ANNALS *of* THE NEW YORK ACADEMY OF SCIENCES

EDITOR-IN-CHIEF
Douglas Braaten

ASSOCIATE EDITOR
Rebecca E. Cooney

PROJECT MANAGER
Steven E. Bohall

EDITORIAL ADMINISTRATOR
Daniel J. Becker

Artwork and design by Ash Ayman Shairzay

The New York Academy of Sciences
7 World Trade Center
250 Greenwich Street, 40th Floor
New York, NY 10007-2157

annals@nyas.org
www.nyas.org/annals

Published by Blackwell Publishing
On behalf of the New York Academy of Sciences

Boston, Massachusetts
2012

ANNALS *of* THE NEW YORK ACADEMY OF SCIENCES

VOLUME 1249

ISSUE

The Year in Ecology and Conservation Biology

ISSUE EDITORS

Richard S. Ostfeld and William H. Schlesinger

Cary Institute of Ecosystem Studies

TABLE OF CONTENTS

Ann. N.Y. Acad. Sci. ISSN 0077-8923

ANNALS OF THE NEW YORK ACADEMY OF SCIENCES
Issue: *The Year in Ecology and Conservation Biology*

Eco-evolutionary dynamics in a changing world

Ilkka Hanski
Department of Biosciences, University of Helsinki, Helsinki, Finland

Address for correspondence: Ilkka Hanski, Department of Biosciences, University of Helsinki, P.O. Box 65, FI-00014 Helsinki, Finland. ilkka.hanski@helsinki.fi

Fast evolutionary changes are common in natural populations, though episodes of rapid evolution do not generally last for long and are typically associated with changing environments. During such periods, evolutionary dynamics may influence ecological population dynamics and vice versa. This review is concerned with spatial eco-evolutionary dynamics with a focus on the occurrence of species in marginal habitats and on metapopulations inhabiting heterogeneous environments. Dispersal and gene flow are key processes in both cases, linking demographic and evolutionary dynamics to each other, facilitating but also constraining the expansion of the current niche and the geographical range of species and determining the spatial scale and pattern of adaptation in heterogeneous environments. An eco-evolutionary metapopulation model helps explain the contrasting responses of species to habitat loss and fragmentation. Eco-evolutionary dynamics may facilitate the persistence of species in changing environments, but typically the evolutionary response only partially compensates for the negative ecological consequences of adverse environmental changes.

Keywords: eco-evolutionary dynamics; spatial dynamics; metapopulation; local adaptation; evolutionary rescue; dispersal

Introduction

Population biology aims at developing a mechanistic and predictive understanding of the dynamics of natural populations. In the context of management and conservation, key questions are why the numbers of individuals fluctuate in time in the manner they do and why the numbers vary from one place to another,[1] which are also questions at the very core of population ecology.[2] Changes in population sizes are due to the many processes that influence births and deaths as well as the movements of individuals into and out of populations. Over the past century, the focus of research has shifted repeatedly to new topics that had been overlooked or were little appreciated by previous researchers. In the middle of the last century, a major concern was the role of density-dependent processes in the demographic dynamics of populations.[3–6] We now view density-dependent population regulation, operating at some though not necessarily at all temporal and spatial scales,[7] as the essential mechanism enhancing the stability of natural populations. Subsequently, however, and to the surprise of many who had learned to associate density dependence with population stability, simple models demonstrated that strong nonlinear density dependence could do just the opposite, lead to wildly and irregularly fluctuating numbers of individuals,[8] which had previously been interpreted as the signature of strong environmental forcing of population dynamics.[9] Interest then shifted from complex temporal dynamics to spatial dynamics, which were seen as the "final frontier for ecological theory."[10,11] But spatial dynamics did not, of course, resolve all the questions. For instance, researchers realized that new understanding of population dynamics could be achieved by taking into account the ubiquitous variation that exists among individuals in life-history traits.[12–15] From here, there is a small step to the idea that genetic variation among individuals, which underlies much of the phenotypic variation, might influence ecological population dynamics.

doi: 10.1111/j.1749-6632.2011.06419.x

The notion that the genetic composition of populations and changes in it (evolutionary dynamics) influence demographic population dynamics is at the same time common sense and unorthodox. Rate of reproduction and the risk of death among individuals vary partly for genetic reasons, and hence the genetic composition of a population should influence its demography. But according to another common wisdom, changes in the genetic composition of populations occur so slowly that the demographic and evolutionary dynamics become effectively decoupled from each other.[16] It is this latter assumption that is being challenged by a body of expanding research on what is now commonly dubbed eco-evolutionary dynamics.[17] I hasten to add that in theoretical population biology, eco-evolutionary dynamics have been an important theme for a long time without the term having been widely used. A classic example is the model by Kirkpatrick and Barton,[18] formalizing the original idea by Haldane [19] and others to explain why species do not constantly expand their geographical ranges through local adaptation. This will be discussed further below, but here I highlight the fact that though certain models combining ecological and evolutionary dynamics have been in the literature for some time, empirical work has been lagging behind.

A notable exception among population ecologists was Dennis Chitty, who articulated already in the 1960s[20,21] a very explicit eco-evolutionary hypothesis about the causes of regular population fluctuations in boreal and arctic small mammals. According to Chitty's hypothesis, cyclic dynamics of voles and lemmings are maintained by high population density selecting for aggressive individuals, which are good competitors but have such a low rate of reproduction that, when their frequency becomes high, the population declines, after which selection starts to favor nonaggressive individuals with a high rate of reproduction, and so on (Fig. 1).[22] The Chitty hypothesis and related hypotheses based on behavioral-endocrine responses[23,24] remained controversial for many reasons, and the Chitty hypothesis was rejected by the 1990s,[25,26] primarily because empirical studies indicated that there is not sufficient heritable genetic variation in the relevant behavioral traits,[27] but also because small mammal population cycles were convincingly explained by other factors, especially the interaction between small mammals and their predators.[28] The failure of the Chitty hypothesis may have discouraged population ecologists from entertaining eco-evolutionary hypotheses until recently, with an upsurge of new interest.[17,29–33] Furthermore, in recent years, the domain of eco-evolutionary dynamics has expanded beyond population dynamics. For instance, researchers have examined the possible influence of genetic variation and evolutionary changes in plant[34] and animal populations[35] on multispecies communities and even ecosystem processes. This paper is, however, restricted to population-level processes, and *ecological dynamics* refer to processes such as changes in population size, dispersal, colonization, and extinction, while *evolutionary dynamics* refer to changes in allele frequency in candidate genes or in heritable phenotypic traits.

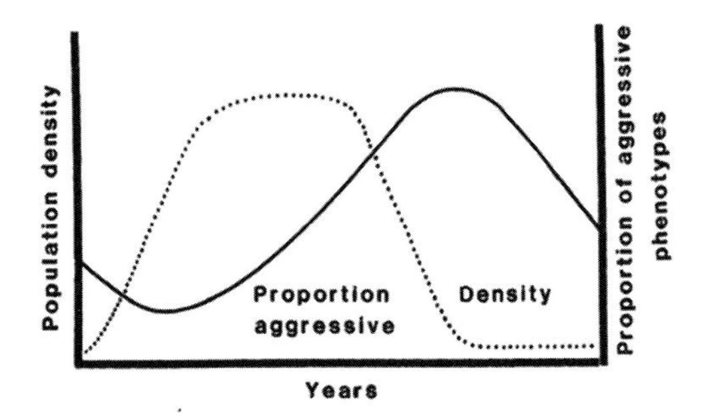

Figure 1. The Chitty hypothesis as interpreted by Ref. 22. Note that the aggressiveness in the population should peak at the end of the decline period in density fluctuations.

Although population ecologists have been slow in broadening their perspective to even consider the possibility of coupled ecological and evolutionary dynamics, the same can be said about population geneticists. Most of the theory in population genetics ignores demographic dynamics and the ecological and environmental context in general,[36] though partly for the understandable reason to keep models tractable for mathematical analysis. However, things are also changing in genetics and evolutionary biology. For instance, Lion *et al.*[37] highlight how one could make progress with the seemingly never-ending debate about the relative merits of kin versus group selection by paying more attention to environmental context and the interplay between demographic and genetic structure and dynamics of populations. Tarnita *et al.*[38] present a general mathematical theory for the reciprocal interaction between the evolutionary dynamics of populations and their spatial population structure. In brief, there is much to be gained for population and evolutionary biologists from broadening our perspectives beyond the conventional subdisciplinary boundaries.

Unidirectional and reciprocal eco-evolutionary feedbacks

Three types of coupling between ecological and evolutionary dynamics are possible: ecological change may influence evolutionary change, evolutionary change may influence ecological change, and there may be reciprocal influences between ecological and evolutionary changes. This paper is concerned with the latter, which represents what might be called the strong form of eco-evolutionary dynamics, or eco-evolutionary feedbacks,[29] but I shall first briefly touch the former two.

Ecological change influences evolutionary change

This is entirely uncontroversial and in the heart of Darwin's thoughts on natural selection and evolution: under particular ecological conditions certain genotypes have higher fitness than others and increase in frequency. It follows that if the ecological conditions change, an evolutionary change is likely to take place—populations become locally adapted. What is not obvious, and where the thinking has shifted over the past decades, is the speed of evolutionary changes. Numerous examples of contemporary (fast) microevolutionary changes[32,39,40] challenge the long-held view of disparate time scales of ecological and evolutionary dynamics.[16]

Evolutionary change influences ecological change

Much of the current research on eco-evolutionary dynamics is concerned with situations where a population's genotypic or phenotypic composition influences ecological change.[41–44] Hairston *et al.*[42] and Ellner *et al.*[45] have developed statistical approaches to partition the ecological and evolutionary contributions to a change in an ecological variable of interest. For example, in the case of the water flea *Daphnia galeata* in Lake Constance, eutrophication increased the abundance of cyanobacteria, a poor-quality food for *Daphnia*.[46] Ellner *et al.*[45] showed that while the ecological contribution (due to change in food quality and quantity) to the change in adult body mass was negative, as juveniles grew poorly on low-quality food, there was a positive evolutionary contribution, juvenile growth rate evolving a higher value under the more eutrophic conditions. The evolutionary contribution was one third in magnitude of the ecological contribution, and thus the former offset a third of the effect of deteriorating food quality on adult body mass. This example highlights the possibly common situation where there is a rapid evolutionary change that is nonetheless difficult to discern because the evolutionary change is countered by the effect of the environmental change.

Reciprocal eco-evolutionary dynamics

Reciprocal influences between ecological and evolutionary changes are more challenging to demonstrate than unidirectional changes, but examples are starting to accumulate. I review some here and more in the subsequent sections on spatial eco-evolutionary dynamics.

Sinervo *et al.*[47] have described an example of reciprocal eco-evolutionary dynamics that fits the scenario originally envisioned by Chitty[20] in the context of vole cycles. The side-blotched lizard (*Uta stansburiana*) exhibits regular 2-year population cycles and has two color morphs of females. One female type is favored at low density due to large clutch size and high rate of reproduction, which leads to overshooting of the carrying capacity and a population crash. Meanwhile, females of the alternative morph produce fewer but larger offspring, and these females are favored at high density. The demographic and evolutionary dynamics become coupled, and selection continues to oscillate between the two alternative life-history syndromes associated with the female color morphs.[47] Another well-studied example of reciprocal eco-evolutionary dynamics involves the interaction between the rotifer *Brachionus calyciflorus* and its prey, obligately asexual green algae *Chlorella vulgaris*.[48,49] When only one clone of the prey is present, the interaction with the rotifer produces typical predator–prey cycles (Fig. 2A–D). These dynamics are, however, significantly modified by the presence of two algal clones, with a trade-off between competitive ability and defense capacity against the rotifer (Fig. 2E–I). This study is especially noteworthy in combining experimental work with mathematical modeling.

A common feature of the above two examples is clonal inheritance of the trait of interest. In the side-blotched lizard, the female color morph has high heritability between dams and daughters,[47] and

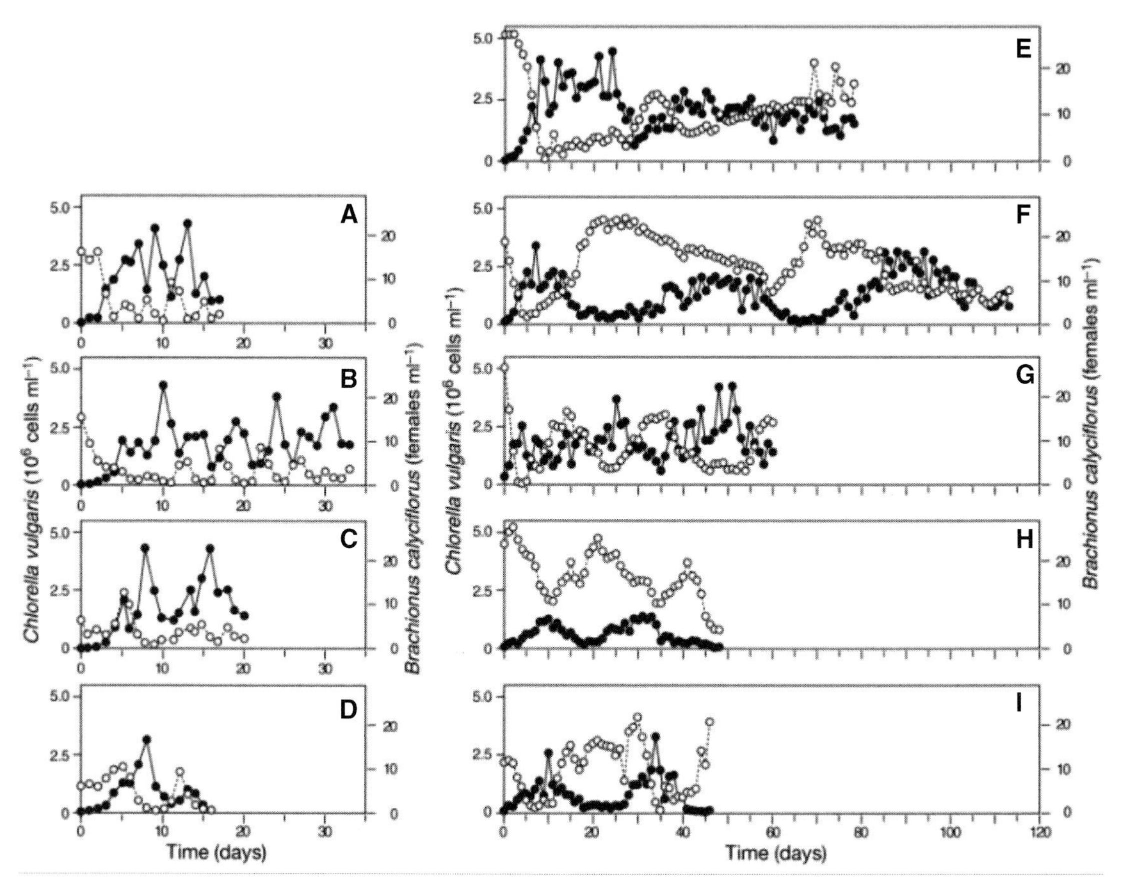

Figure 2. Experimental results showing population cycles in rotifer–algae systems. Filled and open circles give the predator and prey population sizes, respectively. In panels A–D, the algal population consists of a single clone and the dynamics are short-period predator–prey dynamics with the classical phase relations. In panels E–I, the algal populations consist of two clones, and the cycles have long periods with prey and predator oscillations nearly out of phase; such dynamics are not predicted by ecological predator–prey models.[48]

because the male genotype does not appear to make a difference, the population behaves as if it consisted of two clones. Such clonal systems have dynamics that are analogous to the dynamics of multispecies systems. For example, Hanski and Henttonen[50] have analyzed a model of two competing rodent species with a shared mustelid predator, which is comparable to the system studied by Ref. 48, in that one of the rodent species is competitively inferior but less vulnerable to predation. Naturally, however, reciprocal eco-evolutionary dynamics are not restricted to traits with clonal inheritance. The study by Zheng *et al.*[51] on the Glanville fritillary butterfly (*Melitaea cinxia*) exemplifies a sexually reproducing species in which the dynamics of a genetically determined trait (dispersal rate) are closely associated with demographic population dynamics. I shall return to this example later.

Circumstances under which eco-evolutionary dynamics are likely to make a difference

The first point to make is that if selection is weak, the ecological and evolutionary dynamics are likely to be only weakly coupled or not coupled at all. Thus the emerging interest in eco-evolutionary dynamics is based on two other emerging viewpoints in population biology, namely that the rate of evolution is often fast in comparison with the rate of ecological dynamics, which implies strong selection, often on a single gene with a strong effect. Fast rates of evolution appear to be common,[40] though episodes of rapid evolution due to strong selection are likely to be short-lasting and associated with environmental changes. The iconic example is rapid microevolutionary changes in Darwin's finches in response to drought.[52,53] It has been suggested that positive feedbacks during transient periods following an environmental change[54] are especially significant for evolutionary changes, and such situations may often involve the coupling between demographic and evolutionary dynamics.

Recent studies have contributed to an expanding list of examples of strong single gene effects on life-history traits and fitness in natural populations;[55–59] such genes are all good candidates for eco-evolutionary dynamics. Perhaps the best example involves the gene phosphoglucose isomerase (*Pgi*), which encodes for a glycolytic enzyme but may have other functions as well. Classic studies (see Refs. 60–62) established strong links between allozyme phenotypes and individual performance and fitness components in *Colias* butterflies. More recently, similar results have been obtained for the Glanville fritillary,[63–65] other butterflies,[66,67] beetles,[68,69] and other insects.[70] I return below to *Pgi* in the Glanville fritillary.

Based on the observation that environmental changes may often lead to eco-evolutionary dynamics, we might expect such dynamics to be especially prevalent in situations that are characterized by permanent changes. An example is metapopulation dynamics in heterogeneous environments with frequent local extinctions and establishment of new populations by dispersing individuals. Indeed, fast evolutionary changes have been commonly observed in colonizing species and in metapopulations in heterogeneous environments (reviewed by Ref. 71). Colonizations are likely to select for life-history traits that are not selected for in established populations,[72–74] which generates spatial variation in the direction and strength of natural selection among populations with dissimilar demographic histories. In addition, whenever there is spatial variation in habitat type, populations may become locally adapted, in which case gene flow and founder events often involve individuals that are poorly adapted to the environmental conditions that they encounter following dispersal, with likely consequences for both the demographic and evolutionary dynamics of populations. The spatial scale and the amount of dispersal and gene flow will influence local adaptation, but it is also possible that the degree of local adaptation influences local demographic dynamics; hence dispersal may generate reciprocal eco-evolutionary dynamics. In metapopulations with spatio-temporal variation in selection pressures, eco-evolutionary dynamics may not lead to directional evolutionary changes, unless there is a systematic environmental change, but eco-evolutionary dynamics may contribute to the maintenance of genetic variation. I shall return below to eco-evolutionary dynamics in metapopulations in heterogeneous environments.

Seasonality represents another major example of constantly changing environmental conditions, which may lead to eco-evolutionary dynamics in species with multiple generations per year. Year-to-year variation in environmental conditions may lead to reciprocal changes on selection on dormancy in seeds[75,76] and diapause in insects[77,78] and their population dynamics. Such interactions have remained little studied. Another broad class of situations where the biotic environmental conditions are continuously changing involves spatio-temporal dynamics in interacting species. For instance, Vasseur *et al.*[79] have analyzed a model of interspecific competition in which the target and direction of selection on focal individuals depends on whether they are surrounded mostly by conspecific or heterospecific individuals. For a range of parameter values, such "neighbor-dependent selection" allows coexistence because it causes competitive dominance to shift depending on the relative abundances of species within areas of interaction. Empirical studies on *Brassica nigra* present a plausible example, with a trade-off between rapid growth, selected when the focal individual is surrounded by conspecifics, and the production of toxic root exudates, which harm heterospecific competitors and are selected for when the focal individual is mostly surrounded by heterospecifics.[80] Interactions between prey populations and their specialist predators may lead to fluctuating population sizes and fluctuating selection, as originally envisioned by Chitty[20] and exemplified by, for example, epidemic dynamics in *Daphnia dentifera* and its parasite *Metschnikowia bicuspidate*.[81]

Dynamics in marginal habitats

Most species inhabit environments where there is much spatial variation in habitat type and quality. Species and populations have evolved particular ecological requirements, summed up as their niche. But why do species have the niche they have, why do not they constantly expand their niche by becoming better adapted in what used to be low-quality marginal habitats[82] or even sink habitats,[83] in which the intrinsic rate of increase is negative and the population may persist only if there is sufficient immigration from other populations? Similarly, one may ask about the conditions that would allow species to expand their geographical ranges beyond the current range boundaries.[84] These questions have been addressed by an extensive literature in theoretical population biology (for reviews, see Refs. 85–87), which is highly relevant in the present context because the models typically assume close coupling between ecological and evolutionary dynamics.

The dual key process in adaptation in marginal habitats is dispersal and gene flow, which link populations both demographically and genetically. These links are typically asymmetrical, as there is generally more dispersal and gene flow from well-adapted large populations in the core habitats to marginal populations than vice versa. The asymmetry tends to preserve the status quo of the system, making it more difficult for populations in marginal habitats to improve their performance by becoming locally better adapted. A convincing example is presented by studies on the blue tit (*Parus caeruleus*) is southern France, where the species occurs in a patchwork of deciduous and sclerophyllous woodland habitats.[88] Populations in the latter are sinks largely because their breeding phenology is poorly synchronized with local availability of caterpillars. The peak availability of caterpillars occurs about a month earlier in the deciduous woodland, where the populations are well synchronized, whereas in the sclerophyllous habitat, local adaptation is apparently swamped by gene flow from the deciduous habitat.[88] This example is particularly convincing because on the island of Corsica, where the sclerophyllous habitat dominates, the source–sink relationship is reversed.[88] Interestingly, alternative stable source–sink configurations may occur even in the same environment. This is demonstrated by a model of two populations connected by dispersal and gene flow and inhabiting two habitat patches of different type.[89] Assuming that, initially, the population in habitat patch A is well adapted and large but the population in patch B is small, gene flow from A may swamp local adaptation in B, and hence we have a stable source–sink system. However, if the initial condition is reversed, population in B being initially well adapted and large, the system may settle into the alternative source–sink structure. For a range of parameter values, even a third stable configuration exists, in which the species is a generalist, equally adapted to both habitat types, though not as well adapted as a specialist would be in the source habitat.[89] Such complexity is the result of coupling between the demographic and evolutionary dynamics. Boughton[90] describes a plausible example of a butterfly source–sink system with alternative stable states in the same environment.

Asymmetric gene flow from central populations within a species' range may prevent adaptation in marginal populations and thereby prevent range expansion.[18,84] Bridle *et al.*[91] present a putative example on *Drosophila*, comparing adaptation along a steep versus a shallow elevational gradient. There was no cline in the relevant trait on a steep gradient, suggesting gene flow swamping adaptation. Generally, however, there is no strong empirical evidence for range expansion being limited by asymmetric dispersal,[87] and there are many alternative genetic[92] and ecological hypotheses[93] to explain why species' ranges are often restricted without obvious barriers to dispersal.

Gene flow may hinder local adaptation, but gene flow may also facilitate local adaptation by increasing the amount of additive genetic variation, which may otherwise limit evolutionary change in marginal populations.[94] For example, populations of the rainforest-inhabiting *Drosophila birchii* have failed to respond to selection for increased desiccation resistance, and lack of such evolution may prevent the species expanding outside rainforests.[95,96] On the other hand, there are alternative explanations why species may fail to respond to selection, such as complex patterns of pleiotropy and epistasis,[86,97] and it is still an open question how generally marginal populations have particularly low genetic variation.[98,99] In any case, interactions between dispersal and gene flow, local adaptation and population dynamics constitute a very complex process,[86] which implies that one could expect a multitude of outcomes in natural populations.

Eco-evolutionary metapopulation dynamics

A common feature of the models discussed in the previous section is the presence of stable populations that are well adapted to their environment and which send out migrants to less well-adapted and less stable populations, whether they are populations at the range margin or sink populations inhabiting a low-quality habitat type within the range of the species. I shall now turn to metapopulations in which all local populations have a significant risk of extinction either due to ecological factors, such as small population size, or due to maladaptation. Hanski *et al.*[100] have constructed a model for a heterogeneous network of n habitat patches of two or several different habitat types, in which the state of the metapopulation at time t is described with two vectors, one giving the probabilities of occupancy of the n patches and the other one giving the corresponding mean phenotypes conditional on the patch being occupied. When a new population is established, the mean phenotype is determined by the average phenotype of the colonizing individuals. Subsequently, local selection tends to move the mean phenotype towards the local optimal mean phenotype, which depends on habitat type, but changes in mean phenotype are also influenced by gene flow from the surrounding populations. Combining this model of local adaptation with a stochastic patch occupancy metapopulation model leads to a model with several eco-evolutionary feedbacks.[100] For example, the risk of local extinction may depend on the degree of local adaptation, and successful establishment of new populations may depend on the match between the phenotype of the immigrants and the local environmental conditions.[100]

Figure 3 illustrates the model-predicted spatial pattern of adaptation in a network of patches representing two different types. There are four basic patterns depending on the strength of selection, amount of genetic variance, spatial scale of dispersal, and the degree of habitat heterogeneity, though for many parameter combinations the actual pattern is intermediate between the basic patterns. The first two patterns involve populations that become locally adapted (pattern 1 in Fig. 3) or become adapted at the network level (pattern 2), respectively; in the latter case, all local populations, regardless of habitat type, have an intermediate mean phenotype as the long-term equilibrium. Alternatively, all populations across the network may become well adapted to one habitat type only (pattern 3), with the cost of being poorly adapted to the alternative habitat type (habitat specialization). To distinguish this pattern from network adaptation (pattern 2), in an empirical study one would need to know the optimal phenotypes in the different patches. Finally, when the spatial range of dispersal is short, the species may become specialized in different habitat types in different parts of a large network (mosaic specialization; pattern 4). For parameter combinations that yield habitat specialization (pattern 3), the model predicts alternative stable states if the two habitat types are roughly equally common.

The model can be used to explore the consequences of eco-evolutionary dynamics on the ecological viability of metapopulations. At the general level, it can be shown that eco-evolutionary dynamics may both increase and decrease metapopulation size in comparison with demographic dynamics without evolution (Fig. 4). The initial reduction in metapopulation size in Figure 4 with eco-evolutionary dynamics, in comparison with pure ecological dynamics, is due to a shift from network-level adaptation to habitat specialization with increasing fragmentation of habitat and hence with decreasing gene flow between populations (for a comparable result in a two-patch model, see Ref. 89). With further habitat loss and fragmentation, the realized spatial range of dispersal and gene flow become increasingly restricted, and the pattern shifts towards mosaic adaptation. Generally, such shifts in the pattern of adaptation are likely to affect to the commonness of species. For example, the brown argus butterfly (*Aricia agestis*) has been spreading northwards in central England in the past decades apparently due to climate warming.[101] The species used to be restricted to warm habitats with the host plant *Helianthemum chamaecistus*. With warming climate in the past 30 years, previously cooler sites have become more suitable for the butterfly. These sites tend to have alternative host plants, *Geranium* and *Erodium* species, which are used in southern England but were previously not used in central England. Increasing thermal suitability of the cooler sites and the consequent improved demographic performance of the respective local populations have most likely contributed to the observed range expansion in central

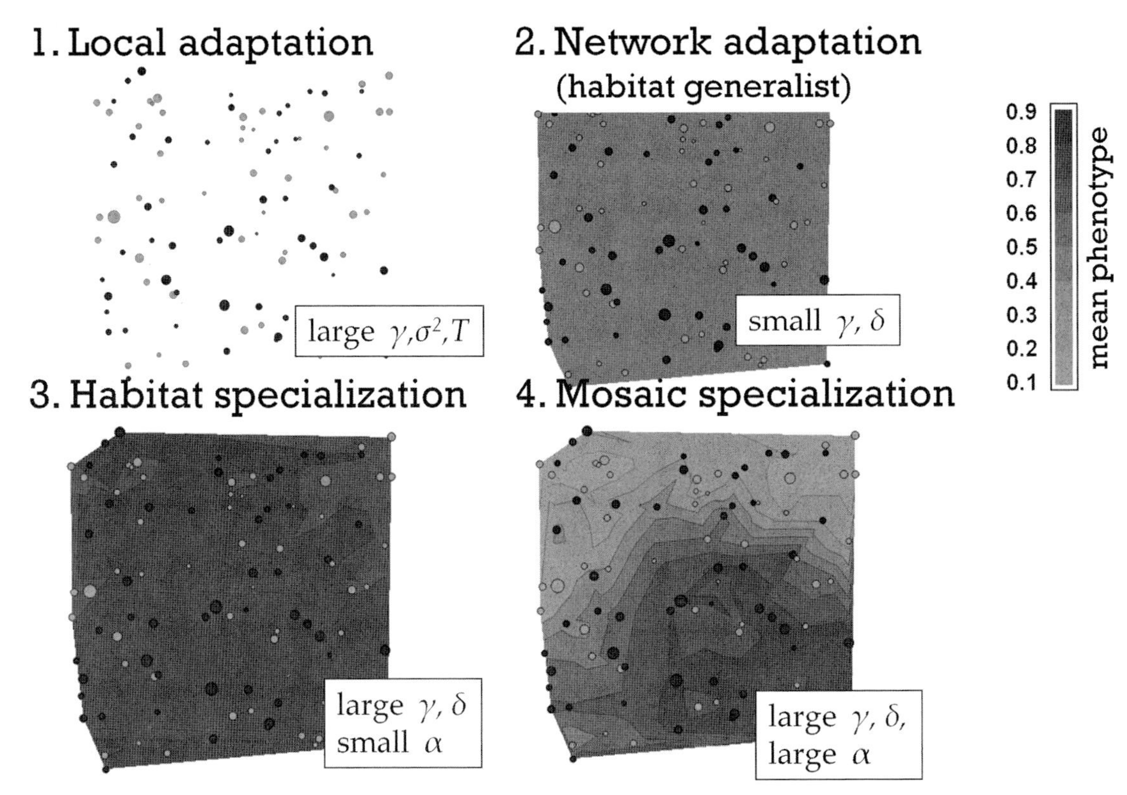

Figure 3. The four basic patterns of network-level adaptation predicted by the eco-evolutionary metapopulation model, which combines a model of local adaptation with an ecological patch occupancy metapopulation model.[100] In each panel, the network is the same, including 100 patches of two different types (black and green). The background color indicates the mean phenotype in the habitat patches across the network, with the exception of panel 1 (local adaptation), in which case the mean phenotype is close to the optimal phenotype, corresponding to the habitat type of each patch. The parameters are γ, strength of selection; σ^2, amount of additive genetic variance; T, expected life-time of local populations; δ, the difference in the optimal phenotypes in different kinds of patches; $1/\alpha$, the spatial range of dispersal and gene flow. The panels indicate the necessary conditions for the different patterns of adaptation in terms of the parameter values. The four patterns are further discussed in the text.

England.[101] However, there is also evidence for a concurrent evolutionary change in host plant preference.[101] The eco-evolutionary metapopulation model discussed above predicts that such a shift from a habitat and host plant specialist (pattern 3 in Fig. 3) to a network-level generalist (pattern 2) leads to a much greater increase in metapopulation size than ecological dynamics alone (see Fig. 6 in Ref. 63).

Range expansion itself, whether owing to climate change or other causes, may select for increased dispersal rate and thereby accelerate range expansion further. This has been demonstrated with models[102–104] and there are also good empirical examples. Thus, Thomas *et al.*[101] showed that in two species of wing dimorphic bush crickets in England, the frequency of the long-winged morph was much higher at the expanding range margin than in more central populations (see also Ref. 105). In two butterfly species in the UK, the speckled wood butterfly (*Pararge aegeria*)[106,107] and the silver spotted skipper butterfly (*Hesperia comma*),[108] results on resource allocation to flight muscles in different populations suggested the same conclusion. In the cane toad (*Bufo marinus*) in Australia, selection at the expanding range margin appears to select for longer legs that allow faster rate of dispersal.[109,110] Direct evidence for genetic differences between expanding and more central populations is still uncommon. One putative example is provided by the European map butterfly.[111] In Finland, the species has expanded its range northwards several hundred

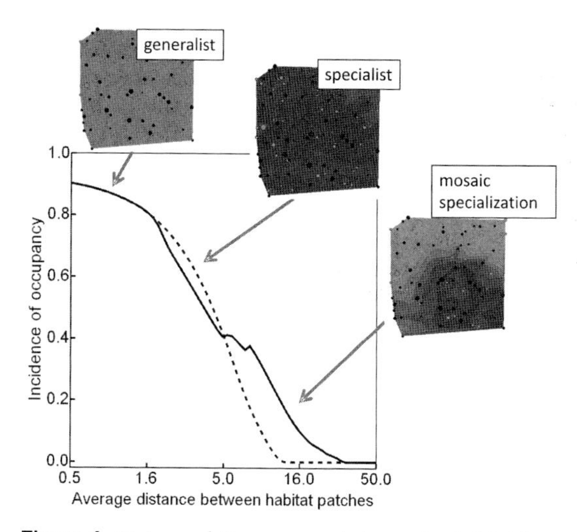

Figure 4. Metapopulation response to increasing habitat fragmentation without (broken line) and with evolution of dispersal rate (continuous line). Habitat fragmentation increases to the right (modeled as increasing distances between a given set of habitat fragments). Eco-evolutionary dynamics make a difference because the network-level pattern of adaptation changes with habitat fragmentation as indicated in the figure and discussed in the text. Modified from Ref. 100.

kilometers in the past 30 years. Mitikka and Hanski[111] found a higher frequency of the phosphoglucose isomerase allozyme phenotype that is associated with higher flight metabolic rate in populations near the range margin than in more central populations.

The example on the map butterfly is consistent with more comprehensive results for the Glanville fritillary butterfly, in which a single nucleotide polymorphism in the phosphoglucose isomerase gene (*Pgi_111*) is strongly associated with flight metabolic rate[112] and dispersal rate in the field,[113] such that the AC heterozygotes in *Pgi_111* fly roughly twice the distance than the AA homozygotes under commonly occurring low ambient temperatures (the CC homozygotes are rare [114]). Hanski and Mononen[116] applied the eco-evolutionary metapopulation model described in the beginning of this section to evolution of dispersal in the Glanville fritillary by defining the mean dispersal phenotype in a particular population as the frequency of the fast-dispersing AC heterozygotes. The model predicts spatially correlated variation in dispersal rate among local populations whenever the range of dispersal is short, as it is in the Glanville fritillary (average dispersal distance around 1 km per generation[115]). The empirical data supported this prediction.[116] At the qualitative level, the model predicts that the long-term frequency of fast-dispersing individuals in a particular habitat patch increases with increasing immigration rate and gene flow (because dispersal is biased towards fast-dispersing individuals), increasing extinction rate (because frequent local extinctions lead to frequent founder events), and the frequency of fast-dispersing individuals among the immigrants. All these factors were significantly related to the frequency of the AC heterozygotes in a set of 97 local populations and together explained 40% of spatial variation in the frequency of the AC heterozygotes among the populations.[116] In addition, as expected, the frequency of the AC heterozygotes was higher in newly established than in old local populations. These results strongly suggest that the demographic metapopulation dynamics and the dynamics of *Pgi* allele frequency are strongly coupled in the Glanville fritillary (see Refs. 51 and 117 for complementary modeling studies).

Spatial dynamics of strongly interacting species may often involve eco-evolutionary feedbacks. For instance, the interaction between the specialist mildew fungus *Podospaera plantaginis* and its host plant *Plantago lanceolata* involve local adaptation in virulence and resistance, leading to spatial eco-evolutionary dynamics,[118] which may commonly characterize plant–pathogen[119] and other host–parasite interactions[120] in general. Gómez and Buckling[121] have reported intriguing experimental studies on coevolutionary dynamics between bacteria and their phages in soil, with never-ending fluctuating selection due to combined spatial demographic and evolutionary dynamics. Thompson's[120] notion about geographic mosaic of coevolution, when applied to a small spatial scale, is essentially a conceptual model of eco-evolutionary dynamics.

Habitat loss and the evolution of dispersal

Habitat loss and fragmentation alter the spatial structure and dynamics of populations, which influences the costs and benefits of dispersal and may therefore affect the evolution of dispersal. Whether habitat loss and fragmentation select for increased,[51,107] decreased,[122,123] or nonmonotonically changing[117,124] rate of dispersal has been much debated.[125,126] Given the multitude of factors affecting dispersal evolution,[127,128] it is not surprising that the evolutionary consequences may be complex. The eco-evolutionary metapopulation model discussed earlier demonstrates that habitat loss and fragmentation may select for either decreased or increased rate of dispersal depending on parameter values.[116] In this model, the long-term equilibrium rate of dispersal in a particular habitat patch depends on the sum of the immigration rate (and gene flow) and extinction rate (and founder effects), and as habitat loss and fragmentation may have opposing effects on these rates, the overall effect depends on quantitative details. For instance, decreasing the areas of habitat patches generally increases extinction rates (because smaller populations typically have a higher risk of extinction than large ones) but decreases immigration rates (because smaller populations typically produce fewer dispersers than large ones).

Empirical studies on the Glanville fritillary and the bog fritillary (*Proclossiana eunomia*) exhibit strikingly different responses to decreasing amount of habitat and increasing fragmentation at the landscape level: in the former, dispersal rate is highest in the most fragmented landscapes, whereas in the bog fritillary, the pattern is just the opposite (see Fig. 3 in Ref. 63). Such contrasting responses appear difficult to explain by for example, the effect of kin competition on dispersal, which has been suggested to be a key process in dispersal evolution in metapopulations.[129] Rather, the explanation may be in the stability of local populations, which is very different in the two species. Small local populations of the Glanville fritillary have a high rate of population extinction (see Fig. 2 in Ref. 63), whereas small populations of the bog fritillary are surprisingly stable.[130] The eco-evolutionary metapopulation model predicts that if local extinctions are uncommon, the dominant effect of habitat loss and fragmentation is reduced immigration rate, which selects for reduced dispersal rate. In contrast, if local extinctions are common, the subsequent founder effects select for increased dispersal, because colonizers are more dispersive than the average individual in the metapopulation, and hence habitat loss and fragmentation increase dispersal rate. The question that remains is why these two fritillary species should have such a big difference in the stability of their local populations? Hanski[63] suggests that the reason is dissimilar egg-laying behavior. The Glanville fritillary lays a small number of large clutches of 150–200 eggs, whereas the bog fritillary lays many small groups of two to four eggs, thereby spreading the risk of mortality among its offspring. This example highlights the potential for critical links between life-history traits, population dynamics, and the evolutionary response of species to changing environmental conditions.

Evolutionary changes may be common, but evolutionary rescue is rare

In the past several decades, human impacts on Earth in the form of land-use changes, climate change, and spread of invasive species represent truly momentous environmental changes. The ecological responses include the decline of populations, many towards imminent extinction,[131,132] although a smaller number of generalist "weedy" species have benefitted from these environmental changes. At the same time, these environmental changes have changed the strength or even the direction of natural selection and caused microevolutionary changes in populations,[133–136] though the extent of evolutionary changes as opposed to plastic phenotypic responses continues to be debated.[137] The question asked in the context of eco-evolutionary dynamics is to what extent, and how, the ecological and evolutionary responses might interact.

The study of the ecological and evolutionary responses of the water flea *Daphnia galeata* to eutrophication,[45] discussed in this paper, highlights a point that is likely to be of general importance. Populations may respond to adverse environmental changes via genetic adaptation, but by definition the direct ecological effect is negative, and therefore the evolutionary response should be strong to entirely compensate for the ecological effect. It is more likely that the evolutionary response will only partly compensate for the negative ecological effect. In the *Daphnia* example, the evolutionary contribution was one third of the ecological contribution to change in adult body mass, and hence eutrophication had an overall adverse effect. Furthermore, there is no guarantee that the evolutionary change will always increase population viability. The example in Figure 4 shows that evolutionary change may both increase and decrease metapopulation size, and a number of theoretical studies have suggested that evolution may even lead to extinction.[138,139]

Figure 5 shows an example of metapopulation viability in response to habitat loss and fragmentation.

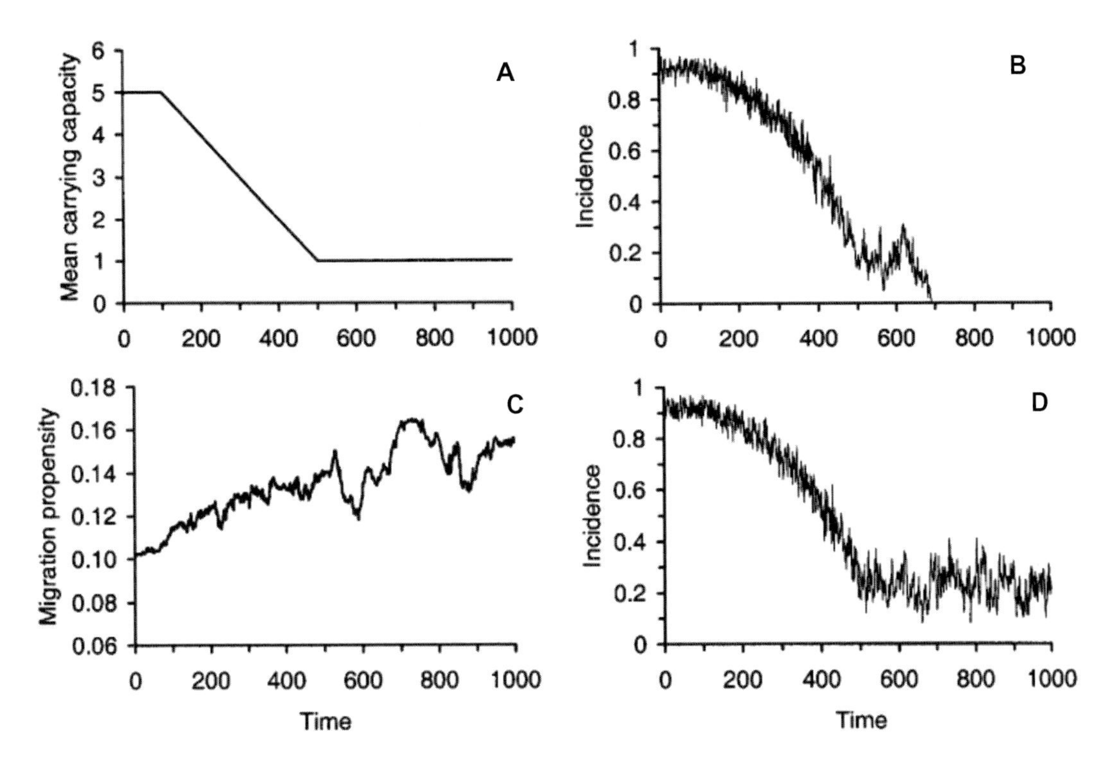

Figure 5. This figure shows the effect of dispersal evolution on metapopulation viability in a changing environment. (A) Habitat is lost during the time interval from 100 to 500 by reducing the carrying capacity of each habitat patch from 5 to 1. (B) Change in metapopulation size (average incidence of patch occupancy) without evolution of dispersal rate. (C) Changes in migration propensity (emigration rate) during and after the period of habitat loss in the model when evolutionary changes are allowed, and (D) the predicted change in metapopulation size with evolution of dispersal rate.[117]

This is a modeling example, but an informative one, as the model was parameterized with a large amount of empirical data for the Glanville fritillary.[117] The example was constructed by assuming a period of habitat loss after which there was no further loss (Fig. 5A). Parameter values were selected in such a manner that without any evolutionary changes in dispersal rate, the initially viable metapopulation went extinct soon after the period of habitat loss (Fig. 5B). In contrast, if dispersal rate was allowed to evolve in response to habitat loss and fragmentation (Fig. 5C), the metapopulation did not go extinct (Fig. 5D). However, just as in the example on *Daphnia*, the direct negative ecological effect of habitat loss and fragmentation dominates: the metapopulation with evolving dispersal rate survived but at a much lower level of habitat occupancy than before habitat loss.

Bell and Gonzalez[140] have described an example of what appears to be a genuine evolutionary rescue. They studied populations of baker's yeast on 96-well plates with a gradient of environmental suitability in the form of salt concentration in the growth medium. In the experiment, they imposed a gradually deteriorating environment by increasing the amount of salt along the gradient and applied either no dispersal, local dispersal, or global dispersal in different replicates. During the experiment, new mutations appeared that facilitated the growth in higher salt concentrations, and the extent of adaptation depended on the history of environmental change during the experiment and on the scale of dispersal among local populations. In this case, evolution truly produced an evolutionary rescue, because specific new mutations allowed efficient use of the growth medium with a higher concentration of salt. Unfortunately, the situation is likely to be more complex in more complex organisms living in more complex environments, where no single mutation can compensate for the consequences of environmental deterioration but instead may only partly offset them, and even that only in the best case. Rapid evolutionary changes and eco-evolutionary feedbacks may be common in natural populations, but we cannot assume that evolution will generally rescue populations that are on decline due to anthropogenic deterioration of their environment.

What next?

The shift in the perception, over the past decades, of how commonly fast microevolutionary changes occur in natural populations has profoundly affected ecology and evolutionary biology. With the recognition of increasingly rapid global environmental changes, which must change the strength and often even the direction of natural selection in countless populations and environments, a new paradigm is emerging. Fast evolutionary changes and local adaptation are not confined to the few iconic examples, such as Darwin's finches[53,141] and heavy metal tolerance in plants growing on mine tailings;[142,143] similar microevolutinary dynamics may characterize many other species in a large array of environments. The challenge is to ask appropriate questions and to employ fitting study approaches. To take a very personal example, I would have been surprised 20 years ago, when I started to work on the Glanville fritillary butterfly, that this study system would provide strong empirical evidence, supported by modeling,[116,144] of reciprocal eco-evolutionary dynamics.[63] I suspect that many ecologists will become similarly surprised in the coming years, and that population geneticists and evolutionary biologists are likely to pay increasing attention to the ecological and environmental context of their study systems. A particular challenge is to develop research programs that take advantage of the very extensive genetic data that will soon become widely available for natural populations of non-model species due to rapidly advancing sequencing technologies.[145–147]

Microevolutionary changes may be rampant especially in changing environments, but at the same time one should not rush to the conclusion that all phenotypic changes are due to evolution. The alternative hypothesis is phenotypic plasticity, which may plausibly explain, for example, temporal changes in body size[134,148] and various other responses to climate change.[137] Realistically, one should expect that plastic phenotypic changes and genetic microevolutionary changes are often mixed in natural populations, leading to new questions about adaptive evolution.[149,150] It should also be noted that even if phenotypic changes would be solely due to plastic responses without any genetic changes, plastic phenotypic changes may significantly interact with ecological dynamics—and generate new questions to study.

More empirical research is needed to charter the extent of eco-evolutionary dynamics, and analogous dynamics based on phenotypic plasticity, in natural populations living under various environmental settings. Exploring the significance of eco-evolutionary dynamics in interspecific interactions and multispecies communities is an especially important challenge that goes beyond the predator–prey and competitive interactions that are familiar from classic ecological studies. Take, for instance, the interactions between insects and their diverse assemblage of primary and secondary endosymbionts[151]—what is the significance of eco-evolutionary dynamics in these systems? As an example, Leonardo and Mondor[152] demonstrate that the facultative bacterial endosymbiont *Regiella insecticola* alters both dispersal and mating in the pea aphid *Acyrthosiphon pisum.* Leonardo and Mondor[152] were interested in the possibility that symbiont-associated changes in dispersal and mating may play a role in the initiation of genetic differentiation and ultimately in speciation, but perhaps an even more likely outcome is coupled evolutionary and ecological dynamics involving both partners in the interaction.

I conclude by highlighting the potential of eco-evolutionary dynamics in addressing one of the major questions in ecology and evolutionary dynamics, namely the maintenance of diversity at molecular, population, and community levels. Of the case studies discussed in this review, the close coupling between the allele frequency dynamics in the gene *Pgi* and the extinction–colonization dynamics in the Glanville fritillary points to the role of eco-evolutionary dynamics in maintaining genetic and life-history variation in metapopulations.[63] Similarly, modeling of interspecific competition[79] and the related empirical work on plants[80,153] suggests a role for eco-evolutionary dynamics in facilitating the coexistence of competitors. Genetic variation is a prerequisite for eco-evolutionary dynamics,[49] but if such dynamics generally facilitate the maintenance of genetic diversity, there are interesting questions to be asked about the conditions and extent of such diversity-enhancing processes.

Acknowledgments

I thank two anonymous reviewers for comments and the European Research Council (Advance grant no 232826) and the Academy of Finland (Finnish CoE Programme 2006–2011) for support.

Conflicts of interest

The author declares no conflicts of interest.

References

1. Caughley, G. 1994. Directions in conservation biology. *J. Anim. Ecol.* **63:** 215–244.
2. Begon, M., J.L. Harper & C.R. Townsend. 1996. *Ecology*. Blackwell. Oxford.
3. Hanski, I. 1990. Density dependence, regulation and variability in animal populations. *Phil. Trans. R. Soc. Lond. B Biol. Sci.* **330:** 141–150.
4. Turchin, P. 1995. Population regulation: old arguments and a new synthesis. In *Population Dynamics: New Approaches and Synthesis*. N. Cappuccino & P. Price, Eds.: 19–40. Academic Press. London.
5. Sinclair, A.R.E. 1989. Population regulation in animals. In *Ecological Concepts*. J.M. Cherrett, Ed.: 197–242. Blackwell. Oxford.
6. Dempster, J.P. 1983. The natural control of populations of butterflies and moths. *Biol. Rev.* **58:** 461–481.
7. Hassell, M.P., J. Latto & R.M. May. 1989. Seeing the wood for the trees: detecting density dependence from existing life-table studies. *J. Anim. Ecol.* **58:** 883–892.
8. May, R.M. 1973. *Stability and Complexity in Model Ecosystems*. Princeton University Press. Princeton.
9. Andrewartha, H.G. & L.C. Birch. 1954. *The Distribution and Abundance of Animals*. The University of Chicago Press. Chicago.
10. Kareiva, P. 1994. Space – the final frontier for ecological theory. *Ecology* **75:** 1.
11. Tilman, D. & P. Kareiva. 1997. *Spatial Ecology*. Princeton University Press. Princeton.
12. Sutherland, W.J. 1996. *From Individual Behaviour to Population Ecology*. Oxford University Press. Oxford.
13. Lomnicki, A. 1988. *Population Ecology of Individuals*. Princeton University Press. Princeton.
14. Coffin, J.M. 1995. HIV population dynamics in vivo – implications for genetic variation, pathogenesis, and therapy. *Science* **267:** 483–489.
15. Ottersen, G., B. Planque, A. Belgrano, *et al.* 2001. Ecological effects of the North Atlantic Oscillation. *Oecologia* **128:** 1–14.
16. Slobodkin, L.B. 1961. *Growth and Regulation of Animal Populations*. Holt, Rinehart and Winston. New York.
17. Pelletier, F., D. Garant & A.P. Hendry. 2009. Eco-evolutionary dynamics: introduction. *Phil. Trans. R. Soc. B Biol. Sci.* **364:** 1483–1489.
18. Kirkpatrick, M. & N.H. Barton. 1997. Evolution of a species' range. *Am. Nat.* **150:** 1–23.
19. Haldane, J.B.S. 1956. The relation between density regulation and natural selection. *Proc. R. Soc. B.* **145:** 306–308.
20. Chitty, D. 1960. Population processes in the vole and their relevance to general theory. *Can. J. Zool.* **38:** 99–113.
21. Chitty, D. 1967. The natural selection of self-regulatory behaviour in animal populations. *Proc. Ecol. Soc. Austral.* **2:** 51–78.
22. Krebs, C.J. 1978. A review of the Chitty hypothesis of population regulation. *Can. J. Zool.* **56:** 2463–2480.
23. Christian, J.J. 1978. Neurobehavioral endocrine regulation of small mammal populations. In *Populations of Small Mammals under Natural Conditions*. D.P. Snyder, Ed.: 143–158, Vol. 5. Special Publ., Pymatuning Lab. Ecology, Pittsburgh.
24. Christian, J.J. 1950. The adreno-pituitary system and population cycles in mammals. *J. Mammal.* **31:** 247–259.
25. Hanski, I. & H. Henttonen. 2002. Population cycles of small rodents in Fennoscandia. In *Population Cycles: Evidence for Trophic Interactions*. A. Berryman, Ed.: 44–68. Oxford University Press. New York.
26. Stenseth, N.C. 1999. Population cycles in voles and lemmings: density dependence and phase dependence in a stochastic world. *Oikos* **87:** 427–461.
27. Boonstra, R. & P.T. Boag. 1987. A test of the Chitty hypothesis: inheritance of life-history traits in meadow voles *Microtus pennsylvanicus*. *Evolution* **41:** 929–947.
28. Turchin, P. & I. Hanski. 2001. Contrasting alternative hypotheses about rodent cycles by translating them into parameterized models. *Ecol. Lett.* **4:** 267–276.
29. Post, D.M. & E.P. Palkovacs. 2009. Eco-evolutionary feedbacks in community and ecosystem ecology: interactions between the ecological theatre and the evolutionary play. *Phil. Trans. R. Soc. B Biol. Sci.* **364:** 1629–1640.
30. Carroll, S.P., A.P. Hendry, D.N. Reznick & C.W. Fox. 2007. Evolution on ecological time-scales. *Funct. Ecol.* **21:** 387–393.
31. Kinnison, M.T. & N.G. Hairston. 2007. Eco-evolutionary conservation biology: contemporary evolution and the dynamics of persistence. *Funct. Ecol.* **21:** 444–454.
32. Saccheri, I. & I. Hanski. 2006. Natural selection and population dynamics. *Trends Ecol. Evol.* **21:** 341–347.
33. Schoener, T.W. 2011. The newest synthesis: understanding the interplay of evolutionary and ecological dynamics. *Science* **331:** 426–429.
34. Bailey, J.K., J.A. Schweitzer, F. Ubeda, *et al.* 2009. From genes to ecosystems: a synthesis of the effects of plant genetic factors across levels of organization. *Phil. Trans. R. Soc. B Biol. Sci.* **364:** 1607–1616.
35. Palkovacs, E.P., M.C. Marshall, B.A. Lamphere, *et al.* 2009. Experimental evaluation of evolution and coevolution as agents of ecosystem change in Trinidadian streams. *Phil. Trans. R. Soc. B Biol. Sci.* **364:** 1617–1628.
36. Whitlock, M.C. 2004. Selection and drift in metapopulations. In *Ecology, Genetics, and Evolution of Metapopulations*. I. Hanski & O. Gaggiotti, Eds.: 153–173. Elsevier Academic Press. Amsterdam.

37. Lion, S., V.A.A. Jansen & T. Day. 2011. Evolution in structured populations: beyond the kin versus group debate. *Trends Ecol. Evol.* **26:** 193–201.
38. Tarnita, C.E., T. Antal, H. Ohtsuki & M.A. Nowak. 2009. Evolutionary dynamics in set structured populations. *Proc. Natl. Acad. Sci. USA* **106:** 8601–8604.
39. Thompson, J.N. 1998. Rapid evolution as an ecological process. *Trends Ecol. Evol.* **13:** 329–332.
40. Hendry, A.P. & M.T. Kinnison. 1999. Perspective. The pace of modern life: measuring rates of contemporary microevolution. *Evolution* **53:** 1637–1653.
41. Fussmann, G.F., M. Loreau & P.A. Abrams. 2007. Eco-evolutionary dynamics of communities and ecosystems. *Funct. Ecol.* **21:** 465–477.
42. Hairston, N.G., S.P. Ellner, M.A. Geber, *et al.* 2005. Rapid evolution and the convergence of ecological and evolutionary time. *Ecol. Lett.* **8:** 1114–1127.
43. Pelletier, F., T. Clutton-Brock, J. Pemberton, *et al.* 2007. The evolutionary demography of ecological change: linking trait variation and population growth. *Science* **315:** 1571–1574.
44. Ezard, T.H.G., S.D. Cote & F. Pelletier. 2009. Eco-evolutionary dynamics: disentangling phenotypic, environmental and population fluctuations. *Phil. Trans. R. Soc. B Biol. Sci.* **364:** 1491–1498.
45. Ellner, S.P., M.A. Geber & N.G. Hairston. 2011. Does rapid evolution matter? Measuring the rate of contemporary evolution and its impacts on ecological dynamics. *Ecol. Lett.* **14:** 603–614.
46. Hairston, N.G., C.L. Holtmeier, W. Lampert, *et al.* 2001. Natural selection for grazer resistance to toxic cyanobacteria: evolution of phenotypic plasticity? *Evolution* **55:** 2203–2214.
47. Sinervo, B., E. Svensson & T. Comendant. 2000. Density cycles and an offspring quantity and quality game driven by natural selection. *Nature* **406:** 985–988.
48. Yoshida, T., L.E. Jones, S.P. Ellner, *et al.* 2003. Rapid evolution drives ecological dynamics in a predator-prey system. *Nature* **424:** 303–306.
49. Becks, L., S.P. Ellner, L.E. Jones & N.G. Hairston. 2010. Reduction of adaptive genetic diversity radically alters eco-evolutionary community dynamics. *Ecol. Lett.* **13:** 989–997.
50. Hanski, I. & H. Henttonen. 1996. Predation on competing rodent species: a simple explanation of complex patterns. *J. Anim. Ecol.* **65:** 220–232.
51. Zheng, C., O. Ovaskainen & I. Hanski. 2009. Modelling single nucleotide effects in phosphoglucose isomerase on dispersal in the Glanville fritillary butterfly: coupling of ecological and evolutionary dynamics. *Phil. Trans. R. Soc. B Biol. Sci.* **364:** 1519–1532.
52. Boag, P.T. & P.R. Grant. 1981. Intense natural selection in a population of Darwin finches (Geospizinae) in the Galapagos. *Science* **214:** 82–85.
53. Grant, P.R. & B.R. Grant. 1995. Predicting microevolutionary responses to directional selection on heritable variation. *Evolution* **49:** 241–251.
54. Crespi, B.J. 2004. Vicious circles: positive feedback in major evolutionary and ecological transitions. *Trends Ecol. Evol.* **19:** 627–633.
55. Barrett, R.D.H., S.M. Rogers & D. Schluter. 2008. Natural selection on a major armor gene in threespine stickleback. *Science* **322:** 255–257.
56. Storz, J.F., A.M. Runck, S.J. Sabatino, *et al.* 2009. Evolutionary and functional insights into the mechanism underlying high-altitude adaptation of deer mouse hemoglobin. *Proc. Natl. Acad. Sci. USA* **106:** 14450–14455.
57. Storz, J.F. & C.W. Wheat. 2010. Integrating evolutionary and functional approaches to infer adaptation at specific loci. *Evolution* **64:** 2489–2509.
58. Dalziel, A.C., S.M. Rogers & P.M. Schulte. 2009. Linking genotypes to phenotypes and fitness: how mechanistic biology can inform molecular ecology. *Mol. Ecol.* **18:** 4997–5017.
59. Hanski, I. & I. Saccheri. 2006. Molecular-level variation affects population growth in a butterfly metapopulation. *PLoS Biol.* **4:** 719–726.
60. Watt, W.B. 1983. Adaptation at specific loci. 2. Demographic and biochemical-elements in the maintenance of the *Colias Pgi* polymorphism. *Genetics* **103:** 691–724.
61. Watt, W.B. 2003. Mechanistic studies of butterfly adaptations. In *Ecology and Evolution taking Flight: Butterflies as Model Systems.* C.L. Boggs, W.B. Watt and P. R. Ehrlich, Eds.: 319–352. University of Chicago Press. Chicago.
62. Watt, W.B. 1977. Adaptation at specific loci. 1. Natural selection on phosphoglucose isomerase of *Colias* butterflies – biochemical and population aspects. *Genetics* **87:** 177–194.
63. Hanski, I. 2011. Eco-evolutionary spatial dynamics in the Glanville fritillary butterfly. *Proc. Natl. Acad. Sci. USA* **108:** 14397–14404.
64. Saastamoinen, M. & I. Hanski. 2008. Genotypic and environmental effects on flight activity and oviposition in the Glanville fritillary butterfly. *Am. Nat.* **171:** 701–712.
65. Klemme, I. & I. Hanski. 2009. Heritability of and strong single gene (*Pgi*) effects on life-history traits in the Glanville fritillary butterfly. *J. Evol. Biol.* **22:** 1944–1953.
66. Karl, I., T. Schmitt & K. Fischer. 2008. Phosphoglucose isomerase genotype affects life-history traits and cold stress resistance in a Copper butterfly. *Funct. Ecol.* **22:** 887–894.
67. Karl, I., K.H. Hoffmann & K. Fischer. 2010. Food stress sensitivity and flight performance across phosphoglucose isomerase enzyme genotypes in the sooty copper butterfly. *Popul. Ecol.* **52:** 307–315.
68. Dahlhoff, E.P. & N.E. Rank. 2000. Functional and physiological consequences of genetic variation at phosphoglucose isomerase: heat shock protein expression is related to enzyme genotype in a montane beetle. *Proc. Natl. Acad. Sci. USA* **97:** 10056–10061.
69. Rank, N.E. & E.P. Dahlhoff. 2002. Allele frequency shifts in response to climate change and physiological consequences of allozyme variation in a montane insect. *Evolution* **56:** 2278–2289.
70. Wheat, C.W. 2010. Phosphoglucose isomerase (*Pgi*) performance and fitness effects among Arthropods and its potential role as an adaptive marker in conservation genetics. *Conserv. Genet.* **11:** 387–397.
71. Reznick, D.N. & C.K. Ghalambor. 2001. The population ecology of contemporary adaptations: what empirical

studies reveal about the conditions that promote adaptive evolution. *Genetica* **112–113:** 183–198.
72. Hanski, I., C. Erälahti, M. Kankare, *et al.* 2004. Variation in migration rate among individuals maintained by landscape structure. *Ecol. Lett.* **7:** 958–966.
73. Wheat, C.W., H.W. Fescemyer, J. Kvist, *et al.* 2011. Functional genomics of life history variation in a butterfly metapopulation. *Mol. Ecol.* **20:** 1813–1828.
74. Olivieri, I. & P.-H. Gouyon. 1997. Evolution of migration rate and other traits: the metapopulation effect. In *Metapopulation Biology*. I.A. Hanski & M.E. Gilpin, Eds.: 293–324. Academic Press. San Diego.
75. Pake, C.E. & D.L. Venable. 1996. Seed banks in desert annuals: implications for persistence and coexistence in variable environments. *Ecology* **77:** 1427–1435.
76. Rees, M. 1996. Evolutionary ecology of seed dormancy and seed size. *Phil. Trans. R. Soc. B Biol. Sci.* **351:** 1299–1308.
77. Hopper, K.R. 1999. Risk-spreading and bet-hedging in insect population biology. *Ann. Rev. Entomol.* **44:** 535–560.
78. Hanski, I. 1988. Four kinds of extra long diapause in insects: a review of theory and observations. *Ann. Zool. Fenn.* **25:** 37–53.
79. Vasseur, D.A., P. Amarasekare, V.H.W. Rudolf & J.M. Levine. 2011. Eco-evolutionary dynamics enable coexistence via neighbor-dependent selection. *Am. Nat.* **178:** E96–E109.
80. Lankau, R.A. & S.Y. Strauss. 2007. Mutual feedbacks maintain both genetic and species diversity in a plant community. *Science* **317:** 1561–1563.
81. Duffy, M.A. & L. Sivars-Becker. 2007. Rapid evolution and ecological host-parasite dynamics. *Ecol. Lett.* **10:** 44–53.
82. Holt, R.D. & M.S. Gaines. 1992. Analysis of adaptation in heterogeneous landscapes: implications for the evolution of fundamental niches. *Evol. Ecol.* **6:** 433–447.
83. Pulliam, H.R. 1988. Sources, sinks, and population regulation. *Am. Nat.* **132:** 652–661.
84. Hoffmann, A.A. & M.W. Blows. 1994. Species borders – ecological and evolutionary perspectives. *Trends Ecol. Evol.* **9:** 223–227.
85. Kawecki, T.J. 2004. Ecological end evolutionary consequences of source-sink population dynamics. In *Ecology, Genetics, and Evolution of Metapopulations*. I. Hanski & O.E. Gaggiotti, Eds.: 387–414. Elsevier Academic Press. Amsterdam.
86. Kawecki, T.J. 2008. Adaptation to marginal habitats. *Ann. Rev. Ecol. Evol. Syst.* **39:** 321–342.
87. Gaston, K.J. 2009. Geographic range limits of species. *Proc. R. Soc. B Biol. Sci.* **276:** 1391–1393.
88. Dias, P.C. & J. Blondel. 1996. Local specialization and maladaptation in Mediterranean blue tits *Parus caeruleus. Oecologia* **107:** 79–86.
89. Ronce, O. & M. Kirkpatrick. 2001. When sources become sinks: migrational meltdown in heterogeneous habitats. *Evolution* **55:** 1520–1531.
90. Boughton, D.A. 2000. The dispersal system of a butterfly: a test of source-sink theory suggests the intermediate-scale hypothesis. *Am. Nat.* **156:** 131–144.
91. Bridle, J.R., S. Gavaz & W.J. Kennington. 2009. Testing limits to adaptation along altitudinal gradients in rainforest *Drosophila. Proc. R. Soc. B Biol. Sci.* **276:** 1507–1515.
92. van Heerwaarden, B., V. Kellermann, M. Schiffer, *et al.* 2009. Testing evolutionary hypotheses about species borders: patterns of genetic variation towards the southern borders of two rainforest *Drosophila* and a related habitat generalist. *Proc. R. Soc. B Biol. Sci.* **276:** 1517–1526.
93. Gaston, K.J. 2003. *The Structure and Dynamics of Geographical Ranges*. Oxford University Press. Oxford.
94. Blows, M.W. & A.A. Hoffmann. 2005. A reassessment of genetic limits to evolutionary change. *Ecology* **86:** 1371–1384.
95. Hoffmann, A.A., R.J. Hallas, J.A. Dean & M. Schiffer. 2003. Low potential for climatic stress adaptation in a rainforest *Drosophila* species. *Science* **301:** 100–102.
96. Kellermann, V.M., B. van Heerwaarden, A.A. Hoffmann & C.M. Sgro. 2006. Very low additive genetic variance and evolutionary potential in multiple populations of two rainforest *Drosophila* species. *Evolution* **60:** 1104–1108.
97. Brakefield, P.M. 2006. Evo-devo and constraints on selection. *Trends Ecol. Evol.* **21:** 362–368.
98. Garner, T.W.J., P.B. Pearman & S. Angelone. 2004. Genetic diversity across a vertebrate species' range: a test of the central-peripheral hypothesis. *Mol. Ecol.* **13:** 1047–1053.
99. Jacquemyn, H., K. Vandepitte, R. Brys, *et al.* 2007. Fitness variation and genetic diversity in small, remnant populations of the food deceptive orchid *Orchis purpurea. Biol. Conserv.* **139:** 203–210.
100. Hanski, I., T. Mononen & O. Ovaskainen. 2011. Eco-evolutionary metapopulation dynamics and the spatial scale of adaptation. *Am. Nat.* **177:** 29–43.
101. Thomas, C.D., E.J. Bodsworth, R.J. Wilson, *et al.* 2001. Ecological and evolutionary processes at expanding range margins. *Nature* **411:** 577–581.
102. Travis, J.M.J., T. Munkemuller & O.J. Burton. 2010. Mutation surfing and the evolution of dispersal during range expansions. *J. Evol. Biol.* **23:** 2656–2667.
103. Phillips, B.L., G.P. Brown & R. Shine. 2010. Life-history evolution in range-shifting populations. *Ecology* **91:** 1617–1627.
104. Travis, J.M.J. & C. Dytham. 2002. Dispersal evolution during invasions. *Evol. Ecol. Res.* **4:** 1119–1129.
105. Simmons, A.D. & C.D. Thomas. 2004. Changes in dispersal during species' range expansions. *Am. Nat.* **164:** 378–395.
106. Hughes, C.L., C. Dytham & J.K. Hill. 2007. Modelling and analysing evolution of dispersal in populations at expanding range boundaries. *Ecol. Entomol.* **32:** 437–445.
107. Hill, J.K., C.D. Thomas & D.S. Blakeley. 1999. Evolution of flight morphology in a butterfly that has recently expanded its geographic range. *Oecologia* **121:** 165–170.
108. Hill, J.K., C.D. Thomas & O.T. Lewis. 1999. Flight morphology in fragmented populations of a rare British butterfly, *Hesperia comma. Biol. Conserv.* **87:** 277–283.
109. Phillips, B.L., G.P. Brown & R. Shine. 2010. Evolutionarily accelerated invasions: the rate of dispersal evolves upwards during the range advance of cane toads. *J. Evol. Biol.* **23:** 2595–2601.

110. Phillips, B.L., G.P. Brown, J.K. Webb & R. Shine. 2006. Invasion and the evolution of speed in toads. *Nature* **439:** 803–803.
111. Mitikka, V. & I. Hanski. 2010. *Pgi* genotype influences flight metabolism at the expanding range margin of the European map butterfly. *Ann. Zool. Fenn.* **47:** 1–14.
112. Niitepõld, K. 2010. Genotype by temperature interactions in the metabolic rate of the Glanville fritillary butterfly. *J. Exp. Biol.* **213:** 1042–1048.
113. Niitepõld, K., A.D. Smith, J.L. Osborne, *et al.* 2009. Flight metabolic rate and *Pgi* genotype influence butterfly dispersal rate in the field. *Ecology* **90:** 2223–2232.
114. Orsini, L., C.W. Wheat, C.R. Haag, *et al.* 2009. Fitness differences associated with *Pgi* SNP genotypes in the Glanville fritillary butterfly (*Melitaea cinxia*). *J. Evol. Biol.* **22:** 367–375.
115. Hanski, I. 1999. *Metapopulation Ecology*. Oxford University Press. New York.
116. Hanski, I. & T. Mononen. 2011. Eco-evolutionary dynamics of dispersal in spatially heterogeneous environments. *Ecol. Lett.* **19:** 1025–1034.
117. Heino, M. & I. Hanski. 2001. Evolution of migration rate in a spatially realistic metapopulation model. *Am. Nat.* **157:** 495–511.
118. Laine, A.L. 2005. Spatial scale of local adaptation in a plant-pathogen metapopulation. *J. Evol. Biol.* **18:** 930–938.
119. Burdon, J.J. & P.H. Thrall. 2009. Coevolution of plants and their pathogens in natural habitats. *Science* **324:** 755–756.
120. Thompson, J.N. 2005. *The Geographic Mosaic of Coevolution.* University of Chicago Press. Chicago.
121. Gomez, P. & A. Buckling. 2011. Bacteria-phage antagonistic coevolution in soil. *Science* **332:** 106–109.
122. Cody, M.L. & J. Overton. 1996. Short-term evolution of reduced dispersal in island plant populations. *J. Ecol.* **84:** 53–61.
123. Schtickzelle, N., G. Mennechez & M. Baguette. 2006. Dispersal depression with habitat fragmentation in the bog fritillary butterfly. *Ecology* **87:** 1057–1065.
124. Gandon, S. & Y. Michalakis. 1999. Evolutionary stable dispersal rate in a metapopulation with extinctions and kin competition. *J. Theor. Biol.* **199:** 275–290.
125. Hanski, I. 2005. *The Shrinking World: Ecological Consequences of Habitat Loss.* International Ecology Institute. Oldendorf/Luhe.
126. Ronce, O. & I. Olivieri. 2004. Life history evolution in metapopulations. In *Ecology, Genetics, and Evolution of Metapopulations.* I. Hanski & O.E. Gaggiotti, Eds.: 227–257. Elsevier Academic Press. Amsterdam.
127. Ronce, O. 2007. How does it feel to be like a rolling stone? Ten questions about dispersal evolution. *Ann. Rev. Ecol. Evol. Syst.* **38:** 231–253.
128. Clobert, J., E. Danchin, A.A. Dhont & J.D. Nichols. 2001. *Dispersal.* Oxford University Press. Oxford.
129. Massol, F., A. Duputie, P. David & P. Jarne. 2011. Asymmetric patch size distribution leads to disruptive selection on dispersal. *Evolution* **65:** 490–500.
130. Baguette, M. 2004. The classical metapopulation theory and the real, natural world: a critical appraisal. *Basic Appl. Ecol.* **5:** 213–224.
131. Millenium Ecosystem Assessment. 2005. Retrieved January 17, 2012, from http://www.maweb.org/en/Index.aspx.
132. IUCN. 2011. Retrieved January 17, 2012, from http://www.iucnredlist.org/.
133. Berteaux, D., D. Reale, A.G. McAdam & S. Boutin. 2004. Keeping pace with fast climate change: can arctic life count on evolution? *Integr. Comp. Biol.* **44:** 140–151.
134. Pulido, F. & P. Berthold. 2004. Microevolutionary response to climatic change. In *Birds and Climate Change.* A.P. Møller, W. Fielder & P. Berthold, Eds.: 151–183, Vol. 35, Adv. Ecol. Research. Elsevier. Amsterdam.
135. Davis, M.B., R.G. Shaw & J.R. Etterson. 2005. Evolutionary responses to changing climate. *Ecology* **86:** 1704–1714.
136. Hill, J.K., H.M. Griffiths & C.D. Thomas. 2011. Climate change and evolutionary adaptations at species' range margins. *Ann. Rev. Entomol.* **56:** 143–159.
137. Gienapp, P., C. Teplitsky, J.S. Alho, *et al.* 2008. Climate change and evolution: disentangling environmental and genetic responses. *Mol. Ecol.* **17:** 167–178.
138. Poethke, H.J., C. Dytham & T. Hovestadt. 2011. A metapopulation paradox: partial improvement of habitat may reduce metapopulation persistence. *Am. Nat.* **177:** 792–799.
139. Parvinen, K. 2007. Evolutionary suicide in a discrete-time metapopulation model. *Evol. Ecol. Res.* **9:** 619–633.
140. Bell, G. & A. Gonzalez. 2011. Adaptation and evolutionary rescue in metapopulations experiencing environmental deterioration. *Science* **332:** 1327–1330.
141. Grant, B.R. 1999. *Evology and Evolution of Darwin's Finches [reprinting with new Afterword].* Princeton University Press. Princeton.
142. Wu, L., A.D. Bradshaw & D.A. Thurman. 1975. Potential for evolution of heavy metal tolerance in plants. 3. Rapid evolution of copper tolerance in *Agrostis stolonifera. Heredity* **34:** 165.
143. Gregory, R.P.G. & A.D. Bradshaw. 1965. Heavy metal tolerance in populations of *Agrostis tenuis* Sibth and other grasses. *New Phytol.* **64:** 131.
144. Zheng, C.Z., O. Ovaskainen & I. Hanski. 2009. Modelling single nucleotide effects in phosphoglucose isomerase on dispersal in the Glanville fritillary butterfly: coupling of ecological and evolutionary dynamics. *Phil. Trans. R. Soc. B Biol. Sci.* **364:** 1519–1532.
145. Brautigam, A. & U. Gowik. 2010. What can next generation sequencing do for you? Next generation sequencing as a valuable tool in plant research. *Plant Biol.* **12:** 831–841.
146. Ekblom, R. & J. Galindo. 2011. Applications of next generation sequencing in molecular ecology of non-model organisms. *Heredity* **107:** 1–15.
147. Bellin, D., A. Ferrarini, A. Chimento, *et al.* 2009. Combining next-generation pyrosequencing with microarray for large scale expression analysis in non-model species. *BMC Genomics.* **10**.
148. Ozgul, A., S. Tuljapurkar, T.G. Benton, *et al.* 2009. The dynamics of phenotypic change and the shrinking sheep of St. Kilda. *Science* **325:** 464–467.
149. Ghalambor, C.K., J.K. McKay, S.P. Carroll & D.N. Reznick. 2007. Adaptive versus non-adaptive phenotypic plasticity

and the potential for contemporary adaptation in new environments. *Funct. Ecol.* **21:** 394–407.

150. Chevin, L.M. & R. Lande. 2011. Adaptation to marginal habitats by evolution of increased phenotypic plasticity. *J. Evol. Biol.* **24:** 1462–1476.
151. Feldhaar, H. 2011. Bacterial symbionts as mediators of ecologically important traits of insect hosts. *Ecol. Entomol.* **36:** 533–543.
152. Leonardo, T.E. & E.B. Mondor. 2006. Symbiont modifies host life-history traits that affect gene flow. *Proc. R. Soc. B Biol. Sci.* **273:** 1079–1084.
153. Lankau, R.A., E. Wheeler, A.E. Bennett & S.Y. Strauss. 2011. Plant-soil feedbacks contribute to an intransitive competitive network that promotes both genetic and species diversity. *J. Ecol.* **99:** 176–185.

Ann. N.Y. Acad. Sci. ISSN 0077-8923

ANNALS OF THE NEW YORK ACADEMY OF SCIENCES
Issue: *The Year in Ecology and Conservation Biology*

The influence of species interactions on geographic range change under climate change

Jessica J. Hellmann,[1] Kirsten M. Prior,[2] and Shannon L. Pelini[3]

[1]Department of Biological Sciences, University of Notre Dame, Notre Dame, Indiana. [2]Department of Ecology and Evolutionary Biology, University of Toronto, Toronto, Canada. [3]Harvard Forest, Harvard University, Petersham, Maine

Address for correspondence: Jessica J. Hellmann, Department of Biological Sciences, 100 Galvin Life Science Center, University of Notre Dame, Notre Dame, IN 46556. hellmann.3@nd.edu

The fossil record tells us that many species shifted their geographic distributions during historic climate changes, but this record does not portray the complete picture of future range change in response to climate change. In particular, it does not provide information on how species interactions will affect range shifts. Therefore, we also need modern research to generate understanding of range change. This paper focuses on the role that species interactions play in promoting or preventing geographic ranges shifts under current and future climate change, and we illustrate key points using empirical case studies from an integrated study system. Case studies can have limited generalizability, but they are critical to defining possible outcomes under climate change. Our case studies emphasize host limitation that could reduce range shifts and enemy release that could facilitate range expansion. We also need improvements in modeling that explicitly consider species interactions, and this modeling can be informed by empirical research. Finally, we discuss how species interactions have implications for range management by people.

Keywords: biogeography; case studies; climate change; enemy release; host specialization; management implications

Introduction

The record of past life embedded in rocks and sediment gives scientists an extraordinary perspective of historic life on Earth. We know from this record, for example, that species migrated both long and short distances in response to glacial and interglacial climatic changes, and geographic change likely predominated over evolutionary change at the end of the last ice age.[1] This basic understanding sets the paradigm for our expectations of how life will change in response to modern climate change—that, if able, species are inclined to move or adjust to changing conditions. In fact, recent evidence suggests that species are already on the move. For example, a recent paper reported that rapid latitudinal and elevational shifts of hundreds of species have occurred in the recent past and the largest range changes occurred where levels of warming were highest.[2]

But there are several aspects of geographic range change due to climate change that are not captured in the paleorecord (e.g., the "Quaternary conundrum" of Botkin *et al.*[3]). These limitations suggest we should not assume that all species have equal ability to shift their ranges as the climate changes. Most importantly, there is limited information in the paleorecord about how interacting species responded to historic climate change (but see Kelley *et al.*[4]). For example, those organisms that depend on other species to sustain minimum viable populations or expand into new areas, such as specialized herbivorous insects or mutualists, would be expected to lag behind their hosts in changing their geographic ranges, but the paleorecord can provide little confirmation (or quantification) of this expectation. Three other limits on our knowledge of historic shifts are also important. First, the record is biased toward abundant and widespread species that left behind conspicuous and quantifiable evidence of their presence. Less well known is the response of rarer or narrowly distributed species, some of which may not have shifted as readily and may have gone extinct as the climate changed,[5] and

doi: 10.1111/j.1749-6632.2011.06410.x

these may have had tight associations with other species. Second, the paleorecord does not provide an accurate picture of the entire range of species that persisted through an ice age. Underestimation of historic ranges, particularly in isolated refugia, may lead to greater migration rates than actually occurred,[6] and variation in the importance of interspecific associations is not captured. Third, the paleorecord does not characterize the historic range of species equally well.[7] Forest trees, for example, leave an excellent record of distribution and abundance, but other organisms are likely to respond differently than trees because of different requirements for cohabitants and other factors.

Given these limitations on knowing about the past, we can turn to a growing, empirical literature that examines species' ranges—and species interactions within those ranges—as we know them today. For example, there have been recent theoretical explorations about species interactions and range change[8–10] and a few summaries of relevant concepts.[11,12] Some recent experimental studies also test the importance of species interactions in the potential for range change. For example, a study by Cunningham *et al.*[13] showed that both climate and interspecific competition affect the biomass of salamanders. Our paper focuses on similar empirical anecdotes to ground some key considerations of climate change and species interactions, with an emphasis on changing species' geography. Individual case studies certainly cannot generalize to all species and instances, but they do help define possible outcomes. We also discuss new methods that must be invented to generate predictions of range change for interacting species, and we explore the implications of species interactions for strategies that humans might use to preserve biodiversity under climate change.

There are other reasons, in addition to species interactions, to suspect that geographic range change may not occur under today's climate change. These additional factors include dispersal capabilities of species that are slow in comparison to the rate of human-caused climate change,[14,15] landscapes that are profoundly altered by human activity that significantly reduce organismal dispersal,[16] and geographic structure in genotype and phenotype that enables local adaptation.[1] These factors also can affect a species' relationship with other taxa. For example, one species dependent on another might be able to disperse through modified landscapes, but its partner species cannot. Or local adaptation of an insect specialist to a host plant may slow or preclude the use of alternate hosts in areas that become climatically suitable to that insect. Alternatively, there may be circumstances where species are freed from a controlling species, thus facilitating a climate-driven range change.[17] Weedy species or species that occupy human habitats may also shift readily under modern climate change, particularly if they are generalists or do not depend on other species.

Our discussion emphasizes species that rely on other species for establishment and persistence under strong, often specialized, biotic control. We recognize that most species interact with others, but some of these interactions involve many potential participants where one interacting species could be exchanged for another (e.g., a generalist predator and its prey). Here, we focus on cases where interactions are relatively specialized because these are the most likely to strongly affect geographic responsiveness to climate change. Finally, we emphasize geographic range shifts because this seems to be a key strategy to enable species persistence. That is not to say that any species that does not shift will go extinct. Existing variation among populations for climatic tolerance, including the sensitivity of species interactions to changing climate, may enable species to stay in place and adjust without genetic evolution.[18] Local or regional microclimates may also allow species and their close associates to persist without moving large distances to track-changing climates. A key research objective is distinguishing which species need to move to persist or thrive, and which do not.

The range shift paradigm

To understand how species interactions may affect range change, it is useful to first explain why species shift at all as the climate changes. If a species is a collection of relatively well-mixed genotypes, that gene pool should have a fitness maximum where conditions for the species are best overall.[19] Fitness should then steadily decline across a gradient from the optimal point.[1] If fitness correlates with population density, abundance would be highest near the center point of a species' range and decline with distance from that point.[20]

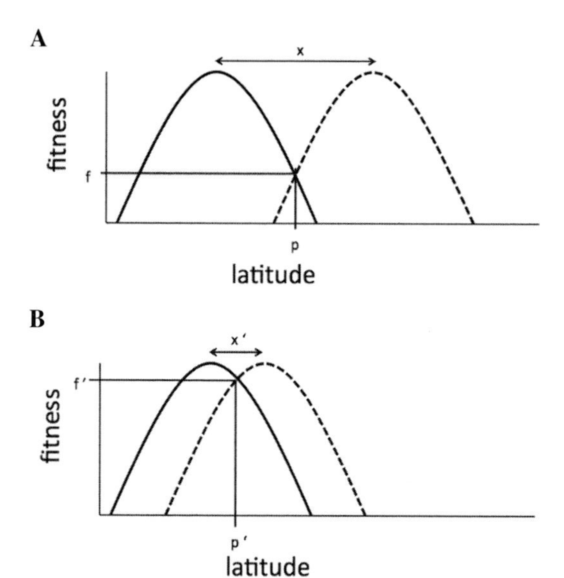

Figure 1. Fitness landscapes and climate change. (A) A peripheral population (p) occurs near the edge of its range with average fitness (f), and the species has a generalized phenotype across its range (solid line). For any change in climate (dashed line) less than or equal to x, the average fitness of p is greater or equal to f. (B) A peripheral population (p′) occurs toward the middle of a local fitness peak with average fitness (f′) (solid line), as in the case of locally adapted genotypes across a species' range. For any change in climate (dashed line) less than or equal to x′, the average fitness of p′ is greater or equal to f′. Comparing (A) and (B), $x' \ll x$.

If we take such a hypothetical range of well-mixed genotypes and start shifting the climate, assuming no evolution and little or no movement of individuals across the climatic gradient, mean population fitness changes so that populations on the side where the optimum is moving away will decline, and populations in the direction of the environmental shift will increase (Fig. 1). In the direction of the environmental change, peripheral populations become less marginal, and we have called this process "peripheral population enhancement."[21] If one then allows for the possibility of colonization or movement of individuals (or their genotypes) into new areas, this enhancement can prime the pump for a poleward range shift.

In reality, there is a great diversity of range sizes and shapes, some with gaps within a species' range where local conditions are unsuitable.[22,23] Further, where a species can live depends not only on abiotic conditions such as climate but also on biotic ones as well, such as available resources and avoidance of predators. These biotic factors also affect density patterns within a range and skew the location of maximum population density. For example, studies on density in peripheral versus central populations of the specialist butterfly, *Erynnis propertius*, showed that higher abundances occurred near the range periphery, with the most northerly populations containing some of the most dense populations.[21,24]

In addition, many species are not genetically well mixed, and this genetic differentiation across the range can promote local adaptation. Gene flow from larger, more centrally located populations are thought to slow or preclude local adaptation at the range periphery, but the balance of gene flow, local selection, and genetic drift all affect the extent to which species tolerances and affinities are preserved over time and space.[19,25,26] The assumed differences in relative density from central to peripheral populations are a factor in this gene swamping process because genes adapted to more central conditions outnumber any genotypes particularly suited for peripheral conditions. Local adaptation has been shown for physiological tolerances in trees[27] and soil microbes,[28] however it can also occur with respect to local interactions. Local adaptation can occur for pathogens, parasitoids, and small herbivores, for example, because many of the conditions for local adaptation, notably strongly local selection, are met in these cases.[29,30]

Local adaptation can also skew the expectation that fitness is high in the range center and lower toward the range periphery, particularly if there are barriers to gene flow from the center to the periphery or if strong selection acts against central traits in some locations. If local adaptation occurs in populations at the poleward (or upward) periphery of a species' range, then fitness and population size may decline in these populations under climate change (Fig. 1). This occurs because the population occupies a local fitness optimum; changing conditions move the population away from that local optimum, causing population declines. This contrasts with a situation of generalized genotypes across the species' range. In this case, climate change would increase the fitness and abundance of peripheral populations because it moves them closer to the global fitness optimum (Fig. 1). For reasons of local adaptation, however, we may need to consider species' responses to climate change at a lower taxonomic resolution.[10,31]

Case studies on interacting species under climate change

To illustrate key issues of species interactions for range change and local adaptation in some of these interactions, we highlight three empirical studies that raise three distinct issues: the lack of hosts for a specialist herbivore in areas that become suitable under climate change; the presence of suboptimal hosts in newly colonized sites under climate change; and the potential for demographic release due to loss of specialized predators under climate change. Two of these show how codependent species can be limited by biotic interactions, even for shifts within a species' range, and the other highlights how escape from antagonistic interactions could facilitate range expansions. Each of these cases emphasizes that although climate may become suitable for a range shift—i.e., a species can physiologically persist outside its current range—species interactions can limit or promote the opportunity for range shifts. In each of these case studies, we draw on a single biome and its associated species where we have extensively studied the potential for changes in geographic distributions as affected by species interactions. We complement these examples with citations to other, similar research.

Case 1: shifts beyond the species' range periphery are hindered by resource availability

A simple reason why a species may not shift its range in response to climate change is because its boundary is set by a specialized resource, often a food source. In the case of specialized herbivorous insects, for example, host plant limitation can prevent range expansion when there are no suitable hosts outside the area of current occupancy. The degree to which insects or other species are host-limited in their geographic range is unknown, but a preliminary analysis performed for butterflies found that 74 butterfly species from 15 subfamilies have a northern range boundary within the United States (where there are good data on boundary location) and use a single host species.[32] On the basis of county-level data on butterfly and host plant occupancy, 46% of these species found their northern boundary within 100 km of the boundary of their host. This means that approximately 35 U.S. species could be limited to shifts less than 100 km without any shift in their host. Any further migration would require a range shift of their host as well or an evolutionary adaptation enabling the use of a novel host plant species. An example of this phenomenon already occurring in nature was observed by Merrill *et al.*[33] regarding elevational range change in a European butterfly. They recorded egg survival at higher elevations with increased warming but found no upward colonization because of a lack of host plants. This implies a positive effect of climate change populations in the direct of climate change, but without a corresponding range shift.

One butterfly that we have studied extensively is another example of a species that is at least partially range-limited by its host. The Propertius duskywing (*E. propertius*) lives throughout coastal, western North America from Baja, Mexico, to Vancouver Island, British Columbia (BC). It feeds on a variety of oak (*Quercus*) species, but in the northern third of its range, only one oak species occurs, Garry oak (*Quercus garryana*).The location of the most northern population of the duskywing is also the most northern population of Garry oak. Garry oak trees live several hundred years and do not reproduce until they are at least 20 years old.[34] Individuals over 60 years old have the highest acorn production. Its acorns are heavy and fall close to the tree, but they are dispersed by animals such as stellar's jays and small mammals.[35,36] Given these life history traits, it is difficult to imagine a large and rapid range shift in Garry oak. In fact, relic populations of Garry oak persisted close to the ice sheet during the last ice age and dispersed less than 300 km to occupy their current range extent today.[37]

It also seems unlikely that the duskywing will evolve new host preferences, as southern BC does not harbor any other species in the Fagaceae family.[38] Thus, the duskywing will likely be prevented from colonizing areas that might become climatically suitable because of its dependence on a slower-migrating host species. This dilemma invites the idea that trees might be planted further to the north for the benefit of duskywing butterfly populations (see management implications below). In fact, this scenario is not implausible because both duskywings and Garry oak are subjects of considerable conservation concern in BC.

Case 2: shifts within the species' range are hindered by resource availability

A second case of species shifts being limited by species interactions is also illustrated by the duskywing butterfly. The geographic range of the duskywing lays over a variety of oak species in western North America, and many of these species have distinct, and sometimes nonoverlapping, geographies. In the southern portion of its range in southern California (CA), duskywings predominately feed on Coast live oak (*Quercus agrifolia*). In the northern part of their range (Oregon [OR], Washington [WA], and BC), they feed exclusively on Garry oak because no other oak occurs. Following on the argument above that shifts in the geographic ranges of trees could lag shifts in the ranges of flying insects, it is likely that duskywing populations—or the genes that compose them—could shift on top of a relatively static distribution of hosts. Specifically, Pelini *et al.*[39] suggested that northward migrating populations (or genes) under climate change could move from a region dominated by Coast live oak to a region dominated by Garry oak, and this novel host could affect insect fitness and abundance for populations that are adapted to more southerly oak species.

To test for fitness effects of novel host plant species, Pelini *et al.*[39] captured duskywings from populations dominated by Garry oak (northern CA and southwest OR) and by Coast live oak (southern CA) and exposed them to their natal and nonnatal host plant. They found local adaptation to Garry oak in the northern populations with reduced survival on Coast live oak. Interestingly, the southern populations did equally well on both hosts (Fig. 2). It seems that southern populations (southern CA) harbor greater host capacity, perhaps associated with greater host diversity. Therefore, its potential movement into a more northerly region (northern CA and OR) where a different host occurs is likely to have minimal fitness effects. If the scenario had been reversed, however, with Garry oak individuals needing to move south into a region of Coast live oak, fitness would likely decline.

The duskywing story is a fortunate one where the capacity for novel host plants seems built in to the populations that are likely to need it as they shift northward. This story speaks, however, to the importance of evolutionary history in species interactions and to the possibility for other species and scenarios where tolerances are reversed or local adaptation is more widespread across populations. Several other studies have demonstrated that geographic turnover in host plant suitability and

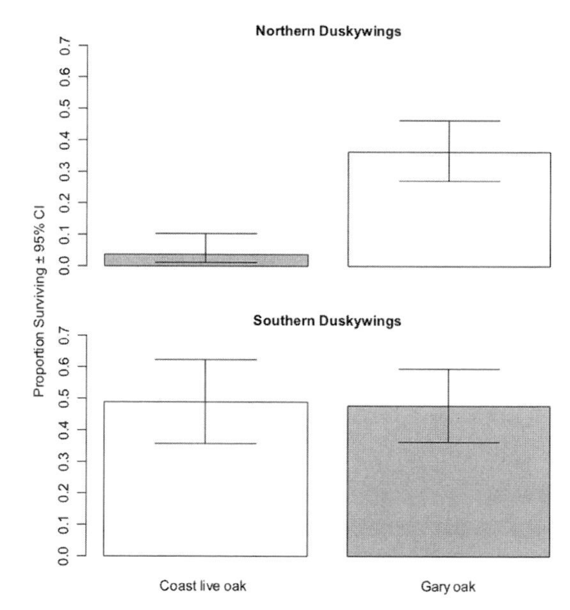

Figure 2. Proportion of northern (top panel) and southern (bottom panel) duskywings (*Erynnis propertius*) surviving (±95% CI) on experimental host plants. Gray bars indicate nonnatal and white bars represents natal *Quercus* hosts. Figure adapted from Pelini *et al.*[39]

availability that fosters local adaptation to different host plants and/or differing in degrees of host specificity across a species' range.[40–42] This "geographic mosaic of coevolution" also has been shown for disease resistance, symbiotic associations, and defense from enemies.[43]

Case 3: shifts beyond the species' range periphery are facilitated by enemy release

A third implication of species interactions for changing biogeography under climate change is that species undergoing range expansions escape from antagonistic interactions, such as those from their specialist enemies. In this case, species' range expansions can be enhanced or facilitated by the loss of interacting species. Enemy release has been demonstrated most frequently for invasive plants that have been introduced over long distances, freed from pathogen or herbivore enemies.[44–47] The role of enemy release in facilitating climate-driven range expansions, however, has yet to be extensively explored and may differ from the intercontinental context.

Species undergoing climate-driven (or short distance) range expansions may receive less benefit from enemy escape opportunities than introduced species that are transported over long distances.[48,49]

This is because species undergoing climate-driven expansions move into adjacent habitats where recipient communities likely contain biologically or taxonomically similar species to native communities. In addition, given the relatively short distance of climate-driven expansions, interacting species will likely be able to more effectively track their range-expanding counterparts. Species transported over long distances, on the other hand, will likely lose more enemies and be introduced into communities with which they share little coevolutionary history. In unrelated communities, fewer enemies may be able to shift to novel species, leading to a higher potential for release.[50] A study by Englekes *et al.*,[51] for example, examined enemy release for species that have moved over different distances. They predicted that plants undergoing intercontinental introductions (i.e., long-distance expansions) would be more released from pathogen and herbivore enemies than plants undergoing intracontinental introductions (i.e., short-distance expansions). Contrary to their predictions, however, they found that both types of range-expanding plants were equally released from enemy control, suggesting that species undergoing climate-driven range expansions can be released from enemies and become invasive.

Another reason that species undergoing poleward range expansions could also benefit from enemy escape opportunities is a decline in enemy richness with latitude.[52] Previous studies of range-expanding insects have found lower parasitoid diversity, changes in composition, and lower parasitoid rates in species' expanded ranges.[17,53,54] However, whether lower parasitoid rates occurred because of a decrease in natural enemy richness towards the poles is unknown in many of these cases. For example, Menéndez *et al.*[17] found that the Brown argus butterfly (*Aricia agestis*), a species that has undergone a climate-driven range expansion, encountered lower parasitoid attack rates, but it experienced similar parasitoid richness in its expanded range. Similar parasitoid species were present in the expanded range, but they attacked an alternative butterfly species. Menéndez *et al.* suggest that the Brown argus butterfly experienced lower attack rates in its expanded range because parasitoids were not locally adapted to the novel species. This study, however, did not explicitly test if enemy reduction translated into demographic release.

Experimental manipulations that measure the effect of enemies on species demographics in native and expanded regions are necessary to uncover if enemy release facilitates range expansions. Consider a recent case study of an herbivorous insect that has undergone an intracontinental introduction, into a similar community. This introduction is analogous to a climate-driven expansion as this species moved into an adjacent habitat where it interacts with many similar community members as it does in its native range. The Jumping gall wasp (*Neuroterus saltatorius*) is an oak gall wasp that is native to oak ecosystems of the western United States including the *Q. garryana* ecosystem described earlier. This species was historically absent from Vancouver Island and neighboring Gulf Islands, BC, however, and it appeared near Victoria, BC, in the early 1980s. It has since spread to all oak habitats on Vancouver Island, where it reaches higher density than in its native range.[55] It was likely brought by car traffic on the ferry from WA. Prior[56] showed that there are fewer parasitoids attacking the gall wasp in BC than in WA, allowing the possibility that release from enemies enables outbreaks to occur in the expanded range (Fig. 3A).

Prior[56] performed an experiment to test if parasitoids control gall abundance in the native range (WA) and if outbreaks are caused by release from parasitoid control in the introduced range (BC). This experiment compared the survivorship of galls reared in the native and introduced range in parasitoid exclosures and open controls. Because the difference in survivorship between parasitoid exclosures and controls was greater in the introduced than in that native range, Prior[56] concluded that enemy release was not a likely driver of outbreaks in the expanded range (Fig. 3B). Instead, the influence of parasitoids was stronger in the introduced area, despite lower parasitoid abundance and lower attack rates. This result suggests that some additional factor controls this species in its native range and facilitates its success in the introduced range, such as host plant suitability or some other factor that modifies the interaction between the gall wasp and its parasitoids. Weather, however, did not significantly differ between study regions.

This case study and other, similar studies suggest that enemies may be lost for species undergoing poleward range expansions, allowing for the

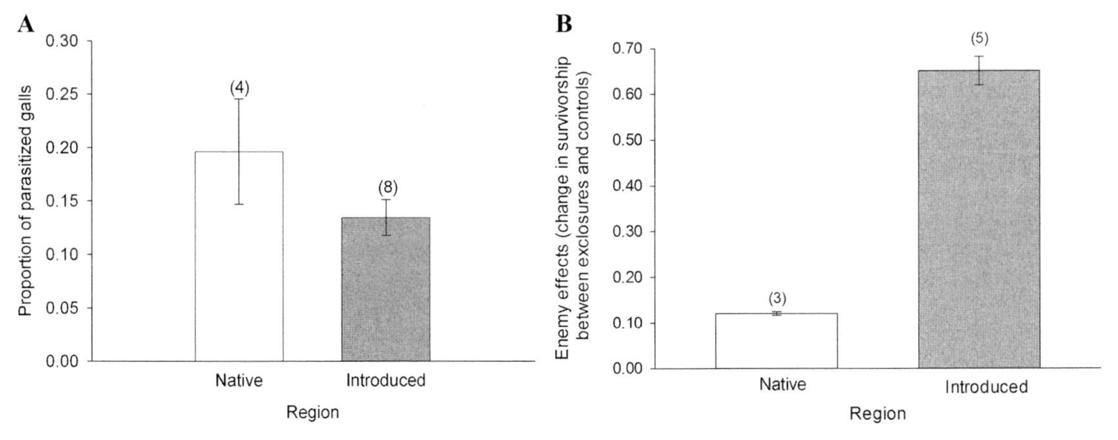

Figure 3. Parasitoid attack rates on *Neuroterus saltatorius* were lower in the introduced region (BC) compared to the native region (WA), suggesting that outbreaks could be a result of enemy release. (A) Mean parasitoid attack rates from galls collected at multiple sites (depicted as numbers above bars) in the native range (white) and the introduced range (gray) (±S.E.). Despite enemy loss, an enemy exclusion experiment found that the effect of enemies on prey performance (gall wasp survivorship) was lower in the native range and higher in the introduced range. (B) Enemy effects were measured as the difference in survivorship between exclosures and controls at multiple sites (depicted above bars) within regions (±S.E.). If enemy release were causing outbreaks, enemy effects would be higher in the native range (depicting control by enemies) and lower in the introduced range (depicting release from enemy control). Figure adapted from Prior.[56]

possibility of outbreaks resulting from enemy release. The existence of outbreaks, and even the pattern of enemy loss, however, does not necessarily imply that enemy release has occurred, and further experiments are needed to determine how frequently enemy release may occur and facilitate range expansions under climate change.

Predicting future distributions of interacting species

Although empirical studies can provide vital insights into our understanding of species interactions and range change, they do not often come with projections of where a species is likely to live in the future. Yet we need to know specifically where to focus our efforts to manage natural resources under climate change.[57] We need to predict not only *how* ranges can change (above), but *where*. At present, climate envelope models, or niche models, fill the role of spatial projection.[58–60] They predict potential future occupancy of a species given data about the conditions where it persists today, and they have been criticized in a number of papers.[61–63]

Of course, projected occupancy is not the same as realized occupancy, and just because a species could theoretically live in an area does not mean that it can or will get there. Species interactions are one reason why future range projections may not be realized, because niche models do not explicitly capture the influence of species' interactions on geographic distribution.[12]

A number of advances to model projections could better account for species interactions. Comodeling of species using existing niche models is a possible start, but few studies have considered the future occupancy of a species and then used that projected occupancy as input to another model. An exception is Araújo and Luoto,[64] who showed that including host plant as an input variable in niche modeling of a butterfly significantly affected both predicted occupancy for the present and the future, relative to a climate-only analysis. Preston *et al.*[65] also incorporated habitat into niche models for a butterfly and a bird that are habitat specialists and found that excluding habitat reduced predicted occupancy by more than 68%.

A second approach is using current occupancy of an interacting species to determine the extent to which a species' current range is determined by association with another organism. This approach could be helpful in instances where the physiology and ecological nature of interacting species may not be well known or well characterized. Many climate envelope models (e.g., Maximum Entropy or

"MaxEnt") generate output on the strength of the relationship between occupancy and other, input variables, and this could be used as a rough measure of strength of interaction.

Third, a coupling of niche models and physiological models could reveal the extent to which species' interactions limit current occupancy. This is because physiological models predict the fundamental niche whereas occupancy data reveal the realized niche. The difference between these two models suggests where species' interactions are limiting today, and information about these locations could be used to adjust niche model projections for the future.[66]

Fourth, advances in "process-based" models also move us closer to incorporating species interactions. These models start from first principles about the factors that affect species occupancy and abundance, often using data collected from experiments.[66,67] For example, Crozier and Dwyer[68] modeled climate-driven range expansion for the butterfly, *Atalopedes campestris*, with population growth rate as a function of temperature, and they estimated model parameters based on field data. In species where growth rate is also a function of another species, a coupled growth model could be used to project future occupancy.

Implications for management under climate change

Species interactions will significantly complicate efforts to advance single-species conservation under climate change, especially as related to geographic range change. Yet conservation for individual species is the main—and probably the only effective—way to intervene on behalf of biodiversity as the climate changes. Although ecosystem- or habitat-level conservation is often cited in the United States as the ideal level for conservation and restoration activities,[69] climate change will alter ecosystem composition as we know it. This will happen because species respond individually, or in close association with a few other species, to climate change. Entire ecosystems do not relocate to new areas as the climate shifts, and the best evidence for this is a comparison of historic communities captured by the paleorecord to modern species assemblages. Many historic communities, such as during the Quaternary, show no analog to modern communities,[70] and projected changes under modern climate change are likely to generate new assemblages with no analogs as well.[71]

Although entire communities do not move together, clearly some species depend on others for their basic livelihood. Such dependent species cannot show up in new areas where codependents are absent (above), and they cannot be placed by people in such areas either. Consider the case of managed relocation, for example, or the idea that humans might assist in the process of species movement under climate change by moving organisms from areas of historic occupancy to places where they are likely to persist in the future. Most of the conversation about managed relocation to date considers single species in isolation,[72–75] but this may be more the exception than the rule. For example, species might be introduced with other species to increase the chances of successful establishment as in the case of the preferred host plant for a specialized butterfly. Mutualistic soil microbes from the native region of a plant are another example, where microbes with historic association may facilitate plant growth more than microbes in the introduction region.[28] The motivation for such multispecies introductions may be particularly strong in the case of highly endangered species where failure to successfully establish a population must be avoided because repeated extraction from source populations is extremely detrimental.

Multiple species introductions might also take place inadvertently via managed relocation. This could occur for pathogens on plant species, for example, and is the basis for much of the plant inspection pursued by USDA APHIS for international transport of plant material. Conversely, managed relocation without specialized predators could lead to outbreaks (see enemy release above). For species with high reproductive rates, biotic checks on population size should be evaluated if managed relocation is pursued. (However, fears of pest outbreaks may be a key reason to avoid managed relocation species with high potential fecundity).

On the other hand, species interactions provide several other opportunities for helping species persist through climate change. Because climate change is likely to alter species interactions, these changes can be exploited for management purposes. For example, studies by Pelini *et al.*[39] demonstrated that fitness of caterpillars on different host plants shifted under simulated warming conditions so that the host plant supporting the highest level of caterpillar growth under current conditions became the least effective under elevated temperature and vice versa for the least effective plant becoming the most effective when temperature was increased. This result implies that management fostering the soon-to-be-preferred plant could directly benefit butterfly populations under climate change, and if these actions were taken in populations near the range periphery, it could affect the probability of poleward colonization by increasing the abundance of potential colonists. Such a strategy does not involve planting new things in new places but instead harnesses an understanding of a climate change impact in one species to cause a benefit in another species.

Conclusion

There is no substitute for good empirical research on species interactions and their responses to climate change in terms of understanding future outcomes, building better models, and designing successful management strategies. We need more of these studies as complex interactions are likely to cause most of the ecological surprises that occur under climate change. A major conundrum, however, is time and resource limitations—there, simply, are not enough experimental ecologists to study global impacts of climate change on all living systems. Instead, we must rely on case studies that highlight issues for consideration (such as those that examine strongly interacting species that are not likely to expand at similar rates), while also enhancing the realism of our ecological models so that we can generate projections with high confidence. At present, inadequate methods are leading us awkwardly into climate-change management. We hope that the next 5–10 years see substantial improvements in ecological modeling, informed by realistic case studies, which will begin to bring us on par with the quantitative projections of global and regional climate models.

The complexities of species interactions also remind us that the only way to maintain biodiversity as we know and appreciate it today is to avoid a climate catastrophe. Respectable scientists and their models suggest that concentrations of CO_2 (or CO_2 equivalents) over 450 ppm could profoundly alter the relatively benign climate that Earth enjoys today. And we are already at 390 ppm as of this writing. This reality should compel us to take strong action today so that nature can take care of itself in the future.

Acknowledgments

Thank you to the Hellmann Lab and two anonymous reviewers for critical comments and to the Office of Science (BER), U.S. Department of Energy Grant DEFG02–05ER for financial support.

Conflicts of interest

The authors declare no conflicts of interest.

References

1. Davis, M.B. & R.G. Shaw. 2001. Range shifts and adaptive responses to Quaternary climate change. *Science* **292:** 673–679.
2. Chen, I.-C. *et al.* 2011. Rapid range shifts of species associated with high levels of climate warming. *Science* **333:** 1024–1026.
3. Botkin, D.B. *et al.* 2010. Forecasting the effects of global warming on biodiversity. *Bioscience* **57:** 227–236.
4. Kelley, P. *et al.* 2003. *Predator-prey Interactions in the Fossil Record.* Kluwer Academic Publishers, New York, NY.
5. Jackson, S.T. & C. Weng. 1999. Late quaternary extinction of a tree species in eastern North America. *Proc. Natl. Acad. Sci. USA* **96:** 13847–13852.
6. McLachlan, J.S. *et al.* 2005. Molecular indicators of tree migration capacity under rapid climate change. *Ecology* **86:** 2088–2098.
7. Kidwell, S.M. & K.W. Flessa. 1995. The quality of the fossil record: populations, species, and communities. *Annu. Rev. Ecol. Evol. Syst.* **26:** 269–299.
8. Münkemüller, T. *et al.* 2009. Disappearing refuges in time and space: how environmental change threatens species coexistence. *Theor. Ecol.* **2:** 217–227.
9. Brooker, R.W. *et al.* 2007. Modelling species' range shifts in a changing climate: the impacts of biotic interactions, dispersal distance and the rate of climate change. *J. Theor. Biol.* **245:** 59–65.
10. Atkins, K.E. & J.M.J. Travis. 2010. Local adaptation and the evolution of species' ranges under climate change. *J. Theor. Biol.* **266:** 449–457.
11. Gilman, S.E. *et al.* 2010. A framework for community interactions under climate change. *Trends Ecol. Evol.* **25:** 325–331.
12. VanderPutten, W.H. *et al.* 2010. Predicting species distribution and abundance responses to climate change: why it is essential to include biotic interactions across trophic levels. *Philos. Trans. R. Soc. Lond. B. Biol. Sci.* **365:** 2025–2034.
13. Cunningham, H.R. *et al.* 2009. Competition at the range boundary in the slimy salamander: using reciprocal transplants for studies on the role of biotic interactions in spatial distributions. *J. Anim. Ecol.* **78:** 52–62.
14. Midgley, G.F. *et al.* 2006. Migration rate limitations on climate change-induced range shifts in Cape Proteaceae. *Diversity Distrib.* **12:** 555–562.
15. Loarie, S.R. *et al.* 2009. The velocity of climate change. *Nature* **462:** 1052–1055.

16. Jules, E.S. & P. Shahani. 2003. A broader ecological context to habitat fragmentation: why matrix habitat is more important than we thought. *J. Veg. Sci.* **14:** 459–464.
17. Menéndez, R. *et al.* 008. Escape from natural enemies during climate-driven range expansion: a case study. *Ecol. Entomol.* **33:** 413–421.
18. Hof, C. *et al.* 2011. Rethinking species' ability to cope with rapid climate change. *Glob. Change Biol.* **17:** 2987–2990.
19. Sexton, J.P. *et al.* 2009. Evolution and ecology of species range limits. *Ann. Rev. Ecol. Evol. Syst.* **40:** 415–436.
20. Brown, J.H. *et al.* 1995. Spatial variation in abundance. *Ecology* **76:** 2028–2043.
21. Pelini, S.L. *et al.* 2009. Translocation experiments with butterflies reveal limits to enhancement of poleward populations under climate change. *Proc. Natl. Acad. Sci. USA* **106:** 11160–11165.
22. Brown, J.H. *et al.* 1996. The geographic range: size, shape, boundaries, and internal structure. *Ann. Rev. Ecol. Evol. Syst.* **27:** 597–623.
23. Sagarin, R.D. *et al.* 2006. Moving beyond assumptions to understand abundance distributions across the ranges of species. *Trends Ecol. Evol.* **21:** 524–530.
24. Hellmann, J.J. *et al.* 2008. The response of two butterfly species to climatic variation at the edge of their range and the implications for poleward range shifts. *Oecologia* **157:** 583–592.
25. Holt, R.D. & R. Gomulkiewicz. 1997. How does immigration influence local adaptation? A reexamination of a familiar paradigm. *Am. Nat.* **149:** 563–572.
26. Alleaume-Benharira, M. *et al.* 2006. Geographical patterns of adaptation within a species' range: interactions between drift and gene flow. *J. Evol. Biol.* **19:** 203–215.
27. Savolainen, O. *et al.* 2007. Gene flow and local adaptation in trees. *Ann. Rev. Ecol. Evol. Syst.* **38:** 595–619.
28. Belotte, D. *et al.* 2003. An experimental test of local adaptation in soil bacteria. *Evolution* **57:** 27–36.
29. Greischar, M. A. & B. Koskella. 2007. A synthesis of experimental work on parasite local adaptation. *Ecol. Lett.* **10:** 418–434.
30. Kawecki, T.J. & D. Ebert. 2004. Conceptual issues in local adaptation. *Ecol. Lett.* **7:** 1225–1241.
31. Zakharov, E.V. & J.J. Hellmann. 2008. Genetic differentiation across a latitudinal gradient in two co-occurring butterfly species: revealing population differences in a context of climate change. *Mol. Ecol.* **17:** 189–208.
32. Pelini, S.L. *et al.* 2009. Climate change and temporal and spatial mismatches in insect communities. In *Climate Change: Observed Impacts on Planet Earth.* T. Letcher, Ed.: 215–231. Elsevier, Amsterdam, the Netherlands.
33. Merrill, R.M. *et al.* 2008. Combined effects of climate and biotic interactions on the elevational range of a phytophagous insect. *J. Anim. Ecol.* **77:** 145–155.
34. Graves, W.C. 1980. Annual oak mast yields from visual estimates. *Proceedings of the Symposium on the Ecology, Management, and Utilization of California Oaks*, 270–274. Claremont, California.
35. Aizen, M.A. & W.A. Patterson. 1990. Acorn size and geographical range in the North American oaks (Quercus L.). *J. Biogeogr.* **17:** 327.
36. Fuchs, M.A. & P.G. Krannitz. 1999. Dispersal of garry oak acorns by steller's jays. *Proceedings of a Conference on the Biology and Management of Species and Habitats at Risk*, 263–266. Kamloops, British Columbia.
37. Marsico, T.D. *et al.* 2009. Patterns of seed dispersal and pollen flow in Quercus garryana (Fagaceae) following post-glacial climatic changes. *J. Biogeogr.* **36:** 929–941.
38. Ehrlich, P.R. & P.H. Raven. 1964. Butterflies and plants: a study in coevolution. *Evolution* **18:** 586–608.
39. Pelini, S.L., J. A. Keppel, A.E. Kelley & J.J. Hellmann. 2010. Adaptation to host plants may prevent rapid insect responses to climate change. *Glob. Change Biol.* **16:** 2923–2929.
40. Fox, L.R. & P.A. Morrow. 1981. Specialization: species property or local phenomenon? *Science* **211:** 887–893.
41. Sword, G.A. & E.B. Dopman. 1999. Developmental specialization and geographic structure of host plant use in a polyphagous grasshopper, Schistocerca emarginata (=lineata) (Orthoptera: Acrididae). *Oecologia* **120:** 437–445.
42. Parry, D. & R.A. Goyer. 2004. Variation in the suitability of host tree species for geographically discrete populations of forest tent caterpillar. *Environ. Entomol.* **33:** 1477–1487.
43. Thompson, J.N. 2005. *The Geographic Mosaic of Coevolution.* University of Chicago Press, Chicago, IL.
44. Prior, K.M. & J.J. Hellmann. 2012. A review of the enemy release hypothesis as an explanation for the success of invasive species in multiple trophic levels and ecosystems. *Invas. Spec. Global. World* In press.
45. Callaway, R.M. *et al.* 2004. Soil biota and exotic plant invasion. *Nature* **427:** 731–733.
46. DeWalt, S.J. *et al.* 2004. Natural-enemy release facilitates habitat expansion of the invasive tropical shrub Clidemia hirta. *Ecology* **85:** 471–483.
47. Williams, J.L. *et al.* 2010. Testing hypotheses for exotic plant success: parallel experiments in the native and introduced ranges. *Ecology* **91:** 1355–1366.
48. Mitchell, C.E. *et al.* 2006. Biotic interactions and plant invasions. *Ecol. Lett.* **9:** 726–740.
49. Mueller, J.M. & J.J. Hellmann. 2008. An assessment of invasion risk from assisted migration. *Conserv. Biol.* **22:** 562–567.
50. Strauss, S.Y. *et al.* 2006. Exotic taxa less related to native species are more invasive. *Proc. Natl. Acad. Sci. USA* **103:** 5841–5845.
51. Engelkes, T. *et al.* 2008. Successful range-expanding plants experience less above-ground and below-ground enemy impact. *Nature* **456:** 946–948.
52. Rosenzweig, M.L. 1995. *Species Diversity in Space and Time.* Cambridge University Press, Cambridge, UK.
53. Schönrogge, K. *et al.* 1998. Invaders on the move: parasitism in the sexual galls of four alien gall wasps in Britain (Hymenoptera: Cynipidae). *Proc. R. Soc. Lond. B. Biol. Sci.* **265:** 1643–1650.
54. Gröbler, B.C. & O.T. Lewis. 2008. Response of native parasitoids to a range-expanding host. *Ecol. Entomol.* **33:** 453–463.
55. Prior, K.M. & J.J. Hellmann. 2010. Impact of an invasive oak gall wasp on a native butterfly: a test of plant-mediated competition. *Ecology* **91:** 3284–3293.
56. Prior, K.M. 2012. Novel community interactions following species' range expansions. Doctoral dissertation. University of Notre Dame, Notre Dame, IN.

57. Sinclair, S.J. *et al.* 2010. How useful are species distribution models for managing biodiversity under future climates? *Ecol. Soc.* **15** (1): 8. Retrieved January 16, 2012, from www.ecologyandsociety.org/vol15/issue1/art8.
58. Roberts, D.R. & A. Hamann. 2011. Predicting potential climate change impacts with bioclimate envelope models: a palaeoecological perspective. *Global Ecol. Biogeogr.* doi:10.1111/j.1466-8238.2011.00657.x
59. Pearson, R.G. & T.P. Dawson. 2003. Predicting the impacts of climate change on the distribution of species: are bioclimate envelope models useful? *Global Ecol. Biogeogr.* **12:** 361–371.
60. Hijmans, R.J. & C.H. Graham. 2006. The ability of climate envelope models to predict the effect of climate change on species distributions. *Glob. Change Biol.* **12:** 2272–2281.
61. Morin, X. & M.J. Lechowicz. 2008. Contemporary perspectives on the niche that can improve models of species range shifts under climate change. *Biol. Lett.* **4:** 573–576.
62. Ibáñez, I. *et al.* 2006. Predicting biodiversity change: outside the climate envelope, beyond the species-area curve. *Ecology* **87:** 1896–906.
63. Wiens, J.A. *et al.* 2009. Niches, models, and climate change: assessing the assumptions and uncertainties. *Proc. Natl. Acac. Sci. USA* **106:** 19729–19736.
64. Araújo, M.B. & M. Luoto. 2007. The importance of biotic interactions for modelling species distributions under climate change. *Global Ecol. Biogeogr.* **16:** 743–753.
65. Preston, K.L. *et al.* 2008. Habitat shifts of endangered species under altered climate conditions: importance of biotic interactions. *Glob. Change Biol.* **14:** 2501–2515.
66. Kearney, M. & W. Porter. 2009. Mechanistic niche modelling: combining physiological and spatial data to predict species' ranges. *Ecol. Lett.* **12:** 334–350.
67. Buckley, L. *et al.* 2011. Does including physiology improve species distribution model predictions of responses to recent climate change? *Ecology* **92:** 2214–2221. doi:10.1890/11-0066.1
68. Crozier, L. & G. Dwyer. 2006. Combining population-dynamic and ecophysiological models to predict climate-induced insect range shifts. *Am. Nat.* **168:** 853–866.
69. Beattie, M. 1996. An ecosystem approach to fish and wildlife conservation. *Ecol. Appl.* **6:** 696–699.
70. Overpeck, J.T. *et al.* 1992. Mapping eastern North American vegetation change of the past 18 ka: no-analogs and the future. *Geology* **20:** 1071–1074.
71. Williams, J.W. *et al.* 2007. Projected distributions of novel and disappearing climates by 2100 AD. *Proc. Natl. Acad. Sci. USA* **104:** 5738–5742.
72. Hoegh-Guldberg, O. *et al.* 2008. Assisted colonization and rapid climate change. *Science* **321:** 345–346.
73. McLachlan, J.S. *et al.* 2007. A framework for debate of assisted migration in an era of climate change. *Conserv. Biol.* **21:** 297–302.
74. Richardson, D.M. *et al.* 2009. Multidimensional evaluation of managed relocation. *Proc. Natl. Acad. Sci. USA* **106:** 9721–9724.
75. Hewitt, N. *et al.* 2011. Taking stock of the assisted migration debate. *Biol. Conserv.* **144:** 2560–2572.

Ann. N.Y. Acad. Sci. ISSN 0077-8923

ANNALS OF THE NEW YORK ACADEMY OF SCIENCES
Issue: *The Year in Ecology and Conservation Biology*

Not by science alone: why orangutan conservationists must think outside the box

Erik Meijaard,[1,2] Serge Wich,[3,4] Marc Ancrenaz,[5] and Andrew J. Marshall[6]

[1]People and Nature Consulting International, Kerobokan, Bali, Indonesia. [2]School of Archaeology and Anthropology, Australian National University, Canberra, Australia. [3]Sumatran Orangutan Conservation Program (PanEco-YEL), Medan, Indonesia. [4]Anthropological Institute and Museum, University of Zurich, Zurich, Switzerland. [5]Kinabatangan Orang-utan Conservation Project, Sandakan, Sabah, Malaysia. [6]Department of Anthropology, Graduate Group in Ecology and Animal Behavior Graduate Group, University of California, Davis, Davis, California

Address for correspondence: Erik Meijaard, People and Nature Consulting International, Country Woods House 306, JL WR Supratman, Pondok Ranji-Rengas, Ciputat, Jakarta 15412, Indonesia. emeijaard@gmail.com

Orangutan survival is threatened by habitat loss and illegal killing. Most wild populations will disappear over the next few decades unless threats are abated. Saving orangutans is ultimately in the hands of the governments and people of Indonesia and Malaysia, which need to ensure that habitats of viable orangutan populations are protected from deforestation and well managed to ensure no hunting takes place. Companies working in orangutan habitat also have to play a much bigger role in habitat management. Although the major problems and the direct actions required to solve them—reducing forest loss and hunting—have been known for decades, orangutan populations continue to decline. Orangutan populations in Sumatra and Borneo have declined by between 2,280 and 5,250 orangutans annually over the past 25 years. As the total current population for the two species is some 60,000 animals in an area of about 90,000 km^2, there is not much time left to make conservation efforts truly effective. Our review discusses what has and has not worked in conservation to guide future conservation efforts.

Keywords: deforestation; great apes; hunting; Indonesia; Malaysia; nongovernmental organizations; orangutan; palm oil; plantations; *Pongo abelii*; *Pongo pygmaeus*

Introduction

Orangutans (*Pongo* spp.) are among the most iconic species in wildlife conservation. Popular television programs such as *Orangutan Island* produced by Animal Planet and *Orangutan Diary* by the British Broadcasting Corporation (BBC) are testimony to the public's interest in these great apes. People's affinity with orangutans is at least partly based on the red ape's behavior, facial expressions, and mannerisms, which can be uncannily human. Indeed, genetic comparisons and morphological, cognitive, and behavioral similarities do indicate the close evolutionary relationship between humans and orangutans.[1–3]

Despite their similarities, the evolutionary paths leading to orangutans and to humans diverged some 9–13 million years ago.[4] After that evolutionary split, a range of orangutan-like taxa or pongins evolved, including *Gigantopithecus*, the largest ape that ever lived. These species primarily occurred in present-day Europe and mainland Asia. Of this diverse lineage, only the genus *Pongo* survives. In this respect, orangutans are not simply two endangered species of great ape, but also represent the sole living representatives of a diverse clade that included dozens of distinct pongin species and genera that occurred for some 10 million years across Asia and Europe.

Although most of pongin evolution occurred in Asia, the human lineage primarily evolved in Africa. Not until the Pleistocene, when species such as *Homo erectus* inhabited Asia, did the orangutan and human lineages meet, and only some 70,000 years ago did our own species, *H. sapiens*, disperse into the orangutan's realm.[5] The co-occurrence of humans

doi: 10.1111/j.1749-6632.2011.06288.x

and orangutans has since then led to the rapid decline of the latter, but it is unlikely that humans were the sole cause of the orangutan's decline. The dramatic environmental changes that occurred in Asia from the late Miocene to the Pleistocene probably contributed to the extinction of many of the Asian apes[6] and may have substantially reduced orangutan population sizes as well. Broadly speaking, climatic changes such as increased aridity, more intense monsoons, and increased frequency and severity of glacial-inter-glacial cycles caused once widespread evergreen rainforests, the habitat of orangutans, to be increasingly restricted to isolated pockets close to the equator.[7,8]

The genus *Pongo* could be found until the Late Pleistocene, or some 40,000 years ago, from as far north of southern China to the island of Java.[9,10] The present habitat of the Sumatran Orangutan (*Pongo abelii*) and Bornean Orangutan (*P. pygmaeus*) is probably less than 5% of the original *Pongo* distribution range. Even on the two islands of Borneo and Sumatra, where the species still occur, their populations had been much reduced by the early 19th century when orangutans became first known to science.[11–13] Orangutans have been hunted by people at least since the Late Pleistocene,[14,15] and a recent study suggests that over-hunting is one of the main reasons for these historic local extinctions of the species.[12] Orangutan populations have declined even more dramatically over the past few decades; both species are considered in danger of becoming extinct in the wild soon.[16,17]

Orangutan distribution and density

In many high-income countries, orangutans are a symbol for the unsustainable management and conversion of tropical rainforests on Borneo and Sumatra, especially where this concerns the development of oil palm plantations (*Elaeis guineensis*), which many consider a major culprit in the orangutan's decline.[18,19] As shown above, the decline of orangutans goes back much further than the last few decades. The southern half of Sumatra had probably lost all its orangutans by the time Alfred Russel Wallace and other early explorers and naturalists first described the orangutan (but see Ref. 13), whereas southeastern and northwestern Borneo, as well as the central highlands of the island, were also devoid of the species. Obviously there is more to declining orangutan populations than simply oil palm expansion, suggesting that better analysis is needed of the threats to orangutans and the orangutans' distribution.

One of the main challenges in orangutan conservation science has been, and continues to be, the development of accurate population estimates so that populations can be monitored. Early estimates in the 1960s and 1970s varied between a few thousand orangutans and 156,000 for all of Borneo[20–22] and between 4,500–10,000 for Sumatra.[23] New approaches based on absence/presence maps,[13] island-wide interview surveys,[24] and aerial surveys[25–27] led to more accurate estimates of at least 54,000 individuals for Borneo.[17] Even this figure should be considered a center point in a wide range of possible real population numbers, because no local densities are known for many parts of the species' range (Fig. 1). The population estimate for Sumatra is *ca.* 6,600, which was recently considered a conservative estimate. What we do know is that the area of orangutan habitats has been in a rapid decline for over 50 years now.[11,13,28]

After decades of survey work, it is reasonably well known where orangutans occur (Fig. 1). In the two Malaysian States (Sabah, Sarawak) on Borneo, and in Sumatra, this knowledge is relatively accurate. Several areas in Kalimantan have also been surveyed quite well, but large areas of unprotected habitat have only been surveyed in a cursory manner. The clumped distribution of orangutans[13] and high spatial variation in density make it impossible to extrapolate density estimates from one area to another. The variation in orangutan population estimates is therefore more due to an uncertainty about local densities than to a lack of information about the extent of the two species' ranges. Many factors have been hypothesized to limit orangutan densities, including the density of fig trees, altitude, forest type, degree of forest degradation, hunting intensity, and possibly the density of salt licks, or even past and current disease outbreaks.[29–34] Measuring these densities, however, is not easy. Orangutans are difficult to survey directly, and accurately estimating densities requires time-consuming surveys of indirect indicators such as their sleeping platforms or "nests."[35–37] This is mostly done on foot, but increasingly remote approaches such as aerial surveys are used.[25,38] These methods have in common that the estimation of the number of orangutans is based on the number and spatial distribution of their nests. This requires

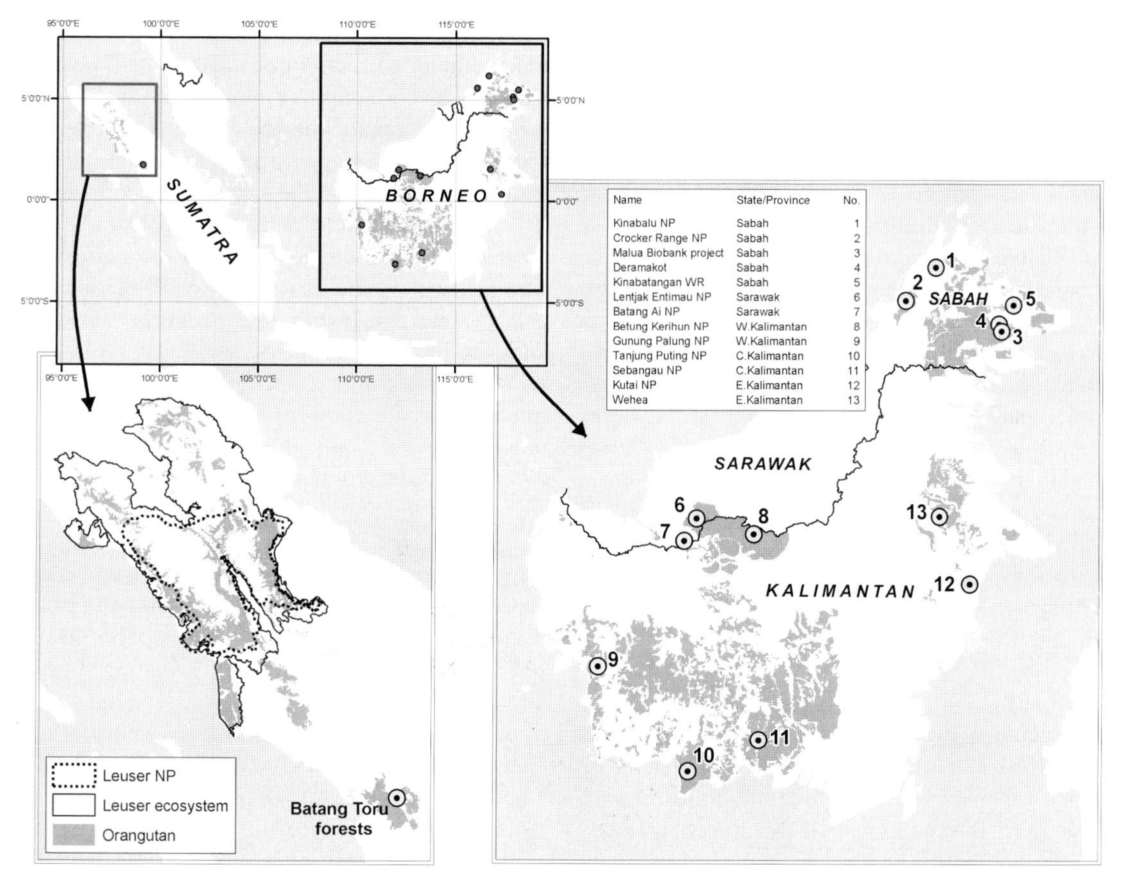

Figure 1. Orangutan distribution based on 2003 surveys, with orangutan populations mentioned in the text indicated on the map.

the determination of two factors: the rate at which orangutans build nests and the rate at which these nests decay over time. Recent studies have shown considerable variation in especially the latter factor, with average nest decay rates varying from 81–602 days, depending on factors such as the species of tree in which the nest is built, height of the nest, as well as the species of orangutans.[39] Nest surveys rarely determine decay rates locally, and the confidence intervals around previous density estimates are therefore large.[40,41] New approaches to nest surveys have been considered,[42] but none effectively overcomes the immense investment of labor and time that would be needed to accurately determine local densities across some 90,000 km^2 of dense, relatively inaccessible rainforest. Until new methods emerge, such as the direct detection from the air of orangutans with a thermal scanner, the best approach appears to be the combined use of on-the-ground nest surveys with remote nest counts, as well as the use of interview surveys, which provide better tools to determine and quantify human threats, such as hunting and agricultural conflict killings.[24] Resources made available for conservation are scarce and should be allocated strategically: assessing spatial variation in threats and ways to manage their impact on major populations is therefore crucial for the survival of the species.

The context of threats

Recent analysis of forest conversion for oil palm suggests that at least 10,000 km^2, or 5% of the total orangutan range, was converted for oil palm between 2000 and 2010,[43] with the high-density orangutan populations in the lowlands of eastern Sabah, a Malaysian state in north Borneo; coastal peat swamps in Sumatra; and peat swamps in the Indonesian province of Central Kalimantan having been especially affected. Oil palm is indeed a considerable threat, but other factors such as unsustainable timber extraction, development of tree plantations for pulp and paper, small-scale community agriculture, mining, and direct orangutan killing are important factors in the orangutan's decline as well. Singling out a particular industry as the main culprit in this process ignores the contribution from others and is unlikely to lead to lasting solutions.

Wild orangutans rarely require species-specific management (e.g., food provisioning or disease treatment), although rope bridges in Kinabatangan to reduce fragmentation effects have been successfully tested. The majority of conservation interventions entail protecting forests and preventing hunting. Despite initial impressions that the species was highly sensitive to forest disturbance,[13,33,44] it is becoming increasingly clear that the species has considerable ecological flexibility, allowing it to cope reasonably well with such disturbances.[45] Some populations even use monocultural plantations, although it is doubtful whether their survival there could be long term without access to more natural forest stands.[46] What is clear, however, is that orangutan populations suffer greatly from hunting. In many parts of their range, orangutans are killed for food, the pet trade, and from conflicts that inevitably occur as natural forests are converted for small and large-scale agriculture. This is a major threat as orangutan reproductive rates are too low to recover from hunting rates >1% females/year.[47] Rates of killing appear to far exceed this: a recent study estimated that annually between 750 and 3,500 orangutans are killed in Kalimantan.[48] Assuming a total population of some 42,500 animals in Kalimantan,[17] this would imply annual female off-take rates between 0.9% and 3.6%. In addition to direct mortality, hunting and deforestation have resulted in a steady flow of rescued and confiscated orangutans into the ten orangutan quarantine, care, and rehabilitation centers in Indonesia and Malaysia.[49]

The different threats to orangutans have related underlying causes. Poor land-use planning allows the degradation of known orangutan habitats. Poor forest management results in degraded, fire-prone forests with limited economic value, providing strong commercial and political incentives to convert these lands to agricultural or silvicultural lands. The resulting fragmentation of orangutan habitats leads to higher human–orangutan conflict and killing rates,[17] which go unpunished because of a lack of law enforcement.[17] In addition, too often the quality of protected area management remains insufficient to prevent forest loss,[50] so even in these areas orangutans are not safe.

In the past, orangutans had significant cultural value among some of the tribes living in or close to forest areas where orangutans occurred.[13] In the face of rapid development and modern consumerism, however, to many local people on Borneo and Sumatra the orangutan has become just another "monkey." In interviews with us, many local people express surprise regarding the amount of international intention for these species while these people themselves are still struggling to survive.[51] A 2008 candidate for the governorship of East Kalimantan (Indonesian Borneo) declared that people should take precedence over orangutans—a politically pragmatic position that is unlikely to change soon in a country where recent figures indicated that 52.4% of the people live on less than US $2 per day.[51] However, losing wild orangutans would have considerable consequences. Orangutans play important ecological roles in forests as ecosystem engineers and seed dispersers[52,53] and also provide a source of income for some forest communities through ecotourism.[54] More importantly, as one of the most recognized global conservation icons, their loss would imply that people are not able to develop life styles that are compatible with the ecological needs of our natural world. Losing a great ape from the wild would be a landmark failure. Not only would we have eradicated one of our closest genetic relatives, we would also substantially affect the forests that depend on orangutans, which would cause declines of many other taxa and erode the environmental services that forests provide and on which many people depend.

Abating the threat to orangutan survival is a political choice requiring societal approval and cooperation for comprehensive solutions that involve strict protection of key habitats, effective law enforcement against harmful illegal activities, and selection of compatible land-use practices. In democratic countries, governments will have to justify their choices in favor of species conservation against the opportunity costs of potentially more profitable land-use practices. In many cases, that justification is ethical rather than economic,[55] and the orangutans' future may depend on whether people are able to prioritize ethics over economics.

Past population trends

Uncertainties in population estimates have translated into widely ranging estimates of population trends, including the prediction in 2007 that orangutans would be virtually extinct in 2012.[56] Recent data allow us to better approximate annual population declines. Loss of orangutan habitat in Sumatra has been estimated at 1–1.5% per year between 1985 and 2001.[28] Based on these rates and the presently remaining area of habitat of 8,641 km^2 (Ref. 57), the estimated losses over the last 25 years range from 2,300 km^2 to 4,000 km^2. These habitat losses primarily occur in lowland forests, which have the highest densities,[31] with a conservative average of 2.5 animals/km^2. This would suggest average annual losses on Sumatra of 230–400 animals over the last 25 years. Such estimates cannot be used to predict future declines, because they are based in forest loss rates in lowland forests, and these rates will likely be lower in remaining upland habitats.

Annual mortality rates of orangutans in Kalimantan are estimated at 750–3,500 per year.[48] In Sabah, the orangutan population is thought to have decreased by a minimum of 95% over the past few centuries based on analysis of population genetics.[11] This is consistent with the decline from a statewide populations estimate in 1987 of 25,000 animals[58] to the presently remaining population size of 10,000 individuals,[45] suggesting annual losses of 1,250 individuals for Sabah only. No estimates are available for Sarawak, where a small, relatively well-protected population remains. We estimate that some 50–100 animals have been lost annually over the past 25 years to hunting, animal trade, or habitat loss.[59,60] This would suggest annual losses for Bornean orangutans over the last 25 years between 2,050 and 4,850. This range fits the estimated annual habitat losses of 3,122 km^2 up to 2004,[28] which, with a density range between 0.5 and 1.5 animals/km^2 (Ref. 31), translates into a similar mortality range.

What has been done in orangutan conservation?

Legally, orangutan conservation started in 1924 when the species was first protected in what is now Indonesia.[13] New laws primarily involved the setting aside of protected areas on Borneo and Sumatra and included prohibition on the then very active zoo trade in orangutans.[61] This prohibition was not enforced, however, and significant trade in orangutans for European and American zoos and for biomedical research continued until at least the 1970s.[62] These issues were brought to the conservation community's attention in the Western world by the efforts of Barbara Harrisson,[63] and her work signifies the start of the serious study of orangutans and international attention to their conservation.[13]

To effectively address threats, orangutan conservation uses four main strategies: management of protected orangutan habitat, combating illegal trade and killing, rehabilitating and reintroducing ex-captives into their wild habitat, and developing orangutan-friendly management in nonprotected areas. These strategies sound relatively straightforward, and, in fact, the principles of orangutan conservation are simple: minimize unnatural deaths and maximize the availability of suitable habitat. We summarize the main issues for each strategy below and explain why things are more complicated than they appear.

Management of protected orangutan habitat

About 25% of all wild orangutans occur in formally protected areas.[17] This does not necessarily mean that these populations are safe, because illegal logging, mining, and poaching continue in many of these protected areas and human–orangutan conflict occurs on their edges.[17,64] Still, these areas provide some level of protection compared to the nonprotected parts of their range.[65]

In Sumatra, orangutans mainly occur in the Leuser Ecosystem, which includes the Gunung Leuser National Park, and in the Batang Toru forest blocks (Fig. 1). Conservation efforts for the Leuser area stem back from the 1930s,[66] but the Batang Toru population was only recently rediscovered,[67,68] and conservation programs therefore began much more recently there. Effective management of the Leuser Ecosystem, which includes the national park and surrounding areas, started in the early 1990s and was boosted by significant European Union and Indonesian government funding and later by funding generated after the 2004 tsunami in Sumatra. We conservatively estimate that a total of US $75 million have been spent on Sumatran orangutan habitat protection from 1990 to 2011. This is a substantial amount of funding and one might expect that there are clear measures that would account for how successful this funding has been for conservation. Unfortunately, this is not as straightforward as it sounds, and we can only approximate the impact of that funding on orangutans. Based on recent forest loss estimates,[69] it has been estimated that between 1985 and 2007, the lowland forests of Aceh and North Sumatra Provinces, where most orangutans occurred, have been reduced by 36% and 61%, respectively.[57] Orangutan density is highest in forests on peatlands,[31,35] and forest loss in those habitats was 33% for Aceh and 78% for North Sumatra. Still, the protected areas of Sumatra do reduce forest loss,[65] and the orangutan's situation would have been worse without them.[70] This is evident in Leuser, even though not all of the Leuser Ecosystem was classified as protected during this period. From 1985 to 2007, 11.7% of forest in the 25,000 km^2 Leuser Ecosystem was converted to other land uses,[57] or up to 20.1% if forest loss in lowland areas is considered. Most of the dryland forest was replaced with agroforestry (31%) and much less by oil palm (19%), but, on peatlands, 79% of the forest that was lost was replaced by oil palm plantations.[57] This is bad for orangutans, but again things would have been worse without the protected areas and the continued conservation efforts for protected and nonprotected areas. For Sumatra, it can therefore be tentatively concluded that protected areas and conservation does work to reduce forests loss, but not to halt it, and that their protected area management and conservation efforts still require much improvement.

Kalimantan (Indonesian Borneo) has more orangutans than any other location. It also has the biggest problems in terms of deforestation; forest degradation and fragmentation; and killing of orangutans for food, pets, and other reasons.[17] Some key populations, such as those in the Gunung Palung, Tanjung Puting, Sabangau, and Betung Kerihun National Parks, are relatively well protected, although forest loss and degradation continue even in these areas. Outside these protected sites, habitat loss and degradation through development of small- and large-scale agricultural and silvicultural plantations is rapidly reducing the areas where viable populations can survive. Forest loss in Borneo between 2000 and 2010 was recently estimated at 500,000 ha/year.[71] Most of this forest loss has occurred in lowland forest on mineral soils and in peat land, both of which are key orangutan habitats. One protected area, Kutai NP (see Fig. 1), has lost much of its forest through illegal activities,[72] although it still maintains a significant orangutan population of over 1,000 animals. Illegal killing of orangutans in Kalimantan appears to impact remaining populations more than it does in Sumatra[73] and the Malaysian side of Borneo.[11] This suggests that protected areas, where the potential for law enforcement is higher than outside them, are especially important in Kalimantan as areas where populations can be protected from hunting.

Sabah, Malaysian Borneo, has lost more than 40% of its forests over the past century, and the majority of its remaining forests are at different stages of degradation and regeneration due to past intense logging activities.[74] Today about 50% of the State is still covered with forest, and 13% of the State protected. Most of Sabah's lowlands used to be occupied by orangutans, and it is estimated that between 50 and 90% of the original orangutan population was lost over the past 100 years.[11] Most orangutan populations are currently found in the eastern side of the state, where nonhunting Muslim communities predominate. Sixty percent of Sabah's orangutans occur outside protected areas,[25] where they are mostly threatened by conflicts, hunting, road kills, lack of food resources, intense population fragmentation, and increased sensitivity of forest blocks to natural and human-made catastrophes. Still, forest loss is slowing in Sabah and protected areas are relatively well managed, although unfortunately most of them were established in marginal orangutan habitat, such as the highland forests of Crocker Range and Kinabalu National Parks (Fig. 1). The hope for Sabah is that relatively low hunting levels and stabilizing land use will allow the development of a network of interconnected, well-protected areas that would allow an orangutan population to survive in a mixed landscape of plantations, timber concessions, and protected areas. In addition, the State is trying to identify mechanisms to secure the long-term sustainable management of some unprotected forests through the development of mechanisms such as biodiversity credits in the Malua Biobank project and forest carbon initiatives. In addition, sustainably managed forests, such as the Deramakot timber concession, which is certified by the Forest Stewardship Council (FSC), are very important orangutan sites that appear to sustain and maintain viable orangutan populations. The State aims to FSC certify all its commercial forests before 2020.

Finally, in Sarawak, Malaysian Borneo, most remaining orangutans occur in two protected areas, Lanjak Entimau and Batang Ai, for which a population of 1,143–1,761 orangutans has been estimated.[17] Little information exists about the management of these populations, because conservation activities by nongovernmental organizations (NGOs) and research are not encouraged in the State.

Law enforcement

Orangutans are fully protected in both Malaysia and Indonesia. Large fines or jail sentences are mandated for those who kill, trade, or illegally keep orangutans or illegally clear their protected habitats. Hunting of orangutans for food, the pet trade, and to avoid crop damage is rampant in Kalimantan,[24] and hunting pressure is the strongest determinant of orangutan density in East Kalimantan Province.[30] Many illegal activities related to orangutans have been reported in the media, but only in 2011 did the first two prosecutions occur in West Kalimantan, Indonesia for people caught trading orangutans.[75] This reportedly resulted in prison sentences of eight months and one and a half years. Malaysia is somewhat stricter in the enforcement of its wildlife laws, but there offenders are rarely prosecuted.

With illegal logging and orangutan killing being among the major threats, improved law enforcement, together with effective public campaigns, are key strategies. Unfortunately, commitment from the government to enforce laws is generally lacking, and few resources have been made available to make law enforcement more effective.[73,76] In addition, few orangutan conservation organizations focus on increasing law enforcement. This might be because it involves the risk of being considered in opposition of people that live alongside the orangutan. In addition, foreign private or government donors shy away from funding law enforcement because of the potential risk of being seen as meddling in internal affairs of another country. Interestingly, some of the best examples of law enforcement are seen among local communities themselves, such as the Honorary Wildlife Warden scheme in Kinabatangan, Sabah, or the Wehea forest in East Kalimantan (Fig. 1), which is managed and patrolled primarily by local community stakeholders.

With regard to law enforcement, most conservation organizations focus on the role of companies in the destruction of orangutan habitats, especially when this is illegal or semilegal. These semilegal cases involve the many gray areas in the legislation related to orangutan conservation. Importantly, among these is the fact that although orangutans are protected this does not mean that it is illegal to destroy their habitat, even though large-scale forest clearing clearly leads to the death of orangutans. For orangutan conservation to function better, governments need to clarify what it means to be a protected species and translate this into meaningful and operational legislation.

Orangutan rehabilitation, reintroduction, and translocation

Orangutan rehabilitation centers were first created in Malaysia in the 1960s, and the earliest efforts in Indonesia date to the 1970s. They were initially set up as law enforcement units in response to the growing number of captive animals seized by law enforcement authorities, with those in Malaysia run by the government and the Indonesian ones by NGO. These centers have been increasingly presented by NGOs and media as the frontline for orangutan conservation in Southeast Asia.[77] Between 1964 and 2008, Indonesian and Malaysian orangutan rehabilitation projects have taken in at least 3,320 animals and released around 1,250,[49] or about one third of the original intake. The intake number is certainly an underestimate, however, because of the high number of mortalities, especially among very young orangutans. Presently, an estimated 2,000 orangutans live in orangutan care centers waiting to be released back into forest habitats. Past efforts to release orangutans have had low success rates, with few orangutans reportedly surviving in one location where over 400 orangutans were released.[78] Mortality rates of reintroduced orangutans vary from 20% to 80%.[49] Assuming that survival rates in these unstudied populations are an average 50%,[49] *ca.* 625 orangutans have been successfully reintroduced over a 45-year period.

Translocation involves the transfer of orangutans from one area, often a forest on the verge of being cut down or an area where orangutans cause conflicts, to another, safer one. There are few published figures on the number of orangutans that have been translocated over the years (but see Ref. 79), nor have follow-up studies been conducted regarding the fate of translocated animals. Similar to reintroductions, the challenge for translocations is to find areas of protected, good habitat into which orangutans can be released. Most of these locations are very remote, making it costly to transport orangutans there and to provide postrelease monitoring and care.

Despite the relatively low success rates, conservation investment in orangutan rehabilitation and reintroduction remains high. The cost of feeding and caring for an orangutan can be estimated from the operational costs of US $3.2 million for about 1,200 orangutans reported by one program in 2007,[80] suggesting annual management cost per orangutan of US $2,670. As this does not take into consideration acquisition and maintenance costs of land, buildings, and vehicles, a minimum cost of US $3,000 per orangutan per year is more likely. In Kalimantan, where many orangutans are released in very remote sites, the one-off logistical cost of releasing one orangutan into the wild, not including postrelease monitoring, is about US $5,000 per animal (M. Desillets, personal communication). With 63–97% of animals arriving in rehabilitation centers being younger than the minimum release age of seven years,[49] mean time from admission and release is likely to exceed three years. A minimum average cost of US $14,000 per animal from rescue to release is realistic. These estimates consider only those animals that are successfully rehabilitated and reach the reintroduction stage. A significant number of orangutans cannot be reintroduced due to diseases or physical impairments and are kept in captivity for life, requiring long-term funding without direct conservation benefits.

Developing orangutan-friendly management in nonprotected areas

With most orangutans occurring outside protected areas and with minimal chances that much more habitat will be formally protected, the inevitable conclusion is that the majority of orangutans will have to be managed in forests that are also used for other purposes. There is hope that well-managed forests, and even plantations, can provide habitat for orangutans, although much more research is needed here to examine which factors determine orangutan survival in multifunctional landscapes.[45,46,81] For such multifunctional landscapes to provide viable habitat for orangutans as well as people, a significant shift in perspective is needed among conservation groups, governments, forest managers, and local communities.[24,82] It will require that we stop seeing conservation in black and white terms of unprotected and protected or natural and unnatural.[83] We need to acknowledge that the orangutans' future will depend on their survival in human-made landscapes and not pristine habitats only.

There is a hugely important role for public awareness campaigns and educational programs to promote peaceful coexistence between people and orangutans. There is also a need to develop clearer management guidelines that prescribe how forests and plantations can best be managed to maximize the survival chances of orangutans (and other wildlife) while minimizing the costs to managers. Governments need to assist this process by making such management guidelines mandatory or to provide financial incentives, such as tax breaks, to offset the extra costs of wildlife-friendly management. Western or local markets could potentially play a role as well by paying higher prices for products harvested from orangutan-friendly sites or refusing to buy those from badly managed ones. This would require some transparent auditing process, however, and the costs of doing this are often higher than the potential gains from premium prices.[84]

The role of research in orangutan conservation

Researchers like orangutans. A search in a scientific literature database (Current Contents) of keywords "Orangutan" and "*Pongo*" revealed about the same number of search hits ($n = 249$) as similar searches for "Tiger" and "*Panthera*" ($n = 284$), "Elephant" and *Elephas*" ($n = 320$), and "Panda" and "*Ailuropoda*" ($n = 226$), all iconic conservation species. Orangutan research covers many fields including evolutionary genetics, ecology, behavior, disease and veterinary medicine, and field survey methodologies. Many orangutan studies justify their research by referring to the species' endangered status, but how relevant are they actually for orangutan conservation?

Although research has provided important data on behavioral ecology, habitat needs, and the genetic structure of populations, which are relevant for orangutan conservation, most research on orangutans provides little specific information or insights that can be directly applied to conservation planning and management.[85,86] For example, even though it has been suspected for a long time that illegal killing of orangutans was a major factor in their conservation,[13,30] the first comprehensive, quantitative data set on this problem was not developed until 2009 and only exists for Kalimantan.[24] Similarly, we are still unable to provide accurate population estimates for the orangutans in the Indonesian part of Borneo, because coordinated research and survey efforts have not been undertaken. Many important practical questions remain unanswered, including the impact of roads and plantations on orangutan dispersal; the nature, causes, and possible solutions of human–orangutan conflicts; effectiveness and efficiency (cost–benefits) of different conservation strategies; orangutan survival in agroforestry or plantation landscapes; or the value of reforestation and corridor development for meta-population dynamics. This is not to say that orangutan researchers do not contribute to orangutan conservation. Often, simply by virtue of their presence, researchers keep hunters and illegal loggers out of areas that would otherwise be unprotected or even play an active role in the development and management of protected areas. Still, this concerns specific, research-intense areas only and not the broader landscape level needed for successful orangutan conservation.

A specific shortcoming of orangutan conservation is that it receives limited input from local scientists. Many Malaysian and Indonesian scientists have been trained in conservation science or specifically worked on orangutans, but their contribution to further orangutan conservation science remains too limited.[87] A search in the scientific literature databases Current Contents and Web of Science for keywords "orangutan/orang-utan" and "conservation" indicated that between 1993 and the present about 1% of these publications were written by Malaysian or Indonesian scientists, as indicated by their position as first author.[87] The scientific literature on orangutan conservation is dominated by English, American, French, Dutch, Australian, Singaporean, and Japanese researchers. The consequence of this is that the role of indigenous researchers in translating research findings into locally relevant policy recommendations or media communications is underused. Successful orangutan conservation will require societal change in Malaysia and Indonesia, and local scientists could play a much more pronounced role in facilitating that change. It is beyond the scope of the present publication to identify the underlying causes of this relatively weak role of indigenous conservation science, but examples from other developing countries indicate that level of economic development may not be a key factor. For example, a similar literature search for "Brazil," "jaguar," and "conservation" indicated that about 50% of the publications were written by Brazilian or Argentinean researchers; about 75% of the publications on "India," "tiger," and "conservation" had a first author of Indian nationality; and about 78% of the publication on "China," "panda," and "conservation" were first-authored by Chinese scientists. There is an urgent need to increase local scientific capacity and active engagement of these scientists in Indonesia and Malaysia, which should eventually result in better societal comprehension of the different aspects of orangutan conservation, improved political uptake of key environmental ideas, and more media attention to these issues supported by local rather than international conservation figureheads.

Finally, conservation involves the manipulation of human–nature interactions with a view towards stabilizing ecological and environmental processes and trends. How these social and ecological systems interact, however, remains poorly understood.[88,89] The lack of progress is largely due to the traditional separation of ecological and social sciences.[90–92] Orangutan researchers who want to justify their work through its supposed positive impacts on orangutan conservation need to do a better job at identifying relevant questions for improving orangutan conservation. This may require a type of science different from the ecology and behavior focus that is normally associated with great ape research. Useful orangutan conservation research needs to delve into disciplines such as economics, political studies, law, sociology, forestry, and agricultural studies—that is, a much more multidisciplinary approach than traditional orangutan science. For conservation science to be useful, the conservation benefits should not be a serendipitous afterthought, but rather a clearly stated primary objective: what research do we need to conduct to achieve significant orangutan conservation outcomes?[86,93]

What could the orangutan's future look like?

The orangutan's extinction is considered to be imminent by some,[56] while others have suggested that the species has good habitat strongholds and also possesses more ecological flexibility to cope with changing environmental conditions than previously thought.[26,28,45,46,81] Despite that apparent flexibility, however, two factors work against the species: their inherently low reproductive rates and the ongoing high mortality rates caused by conflict killings, hunting, and habitat destruction. These factors, in addition to the continuing habitat conversion, make it difficult to predict whether any viable populations will remain a few decades from now. Key factors that will determine the outcome of this are the effectiveness of law enforcement that is needed to reduce orangutan killings and increased awareness leading to cultural change regarding land management and the treatment of protected species. The value of the latter is obvious in areas such as the Kinabatangan region in Malaysian Sabah, where traditionally people do not hunt orangutans and where despite severe fragmentation and degradation of original habitats, high densities of orangutans remain.[11,26] The situation in this part of Malaysia indicates that orangutans and people can potentially coexist and that orangutan populations can be sustained if large enough forest areas are connected to permit dispersal. This will, however, require significant changes in the attitude of politicians; land-use planners; large-land holders such as plantations, timber operations, and mining companies; and the millions of rural people that share ever-diminishing forest resources with the remaining orangutan populations.

Population viability analyses indicate that, in moderate quality habitats, orangutan populations starting at 500 individuals can maintain sufficient size and genetic diversity to persist for hundreds of years.[47] At least on Borneo, orangutan population densities rarely exceed two individuals/km^2,[31] meaning that at least 250 km^2 is required to provide a reasonable chance of long-term persistence. This assumes that no other factors such as hunting or disease cause unnatural deaths. There is more to this story though, as is becoming clear in Kinabatangan, where populations have shown slow but steady decreases that might be caused by unflanged males leaving their natal population to look for other areas in which to settle. To facilitate this interpopulation dispersal, habitat connectivity between major orangutan populations should be maintained or reestablished whenever possible.

We expect that the future of orangutans will very much depend on the long-term security of large, strictly protected areas where illegal logging and hunting are effectively controlled and orangutan populations are large enough to cope with potential catastrophic events such as fires or disease outbreaks. In addition, these areas need to contain ecological gradients that allow orangutans and their key resource requirements to adapt to more gradual changes, such as those brought about by climate change. Ideally, the core-protected parts of the orangutan range should remain connected with other forest areas. These would not necessarily have to be totally protected, but could also be used for commercial extraction purposes, at least for the orangutan subspecies with the ecological resilience to survive in such used forests.[45] Such an ecological network of protected areas, interconnected via forested watersheds, could in turn be buffered by low-intensity plantations, such as those used for pulp and paper and possibly also large-scale oil palm. This would then border on the high intensity-use areas where most people live, where infrastructure such as roads is concentrated, and where small-scale agriculture and silviculture are concentrated. Such a landscape would have ample room for economic development while optimizing the use of ecosystem services such as prevention of soil erosion, regulation of hydrology, and storage of carbon.[57]

The above idealized picture of ecologically connected networks remains very far removed from the present paradigm of rapid economic development through exploitation of natural resources and conversion of forest landscapes to nonforest ones without large-scale future land-use planning. The type of planning needed to retain or regain ecological connectivity on Borneo and Sumatra urgently requires that political decision makers recognize the value of such landscapes for the long-term socioeconomics of these lands. For this, environmental values (economic, ethical, and legal) need to be considered much more specifically in the land-use planning process. Island-wide planning frameworks need to be created that show different tradeoff scenarios between short-term economic gains from land development and longer-term benefits from maintaining sustainably managed forest environments. Based on this, governments can make an informed choice on how they envisage the long-term development of Borneo and Sumatra.

The way forward in orangutan conservation

Despite five decades of conservation attention for the orangutan, there has been frustratingly little progress. Even though we do not know how many orangutans existed some 50 years ago, we do know that every year large areas of orangutan habitat are degraded or lost at a rate of some 4,000 km^2 per year in Borneo and Sumatra.[28] This suggests that unless deforestation and hunting trends improve significantly, most wild orangutans will perish by 2025 and some important populations such as those in Tripa, Sumatra could disappear as early as 2015/16.[57] Even in protected areas, forests are not safe from illegal logging,[50,94,95] and we are a long way away from the goal of the Indonesian government to stabilize all wild orangutan populations in Indonesia by 2017.[96] The obvious conclusion is that, despite local progress and conservation success, the orangutan conservation movement has not done enough to turn the tide of the orangutan's fate.

This is not to say there has been no success. Many protected and relatively well managed areas would not exist or would have lost most their forest without the major lobbying and management efforts of people or groups highly committed to the orangutans survival: Kinabatangan, the Leuser Ecosystem, Gunung Palung, Tanjung Puting, and Sabangau are good examples. Other areas, such as the Wehea community forest and the FSC-certified Deramakot timber concession show that forests do not have to be legally protected for orangutans to have a safe home. We should learn from these successes and replicate them where possible. At the same time, we should not lose sight of the fact that for most wild orangutans living outside these relatively protected areas (comprising some 75% of the total population), the immediate future remains grim if overexploitation of forests and conversion of natural forests to planted or nonforest continues. Minimizing losses among these populations requires a major strategic change in how orangutan conservation is done.

The two most important groups involved in conservation are the governments of Indonesia and Malaysia and a range of local and international NGOs. Governments are responsible for policy development, land-use planning, law enforcement, and conservation management in protected areas. It is obvious that, despite orangutans being fully protected, existing legislation and policy do not provide sufficient protection for them. Governments need to develop much more holistic policies that not only target economic progress, but also balance that progress with ecological and environmental sustainability. The Indonesian government has shown its commitment to this in its national orangutan action plan and its low carbon growth objectives,[97] but this general commitment needs to be translated into new land-use plans and new policies on land use that integrate protected area management with broader landscape-level management. Socio-ecological sustainability should become a general policy principle if orangutans outside protected areas are to survive. Governments also need to recognize the spatial heterogeneity of threats and design area-specific plans to reduce these threats. This links to the law enforcement and protected area management roles of the government, neither of which is up to standards at the moment. Governments need to seriously consider how they can improve their effectiveness in both these roles. Strengthening law enforcement needs to happen by ensuring that those that break the law are actually caught and prosecuted. This will require training of police, judges, and conservation authorities; effectively combating corruption in government offices; and also ensuring that the public understands why these laws are now taken seriously. The latter requires effective campaigns informing the public. This is needed because a recent study suggested that 27% of the people in Kalimantan did not know that orangutans are legally protected.[48] The Indonesian government has voiced the idea to privatize some of these functions, specifically aiming to outsource the management of national parks to the private sector.[98] How protected area management could be financially attractive to the private sector is unclear, however, and privatizing conservation management is unlikely to be a panacea.

Many NGOs are involved in orangutan conservation. These range from local activists groups that use public protests and graphic images of orangutan suffering to engender public awareness and compassion to big global conservation organizations. What many of these organizations have in common is that they work by themselves on their own projects and rarely coordinate their activities with those of other groups. This lack of cooperation and absence of coherent joint planning is a significant weakness of the orangutan conservation community.[99] Attempts have been made to improve this, and the Indonesian orangutan conservation action plan of 2007 was a good example of different NGOs working together to effectively push for new legislation. This happens too rarely though, leaving much conservation action uncoordinated and not adding to an overall conservation framework. This is not to say that all conservation activities should be coordinated—great conservation gains have been made by people or organizations working on their own. The key, however, is to collaborate and team up with others when broader issues are at stake, such as the development of new legislation or the implementation of country-level management plans for orangutans, which requires coherent and generally supported lobbying effort to influence government.

Companies play an increasingly important role in orangutan conservation. Many of them have a number of key characteristics that would permit them to effectively address conservation goals, attributes, which are not always present in governmental and NGOs. Companies generally have well-trained staff and strong operational procedures for their management—slotting in orangutan-specific management should be relatively easy. Companies also tend to have significant financial resources at their disposal. The key question to them is whether the cost of investing in good environmental and species management is offset by the benefits. Those benefits can be a direct financial one (e.g., when markets pay premium prices for products from well-managed companies), but more often they are associated with having a green image, which in today's world can be a significant indirect financial benefit for companies. With most of the land outside the protected areas being managed by companies, there are obvious needs to involve this sector more in orangutan habitat management. Whether or not such companies are interested will depend greatly on regulatory requirements, the companies' sensitivity to public opinion, and the potential financial gains.

The most important and least engaged sector are the estimated 36.5 million people in Borneo and northern Sumatra that live alongside the approximately 60,000 orangutans (i.e., 605 people per orangutan). The day-to-day needs of these people and their aspirations in life play a hugely important role in what will happen to remaining orangutan habitats. The extent to which these people are willing to live next to orangutans might be the deciding factor for the great ape's future outside protected areas.[24,81,82] Encouraging these people to support the principles of environmental conservation and sustainable development and be actively responsible for the management of their resources is therefore a crucial requirement for successful orangutan conservation.[54] Conservation groups need to think really hard about how this issue can be more effectively addressed. It would likely include the continuation of various small-scale education programs, but it would also require broader, regional, or national level campaigns with messages that do not just call for the protection of orangutans but challenge people to think of conservation in the context of their own lives. Linking conservation to other aspects of people's lives such as more efficient agricultural methods or health programs[100] might be more effective than simply telling people that they cannot harm orangutans. After all, people are sometimes confused. Why does the West pay so much attention to the protection of orangutans (i.e., "people of the forest" in the Malay language), while local communities consider themselves to be the people of the forest?[51] Conservation needs to make sure that double standards and different viewpoints about conservation are clarified and addressed where possible.[58]

Finally, who is going to pay for conservation? Orangutan conservation costs money. At least as long as markets are not paying for environmental services provided by forests, exploiting these forests or replacing them with more productive uses is always going to generate more revenue than conserving them. Some of these opportunity costs can be offset by legislation or ethics—we are legally obliged to protect these forests, or we feel it is the right thing to do—but in most situations the protection or sustainable management of orangutan habitats will have to be paid for. Carbon markets may provide some of these financial means required for this,[57,101] while other sources of income could be from tourism, payments for water, and other forest-related economic activities. Companies could also fund a significant part of the total financial requirements by implementing orangutan-friendly management on their land. Ultimately, however, the main responsibility for orangutan conservation lies with the Indonesian and Malaysian governments, which need to develop the financial and economic tools that allow their countries to continue economic development, but not at the expense of the few remaining orangutan habitats.

Acknowledgments

We thank the Arcus Foundation for their support. We thank Riswan and Graham Usher for their help in making the map and two anonymous reviewers for providing feedback about our manuscript.

Conflicts of interest

The authors declare no conflicts of interest.

References

1. Harrison, T. 2010. Apes among the tangled branches of human origins. *Science* **327:** 532–534.
2. Wood, B. & T. Harrison. 2011. The evolutionary context of the first hominins. *Nature* **470:** 347–352.
3. Grehan, J.R. & J.H. Schwartz. 2009. Evolution of the second orangutan: phylogeny and biogeography of hominid origins. *J. Biogeogr.* **36:** 1823–1844.
4. Hobolth, A., J.Y. Dutheil, J. Hawks, *et al.* 2011. Incomplete lineage sorting patterns among human, chimpanzee, and orangutan suggest recent orangutan speciation and widespread selection. *Genome Res.* **21:** 349–356. doi:10.1101/ gr.114751.114110.
5. Petraglia, M.D., M. Haslam, D.Q. Fuller, *et al.* 2010. Out of Africa: new hypotheses and evidence for the dispersal of Homo sapiens along the Indian Ocean rim. *Ann. Hum. Biol.* **37:** 288–311.
6. Jablonski, N.G. 1993. Quaternary environments and the evolution of primates in East Asia, with notes on two new specimens of fossil Cercopithecidae from China. *Folia Primatol.* **60:** 118–132.
7. Patnaik, R. & P. Chauhan. 2009. India at the cross-roads of human evolution. *J. Biosc.* **34:** 729–747.
8. Barry, J.C., M.L.E. Morgan, L.J. Flynn, *et al.* 2002. Faunal and environmental change in the late Miocene Siwaliks of northern Pakistan. *Paleobiol.* **28:** 1–71.
9. Kahlke, H.D. 1972. A review of the Pleistocene history of the orangutan. *Asian Perspect.* **15:** 5–15.
10. Steiper, M.E. 2006. Population history, biogeography, and taxonomy of orangutans (Genus: *Pongo*) based on a population genetic meta-analysis of multiple loci. *J. Hum. Evol.* **50:** 509–522.
11. Goossens, B., L. Chikhi, M. Ancrenaz, *et al.* 2006. Genetic signature of anthropogenic population collapse in orangutans – art. no. e25. *PLoS Biol.* **4:** 285–291.
12. Meijaard, E., A. Welsh, M. Ancrenaz, *et al.* 2010. Declining orangutan encounter rates from Wallace to the present suggest the species was once more abundant. *Plos ONE.* **5:** e12042.
13. Rijksen, H.D. & E. Meijaard. 1999. Our vanishing relative. In *The Status of Wild Orang-Utans at the Close of the Twentieth Century*. Kluwer Academic Publishers. Dordrecht, The Netherlands.
14. Hooijer, D.A. 1960. The orang utan in Niah Cave prehistory. *Sarawak Mus. J.* **9:** 408–419.
15. Piper, P.J. & R.J. Rabett. 2009. Hunting in a tropical rainforest: evidence from the terminal pleistocene at Lobang Hangus, Niah Caves, Sarawak. *Int. J. Osteoarchaeol.* **19:** 551–565.
16. IUCN. 2010. IUCN Red List of Threatened Species. Version 2010.2. Available at: www.iucnredlist.org, Accessed 24/8/2010.
17. Wich, S.A., E. Meijaard, A.J. Marshall, *et al.* 2008. Distribution and conservation status of the orang-utan (*Pongo* spp.) on Borneo and Sumatra: how many remain? *Oryx.* **42:** 329–339.
18. Nantha, H.S. & C. Tisdell. 2009. The orangutan-oil palm conflict: economic constraints and opportunities for conservation. *Biod.Conserv.* **18:** 487–502.
19. WWF. 2011. Orangutans and oil palm plantations. Available at: http://wwf.panda.org/about_our_earth/about_forests/deforestation/forest_conversion_agriculture/orang_utans_palm_oil./. Accessed 15/2/2011.
20. Reynolds, V.A. 1967. *The Apes. The Gorilla, Chimpanzee, Orangutan, and Gibbon. Their History and their World.* Harper & Row. New York.
21. Schaller, G.B. 1961. The orangutan in Sarawak. *Zoologica* **46:** 73–82.
22. MacKinnon, K.S. 1986. Conservation status of Indonesian primates. *Primate Eye* **29:** 30–35.
23. Borner, M. 1976. Sumatra's orang-utans. *Oryx* **13:** 290–293.
24. Meijaard, E., K. Mengersen, D. Buchori, *et al.* 2011. Why don't we ask? A complementary method for assessing the status of great apes. *PloS One* **6:** e18008.
25. Ancrenaz, M., O. Gimenez, L. Ambu, *et al.* 2005. Aerial surveys give new estimates for orangutans in Sabah, Malaysia. *PLoS Biol.* **3:** e3.
26. Ancrenaz, M., B. Goossens, O. Gimenez, *et al.* 2004. Determination of ape distribution and population size using ground and aerial surveys: a case study with orang-utans in lower Kinabatangan, Sabah, Malaysia. *Anim. Conserv.* **7:** 375–385.
27. Ancrenaz, M. & I. Lackman-Ancrenaz. 2004. Orang-utan status in Sabah: distribution and population size. Kinabatangan Orang-utan Conservation Project. Kota Kinabalu, Sabah, Malaysia.
28. Meijaard, E. & S. Wich. 2007. Putting orang-utan population trends into perspective. *Curr. Biol.* **17:** R540.
29. Wich, S. A., R. Buij & C. P. van Schaik. 2004. Determinants of orangutan density in the dryland forests of the Leuser Ecosystem. *Primates* **45:** 177–182.
30. Marshall, A.J., Nardiyono, L.M. Engstrom, *et al.* 2006. The blowgun is mightier than the chainsaw in determining population density of Bornean orangutans (*Pongo pygmaeus*

morio) in the forests of East Kalimantan. *Biol. Conserv.* **129:** 566–578.
31. Husson, S.J., S.A. Wich, A.J. Marshall, *et al.* 2009. Orangutan distribution, density, abundance and impacts of disturbance. In *Orangutans: Geographic Variation in Behavioral Ecology and Conservation.* S.A. Wich, S.U. Atmoko, T.M. Setia & C.P. van Schaik, Eds.: 77–96. Oxford University Press. Oxford, UK.
32. Matsubayashi, H., A.H. Ahmad, N. Wakamatsu, *et al.* 2011. Natural-licks use by orangutans and conservation of their in Bornean tropical production forest. *Raff. Bull. Zool.* **59:** 109–115.
33. Knop, E., P.I. Ward & S.A. Wich. 2004. A comparison of orang-utan density in a logged and unlogged forest on Sumatra. *Biol. Conserv.* **120:** 187–192.
34. Payne, J. 1990. Rarity and extinctions of large mammals in Malaysian rainforests. Proceedings of the International Conference on Tropical Biodiversity, "In Harmony with Nature." Y.S. Kheong & L. S. Win, Eds.: 310–320. 12–16 June 1990, Kuala Lumpur, Malaysia.
35. van Schaik, C.P., A. Priatna & D. Priatna. 1995. Population estimates and habitat preferences of orang-utans based on line transects of nests. In *The Neglected Ape.* R.D. Nadler, B.F.M. Galdikas, L.K. Sheeran, & N. Rosen, Eds.: 109–116. Plenum Press. New York.
36. van Schaik, C.P., S.A. Wich, S.S. Utami, *et al.* 2005. A simple alternative to line transects of nests for estimating orangutan densities. *Primates* **46:** 249–254.
37. Buij, R., I. Singleton, E. Krakauer, *et al.* 2003. Rapid assessment of orangutan density. *Biol. Conserv.* **114:** 103–113.
38. Ancrenaz, M. 2007. Orangutan aerial survey in Sebangau National Park, Central Kalimantan, Indonesia. KOCP. Kota Kinabalu, Malaysia.
39. Mathewson, PD., S.N. Spehar, E. Meijaard, *et al.* 2008. Evaluating orangutan census techniques using nest decay rates: implications for population estimates. *Ecol. Appl.* **18:** 208–221.
40. Boyko, R.H. & A.J. Marshall. 2010. Using simulation models to evaluate ape nest survey techniques. *PloS One* **5:** e10754.
41. Marshall, A.J. & E. Meijaard. 2009. Orangutan nest surveys: the devil is in the details. *Oryx* **43:** 416–418.
42. Spehar, S.N., P.D. Mathewson, Nuzuar, *et al.* 2010. Estimating orangutan densities using the standing crop and marked nest count methods: lessons learned for conservation. *Biotropica* **42:** 748–757.
43. Koh, L.P., J. Miettinen, S.C. Liew, *et al.* 2011. Remotely sensed evidence of tropical peatland conversion to oil palm. *Proc. Natl. Acad. Sci. USA* **108:** 5127–5132.
44. Felton, A.M., L.M. Engstrom, A. Felton, *et al.* 2003. Orangutan population density, forest structure and fruit availability in hand-logged and unlogged peat swamp forests in West Kalimantan, Indonesia. *Biol. Conserv.* **114:** 91–101.
45. Ancrenaz, M., L. Ambu, I. Sunjoto, *et al.* 2010. Recent surveys in the forests of Ulu Segama Malua, Sabah, Malaysia, show that orang-utans (*P. p. morio*) can be maintained in slightly logged forests. *PloS One* **5:** e11510.
46. Meijaard, E., G. Albar, Y. Rayadin, *et al.* 2010. Unexpected ecological resilience in Bornean Orangutans and implications for pulp and paper plantation management. *PloS One* **5:** e12813.
47. Marshall, A.J., R. Lacy, M. Ancrenaz, *et al.* 2009. Orangutan population biology, life history, and conservation. Perspectives from population viability analysis models. In *Orangutans: Geographic Variation in Behavioral Ecology and Conservation.* S.A. Wich, S.U. Atmoko, T.M. Setia & C.P. van Schaik, Eds.: 311–326. Oxford University Press. Oxford, UK.
48. Meijaard, E., D. Buchori, Y. Hadiprakoso, *et al.* 2011. Quantifying killing of orangutans and human-orangutan conflict in Kalimantan, Indonesia. *PLoS One*. In press.
49. Russon, A.E. 2009. Orangutan rehabilitation and reintroduction. In *Orangutans. Geographic Variation in Behavioral Ecology and Conservation.* S.A. Wich, S.U. Atmoko, T.M. Setia & C.P. van Schaik, Eds.: 327–350. Oxford University Press. Oxford, UK.
50. Curran, L.M., S.N. Trigg, A.K. McDonald, *et al.* 2004. Lowland forest loss in protected areas of Indonesian Borneo. *Science* **303:** 1000–1003.
51. Meijaard, E. & D. Sheil. 2008. Cuddly animals don't persuade poor people to back conservation. *Nature* **454:** 159.
52. Ancrenaz, M., I. Lackman-Ancrenaz & H. Elahan. 2006. Seed spitting and seed swallowing by wild orang-utans (*Pongo pygmaeus morio*) in Sabah, Malaysia. *J. Trop. Biol.Conserv.* **2:** 65–70.
53. Galdikas, B.M.F. 1982. Orang utans as seed dispersers at Tanjung Puting, Central Kalimantan: Implications for conservation. In *The Orang Utan: Its Biology and Conservation.* L.E.M. de Boer Ed.: 285–298. Springer. New York.
54. Ancrenaz, M., L. Dabek & S. O'Neil. 2007. The costs of exclusion: recognizing a role for local communities in biodiversity conservation. *PLoS Biol.* **5:** 2443–2448.
55. Meijaard, E. & D. Sheil. 2011. A modest proposal for wealthy countries to reforest their land for the common good. *Biotropica* **43:** 544–548.
56. Williams, N. 2007. Orang-utan extinction threat shortens. *Curr. Biol.* **17:** R261.
57. Wich, S., Riswan, J. Jenson, *et al.* 2011. *The Orangutan and the Economics of Sustainable Forest Management in Sumatra.* UNEP/GRASP/PanEco/YEL/ICRAF/GRID-Arendal.
58. Payne, J. 1987. Surveying orang-utan populations by counting nests from a helicopter: a pilot survey in Sabah. *Primate Conserv.* **8:** 92–103.
59. Blouch, R.A. 1997. Distribution and abundance of orangutan (*Pongo pygmaeus*) and other primates in the Lanjak Entimau wildlife reserve, Sarawak, Malaysia. *Trop. Biodivers.* **3:** 259–274.
60. Meredith, M. 1993. *A Faunal Survey of Batang Ai National Park, Sarawak-Malaysia.* The Wildlife Conservation Society. New York, NY.
61. Jones, M.L. 1982. The orang-utan in captivity. In *The Orangutan: Its Biology and Conservation*: 17–37. Dr. W. Junk Publishers. The Hague, Netherlands.
62. Bourne, G.H. 1971. *The Ape People.* Putnam. London.
63. Harrisson, B. 1961. Orang utan: what chances of survival. *Sarawak Mus. J.* **10:** 238–261.
64. Nellemann, C., L. Miles, B.P. Kaltenborn, *et al.* 2007. *The Last Stand of the Orangutan. State of Emergency: Illegal*

Logging, Fire, and Palm Oil in Indonesia's National Parks. United Nations Environment Programme-GRID. Arendal, Norway.

65. Gaveau, D.L.A., J. Epting, O. Lyne, *et al*. 2009. Evaluating whether protected areas reduce tropical deforestation in Sumatra. *J. Biogeogr.* **36:** 2165–2175.
66. Wind, J. 1996. Gunung Leuser National Park: History, Threats and Options. In *Leuser. A Sumatran Sanctuary*. C.P. van Schaik & J. Supratna, Eds.: 4–27. Perdana Ciptamandiri. Jakarta, Indonesia.
67. Wich, S.A., I. Singleton, S.S. Utami-Atmoko, *et al*. 2003. The status of the Sumatran orang-utan *Pongo abelii*: an update. *Oryx* **37:** 49–54.
68. Meijaard, E. 1997. *A Survey of Some Forested Areas in South and Central Tapanuli, North Sumatra; New Chances for Orangutan Conservation*. Tropenbos and the Golden Ark. Wageningen.
69. WWF. 2010. *Sumatra's Forests, their Wildlife and the Climate. Windows in Time: 1985, 1990, 2000 and 2009*. A Quantitative Assessment of Some of Sumatra's Natural Resources submitted as Technical Report by Invitation to the National Forestry Council (DKN) and to the National Development Planning Agency (BAPPENAS) of Indonesia. WWF-Indonesia and BAPPENAS. Jakarta, Indonesia.
70. Gaveau, D.L.A., S. Wich, J. Epting, *et al*. 2009. The future of forests and orangutans (*Pongo abelii*) in Sumatra: predicting impacts of oil palm plantations, road construction, and mechanisms for reducing carbon emissions from deforestation. *Env. Res. Lett*. **4:** 34013.
71. Miettinen, J., C. Shi & S. C. Liew. 2011. Deforestation rates in insular Southeast Asia between 2000 and 2010. *Glob. Change Biol.* **17:** 2261–2270.
72. Soehartono, T. & A. Mardiastuti. 2001. Kutai National Park: where to go. *Trop. Biodivers.* **7:** 83–101.
73. Robertson, J.M.Y. & C.P. van Schaik. 2001. Causal factors underlying the dramatic decline of the Sumatran orangutan. *Oryx* **35:** 26–38.
74. McMorrow, J. & M.A. Talip. 2001. Decline of forest area in Sabah, Malaysia: relationship to state policies, land code and land capability. *Glob. Env. Change-Hum. Policy Dimens.* **11:** 217–230.
75. Berita Polhut. 2011. Vonis Pertama Kasus Perdagangan Orangutan di Indonesia. In *Indonesian Forest Ranger's Blog*. Available at http://polhut08.wordpress.com/2011/06/28/vonis-pertama-kasus-perdagangan-orangutan-di-indonesia/, Accessed 28/06/2011.
76. Gaveau, D.L.A., M. Linkie, Suyadi, *et al*. 2009. Three decades of deforestation in southwest Sumatra: effects of coffee prices, law enforcement and rural poverty. *Biol. Conserv.* **142:** 597–605.
77. Suryakusuma, J. 2009. Monkeying with the environment? *The Jakarta Post*. Available at http://www.thejakartapost.com/news/2009/12/02/monkeying-with-environment.html, Accessed 2/12/2009.
78. Grundmann, E. 2006. Back to the wild: will reintroduction and rehabilitation help the long-term conservation of orang-utans in Indonesia? *Social Sci. Inf.* **45:** 265–284.
79. Andau, P.M., L.K. Hiong & J.B. Sale. 1994. Translocation of pocketed orang-utans in Sabah. *Oryx* **28:** 263–268.
80. Borneo Orangutan Survival Foundation. 2008. *Independent Auditors' Report on Financial Transactions for the Years Then Ended December 31, 2007 and 2006*. Borneo Orangutan Survival Foundation. Balikpapan, Indonesia.
81. Campbell-Smith, G., M. Campbell-Smith, I. Singleton, *et al*. 2011. Apes in space: saving an imperilled orangutan population in Sumatra. *PLoS One* **6:** e17210.
82. Campbell-Smith, G., H.V.P. Simanjorang, N. Leader-Williams, *et al*. 2010. Local attitudes and perceptions toward crop-raiding by Orangutans (*Pongo abelii*) and other nonhuman primates in Northern Sumatra, Indonesia. *Am. J. Primatol.* **72:** 866–876.
83. Sheil, D. & E. Meijaard. 2010. Purity and prejudice: deluding ourselves about biodiversity conservation. *Biotropica* **42:** 566–568.
84. Zagt, R., D. Sheil & F.E. Putz. 2010. Biodiversity conservation in certified forests: An overview. In *Biodiversity Conservation in Certified Forests*. D. Sheil, F.E. Putz & R. Zagt, Eds.: V–XXVIII. Tropenbos International. Wageningen, The Netherlands.
85. Whitten, T., D. Holmes & K. MacKinnon. 2001. Conservation biology: a displacement behavior for academia? *Conserv. Biol.* **15:** 1–3.
86. Meijaard, E. & D. Sheil. 2007. Is wildlife research useful for wildlife conservation in the tropics? A review for Borneo with global implications. *Biod. Cons.* **16:** 3053–3065.
87. Meijaard, E. 2011. Indonesia Has Its Share Of Scientists, So Where's the Science? *The Jakarta Globe*. Available at http://www.thejakartaglobe.com/opinion/indonesia-has-its-share-of-scientists-so-wheres-the-science/430931. Accessed 24/3/2011.
88. Ostrom, E. & M. Cox. 2010. Moving beyond panaceas: a multi-tiered diagnostic approach for social-ecological analysis. *Environv. Conserv.* **37:** 451–463.
89. Liu, J.G., T. Dietz, S.R. Carpenter, *et al*. 2007. Complexity of coupled human and natural systems. *Science* **317:** 1513–1516.
90. Rosa, E.A. & T. Dietz. 1998. Climate change and society – Speculation, construction and scientific investigation. *Int. Sociol.* **13:** 421–455.
91. Reyers, B., D.J. Roux & P.J. O'Farrell. 2010. Can ecosystem services lead ecology on a transdisciplinary pathway? *Environ. Conserv.* **37:** 501–511.
92. Gowdy, J., C. Hall, K. Klitgaard, *et al*. 2010. What every conservation biologist should know about economic theory. *Conserv. Biol.* **24:** 1440–1447.
93. Caro, T. & P.W. Sherman. 2011. Endangered species and a threatened discipline: behavioural ecology. *Trends Ecol. Evol.* **26:** 111–118.
94. van Schaik, C.P., K.A. Monk & J.M.Y. Robertson. 2001. Dramatic decline in orang-utan numbers in the Leuser Ecosystem, Northern Sumatra. *Oryx* **35:** 14–25.
95. Fuller, D.O., T.C. Jessup & A. Salim. 2004. Loss of forest cover in Kalimantan, Indonesia, since the 1997–1998 El Niño. *Conserv. Biol.* **18:** 249–254.
96. Departmen Kehutanan. 2007. Strategi dan rencana aksi konservasi orangutan Indonesia 2007–2017. *Direktorat Jenderal Perlindungan Hutan Dan Konservasi Alam,*

Departemen Kehutanan (Indonesian Ministry of Forestry). Jakarta, Indonesia.

97. Elson, D. 2011. *Cost-Benefit Analysis of a Shift to a Low Carbon Economy in the Land Use Sector in Indonesia.* UK Climate Change Unit of the British Embassy. Jakarta, Indonesia.
98. Simamora, A.P. 2011, 24 March. Govt wants national parks to become 'profit centers'. *The Jakarta Post*. Available at http://www.thejakartapost.com/news/2011/03/24/govt-wants-national-parks-become-%E2%80%98profit-centers%E2%80%99.html. Accessed on 25/3/2011.
99. Meijaard, E. 2007. We must act now to save orangutans. *The Jakarta Post.* Available at http://www.thejakartapost.com/news/2007/01/18/we-must-act-now-save-orangutan.html. Accessed 8/1/2007.
100. Ali, R. & S.M. Jacobs. 2007. Saving the rainforest through health care: medicine as conservation in Borneo. *Int. J. Occup. Environm. Health* **13:** 295–311.
101. Venter, O., E. Meijaard, H.P. Possingham, *et al.* 2009. Carbon payments as a safeguard for threatened tropical mammals. *Conserv. Lett.* **2:** 123–129.

Ann. N.Y. Acad. Sci. ISSN 0077-8923

ANNALS OF THE NEW YORK ACADEMY OF SCIENCES
Issue: *The Year in Ecology and Conservation Biology*

Ecology and management of white-tailed deer in a changing world

William J. McShea

Center for Conservation Ecology, Smithsonian Conservation Biology Institute, Front Royal, Virginia

Address for correspondence: William J. McShea, Center for Conservation Ecology, Smithsonian Conservation Biology Institute, 1500 Remount Rd., Front Royal, VA 22630. mcsheaw@si.edu

Due to chronic high densities and preferential browsing, white-tailed deer have significant impacts on woody and herbaceous plants. These impacts have ramifications for animals that share resources and across trophic levels. High deer densities result from an absence of predators or high plant productivity, often due to human habitat modifications, and from the desires of stakeholders that set deer management goals based on cultural, rather than biological, carrying capacity. Success at maintaining forest ecosystems require regulating deer below biological carrying capacity, as measured by ecological impacts. Control methods limit reproduction through modifications in habitat productivity or increase mortality through increasing predators or hunting. Hunting is the primary deer management tool and relies on active participation of citizens. Hunters are capable of reducing deer densities but struggle with creating densities sufficiently low to ensure the persistence of rare species. Alternative management models may be necessary to achieve densities sufficiently below biological carrying capacity. Regardless of the population control adopted, success should be measured by ecological benchmarks and not solely by cultural acceptance.

Keywords: forest; *Odocoileus virginianus*; population control; hunting; carry capacity

Introduction

Large herbivores have a central role in the functioning of many terrestrial ecosystems.[1–3] Large carnivores are viewed as keystone species within terrestrial ecosystems primarily because of their role in regulating the numbers of herbivores, which convert plant material into energy and nutrients that are assessable to other animals.[4–6] By shaping plant communities and supporting apex predators, most forest ecosystems are structurally and compositionally different depending on whether or not large herbivores are present.[7–9] Whether ecosystems are regulated by top-down processes or bottom-up processes, significant energy and nutrients flow through the large herbivore community, and this feature makes understanding their ecology important for understanding ecosystem structure and functioning.[2,10]

White-tailed deer (*Odocoileus virginianus*) are the largest herbivore in many forested ecosystems in the eastern United States. In many forests, deer densities are chronically above historical levels.[11] White-tailed deer are not the only ungulates in eastern North America, with the recolonization of moose (*Alces alces*) to many areas of the Northeast and the reintroduction of elk (*Cervus elaphus*) to reclaimed mining areas in the Appalachians.[12,13] The larger body size of these ungulates exacerbates animal–human conflicts such as damage from collisions with vehicles and crop loss, but their overall densities and distributions have yet to exceed historic levels. For moose and elk, the ecological principles and management options are generally the same as those outlined here, but I will not deal with them directly in this paper. I will also focus my review primarily on forests east of the Great Plains due to space limitations, depth of the literature, and commonality of habitat.

As opposed to most species in the eastern forests, expertise, manpower, and bureaucracy are in place to manage deer populations across its range. With

doi: 10.1111/j.1749-6632.2011.06376.x

white-tailed deer, there is a dedicated management structure at all levels of government that can enact recommendations based on a public mandate. The critical junctures are often informing and engaging the public on the need for management, and managers and balancing the sometimes competing interests of stakeholders.[14,15] However, deer management is one of the few instances where citizens have an active and pivotal role. Whether the land is in public stewardship or is privately owned, governments rely primarily on citizens to enact deer management. This offers an opportunity to engage the public in ecosystem management, but it can also lead to conflicts between land manager and hunter goals and to uneven management across a landscape.[16]

The purpose of this paper is to review in brief the role of white-tailed deer (hereafter deer) in ecosystems in eastern North America and their impact on human communities, and then outline mechanisms and management strategies for controlling deer populations.

Ecological impacts

For most of eastern North America, the climax terrestrial community is forest.[17,18] In the most southeastern and northern climes, forests are primarily coniferous, but deer primarily exist within forests composed of diverse deciduous tree species that occur along a gradient of moisture and soil nutrients. Rare herbaceous plants or trees are impacted by deer in northern coniferous forests,[19–23] but it is primarily within deciduous forests that deer reach their highest densities and have been documented to have profound and consequential impacts on plant species.[24–27]

Deer are primarily browsers.[28] Their diet consists of buds and young leaves and branches, as well as forbs that occur within forests.[29] These forage items are not generally abundant in mature forests, except in temporary canopy gaps or along natural edges.[30,31] In human-modified forests, deer can increase their access to forage by moving between forested and human landscapes.[32–34] Agricultural crop damage is highest along the forest boundary, and these crops enhance the productivity of the landscape.[35,36] Forestry practices create large patches of early successional trees that are readily fed upon by deer.[37,38] Mature forest productivity alone likely would not support high deer densities, but seasonal access to human-added productivity results in seasonal bouts of heavy browsing pressure on natural systems.[19] No high-density populations have been reported outside of this human–natural system dynamic, with the exception of deer isolated from predators due to natural or human-made barriers.[27]

As its primary forage item, the abundance and distribution of woody seedlings and saplings can be significantly impacted by deer.[21,39–41] Studies of canopy gaps, logging operations, and mature forest find that deer browsing can shift woody plant composition toward unpalatable species or toward low species richness or density.[21,39,42] Shifts in plant composition toward unpalatable invasive species can alter forest succession by reducing light levels on forest floor, and deer herbivory after canopy tree defoliation can change successional patterns.[43–45] At the highest deer densities, forest succession is halted, and natural disturbance or timber harvest transitions the forest into an alternative stable or climax community of open woodland with a grass, fern, or exotic forb ground cover.[46–49]

Herbaceous plants are a rich species component of eastern deciduous forests.[17] Species extirpation has been postulated for midwestern forest patches, and overall diversity and density measures for forbs are generally lower when deer densities are high.[19,24,25,50,51] Plant competitive interactions are altered by preferential browsing by deer, with nonpalatable species (including exotic species) becoming dominant.[26,52] The spread of invasive plants into a forest understory also can be facilitated by deer, both through transport of seeds and through altering forest floor conditions.[52,53] Deer browsing may not cause plant morality, as many perennial forest forbs store significant resources in belowground roots, but does decrease plant growth during that year and lowers rates of flowering and fruit production.[54,55] Deer do have preferred browse species, but even unpalatable species can be impacted at high deer densities through changes in soil compaction and possibly nutrients.[56]

If deer can shape the diversity and structure of plant communities, then possibly this foraging will impact other species within the ecosystem.[26,27,57] These impacts fall into two categories: food web impacts for species that consume the same food resources or trophic level effects where shifts in resources at one trophic level have significant impacts at multiple levels. Consumption of key resources, such as acorns for small mammals and plant biomass for insect densities, leads to lower densities of these animals and are indications of direct impacts on food webs.[58–60] As with other large herbivore systems, these direct impacts have consequences at multiple trophic levels within forest systems, with changes in bird communities and both insect and disease outbreaks.[61–64] Most trophic level and food web interactions are only obvious with the addition or removal of apex predators or the exclusion of deer from small areas.[5,6,62]

An important consideration is that deer impacts on vegetation are not proportional to their density. For most large herbivores, the shape of the functional response curve to plant biomass depends on relative forage preference, the animal's nutritional state, and predation risks.[6,65,66] Augustine *et al.* demonstrated a Hollings Type II functional response curve for a forest herb, *Laportea canadensis*, in Wisconsin forests.[67] Elk browsing in riparian areas of Yellowstone ecosystem shifted in response to the arrival of wolves (*Canis lupus*), in the absence of significant changes in elk density aspen regeneration increased.[68] In addition to lower herbivore densities, this spatial and temporal variability results in a heterogeneous distribution of plants.[66] As examples of this effect, Royo *et al.* found that the low levels of deer browsing increased forest forb diversity, and Parker *et al.* reported that intermediate deer browsing on a herbaceous plant (*Oenothera biennis)* increased the genetic diversity within the population and thereby reduced overall damage by the main herbivore, *Microtus pennsylvanicus*.[31,69] Although these considerations are important at low or intermediate deer densities, high deer densities (i.e., approaching carrying capacity) result in homogenization of forest understory communities through chronic heavy browsing.[70]

Human impacts

In addition to the significant ecological impact of deer, it would be remiss to review the species without detailing the economic benefits, and both economic and health risks, which are also a product of their abundance on the landscape. Approximately 12.5 million Americans hunt and 25 billion dollars are spent each year on hunting activities in the United States, with deer hunting being the dominant activity.[71] This revenue comes through three main avenues: sale of licenses to hunters, leasing of land by landowners, and purchase of hunting gear and logistics around the actual hunt (e.g., hotel, travel, guides). License sales are directly related to deer management; for example, Virginia sells over 250,000 hunting licenses annually, which are the primary means of supporting their wildlife department.[72] In addition, the federal tax on hunting gear annually returns 265 million dollars (2007 estimates) to states for wildlife conservation through the Pittman–Robertson Federal Aid in Wildlife Restoration Program.[71] Maintaining these revenue sources is critical for many states.

The economic benefits of deer are countered by costs incurred by multiple segments of the community. In 2009, the insurance industry estimated that 2.4 million deer–vehicle collisions had occurred over the previous 24 months, with an estimated cost of over 7 billion dollars and 300 human fatalities.[73] These collisions increased 18% over the previous five years, although an unknown portion of this increase is due to better record keeping.[73] These numbers are alarmingly high from an economic standpoint, but removing 1.2 million deer per year from a national population that exceeds 25 million is well below the annual recruitment rate. For example, the legal harvest and vehicle collisions of deer was followed for two years within a rural county in Virginia, and the combined annual mortality did not exceed 20% of the estimated population.[74] Deer are estimated to cause more damage to agricultural crops than any other wildlife species.[75] Drake *et al.* estimated 94 million dollars in annual vegetable crop damage and 74 million dollars in grain crop damage for 13 northeastern states.[76] The same study estimated annual residential and commercial ornamental damage at 49 million dollars.[76] These large losses have consequences on landowner attitudes toward deer. For Virginia farmers, the percentage desiring lower deer numbers across the landscape increased from 50% to 93% if they had experienced crop damage in the last year.[77] The industrial forest community has long advocated lower deer densities due to the ability of deer to halt regeneration of valuable timber species.[78,79] Direct estimates of forestry losses due to deer browsing are difficult to determine, but a small subset of forest practices (nurseries) estimated their annual stock damage at 27 million dollars for 13 northeastern states.[76]

Disease transmission between deer and livestock is a consideration for both deer population regulation and economic costs incurred by high deer densities. Bovine tuberculosis (*Mycobacterium bovis*) and brucellosis (*Brucella arbortus*) are bacterial diseases capable of moving between livestock and deer, and, where the disease is present in wild deer, deer likely are a reservoir for diseases they contracted from livestock.[80] Bovine tuberculosis has been found in deer in Michigan and, although no transmission to livestock has been documented, control measures do cost state agencies.[80] Chronic wasting disease is a transportable spongiform encephalopathy, or a prion disease, that is specific to deer and is not a concern from transmission to livestock, except farmed deer, but is a concern if transmitted to the wild deer population.[80]

One cost of having wildlife on the landscape is the transmission of zoonoses, and deer are not different from other abundant wildlife. Deer serve as an intermediate host for several diseases that are transmitted to humans through ticks (*Ixodes* sp.), such as Rocky Mountain spotted fever (*Rickettsia rickettsii*) and Lyme disease (*Borrelia burgdorferi*). Lyme disease is found in 12 eastern states and was found to affect over 38,000 people in 2009.[81] Deer are the primary host for adult ticks, but risk factors for the disease are better predicted by knowing small mammal abundance (the host for intermediate stages) than the abundance of deer.[82] The primary role of deer in tick-borne diseases is transporting ticks across the landscape, through their propensity to move in response to variable mast production and other shifting food resources.[83] Whether deer herds can be reduced sufficiently to reduce transmission rates is unclear and doubtful. Application of an acaricide to deer can reduce the prevalence of tick-borne diseases.[84] However, logistical concerns, which include baiting deer, limit its applicability.

I have listed some direct economic costs and benefits of deer, but most estimates are rough approximations. Estimating economic costs for items with known value (e.g., automobiles, crops) is relatively easy compared to estimating ecological costs, which I have not attempted. The key point is to view each cost and benefit as representing a strong stakeholder group that has a voice in deer management.

Carrying capacity

The impacts of deer may be significant, but they are not an invasive or exotic species; their removal from an ecosystem does not "restore" natural conditions. Part of the management conflict with deer is that many present-day forests were initiated after logging activities in the first half of 20th century, when deer were absent or at much lower densities on the landscape, and these forests are currently difficult to restore after harvest.[85,86] For example, oak forest reestablishment after harvest depends on relatively low deer densities, but is only successful in conjunction with other factors, such as fire.[87,88] Reducing deer numbers does not always achieve objectives, as herbaceous plant recovery depends partially on soil and seed bank conditions that may no longer support rich communities.[89] Deer are an adaptive, prolific species, whose selective browsing has ramifications that are important for forest managers, not because they are exotic, but because of their sheer numbers. The question is when managers should regulate deer.

How and when to regulate deer herds is tied to the concept of carrying capacity. Carrying capacity is the sustainable biological limit of a population with its environment; a sum total of mortality and reproduction rates that will fluctuate over time as the environment changes.[90,91] As a population approaches carrying capacity, recruitment is limited and adult mortality increases.[92] For deer managers, there are two important population levels: when numbers equal those that can be supported by the plant productivity (i.e., carrying capacity), and the point where the annual mortality of deer (both harvest and natural) equals the annual recruitment of deer, which is referred to as maximum sustainable yield (MSY). Agencies and landowners interested in maximizing hunting opportunities manage for populations approaching MSY.[91–93] Carrying capacity is one of the oldest concepts in wildlife management.[94–96] It is a wonderful theory for explaining the limit of environments, but it is nearly impossible to calculate for a specific area without extensive data.[92] With regard to MSY, deer harvest and other sources of mortality that do not exceed annual fawn production will not reduce deer populations over the long term. Many control efforts remove animals from the population, but few reduce numbers sufficiently to counter annual recruitment of this fecund animal.

For managed wildlife populations such as deer, the concept of a biological carrying capacity is often replaced by the concept of a cultural carrying capacity.[16,91] The cultural carry capacity is based on a political process among community stakeholders.[14,15] Cultural carrying capacity is usually below biological carrying capacity, but depends less on the attributes of the habitat and more on the views of the stakeholders. Many states have adopted a stakeholder approach to managing deer to a cultural carrying capacity for each specific community.[14,72]

Ecological carrying capacity is the primary concept for protected areas that have management goals based on biodiversity or on endangered species that are impacted by deer browsing.[26,97] The functional foraging response of deer means they will not select forage randomly, but will preferentially browse on specific plant species.[65,98] These preferred species will decline or disappear long before the deer population is limited by plant productivity and probably before the limits imposed by cultural carrying capacity. This functional response is exacerbated by productivity inputs from humans that increase biological carrying capacity but do not change the browsing preferences of the deer. Some preferred browse species can be impacted at densities of 3 deer/km^{2},[21] deer densities that are well below both cultural carry capacity and achievable goals for state management agencies.[99] Therefore, it is hard to manage for rare species on private or public lands while staying within the strictures of public hunting.[100]

These differing concepts of carrying capacity do not impact how the deer are managed but do impact how management success is measured. Biological carrying capacity is a quantifiable measure based on deer population metrics. Ecological carrying capacity is based on deer impacts to a single species or guild of species that can be measured directly on the landscape. Cultural carrying capacity is derived from stakeholder meetings and is measured through feedback from the constituent groups involved. A deer management program adopts one of these measures and proceeds to limit the deer population according to the metrics adopted.

Managing deer densities that exceed carry capacity

Deer population size is determined by reproduction and mortality, and control is focused on impacting at least one of those demographic traits.

Reproduction

In the absence of major predators, the case for much of the eastern United States, the primary limit to deer numbers is access to plant productivity.[27,92,101] White-tailed deer are one of the most fecund deer species in the world, with females in unhunted populations capable of producing 30 offsprings in their lifetime.[28,102] High lifetime fecundity means deer numbers can change quickly to fluctuations in forage availability or predation. Island populations of deer are good examples of the potential for rapid increases.[63,103] Forestry practices in Pennsylvania during the early 20th century shifted forests dramatically to younger age classes, which coincided with rapid increases in deer numbers.[79,97]

As discussed throughout this review, landscape productivity is the key to deer population growth and reduced productivity, or access to productivity, will lower deer densities. Reduced productivity occurs as human landscapes transition from rural to suburban to urban. Reduced access to productivity can occur as major roads bisect deer ranges or fences restrict movement of deer across landscapes. Agricultural shifts from edible crops to biofuel have the potential to lower landscape productivity.[104] Lower palatability of invasive plants may initially shift browsing to native species,[105] but ultimately these exotics will lower habitat productivity for deer. These are all unintentional consequences of human development that might ultimately reduce deer densities in many regions where densities are currently high. Intentional reductions in habitat productivity usually entail limiting access through repellents, fences, or dogs.[16] These remedies may work for individual landowners or small landholdings, but are not effective across landscapes or away from human habitation. A subset of this approach is to shift productivity in an effort to shift the browsing pressure of deer. Foresters have had success shifting deer through placement of food plots or precuts away from valuable timber stands before harvest.[38] This short-term relief from browsing pressure will be counter-productive over the long term, as it raises overall landscape productivity, but it can achieve immediate goals.

Rather than reducing productivity, it might be possible to reduce the ability of deer to utilize plant energy by limiting their reproduction. Extensive research has gone into developing contraceptives that limit reproduction in female deer and, if applied properly, could limit deer population growth.[106–109] GonaCon™ (U.S. Dept. of Agriculture, Animal and Plant Health Inspection Service, Fort Collins, CO, USA) is currently the sole contraception approved by USDA for commercial use, and an intramuscular application results in female deer producing GnRH antibodies, which prevent development of a corpus luteum in the ovaries and thereby eliminates mating behavior and ovulation.[108,110] A present limitation is that the contraceptive must be hand-injected, entailing capture of individuals. Development of an oral or remote-delivered contraceptive will remove this limitation, but contraception does not directly reduce the population number, only the recruitment rate. Used in conjunction with increased mortality, it has potential for limiting populations around human development if sufficient females can be maintained in a contraceptive state.[111]

Mortality

Limiting habitat productivity will not only limit reproduction, but also can increase mortality, as food restrictions can increase overwinter mortality and disease susceptibility.[92,112] Most overwinter mortality is confined to fawns of the year, although severe winter weather can impact adults.[92] Malnourished deer do suffer higher parasite rates that may increase their mortality rate.[112] Viral diseases, such as bluetongue and various hemorrhagic diseases, are episodic, but more prevalent in high-density populations of deer, and might reduce populations by 15% in a single year.[112] Neither these diseases nor hunger, however, will regulate deer numbers at densities well below biological carrying capacity. Both bovine tuberculosis and chronic wasting disease are transmitted by contact with infected individuals or materials, and transmission should increase with density.[80] Theoretically, chronic wasting disease is more likely to persist in deer populations where the carrying capacity has been increased through human modification of habitat.[113] However, there is yet no empirical evidence that either of these diseases limit deer populations.[80]

Predation has been shown to significantly reduce deer populations. The reintroduction of wolves into both western ecosystems have changed both the behavior and the number of large herbivores.[5,114] Large predators accomplish many of the goals of ecological carrying capacity by both reducing overall numbers and increasing the perceived predation risk of deer.[114] In the case of the Yellowstone system, the reintroduction of wolves caused elk and bison to spend less time in open riparian areas and less time feeding overall, which resulted in increased stem density of aspen within riparian areas.[5] Increased predation risks also lowered reproduction in elk through increased glucosteroid stress hormones, reducing fawn production.[115,116] In northern Minnesota, wolves were the main source of mortality for female deer within five years of their arrival.[117] Wolves were historically part of eastern forests, but the politics of their return in significant numbers is problematic.

Besides wolves, there is evidence that cougars (*Puma concolor*) limit deer populations in western states, but eastern cougar populations only occur in Florida.[118] Coyotes (*Canis latrins*) do reduce deer numbers in eastern Canada and are postulated to be able to reduce southern deer populations.[119,120] There is, however, limited evidence that deer densities are lower throughout the expanding range of coyotes. Introduction of bobcats to a South Carolina island did reduce deer numbers, whereas other forest carnivores seem to be incidental predators on fawns.[16]

Extirpated predators should be reintroduced where possible to both reduce numbers and change the functional foraging of deer, but this option will not be always be viable. Hunting is currently the primary tool for deer management in the United States. Nationwide in 2006 (the most current year with summary statistics), 10.7 million hunters harvested 6.2 million deer.[121] Deer herds can be reduced when exposed to hunters, and indexes show lower deer damage when herds are newly harvested.[122–124] Hunters do not mimic predators, as they only impact the number of deer and not their preferential browsing.[125] It has not been demonstrated that hunters with restricted seasons, locations, and hours can duplicate the presence of apex predators on the landscape. Hunting, however, is the sole tool currently available that can significantly reduce deer numbers at limited cost and has the potential to achieve management goals.

Managing hunters in North America

It is difficult to generalize the current densities of deer in eastern United States and their impact on forest resources. Unhunted population in moderate or highly productive landscapes are found in densities of 30–50 deer/km^2, with isolated examples of >100 deer/km^2.[24,26,27,32,35,62] Hunted populations generally are in a range of 15–30 deer/km^2, and preferred browse species have been demonstrated to be impacted at 3–10 deer km^2.[21,24,26,32,40] Without affecting functional foraging responses, it will be difficult to maintain preferred browse species through use of hunters unless deer densities are reduced to numbers significantly below biological carrying capacity. Achieving these low densities through hunting is a matter of managing hunters and their behavior.[126]

Hunting policy in the United States is unlike hunting in most of the world. In most developing countries, hunting is banned because of inadequate enforcement and low wildlife densities. In Europe, wildlife belongs to the landowner.[127,128] A landowner can manage their wildlife as they would their livestock, setting their own limits and rules, and meat and wildlife products can be sold to restaurants and shops.[129] Government funds are limited for wildlife management and focused on conservation of rare species, as most management is under private control.[128] In the United States, wildlife does not belong to the landowner, but the citizen, and in most states the landowner cannot restrict the movement of wildlife across their land.[130] Landowners may try to entice wildlife by planting food plots or bait piles, but they can only harvest the animals at the discretion of the state. Any game, or its product, killed by the landowner cannot be sold; meat not used for personal consumption can be donated to public food banks or institutions. Management of deer herds is a state function and deer managers set permit levels at a county or regional scale with limited attention to the local property.[72,99] An exception is the issuance of damage control permits for landowners that can demonstrate economic losses due to deer, and these allow for harvest outside of standard regulations.[72] Public lands that wish to engage in hunting must conduct lengthy sessions with public stakeholders and state agencies, and national lands must allow input from citizens and organizations throughout the country.[14] As mentioned earlier, revenues generated from sale of licenses and taxes placed on hunting equipment are used to manage the wildlife and in some states are the sole source of wildlife agency funds.[71]

The reliance on citizen hunters to achieve management goals has come to be called "The North American model."[130] This model has seven tenets that call for ethical hunting of a shared resource for sport and personal consumption. A modification on this model is "Quality Deer Management," which engages the public more directly in population management by encouraging relatively low deer densities through high harvest rates on females, thereby allowing males to reach older age classes under optimal forage conditions.[131]

The North American model has been credited with expanding game populations in North America and creating a system of forest land that is accessible to the public.[130] It should be noted that game populations have also increased throughout Europe without the benefit of the model.[129] Wildlife managers and researchers have noted that the model has problems with expanding use of its revenues and effort beyond game species, developing a strong role for the nonhunting citizen, and replacing an aging constituency of hunters.[132,133] These socioeconomic changes, and safety concerns, result in increasing portions of private land in exurbia being closed to hunting either by individual landowners or homeowner associations.[134] The reliance on a volunteer hunter limits a manager's ability to target deer harvest to specific forests, but some success can be achieved through incentive programs to encourage increased harvest of females.[126] Hunters go where they have the highest probability of obtaining a quality deer, even when they know the management intent is to reduce deer density.[135] Quality Deer Management guidelines do encourage lower densities, but it is limited to cooperatives where hunters agree to shared quotas.[131] Many suburban communities have gone to sharpshooters, usually professional companies, to accomplish goals due to safety concerns.[121]

A major limit to managing the impact of deer herds on forests in North America is that the multiple constituencies involved with deer management who do not all view the ecological role of white-tailed deer as their highest priority.[14] Whereas ecological damage and disease spread may be a direct function of high densities, states have not been able to reduce deer populations across a broad landscape. If state-wide reduction is not possible, then a primary concern of ecologists is that high densities of deer do not result in the homogenization of forests.[70,136] To satisfy stakeholders demanding higher densities, specific areas may be able to "survive" high deer densities if we can shift these areas over time and allow plant communities a periodic release from heavy browsing pressure. In rangelands, livestock-grazing systems that promote both temporal and spatial variation have been shown to increase plant and bird biodiversity.[137,138] In unharvested forests, an effective stocking rate for oak seedlings can be achieved in three years of low deer density, and subsequent canopy disturbance would release seedlings to reach sufficient heights to escape damaging deer browsing within an additional five years.[87,88] Responses within harvested forests would be quicker, assuming seed banks are still viable.[22] It might be possible to create a three-tiered system of deer management for public lands composed of areas with >30, 15–30, and 5–15 deer/km^2. Hunters could have access to the first two tiers, and the third tier would be maintained at lower densities through targeted management by either commercial or professional staff. Converting to a hybrid model of deer management, where citizen hunters are initially allowed to harvest deer under standard regulations, followed by subsequent years where regulated sale of meat from harvested deer is allowed, might provide incentive to lower deer densities into the third tier and below ecological carrying capacity. Without the economic incentive to remove deer from already low-density populations, managers probably will have to expend funds to recruit hunters. The goal of such management would be to bring deer densities as low as possible in the focal areas and create a heterogeneous deer density across the landscape.

Several researchers argue for creating ecological benchmarks and managing for impact rather than deer density.[139–141] Transition points between the three tiers of deer density outlined above can be converted to benchmarks, which are easier to measure than deer density and which trigger shifts in management prescription. These measures would manage deer on an ecological basis rather than on cultural carrying capacity. The support of conservationists and ecologists for deer management would be stronger if management was based on ecological principles.

Management conclusions

The ecological evidence is compelling that deer populations in eastern North America need to be managed significantly below biological carrying capacity to maintain intact, diverse forested ecosystems, but that this regulation is not likely to be accomplished under the present suite of natural predators or through significant habitat modification. For the immediate future, managers must rely on hunters to reduce deer populations. Two issues hinder the ability of managers to achieve their goals. First, the current wildlife management system (i.e., the North American model) was developed to grow wildlife populations and may not have enough incentives to meet the current challenges of reducing deer populations. Second, state wildlife managers have adopted a paradigm of cultural carry capacity for setting population levels, and this qualitative measure does not insure densities below biological carry capacity. The primary function of management should be stewardship of the public's natural resources, and any system not based on quantifiable measures will not be able to withstand careful scrutiny by opposing groups. Adoption of a management plan based on biological carrying capacity relies on cross-agency cooperation and buy-in by stakeholder groups, which includes the continued support of the citizen hunter and by gaining the support of other conservationists.

Conflicts of interest

The author declares no conflicts of interest.

References

1. McNaughton, S.J., D.A. Oesterheld, D.A. Frank & K. J. Williams. 1989. Ecosystem-level patterns of primary productivity and herbivory in terrestrial habitat. *Nature* **341:** 142–144.
2. Hobbs, N.T. 1996. Modification of ecosystems by ungulates. *J. Wildl. Manage.* **60:** 695–713.
3. Danell, K., R. Bergstrom, P. Duncan & J. Pastor, eds. 2006. *Large Herbivore Ecology, Ecosystem Dynamics and Conservation.* Cambridge University Press. Cambridge, UK.
4. Ray, J.C., K.H. Redford, R.S. Stenbeck & J. Berger, eds. 2005. *Large Carnivores and the Conservation of Biodiversity.* Island Press. Washington, DC.
5. Beschta, R.L. & W.J. Ripple. 2009. Large predators and trophic cascades in terrestrial ecosystems of the western United States. *Biol. Conserv.* **142:** 2401–2414.
6. Terbourgh, J. & J. A. Estes. 2010. *Trophic Cascades: Predators, Prey and the Changing Dynamics of Nature.* Island Press. New York, NY.
7. Hester, A.J., M. Bergman, G.R. Iason & J. Moen. 2006. Impacts of large herbivores on plant community structure and dynamics. In *Large Herbivore Ecology, Ecosystem Dynamics and Conservation.* K. Danell, R. Bergstrom,

P. Duncan & J. Pastor, Eds.: 97–141. Cambridge University Press. Cambridge, UK.

8. Karath, K.U., J.D. Nichols, N.S. Kumar & J.E. Hines. 2004. Tigers and their prey; predicting carnivore densities from prey abundance. *Proc. Natl. Acad. Sci. USA* **101:** 4854–4858.
9. Vera, F.W.M., E.S. Bakker & H. Olff. 2006. Large herbivores: missing patterns of western European light-demanding tree and shrub species. In *Large Herbivore Ecology, Ecosystem Dynamics and Conservation.* K. Danell, R. Bergstrom, P. Duncan & J. Pastor, Eds.: 203–231. Cambridge University Press. Cambridge, UK.
10. Pastor, J. & Y. Cohen. 1997. Herbivores, the functional diversity of plant species and the cycling of nutrients in ecosystems. *Theor. Popul. Biol.* **51:** 165–179.
11. McCabe, T.R. & R.E. McCabe. 1997. Recounting whitetails past. In *The Science of Overabundance: Deer Ecology and Population Management*. W.J. McShea, H.B. Underwood & J.H. Rappole, Eds.: 11–26. Smithsonian Institution Press. Washington, DC.
12. Hickey, L. 2008. Assessing re-colonization of moose in New York with HIS models. *Alces* **44:** 117–126.
13. Larkin, J.L., R.A. Grims, L. Cornicelli, *et al.* 2001. Returning elk to Appalchia: foiling Murphy's law. In *Large Mammal Restoration; Ecological and Sociological Challenges in the 21st Century.* D.S. Maehr, R.F. Noss & J.L. Larkin, Eds.: 101–117. Island Press. Washington, DC.
14. Leong, K.M., D.J. Decker, T.B. Lauber, *et al.* 2009. Overcoming jurisdictional boundaries through stakeholder engagement and collaborative governance: lessons learned from white-tailed deer management in the U.S. *Res. Rural Sociol. Dev.* **14:** 221–247.
15. Ruggiero, L.F. 2010. Scientific independence and credibility in sociopolitical processes. *J. Wildl. Manage.* **74:** 1179–1182.
16. Warren, R.J. 2011. Deer overabundance in the USA: recent advances in population control. *Anim. Prod. Sci.* **51:** 259–266.
17. Braun, E.L. 1950. *Deciduous Forests of Eastern North America.* The Blakiston Co. Philadelphia, PA.
18. Bailey, R. G. 2009. *Ecosystem Geography*. Springer-Verlag. New York, NY.
19. Augustine, D.J. & P.A. Jordon. 1998. Predictors of white-tailed deer grazing intensity in fragmented deciduous forests. *J. Wildl. Manage.* **62:** 1076–1085.
20. Morisette, E.M., C. Lavoie & J. Huot. 2009. Fairy slipper (*Calypso bulbosa*) on Anticosti Island: the occurrence of a rare plant in an environment strongly modified by white-tailed deer. *Botany* **87:** 1223–1231.
21. Alverson, W.S. & D.M. Waller. 1997. Deer population and widespread failure of hemlock regeneration in northern forests. In *The Science of Overabundance: Deer Population Ecology and Management*. W.J. McShea, H.B. Underwood & J.H. Rappole, Eds.: 280–297. Smithsonian Institution Press. Washington, DC.
22. Mallik, A.U. 2003. Conifer regeneration problems in boreal and temperate forests with ericaceous understory: role of disturbance, seedbed limitation, and keystone species change. *Crit. Rev. Plant Sci.* **22:** 341–366.
23. Sauve, D.G. & S.D. Cote. 2010. Winter foraging selection in white-tailed deer at high densities: balsam fir is the best of a bad choice. *J. Wildl. Manage.* **71:** 911–914.
24. Rooney, T.P. 2001. Deer impacts on forest ecosystems: a North American perspective. *Forestry* **74:** 201–208.
25. Rooney T.P. & D.M. Waller. 2003. Direct and indirect effects of white-tailed deer in forest ecosystems. *For. Ecol. Manage.* **181:** 165–176.
26. Cote, S.D., T.P. Rooney, J. Tremblay, *et al.* 2004. Ecological impacts of deer overabundance. *Annu. Rev. Ecol. Evol. Syst.* **35:** 113–147.
27. McShea, W.J. 2005. Forest ecosystems without carnivores: when ungulates rule the world. In *Large Carnivores and the Conservation of Biodiversity*. J.C. Ray, K. Redford, R.S. Stenbeck & J. Berger, Eds.: 138–153. Island Press. Washington, DC.
28. Geist, V. 1998. *Deer of the World: Their Evolution, Behavior and Ecology*. Stackppole Books. Mechanicsburg, PA.
29. McCaffery, K.R., J. Tranetzki & T. Piechura, Jr. 1974. Summer food of deer in Northern Wisconsin. *J. Wildl. Manage.* **38:** 215–219.
30. Runkle, J.R. 1981. Gap regeneration in some old-growth forests of the eastern United States. *Ecology* **62:** 1041–1051.
31. Royo, A.R., R. Collins, M.B. Adams, *et al.* 2010. Pervasive interactions between ungulate browsers and disturbance regimes promote temperate forest herbaceous diversity. *Ecology* **91:** 93–105.
32. Hansen, L.P., C.M. Nixon & J. Beringer. 1997. Role of refuges in the dynamics of outlying deer populations. In *The Science of Overabundance: Deer Population Ecology and Management*. W.J. McShea, H.B. Underwood & J.H. Rappole, Eds.: 327–345. Smithsonian Institution Press. Washington, DC.
33. Etter, D.R., K.M. Hollis, T.R. Van Deelen, *et al.* 2002. Survival and movements of whitetailed deer in suburban Chicago, Illinois. *J. Wildl. Manage.* **66:** 500–510.
34. Storm, D.J., C.K. Nielson, E.M. Schauber & A. Woolf. 2007. Deer–human conflict and hunter access in an exurban landscape. *Hum. Wildl. Inter.* **1:** 53–59.
35. Stewart, C.M., W.J. McShea & B.P. Piccolo. 2007. The impact of white-tailed deer foraging on agricultural resources at 3 National Historical Parks in Maryland. *J. Wildl. Manage.* **71:** 1525–1530.
36. Retamosa, M. I., L.A. Humberg, J.C. Beasley & O.E. Rhodes, Jr. 2008. Modeling wildlife damage to crops in northern Indiana. *Hum. Wildl. Inter.* **2:** 225–239.
37. Meier, A.J., S.P. Bratton & D.C. Duffy. 1995. Possible ecological mechanisms for loss of vernal-herb diversity in logged eastern deciduous forests. *Ecol. Appl.* **5:** 935–946.
38. Miller, B.F., T.A. Campbell, B.R. Laseter, *et al.* 2009. White-tailed deer herbivory and timber harvesting rates: implications for regeneration success. *For. Ecol. Manage.* **258:** 1067–1072.
39. Tilghman, N.G. 1989. Impacts of white-tailed deer on forest regeneration in northwestern Pennsylvania. *J. Wildl. Manage.* **53:** 524–532.
40. Anderson, R.C. & A.J. Katz. 1993. Recovery of browse-sensitive tree species following release from white-tailed deer (*Odocoileus virginianus* Zimmerman) browsing pressure. *Biol. Conserv.* **63:** 203–208.

41. Russell, F.L., D.B. Zippin & N.L. Fowler. 2001. Effects of white-tailed deer (*Odocoileus virginianus*) on plants, plant populations and communities: a review. *Am. Midl. Nat.* **146:** 1–26.
42. Pedersen, B.S. & A.M. Wallis. 2004. Effects of white-tailed deer herbivory on forest gap dynamics in a wildlife preserve, Pennsylvania, USA. *Nat. Areas J.* **24:** 82–94.
43. Royo, A.A. & W.P. Carson. 2006. On the formation of dense understory layers in forests worldwide: consequences and implications for forest dynamics, biodiversity, and succession. *Can. J. For. Res.* **36:** 1345–1362.
44. Huebner, C.D., K.W. Gottschalk, G.W. Miller & P.H. Brose. 2011. Restoration of three forest herbs in the Liliaceae family by manipulating deer herbivory and overstorey and understorey vegetation. *Plant Ecol. Divers.* **3:** 259–272.
45. Eschtruth, A.K. & J.J. Battles. 2008. Deer herbivory alters forest response to canopy decline caused by an exotic insect pest. *Ecol. Appl.* **18:** 360–376.
46. Schmitz, O.J. & A.R.E. Sinclair. 1997. Rethinking the role of deer in forests. In *The Science of Overabundance: Deer Ecology and Population Management*. W.J. McShea, H.B. Underwood & J.H. Rappole, Eds.: 201–223. Smithsonian Institution Press. Washington, DC.
47. Stromayer, K.A.K. & R.J. Warren. 1997. Are overabundant deer herds in the eastern United States creating alternate stable states in forest plant communities? *Wildl. Soc. Bull.* **25:** 227–34.
48. Horsley, S.B., S.L. Stout & D.S. DeCalesta. 2003. Whitetailed deer impact on the vegetation dynamics of a northern hardwood forest. *Ecol. Appl.* **13:** 98–118.
49. Tremblay, J.P., J. Huot & F. Potvin. 2006. Divergent nonlinear responses of the boreal forest field layer along an experimental gradient of deer densities. *Oecologia* **150:** 78–88.
50. Webster, C.R., M.A. Jenkins & J.H. Rock. 2005. Longterm response of spring flora to chronic herbivory and deer exclusion in Great Smoky Mountains National Park, USA. *Biol. Conserv.* **125:** 297–307.
51. Taverna, K., R.K. Peet & L.C. Phillips. 2005. Long-term change in ground-layer vegetation of deciduous forests of the North Carolina Piedmont, USA. *J. Ecol.* **93:** 202–213.
52. Knight, T.M., J.L. Dunn, L.A. Smith, *et al.* 2009. Deer facilitate invasive plant success in a Pennsylvania forest understory. *Nat. Areas J.* **29:** 110–116.
53. Myers, J.A., M. Vellend, S. Gardescu & P.L. Marks. 2004. Seed dispersal by white-tailed deer: implications for long distance dispersal, invasion, and migration of plants in eastern North America. *Oecologia* **139:** 35–44.
54. Fletcher, J.D., W.J. McShea, L.A. Shipley & D. Shumway. 2001. Use of common forest forbs to measure browsing pressure by white-tailed deer (*Odocoileus virginianus* Zimmerman) in Virginia, USA. *Nat. Areas J.* **21:** 172–176.
55. Augustine, D.J. & D. deCalesta. 2003. Defining deer overabundance and threats to forest communities: from individual plants to landscape structure. *Ecoscience* **10:** 472–486.
56. Heckel, C.D., N.A. Bourg, W.J. McShea & S. Kalisz. 2010. Nonconsumptive effects of a generalist ungulate herbivore drive decline of unpalatable forest herbs. *Ecology* **91:** 319–326.
57. Waller, D.M. & W.S. Alverson. 1997. The white-tailed deer: a keystone herbivore. *Wildl. Soc. Bull.* **25:** 217–226.
58. Ostfeld, R.S., C.G. Jones & J. O. Wolff. 1996. Of mice and mast; ecological connections in eastern deciduous forests. *Bioscience* **46:** 323–330.
59. McShea, W.J. 2000. The influence of acorn crops on annual variation in rodent and bird populations. *Ecology* **81:** 228–238.
60. Martin, J.L., S.A. Stockton, S. Allombert & A.J. Gaston. 2010. Top-down and bottom-up consequences of unchecked ungulate browsing on plant and animal diversity in temperate forests: lessons from a deer introduction. *Biol. Invasions* **12:** 353–371.
61. Pringle, R.M., T.P. Young, D.I. Rubenstein & D.J. McCauley. 2007. Herbivore-initiated interaction cascades and their modulation by productivity in an African savanna. *Proc. Natl. Acad. Sci. USA* **104:** 193–197.
62. McShea, W.J. & J.H. Rappole. 2000. Managing the abundance and diversity of breeding bird populations through manipulation of deer populations. *Conserv. Biol.* **14:** 1161–1170.
63. Martin, T.G., P. Arcese & N. Scheerder. 2011. Browsing down our natural heritage: deer impact on vegetation structure and songbird populations on across an island archipelago. *Biol. Conserv.* **144:** 459–469.
64. Jones, C.G., R.S. Ostfeld, M.P. Richard, *et al.* 1998. Chain reactions linking acorns to gypsy moth outbreaks and Lyme disease. *Risk Sci.* **279:** 1023–1026.
65. Illius, A.W. 2004. Linking functional responses and foraging to population dynamics. In *Large Herbivore Ecology, Ecosystem Dynamics and Conservation*. K. Danell, R. Bergstrom, P. Duncan & J. Pastore, Eds.: 71–96. Cambridge University Press. Cambridge, UK.
66. Ward, D. 2006. Long-term effects of herbivory on plant diversity and functional types in arid ecosystems. In *Large Herbivore Ecology, Ecosystem Dynamics and Conservation*. K. Danell, R. Bergstrom, P. Duncan & J. Pastor, Eds.: 142–169. Cambridge University Press. Cambridge, UK.
67. Augustine, D.J., L.E. Frelich & P.A. Jordon. 1998. Evidence for two alternative stable states in an ungulate grazing system. *Ecol. Appl.* **8:** 1260–1269.
68. Ripple, W.J. & R.L. Beschta. 2003. Wolf reintroduction, predation risk, and cottonwood recovery in Yellowstone National Park. *For. Ecol. Manage.* **184:** 299–313.
69. Parker, J.D., J-P Salminen & A.A. Agrawal. 2010. Herbivory enhances positive effects of plant genetic diversity. *Ecol. Lett.* **13:** 553–563.
70. Rooney, T.P., S.M. Wiegmann, D.A. Rogers & D.M. Waller. 2004. Biotic impoverishment and homogenization in unfragmented forest understory communities. *Conserv. Biol.* **18:** 787–798.
71. Southwick, R. 2009. The economic contributions of hunting in the United States. 199–211 In *Transactions of the Seventy-third North American Wildlife and Natural Resources Conference*, Wildlife Management Institute. Washington, DC.
72. Virginia Department of Game and Inland Fisheries. 2007. Virginia Deer Management Plan 2006–2015. Wildl. Inform. Publ. 07–1. Richmond, VA.

73. State Farm. 2009. Deer- vehicle collision frequency jumps 18 percent in five years. Available at: http://www.statefarm.com/aboutus/_pressreleases/2009/deer_vehicle_collision_frequency_jumps.asp accessed Dec. 29, 2011.
74. McShea, W.J., C.M. Stewart, L.J. Kearns, *et al.* 2008. Factors affecting autumn deer/vehicle collisions in a rural Virginia county. *Hum. Wildl. Confl.* **2:** 110–121.
75. Conover, M.R. & D.J. Decker. 1991. Wildlife damage to crops: perceptions of agricultural and wildlife professionals in 1957 and 1987. *Wildl. Soc. Bull.* **19:** 46–52.
76. Drake, D., J.B. Paulin, P.D. Curstis, *et al.* 2005. Assessment of negative economic impacts from deer in the northeastern United States. *J. Ext.* **43.** Available at: http://www.joe.org/joe/2005february/rb5.php accessed Dec. 29, 2011.
77. West, B.C. & J.A. Parkhurst. 2002. Interactions between deer damage, deer density, and stakeholder attitudes in Virginia. *Wildl. Soc. Bull.* **30:** 139–147.
78. Marquis, D.A. 1974. *The Impact of Deer Browsing on Allegheny Hardwood Regeneration.* USDA Forest Service Research Paper NE-308, Northeastern Forest Experiment Station, Upper Darby, PA.
79. Horsely, S.B. & D.A. Marquis. 1983. Interference of deer and weeds with Allegheny hardwood reproduction. *Can. J. For. Res.* **13:** 61–69.
80. Conner, M.M., M.R. Ebinger, J. A. Blanchong, & P.C. Cross. 2008. Infectious disease in cervids of North America. *Ann. N. Y. Acad. Sci.* **1134:** 146–172.
81. Centers for Disease Control and Prevention. 2011. *Summary statistics.* Available at: http://www.cdc.gov/lyme/stats/index.html accessed Dec. 29, 2011.
82. Ostfeld, R.S., C.D. Canham, K. Oggenfuss, *et al.* 2006. Climate, deer, rodents, and acorns as determinants of variation in Lyme-disease risk. *PloS Biol.* **4:** 1058–1068.
83. Ostfeld, R.S., F. Keesing, C.G. Jones, *et al.* 1998. Integrative ecology and the dynamics of species in oak forests. *Inter. Biol.* **1:** 178–186.
84. Fish, D. & J.E. Childs. 2009. Community-based prevention of Lyme disease and other tick-borne diseases through topical application of acaricide to white-tailed deer: background and rational. *Vector-borne Zoonotic Dis.* **4:** 357–364.
85. Abrams, M.D. 2003. Where have all the white oak gone? *Bioscience* **53:** 927–939.
86. McShea, W.J., W.M. Healy, P. Devers, *et al.* 2007. Forestry Matters: decline of oaks will impact wildlife in hardwood forests. *J. Wildl. Manage.* **71:** 1717–1728.
87. Dey, D. 2002. The ecological basis for oak silviculture in eastern North America. In *Oak Forest Ecosystems; Ecology and Management for Wildlife.* W.J. McShea & W.M. Healy, Eds.: 60–79. Johns Hopkins Press. Baltimore, MD.
88. Healy, W.M. & W.J. McShea. 2002. Goals and guidelines for ecosystem management of oak forests. In *Oak Forest Ecosystems: Ecology and Management for Wildlife.* W.J. McShea & W.M. Healy, Eds.: 33–340. John Hopkins University Press. Baltimore, MD.
89. Suding, K.N., K.L. Gross & G.R. Houseman. 2004. Alternative states and positive feedbacks in restoration ecology. *Trends Ecol. Evol.* **19:** 46–53.
90. Caughley, G. 1976. Wildlife management and the dynamics of ungulate populations. In *Applied Biology.* T.H. Coaker, Eds.: 183–246. Academic Press. New York, NY.
91. Sinclair, A.RE. 1997. Carrying capacity and the overabundance of deer. In *The Science of Overabundance: Deer Ecology and Population Management.* W.J. McShea, H.B. Underwood & J.H. Rappole, Eds.: 80–394. Smithsonian Institution Press. Washington, DC.
92. McCullough, D.R. 1979. *The George Reserve Deer Herd.* Michigan State University Press. Ann Arbor, MI.
93. Lancia, R.A., K.H. Pollock, J.W. Bishir & M.C. Conner. 1988. A white-tailed deer harvest strategy. *J. Wildl. Manage.* **52:** 589–595.
94. Leopold, A. 1943. Deer irruptions. *Trans. Wisc. Acad. Sci. Arts Lett.* **35:** 351–366.
95. Leopold, A., L.K. Sowls & D.L. Spencer. 1947. A survey of over-populated deer ranges in the U.S. *J. Wildl. Manage.* **11:** 162–177.
96. McNab, J. 1985. Carrying capacity and related slippery shibboleths. *Wildl. Soc. Bull.* **13:** 403–410.
97. Latham, R.E., J. Beyea, M. Benner, *et al.* 2005. *Managing white-tailed deer in forest habitat from an ecosystem perspective: Pennsylvania case study.* Audubon Pennsylvania and the Pennsylvania Habitat Alliance, Harrisburg, PA.
98. Banta, J.A., A.A. Royo, C. Kirschbaum & W.P. Carson. 2005. Plant communities growing on boulders in the Allegheny National Forest: evidence for boulders as refugia from deer and as a bioassay of overbrowsing. *Nat. Areas J.* **25:** 10–18.
99. Knox, M.W. 1997. Historical changes in the abundance and distribution of deer in Virginia. In *The Science of Overabundance: Deer Ecology and Population Management.* W.J. McShea, H.B. Underwood & J.H. Rappole, Eds.: 27–36. Smithsonian Institution Press. Washington, DC.
100. McGraw, J.B. & M.A. Furedi. 2004. Deer browsing and population viability of a forest understory plant. *Science* **307:** 920–922.
101. Nixon, C.M., L.P. Hansen, P.A. Brewer & J.E. Chelsvig. 1991. Ecology of white-tailed deer in an intensively farmed region of Illinois. *Wildl. Monogr.* **188:** 1–77.
102. Nowak, R.M. 1999. *Walker's Mammals of the World.* 6th ed. Johns Hopkins Press. Baltimore, MD.
103. Cote, S.D. 2005. Extirpation of a large black bear population by introduced white-tailed deer. *Conserv. Biol.* **19**: 1668–1671.
104. Walter, W.D., K.C. VerCauteren, J.M. Gilsdorf, & S.E. Hynstrom. 2009. Crop, native vegetation, and biofuel: response of white-tailed deer to changes in management priorities. *J. Wildl. Manage.* **73:** 339–344.
105. de la Cretaz, A.L. & M.J. Kelty. 1999. Establishment and control of hay-scented fern: a native invasive species. *Biol. Invasions* **1:** 223–236.
106. McShea, W.J., S.L. Monfort, S. Hakim, *et al.* 1997. The effect of immunocontraception on the behavior and reproduction of white-tailed deer. *J. Wildl. Manage.* **41:** 560–569.
107. Fraker, M.A., R.G. Brown, G.E. Grant, *et al.* 2002. Long-lasting, single-dose immunocontraception of feral fallow deer in British Columbia. *J. Wildl. Manage.* **66:** 1141–1147.
108. Miller, L.A., J.P. Gionfriddo, K.A. Fagerstone, *et al.* 2008. The single-shot GnRH immunocontraceptive vaccine (GonaCon) in whitetailed deer: comparison of several GnRH preparations. *Am. J. Reprod. Immunol.* **60:** 214–23.

109. Rutberg, A.T. & R.E. Nagle. 2007. Population-level effects of immunocontraception in white-tailed deer (*Odocoileus virginianus*). *Wildl. Res.* **35:** 494–501.
110. Gionfriddo, J.P., J.D. Eisemann, K.J. Sullivan, *et al.* 2009. Field test of a single-injection gonadotrophin releasing hormone immunocontraceptive vaccine in female white-tailed deer. *Wildl. Res.* **36:** 177–84.
111. Porter, W.F., H.B. Underwood & J.L. Woodward. 2004. Movement behavior, dispersal, and the potential for localized management of deer in a suburban environment. *J. Wildl. Manage.* **68:** 247–256.
112. Davidson, W.R. & G.L. Doster. 1997. Health characteristics and white-tailed deer populations in the southeastern United States. In *The Science of Overabundance: Deer Ecology and Population Management*. W.J. McShea, H.B. Underwood & J.H. Rappole, Eds.: 164–184. Smithsonian Institution Press. Washington, DC.
113. Sharp, A. & J. Pastor. 2011. Stable limit cycles and the paradox of enrichment in a model of chronic wasting disease. *Ecol. Appl.* **21:** 1024–1030.
114. Laundré, J.W., L. Hernandez & K.B. Altendorf. 2001 Wolves, elk, and bison: reestablishing the 'landscape of fear' in Yellowstone National Park, USA. *Can. J. Zool.* **79:** 1401–1409.
115. Creel S., D. Christianson, S. Liley & J.A. Winnie. 2007. Predation risk affects reproductive physiology and demography of elk. *Science* **315:** 960.
116. Creel S., J.A. Winnie & D. Christianson. 2009. Glucocorticoid stress hormones and the effect of predation risk on elk reproduction. *Proc. Natl. Acad. Sci. USA* **106:** 388–393.
117. DelGiudice, G.D., M.R. Riggs, P. Joly & W. Pan. 2002. Winter severity, survival and cause of specific mortality in female white-tailed deer in north-central Minnesota. *J. Wildl. Manage.* **66:** 698–717.
118. Ripple, W.J. & R.L. Beschta. 2008. Trophic cascade involving cougar, mule deer and black oaks in Yosemite National Park. *Biol. Conserv.* **141:** 1249–1256.
119. Ballard, W.B., H.A. Whitlaw, S.J. Young, *et al.* 1999. Predation and survival of white-tailed deer fawns in north central New Brunswick. *J. Wildl. Manage.* **63:** 574–579.
120. Kilgo, J.C., H.S. Ray, C. Ruth & K.V. Miller. 2010. Can coyotes affect deer populations in southeastern North America? *J. Wildl. Manage.* **74:** 929–933.
121. Fish and Wildlife Service. 2011. *Hunting Summary Statistics*. Available at: http://www.fws.gov/hunting/huntingstats.html accessed Dec. 29, 2011.
122. Killmaster, C.H., D.A. Osborn, R.J. Warren, *et al.* 2007. Deer and understory plant responses to a large-scale herd reduction on a Georgia state park. *Nat. Areas J.* **27:** 161–168.
123. DeNicola, A.J. & S.C. Williams. 2008. Sharpshooting suburban white-tailed deer reduces deer-vehicle collisions. *Hum. Wildl. Confl.* **2:** 28–33.
124. Winchcombe, R.J. 2010. Hunting for balance: a long-term effort to control local deer abundance. *Wildl. Prof.* **4:** 48–50.
125. Berger J. 2005. Hunting by carnivores and humans: does functional redundancy occur and does it matter? In *Large Carnivores and the Conservation of Biodiversity*. J.C. Ray, K.H. Redford, R.S. Steneck & J. Berger, Eds.: 315–341. Island Press. Washington, DC.
126. Van Deelen, T.R., B.J. Dhuey, C.N. Jacques, *et al.* 2010. Effects of earn-a-buck and special antlerless-only season on Wisconsin's deer herd. *J. Wildl. Manage.* **74:** 1693–1700.
127. Apollonio, M., R. Anderson & R. Putman. 2010. *European Ungulates and their management in the 21st Century*. Cambridge University Press. New York, NY.
128. Apollonio, M., R. Anderson & R. Putman. 2010. Recent status and future challenges for European ungulate management. In *European Ungulates and their management in the 21st Century*. M. Apollonio, R. Anderson & R. Putman, Eds.: 578–604. Cambridge University Press. New York, NY.
129. Brainerd, S.M. & B. Kaltenborn. 2010. The Scandinavian model: a different path to wildlife management. *Wildl. Prof.* **4:** 52–55.
130. Organ, J.F., S.H. Mahoney & V. Geist. 2010. Born in the hands of hunters: the North American model of wildlife conservation. *Wildl. Prof.* **4:** 22–27.
131. Miller, K.V. & R.L. Marchinton, eds. 1995. *Quality Whitetails: The Why and How of Quality Deer Management*. Stackpole Books. Mechanicsburg, PA.
132. Dratch, P. & R. Kahn. 2011. Moving beyond the model: our ethical responsibility as the top trophic predators. *Wildl. Prof.* **5:** 58–60.
133. Nelson, M.P., J.A. Vucetich, P.C. Paquet & J.K. Bump. 2011. An inadequate construct? North American model: what's flawed, what's missing, what's needed. *Wildl. Prof.* **5:** 58–60.
134. Harden, C.D., A. Woolf & J. Roseberry. 2005. Influence of exurban development on hunting opportunity, hunter distribution, and harvest efficiency of white-tailed deer. *Wildl. Soc. Bull.* **33:** 233–242.
135. Blanchong, J.A., D.O. Joly, M.D. Samuel, *et al.* 2006. White-tailed deer harvest from the chronic wasting disease eradication zone in South-Central Wisconsin. *Wildl. Soc. Bull.* **34:** 725–731.
136. McKinney, M.L. & J.L. Lockwood. 1999. Biotic homogenization: a few winners replacing many losers in the next mass extinction. *Trends Ecol. Evol.* **14:** 450–453.
137. Derner, J.D., W.K. Lauenroth, P. Stapp & D.J. Augustine. 2009. Livestock as ecosystem engineers for grassland bird habitat in the Western Great Plains of North America. *Range. Ecol. Manag.* **62:** 111–118.
138. Nelson, K.S., E.M. Gray & J.R. Evans. 2011. Finding solutions for bird restoration and livestock management: comparing grazing exclusion levels. *Ecol. Appl.* **21:** 547–554.
139. Morellet, N., J.M. Gaillard, A.J.M. Hewison, *et al.* 2007. Indicators of ecological change: new tools for managing populations of large herbivores. *J. Appl. Ecol.* **44:** 634–643.
140. deCalesta, D.S. & S.L. Stout. 1997. Relative deer density and sustainability: a conceptual framework for integrating deer management with ecosystem management. *Wildl. Soc. Bull.* **25:** 252–258.
141. Tierney, G.L., D. Faber-Langendoen, B.R. Mitchell, *et al.* 2009. Monitoring and evaluating the ecological integrity of forest ecosystems. *Front. Ecol. Environ.* **7:** 308–316.

Ann. N.Y. Acad. Sci. ISSN 0077-8923

ANNALS OF THE NEW YORK ACADEMY OF SCIENCES
Issue: *The Year in Ecology and Conservation Biology*

Dropping dead: causes and consequences of vulture population declines worldwide

Darcy L. Ogada,[1] Felicia Keesing,[3] and Munir Z. Virani[1,2]

[1]The Peregrine Fund, Boise, Idaho. [2]National Museums of Kenya, Ornithology Section, Nairobi, Kenya. [3]Bard College, Annadale-on-Hudson, New York

Address for correspondence: Darcy L. Ogada, P.O. Box 1629, 00606 Nairobi, Kenya. darcyogada@yahoo.com

Vultures are nature's most successful scavengers, and they provide an array of ecological, economic, and cultural services. As the only known obligate scavengers, vultures are uniquely adapted to a scavenging lifestyle. Vultures' unique adaptations include soaring flight, keen eyesight, and extremely low pH levels in their stomachs. Presently, 14 of 23 (61%) vulture species worldwide are threatened with extinction, and the most rapid declines have occurred in the vulture-rich regions of Asia and Africa. The reasons for the population declines are varied, but poisoning or human persecution, or both, feature in the list of nearly every declining species. Deliberate poisoning of carnivores is likely the most widespread cause of vulture poisoning. In Asia, *Gyps* vultures have declined by >95% due to poisoning by the veterinary drug diclofenac, which was banned by regional governments in 2006. Human persecution of vultures has occurred for centuries, and shooting and deliberate poisoning are the most widely practiced activities. Ecological consequences of vulture declines include changes in community composition of scavengers at carcasses and an increased potential for disease transmission between mammalian scavengers at carcasses. There have been cultural and economic costs of vulture declines as well, particularly in Asia. In the wake of catastrophic vulture declines in Asia, regional governments, the international scientific and donor communities, and the media have given the crisis substantial attention. Even though the Asian vulture crisis focused attention on the plight of vultures worldwide, the situation for African vultures has received relatively little attention especially given the similar levels of population decline. While the Asian crisis has been largely linked to poisoning by diclofenac, vulture population declines in Africa have numerous causes, which have made conserving existing populations more difficult. And in Africa there has been little government support to conserve vultures despite mounting evidence of the major threats. In other regions with successful vulture conservation programs, a common theme is a huge investment of financial resources and highly skilled personnel, as well as political will and community support.

Keywords: scavenger; condor; ecosystem services; carcass decomposition; disease transmission; vulture decline; poisoning; persecution; Africa; Asian vulture crisis; vulture conservation; diclofenac; furadan

Introduction

Charles Darwin thought vultures were "disgusting."[1] From a human perspective, perhaps they are, but vultures are nature's most successful scavengers, and they provide us with an extensive array of ecological, economic, and cultural services. Most notably, vultures dispose of carrion and other organic refuse, providing a free and highly effective sanitation service. The vulture-governed cleaning service protects the health of humans, domesticated animals, and wildlife because the abundance of other scavengers, some of which are well-known disease reservoirs, increases substantially at carcasses without vultures.[2,3,4] Scavenging of carcasses by vultures promotes the flow of energy through food webs,[5,6] and vultures have been shown to facilitate African predators, such as lions and hyenas, in locating food resources.[7,8]

In this review, we highlight the unique adaptations of vultures to scavenging. We then describe the dramatic recent and historic declines in many

doi: 10.1111/j.1749-6632.2011.06293.x

vulture species worldwide. We explore the apparent causes and consequences of these declines in different regions, and we conclude by characterizing the elements that appear to be necessary for successful vulture conservation programs.

Taxonomy, distribution, and unique adaptations to scavenging

Globally, there are 23 species of vultures (including condors), of which the majority ($n = 16$) occur in the Old World and within the family Accipitridae. The remaining seven species comprise the New World Cathartidae family. Most species ($n = 15$) occupy a range within one continent comprised of two or more countries. Four species, the Griffon vulture (*Gyps fulvus*), Bearded vulture (*Gypaetus barbatus*), Egyptian vulture (*Neophron percnopterus*), and Cinereous vulture (*Aegypius monachus*), have or historically had large ranges that span three continents. Two species, Turkey (*Cathartes aura*) and Black vultures (*Coragyps atratus*), range widely within both North and South America. Cape vultures (*G. coprotheres*) in southern Africa and California condors (*Gymnogyps californianus*) in North America have historically small ranges,[9,10] though fossil evidence suggests that California condors were once found throughout the United States, southern Canada, and northern Mexico.[10] Vulture-rich regions include Central and South America ($n = 6$ spp.), South Asia ($n = 10$ spp.), and Africa ($n = 11$ spp.).

Outside of the oceans, vultures are the only known obligate scavengers.[11] They are uniquely adapted to exploit a transient food source that occurs intermittently over large areas.[11,12] Using gliding flight, vultures take advantage of upward air movements that enable them to travel rapidly over long distances with relatively little energy expenditure.[13] This allows them to search for food efficiently. They can also search communally by observing other birds from the air. Aerial searching also gives them a considerable advantage over terrestrial scavengers because the latter have limited feeding ranges, higher energy expenditures to locate carcasses, and comparatively poor visibility from the ground.[14,15] The superior foraging efficiency of avian scavengers is nowhere more apparent than in the Serengeti, where only vultures have the ability to follow migratory ungulates over vast distances and benefit from heavy mortality that occurs along the way.[14] It has been estimated that vultures in the Serengeti consume more meat than all the other mammalian carnivores combined.[12]

All vultures locate carcasses using keen eyesight. New World *Cathartes* vultures also have a well-developed sense of smell that is used for locating food in forested areas.[16,17,18] Once they have located food, they can travel quickly to reach it, avoiding displacement by larger terrestrial scavengers.[12] For example, vultures in the Serengeti entirely consumed 84% of experimentally placed carcasses before any mammalian scavengers appeared.[8] Facultative scavengers, such as hyenas and especially lions, use the activity of vultures to detect carrion,[7] but vultures more than compensate for this competition by arriving rapidly and in large numbers.[12]

Vultures are among the largest of flying birds. Their size allows them to consume more food at each carcass discovery and to carry greater body reserves, which is important given their erratic food supply. A large body also helps them to outcompete smaller scavengers at carcasses and because flight speed is largely determined by body mass;[11] it increases the area that they can search each day. A recent study using satellite tracking devices determined that the mean foraging range for two immature Cape vultures (one was likely a Cape x White-backed (*G. africanus*) hybrid vulture) was an astounding 480,000 km^2 over an eight-month period.[19] Physiologically, vultures have low pH levels in their digestive tracts (pH 1–2); this destroys most microscopic organisms and greatly reduces the probability that vultures act as sources of infection at carcasses.[20] Finally, vulture life history is characterized by delayed maturity, low productivity, and relatively high adult survivorship. Vultures and especially condors have some of the lowest reproductive rates among birds, and their populations are particularly vulnerable to high mortality, whether by natural or human causes.[21]

Historical and recent vulture population trends

Vulture population declines in Europe and North American likely began as early as the mid-19th century.[22,23] One hundred years later, some populations of Bearded vultures in Europe and the California condor in North America were already nearing extinction.[22,23] Further reports of population

Table 1. Overview of historical vulture declines

Species	Range	Area declining	Peak of declines	Causes of decline	References
Egyptian vulture	Africa, Asia, Europe	Europe (first), Asia, Africa	20th century	Persecution through poisoning and hunting	26
Bearded vulture	Africa, Asia, Europe	Europe (first), Africa	Began 1860 through 1900 when nearly extirpated	Persecution, poisoning, also electrocution	22, 27
California condor	North America	North America	19th century through 1937	Lead poisoning, persecution, collision with overhead wires, secondary poisoning	28, 29
Cape vulture	Southern Africa	Southern Africa	1900	Poisoning, decline in food supply, electrocution, persecution, disturbance at breeding sites	9
White-rumped, Slender-billed, Red-headed vultures	Asia	South Asia (Malaysia, Laos, Cambodia, Thailand rare if not extinct) and Burma, Bangladesh and Pakistan seriously declined, with exception of India	1950–1970s	Decline in food supply due to overhunting and changes in livestock husbandry practices (primarily) also persecution and poisoning	3, 24, 25

declines of Cape vultures and of vultures in South Asia testify to the global nature of declines that had already begun prior to the mid-20th century (Table 1 and Refs. 9, 24, and 25).

In few areas, if any, have vulture populations maintained historical levels of abundance, but there are substantial differences in the numbers of species declining and in population trends among regions (Table 2). In the Middle East, populations of three species are reported to be in decline in the United Arab Emirates,[51] and five species present in Israel are similarly declining.[52] In Europe and North America, which have historically recorded large population declines,[22,23,27] the majority of vulture populations are now increasing or stable (Table 2). In vulture-rich regions, large population declines have occurred in recent decades, particularly in Asia and Africa.

Within the Central and South American region, half of the vulture species are estimated to be in decline, though the region has comparatively little published research on vulture populations apart from that on the Andean condor (*Vultur gryphus*).[31,53,54,55]

Table 2. Conservation status of the world's vultures. References are listed in respect to recent status overviews where available; these are indicated by "and references therein." For species lacking a recent status overview, important historical and recent references are listed, with BirdLife International (2011) as a general source.

Species	Scientific name	Region(s)	IUCN Red List category[a]	Current status	References
California condor	*Gymnogyps californianus*	North America	CR	Small population increasing, though lead shot in carcasses remains a threat	30 and references therein
Turkey vulture	*Cathartes aura*	Americas	LC	Population increasing	31–33 and references therein
Black vulture	*Coragyps atratus*	Americas	LC	Population increasing	31–33 and references therein
Andean condor	*Vultur gryphus*	South America	NR	Moderately rapid decline	31, 34 and references therein
Lesser Yellow-headed vulture	*Cathartes burrovianus*	Central and South America	LC	Apparently stable	31
Greater Yellow-headed vulture	*Cathartes melambrotus*	South America	LC	Slow decline	31
King vulture	*Sarcoramphus papa*	Central and South America	LC	Slow decline	31
Egyptian vulture	*Neophron percnopterus*	Europe, Africa, Middle East, Asia	EN	Declining significantly throughout most of its range. Rapid declines in India	26, 31, 35, 36
Bearded vulture	*Gypaetus barbatus*	Europe, Africa, Middle East, Asia	LC	Slow decline throughout much of range	27, 31 and references therein
Griffon vulture	*Gyps fulvus*	Europe, Africa, Middle East, Asia	LC	Increasing in Europe, stable in central Asia and decreasing in North Africa and Turkey	31
Cinereous vulture	*Aegypius monachus*	Europe, Africa,[b] Middle East, Asia	NT	Slow to moderate decline	31, 37 and references therein

Continued

Table 2. *Continued*

Species	Scientific name	Region(s)	IUCN Red List category[a]	Current status	References
Hooded vulture	*Necrosyrtes monachus*	Africa	EN	Rapid decline of at least 50% over three generations	38 and references therein
White-backed vulture	*Gyps africanus*	Africa	NT	Large declines in West Africa, but apparently stable in other parts of Africa	16, 31, 39–42
Rüppell's vulture	*Gyps rueppellii*	Africa	NT	Extremely rapid declines in West Africa and moderate declines elsewhere	16, 31, 39–42
Lappet-faced vulture	*Torgos tracheliotos*	Africa, Middle East	VU	Moderately rapid decline	16, 31, 39–42
White-headed vulture	*Trigonoceps occipitalis*	Africa	VU	Declining at slow to moderate rate	16, 31, 39–42
Cape vulture	*Gyps coprotheres*	Africa	VU	Declining at ~20% over three generations	9, 31, 43–45
Palm-nut vulture	*Gypohierax angolensis*	Africa	LC	Apparently stable	31
Indian vulture	*Gyps indicus*	Asia	CR	Extremely rapid declines of >97% over 10–15 years	4, 31, 46–49
Himalayan vulture	*Gyps himalayensis*	Asia	LC	Apparently stable	31
White-rumped vulture	*Gyps bengalensis*	Asia	CR	Extremely rapid decline of >99% over 10–15 years	4, 31, 46–50
Red-headed vulture	*Sarcogyps calvus*	Asia	CR	Rapid decline in excess of 90% over 10 years in India	31, 35
Slender-billed vulture	*Gyps tenuirostris*	Asia	CR	Extremely rapid declines	4, 31

[a]IUCN Red List category abbreviations are as follows: LC, least concern; NT, near threatened; V, vulnerable; EN, endangered; CR, critically endangered.
[b]Considered extinct in the region.[16]

Vulture populations in South Asia have incurred the most precipitous and rapid declines ever recorded. Prakash[48] first reported population declines of >95% for *Gyps* vultures in Keoladeo National Park, India, which occurred within a 10-year period. Subsequent surveys over the next decade confirmed massive declines (>96%) of three species of *Gyps* vultures in India.[4,49] Rapid declines have been similarly noted in Pakistan[46,47,50] and Nepal.[56,57] More recent declines in Egyptian and Red-headed vultures (*Sarcogyps calvus*) have been documented throughout India.[35]

In Africa, recent population collapses have been recorded, particularly in West and East Africa. In West Africa, populations of all vultures except the Hooded vulture (*Necrosyrtes monachus*) have declined by an average of 95% in rural areas over the last 30 years.[40,41,58] In protected areas of the Sudanese zone, their collective populations fell by an average of 42% over the same period.[40,41,58] In East Africa, vulture declines of 70% were recently recorded over a three-year period in north-central Kenya.[59] Even the wildlife-rich Masai Mara region has lost an average of 62% of its vultures over the past 30 years, and annual vulture mortality as high as 25% has recently been recorded.[42,60] The situation for vultures in North Africa is dire, particularly in Morocco, where two species, Cinereous and Lappet-faced vultures (*Torgos tracheliotos*), have been extirpated.[61] Others are predicted to follow, and the rest of the region offers little hope for long-term vulture conservation.[61] Vulture research and conservation have a relatively long history in southern Africa, beginning with the formation of the Vulture Study Group in 1977. That group produced the seminal book *The Vultures of Africa* and ingrained vulture research within the ornithological community. Vulture populations in the region continue to be the best studied in Africa and rival the level of study of those in Europe and North America. However, there are a few species whose populations continue to decline, and one, the Egyptian vulture, is believed to be extinct as a breeding species in southern Africa.[62] Throughout Africa, the Hooded vulture, a widespread human commensal, has declined by an average of 62% over the past 40–50 years, and in some areas the decline has been much more rapid.[38]

Causes of declines

There are numerous reasons behind the global vulture decline. However, either poisoning or human persecution, or both, feature in the list of nearly every declining vulture population. We define poisoning as the *unintentional* killing or harming of vultures through consumption of contaminated carcasses or remains. Human persecution refers to the *intentional* killing or disturbance of vultures through actions such as shooting, harassment, and deliberate poisoning.

Vultures are particularly vulnerable to toxic substances due to a combination of foraging behaviors and life history traits found collectively only in vultures.[63] First, most vultures are obligate scavengers that rely on eating dead animals or waste products; this may increase their likelihood of exposure to contaminants. Second, because vultures feed communally, large numbers can be poisoned at a single carcass. Finally, vultures are very long lived and at a high trophic level, which increases their vulnerability to bioaccumulation. Bioaccumulation may have sublethal effects on reproductive success, behavior, immune response, and physiology.[64]

The deliberate poisoning by humans of carnivores, which kills scavengers as well as the intended victims, is likely the most widespread cause of vulture poisoning. Carnivore poisoning continues to be common, especially in Europe and Africa.[65–75] In Europe, poisoning is used to kill predators of game animals (e.g., rabbits, pheasants, and partridges) because hunters believe carnivores such as foxes and mongooses reduce their hunting success.[76] In both Europe and Africa, poisoning is used to "protect" livestock from predators. In Europe, it is regarded as the first option to deter carnivores from attacking livestock, while in Africa poisoning is largely used to avenge the killing of livestock (D. Ogada, pers. obs.[76]).

Some European countries (e.g., Spain and UK) have reduced poisoning incidents through more stringent penalties; tougher restrictions on the use of toxic chemicals; increased public awareness; and cooperation between government ministries, landowner associations, and nongovernmental organizations.[76,77] Spain has taken an innovative approach to tackling this problem by training dogs to detect specific poisons (baits and carcasses of poisoned animals) that are most commonly used against wildlife. Dogs detected 70% more poisoned baits than did specialized (human) detection teams. Funded by the regional government, the canine unit assists in the discovery of offenders as well as in dissuading poisoning through routine inspections in known hotspots.[78] In Africa, interventions are largely spearheaded by nongovernmental organizations, such as the Endangered Wildlife Trust in South Africa and WildlifeDirect in Kenya. While some success has been achieved, there is a need for greater financial support and more training and enforcement if prevention programs are to be sustainable and effective (D. Ogada, pers. obs.[72]).

Poisoning of vultures by ingestion of nonsteroidal anti-inflammatory drugs, particularly diclofenac-sodium, has caused rapid and severe population declines in Asian *Gyps* vultures. During the 1990s, researchers first noted high mortality rates in three *Gyps* species. Subsequent postmortem analyses indicated the vultures died of renal failure triggered by diclofenac residues found in livestock carcasses.[46,47,79] In South Asia, diclofenac was a widely used veterinary analgesic in livestock that typically is consumed by scavengers if they die of disease or injury. Studies showed that diclofenac concentrations in treated livestock were sufficient to cause the high levels of vulture mortality that had been recorded.[79] Contaminated carcasses were subsequently linked to vulture population declines across the subcontinent, and models further confirmed the correlation between vulture declines and diclofenac residues in carcasses.[80,81] In response to the vulture crisis, the governments of three of the most affected countries—India, Pakistan, and Nepal—withdrew manufacturing licenses for veterinary diclofenac in 2006. Recent surveys in India indicate that the ban on veterinary use of diclofenac has markedly reduced its presence in livestock carcasses to levels almost half of what they were prior to and immediately after the ban.[82,83] Despite the reduction of diclofenac in carcasses, prevalence levels still remain sufficiently high to continue to cause a rapid rate of vulture decline that has been estimated at 18% year for Oriental White-backed vultures (*G. bengalensis*).[83]

Lead poisoning through the ingestion of pellets or fragments of lead-based bullets in hunter-killed carcasses is a serious threat to some vulture populations, most notably the California condor. Lead poisoning has been implicated as the leading cause of death in the Arizona population of the condor (~40% of the worldwide population), and it had been estimated that without the intensive intervention of veterinarians, this population would not be self-sustaining in the wild.[29,84–86] However, in 2005 the Arizona Game and Fish Department implemented a hunter-education campaign and provided free, nonlead ammunition for hunters in condor habitats. With >80% participation by hunters, the severity of lead exposure during 2007 was low and there were no lead-related condor deaths.[87] However, there were three lead-related condor deaths after the 2009 hunting season.[30]

In contrast to unintentional poisoning, deliberate persecution of vultures has occurred for centuries. Most vultures have been victimized due to people's ignorance, superstition, wantonness, and retaliation.[9] Vultures have been believed to spread diseases such as anthrax, to be responsible for blowfly plagues, to foul drinking water provided for livestock, and to represent evil spirits, among other beliefs.[9,88]

Shooting and intentional poisoning are the two main forms of persecution. Shooting vultures has long been documented in the United States, Europe, and North Africa, where the activity appears to be largely for sport.[22,28,29,61] Documented cases of intentional poisoning of vultures have been noted as being in retaliation for the suspected killing of newborn lambs, to disguise the locations of poachers' activities, and to obtain vulture parts for traditional medicine.[65,72,87,89,90] The acquisition of vulture parts for traditional medicine has been documented in West and southern Africa[72,91–93] and is suspected in parts of East Africa (N. Baker, pers. comm.). Though it is likely to be a substantial threat to vultures in those areas, there have been no documented efforts to tackle the problem. Interventions by local governments appear unlikely due to cultural beliefs and practices that remain deeply entrenched in African societies.[91,94] In addition to use in traditional medicines, vultures, and in particular the Hooded vulture, are hunted for food in West Africa.[40,41] Though it is difficult to ascertain population-level effects of persecution on individual species, it is thought to be a significant cause of mortality for some species and populations, including European Bearded vultures, Cape griffons in South Africa, Hooded vultures in West Africa, and large vultures in Nigeria.[22,27,40,41,91,92]

Poisoning and persecution are not the only threats to vulture survival. Electrocution and collision with power lines have also caused significant levels of vulture mortality,[95–99] and the recent proliferation of wind farms as a source of green energy production also has had adverse effects. In recent studies, Griffon vultures suffered among the highest levels of mortality after collision with wind turbine blades.[100,101] Given the rapid increase in the development of "green" technology and electricity infrastructure worldwide, these threats are likely to increase in coming decades. Mass drowning of vultures (of up to 38 birds at a time) has also been reported, particularly in semiarid areas of South Africa where the birds enter artificial reservoirs presumably to bathe or cool off, only to be unable to extricate themselves due to vertical reservoir walls.[67,102]

Other threats to vultures mentioned in the literature include habitat changes and food shortage. Habitat loss and degradation are suspected to have played roles in the dramatic declines (>98%) of large vultures outside of protected areas in West Africa where human population growth has been very rapid.[41,58] Declines in large game birds and mammals have also been noted.[41,58] Lack of food—due to overhunting or changes in livestock husbandry—could have major impacts on vultures[18] and is thought to have contributed to large-scale vulture declines in West Africa,[40,58] Southeast Asia,[24] and Europe.[36] More recently, an outbreak of bovine spongiform encephalopathy in Europe led to the passing of sanitary legislation that restricted the use of carcasses and animal by-products.[103] Vultures and other avian scavengers have been hard hit by the ensuing food shortage, which has been linked to lower breeding success and higher mortality in juvenile vultures.[104]

Consequences of declines

Given the rate at which vultures are declining, there have been surprisingly few studies about the ecological consequences of the widespread disappearance of these scavenging birds.[105–107] Communities of facultative scavengers are highly structured (not random) and complex, and birds contribute most to this structure because they are the most specialized scavengers.[108] For example, in a Polish forest, well-adapted scavengers such as ravens and foxes were nearly ubiquitous at carcasses, whereas rare or sporadic scavengers associated with the main scavenger species and did not scavenge randomly, but rather on those carcasses where carrion specialists were present.[108] As carrion specialists, the absence of vultures from carcasses may affect the community composition of scavengers at carcasses, which could alter scavenging rates for individual species.

In localized regions where vultures are functionally extinct, such as in India, the absence of vultures at carcasses appears to have driven a rapid increase in the abundance of opportunistic species such as feral dogs and rats (*Rattus rattus*).[3,4] Feral dogs have been shown to compete directly with vultures for food and are capable of displacing vultures from carcasses. In areas adjacent to communal lands in Zimbabwe, feral dogs dominated carcasses, but inside protected reserves, vultures were the major scavengers at carcasses.[109] Both feral dogs and rats are well-known disease reservoirs, and their increase at carcasses in the absence of vultures may increase rates of infectious disease transmission to other species. Diseases such as rabies and bubonic plague, for which dogs and rats respectively are the primary reservoirs, may increase as a consequence of vulture declines. Wildlife and livestock could also be at increased risk from dog- and rat-borne pathogens, including canine distemper virus, canine parvovirus, and *Leptospira* spp. bacteria.[3] In India, rising cases of human anthrax due to handling infected carcasses or consuming poorly cooked meat of infected livestock are believed to be linked to the precipitous decline of vultures.[110]

A recent confirmed the important role of vultures in decomposing carcasses, maintaining community structure, and moderating contact between mammalian scavengers at carcasses.[2] In Kenya, in the absence of vultures, carcass decomposition time nearly tripled, and both the number of scavenging mammals and the time they spent at carcasses increased threefold. Further, there was a nearly threefold increase in the number of contacts between mammalian scavengers at carcasses without vultures, suggesting that the demise of vultures could facilitate disease transmission at carcasses.[2]

In addition to ecological impacts, vulture declines have already had socioeconomic and cultural impacts in South Asia, where they hold important social significance. Amongst the Zoroastrian-practicing Parsi community in India, who believe that fire, water, air, and earth are pure elements that need to be preserved, the dead are laid in "Towers of Silence." Built on hilltops or low mountains, these circular pillars of stone enclose corpses that are left in the open to be disposed of by scavengers, principally vultures.[111,112] This ancient custom, also known as sky burial and similarly practiced by Tibetan Buddhists, has come to an abrupt end in the last decade due to the collapse of vulture populations in the region. The Parsi community has unsuccessfully turned to solar reflectors in the hopes of hastening decomposition.[112,113]

Economic consequences of vulture declines include increased costs to human health. Estimates of the human health costs of the loss of vultures and subsequent increases in dogs and rabies are about $1.5 billion annually in India.[111] In Nepal, the economic benefits of conserving vultures were estimated at $6.9 million, and interviews in communities within Important Bird Areas showed they were considered significantly beneficial to humans.[114] In Uganda, Hooded vultures consume primarily internal organs from diseased animals thrown from abbatoirs, thereby saving local councils the expense of a more sophisticated system of collection and disposal of refuse.[115]

Comparison of the situation in Asia versus Africa

The catastrophic declines in vulture populations in South Asia have generated considerable attention from the media and the scientific community, which has catalyzed vulture research, particularly in Africa.[116,117] There are many similarities between Africa and South Asia. In particular, both regions harbor human populations that are mostly rural and poor, growing rapidly, and divided into many ethnic groups. Although the causes may differ, there have also been similar levels of decline in vulture populations in both regions.[4,40–42,46,47,59]

The prospects for vultures and their continued survival, however, differ between the two regions. In particular, the regions could not be more disparate in the level of attention given to the issue from local governments, the international scientific and donor communities, and the media. There are undoubtedly many reasons for this, but the fact that the Asian vulture crisis has been largely pinpointed to a single source—diclofenac poisoning—may be crucial. The discovery of diclofenac has focused conservation efforts (and hence funding) and media attention on a single target, and it has provided a mechanism whereby success of conservation interventions can be directly measured.

Although the situation for Asian vultures is critical and it will likely remain so over coming decades, the governments within the region have played a crucial role in halting the declines and protecting and bolstering remaining populations.[82] Indeed, evidence from four years since the ban on veterinary diclofenac has shown that productivity of Indian vultures (*G. indicus*) in southeast Pakistan and in the Indian states of Rajasthan and Madhya Pradesh has already begun to increase.[118,119] The identification of veterinary diclofenac as the primary cause of vulture population declines in South Asia took at least four years of intensive field and diagnostic research.[82,120,121] The results of this work were published in a prominent journal[79] and presented to South Asian government authorities at a Kathmandu Summit Meeting organized by The Peregrine Fund in February 2004. The dissemination of scientific research to government authorities plus advocacy work can play a crucial role in achieving conservation results. The key was getting governments to listen and to understand the gravity of the vulture situation and its consequences should conservation interventions fail. The role of conservation organizations in India, Nepal, and Pakistan was critical in advocating for a ban on the manufacture, sale, and import of veterinary diclofenac. In addition, conservation organizations in South Asia, along with their overseas partners embarked on large-scale education and outreach programs to ensure that rural people understood the lethal impacts of administering veterinary diclofenac to livestock. The Indian government in particular was extremely supportive in providing the necessary framework for legislation, financial support, and assistance with developing vulture captive breeding and restoration efforts (MZV, pers. obs.). Also significant to the conservation of vultures on the subcontinent is the role of vultures in Hindu mythology where they are highly esteemed, as evidenced by the existence of a vulture god, Jatayu, which is regarded as a holy bird.[111,113]

The situation in Africa is vastly different because there are many significant threats to vultures, not just one. In addition, the threats vary by region. For example, the cause(s) of vulture disappearances in West Africa remain unclear, though many suggestions have been made, including food shortage, poisoning, and persecution for traditional medicine and food.[40,41] Whereas in South Asia recent population surveys have indicated that declines have slowed or ceased in some areas and the threat from diclofenac poisoning has reduced substantially,[83,118,119] in Africa there is little similar evidence.[122,123]

In addition, the vast majority of African governments, with the exception of South Africa, have provided little, if any, support for vulture conservation or have attempted to resolve known vulture threats. In West Africa, studies of vultures over the past 40 years are virtually nonexistent[122] though seven years have passed since the first reports of massive vulture population declines there were first published.[40] Kenya and Uganda are the only two East African countries where some populations of vultures have been studied and monitored for many decades (S. Thomsett unpub. data,[42,115,124]). In Kenya, two recent studies have shown large declines in most species, and there exist both scientific and anecdotal reports linking the declines to poisoning, primarily to the agricultural pesticide Furadan.[42,59,75,125] In addition to over 350 vultures known to have been poisoned in the past seven years, Kenya's lion population has similarly plummeted, largely due to illegal poisoning.[126] Indeed, wildlife poisoning is so rife in Kenya that a national taskforce, first spearheaded by conservationists and later incorporated under the Ministry of Agriculture, was initiated in 2008 to tackle the problem. The primary aim of conservationists engaged in the Stop Wildlife Poisoning Taskforce was to implore the Kenyan government to ban the registration of carbamate pesticides, of which Furadan is the most notorious, for sale in Kenya.[127] In lieu of a ban on carbamate pesticides, the taskforce also lobbied for tighter restrictions on the sale of Furadan in Kenya, as it is used to poison both wildlife and feral dogs. These pesticides are affordable, highly effective, and easily accessible.[128,129] Apart from reports of wildlife poisoning, Furadan is known to be used to kill birds and fish that are later sold for human consumption.[130–132] Despite the raft of evidence provided to Kenyan government officials and a report on the situation as requested by the Prime Minister's office,[131] the Kenyan government has yet to take any action restricting its use or sale in the country despite the existence of pesticides that are purported to be less toxic to wildlife.[128]

Overview of successful conservation programs and species

While the Asian vulture crisis is still unfolding and the full results of vulture conservation efforts will not be known for some years to come, in Europe and the United States a number of conservation programs have already succeeded in bolstering dwindling vulture populations. Three of the most successful programs involved the reintroduction and supplementary feeding of California Condors in the United States, Griffon vultures in the Massif Central region of France, and Bearded vultures in the European Alps.[133] In addition to releasing and feeding vultures, intensive public education programs were launched that targeted hunters, farmers, and livestock keepers, and all released individuals have been marked and in some cases are intensively monitored.[134–136] Common themes among these successful conservation efforts are a huge investment, at both national and international levels, of financial resources and highly skilled personnel, as well as political will and community support.

While reintroductions have boosted populations of some species, populations of Turkey and Black vultures have experienced major increases throughout much of their ranges in North America since the mid-1970s without any human intervention.[32,137] Although the reasons for their population increases have not been well studied, a number of suggestions have been made, including substantial decreases in persecution, increases in food availability due to the large resurgence of White-tailed deer populations (*Odocoileus virginianus*), and decreases in secondary poisoning, especially exposure to DDE (through DDT spraying), which induced eggshell thinning in many raptor species.[32] While both species are susceptible to the effects of DDT spraying, their susceptibility to other toxins is mixed. Turkey vultures have been shown to be relatively insensitive to some rodenticides but quite sensitive to strychnine and cyanide. It has also been suggested that both species may be particularly sensitive to lead poisoning.[32] A review of poisoning cases in the United States[138] suggests that both Turkey and Black vultures may be less susceptible to poisoning than some other raptor species. In general, raptors are more susceptible to poisoning compared to birds of a similar size; their stomachs have low pH levels that increase their susceptibility to lead poisoning.[138,139] A review of data presented by Mineau *et al.*[138] shows that 181 raptor deaths were attributed to the labeled use of pesticides, implying that the raptor deaths were accidental and were not as a result of persecution. No mortality of Turkey or Black vultures was recorded, despite mortality of scavenging Red-tailed hawks, Bald eagles, and Swainson's hawks, representing 31%, 17%, and 11% of cases, respectively.[138] While the results are only suggestive that both species may exhibit a reduced susceptibility to poisons, they do not suggest whether physiological and/or behavioral traits may play a greater role. In a recent study, Turkey vultures showed no signs of toxicity after being dosed with diclofenac at >100 times the lethal dose for *Gyps* vultures.[140] Turkey vultures have also been shown to have exceptional immune function.[141] Behaviorally, vultures may limit their susceptibility to toxins by foraging in smaller group sizes. The mean number of vultures poisoned per carcass appears to be proportional to their foraging group size (data from Ref. 16). In addition to physiological or behavioral adaptations that may benefit populations of Turkey and Black vultures, we posit that the governments of some countries in the Americas have stronger conservation policies, more political will, and more resources to conserve vultures as compared to countries in Asia and Africa.

Other ongoing and potentially important conservation initiatives consist of public involvement in vulture research or citizen science programs and increasingly through public education and outreach programs. Citizen science programs have largely focused on creating awareness of wing-tagged vultures and requesting the public to report sightings of marked individuals. The use of vulture restaurants could also be particularly useful to obtain reports of wing-tagged vultures from citizen scientists. Results from South Africa indicate that 4,443 resightings from a sample of 120 vultures have been reported, largely by citizen scientists at vulture restaurants over a five-year period (A. Botha, pers. comm.). Public education focused on the plight and importance of vultures has recently come to the forefront with the recognition of International Vulture Awareness Day (www.vultureday.org). International Vulture Awareness Day aims to have conservation organizations worldwide, including zoos and local government agencies, organize a one-day celebration of vultures that publicizes the conservation of vultures to a wider audience and highlights important work being carried out by vulture conservationists throughout the world.

Conclusions

A summary of the available evidence has shown that scavenging birds, and vultures in particular, are the most threatened group of birds in the world, and presently 61% are listed on the IUCN Red List of Threatened Species.[31,106] As a consequence of being highly specialized, vultures are particularly prone to poisoning, which has led to widespread and serious population declines in many species. Though the long-term impacts of these declines are not fully understood, ecosystem services provided by vultures have already plummeted, and novel research has shown that disease transmission at carcasses may increase. The two most important regions for Old World vultures—Asia and Africa—have witnessed recent catastrophic population declines in most species. While the declines in Asia have been linked to poisoning by the veterinary drug diclofenac, the reasons for the declines across Africa remain little understood. The Asian vulture crisis has highlighted the importance of collaboration between scientists, regional governments, donors, and the media to effectively conserve vultures over the vast areas where they range. In many African countries, vulture populations remain little known and even less is being done on the ground to ensure their survival. In countries such as South Africa and Kenya, where vulture conservation programs are ongoing, efforts are largely driven by or with the support of local conservation organizations. As the Asian vulture crisis has shown, without support and backing from national governments and local communities, any conservation efforts are likely to be met with limited success over the long term.

Acknowledgments

We thank N. Richards for providing expertise on poisoning issues and A. Botha, C. Whelan, and one anonymous reviewer for their comments on an earlier draft of the manuscript.

Conflicts of interest

The authors declare no conflicts of interest.

References

1. Houston, D.C. 2001. *Condors and Vultures*. Voyageur Press, Inc., Stilwater, Minnesota, USA.
2. Ogada, D.L., M.E. Torchin, M.F. Kinnaird & V.O. Ezenwa. In press. Ecological consequences of vulture declines on facultative scavengers and potential implications for mammalian disease transmission in Kenya. *Conserv. Biol.*
3. Pain, D.J., A.A. Cunningham, P.F. Donald, *et al.* 2003. Causes and effects of temporospatial declines of *Gyps* vultures in Asia. *Conserv. Biol.* **17:** 661–671.
4. Prakash, V., D.J. Pain, A.A. Cunningham, *et al.* 2003. Catastrophic collapse of Indian white-backed *Gyps bengalensis* and long-billed *Gyps indicus* vulture populations. *Biol. Conserv.* **109:** 381–390.
5. DeVault, T.L., O.E. Rhodes & J.A. Shivik. 2003. Scavenging by vertebrates: behavioural, ecological, and evolutionary

perspectives on an important energy transfer pathway in terrestrial ecosystems. *Oikos* **102:** 225–234.
6. Wilson, E.E. & E.M. Wolkovich. 2011. Scavenging: how carnivores and carrion structure communities. *Trends Ecol. Evol.* **26:** 129–135.
7. Schaller, G.B. 1972. *The Serengeti Lion.* University of Chicago Press. Chicago.
8. Houston, D.C. 1974a. Food searching behaviour in griffon vultures. *East Afr. Wildl. J.* **12:** 63–77.
9. Boshoff, A.F. & C.J. Vernon. 1980. Past and present distribution and status of the Cape vulture in Cape Province. *Ostrich* **51:** 230–250.
10. Snyder, N.F. & N.J. Schmitt. 2002. California Condor (*Gymnogyps californianus*). In *The Birds of North America Online* A. Poole, Ed. Ithaca: Cornell Lab of Ornithology; Retrieved from the Birds of North America Online: http://bna.birds.cornell.edu.proxy2.library.illinois.edu/bna/species/610
11. Ruxton, G.D. & D.C. Houston. 2004. Obligate vertebrate scavengers must be large soaring fliers. *J. Theor. Biol.* **228:** 431–436.
12. Houston, D.C. 1979. The adaptations of scavengers. In *Serengeti: Dynamics of an Ecosytem.* A.R.E. Sinclair & M. Norton-Griffiths Eds.: 263–286. Cambridge University Press. Cambridge, UK.
13. Pennycuick, C.J., 1972. Soaring behaviour and performance of some African birds observed from a motor-glider. *Ibis* **114:** 178–218.
14. Houston, D.C. 1974b. The role of Griffon Vultures *Gyps africanus* and *Gyps ruppellii* as scavengers. *J. Zool.* **172:** 35–46.
15. Shivik, J.A. 2006. Are vultures birds, and do snakes have venom, because of macro- and microscavenger conflict? *BioScience* **56:** 819-823.
16. Mundy, P., D. Butchart, J. Ledger & S. Piper. 1992. *The Vultures of Africa.* Academic Press. London.
17. Houston, D.C. 1986. Scavenging efficiency of turkey vultures in tropical forest. *Condor* **88:** 318–323.
18. Houston, D.C. 1987. The effect of reduced mammal numbers on *Cathartes* vultures in neotropical forests. *Biol. Conserv.* **41:** 91–98.
19. Bamford, A.J., M. Diekmann, A. Monadjem & J. Mendelsohn. 2007. Ranging behaviour of Cape Vultures (*Gyps coprotheres*) from an endangered population in Namibia. *Bird. Conserv. Int.* **17:** 331–339.
20. Houston, D.C. & J.E. Cooper. 1975. The digestive tract of the Whiteback Griffon Vulture and its role in disease transmission among wild ungulates. *J. Wildl. Dis.* **11:** 306–313.
21. Wynne-Edwards, V.C. 1955. Low reproductive rates in birds, especially sea-birds. *Acta XI Congressus Internationalis Ornithologici.* (Basel) 540–547.
22. Mingozzi, T. & R. Estève 1997. Analysis of a historical extirpation of the Bearded vulture *Gypaetus barbatus* (L.) in the Western Alps (France-Italy): former distribution and causes of extirpation. *Biol. Conserv.* **79:** 155–171.
23. Snyder, N.F.R. 1983. California Condor reproduction, past and present. *Bird Conserv.* **1:** 67–86.
24. Hla, H., N.M. Shwe, T.W. Htun, *et al.* 2011. Historical and current status of vultures in Myanmar. *Bird. Conserv. Int.* **21:** 376–387. doi:10.1017/S0959270910000560.
25. Thiollay, J.M. 2000. Vultures in India. *Vulture News* **42:** 36–38.
26. Donazar, J.A., 1994. Egyptian vulture Neophron percnopterus. In *Birds in Europe: Their Conservation Status.* BirdLife Conservation Series No. 3. Tucker, G.M. & Heat, M.F., Eds.: 154–155. BirdLife International. Cambridge.
27. Margalida A., R. Heredia, M. Razin & M. Hernandez. 2008. Sources of variation in mortality of the Bearded vulture *Gypaetus barbatus* in Europe. *Bird Conserv. Int.* **18:** 1–10.
28. Miller, A.H., I. McMillan & E. McMillan. 1965. The current status and welfare of the California Condor. National Audubon Research Report No. **6:** 1–61.
29. Hunt, W.G., C.N. Parish, K. Orr, & R.F. Aguilar. 2009. Lead poisoning and the reintroduction of the California condor in Northern Arizona. *J. Avi. Med. Surg.* **23:** 145–150.
30. The Peregrine Fund. 2011. http://www.peregrinefund.org/condor. Accessed 27 July, 2011.
31. BirdLife International. 2011. IUCN Red List for birds. http://www.birdlife.org. Accessed 9 May, 2011.
32. Kiff, L.F. 2000. The current status of North American vultures. In *Raptors at Risk.* R.D. Chancellor, & B.-U., Meyburg, Eds.: 175–189. WWGBP/Hancock House. Berlin and Surrey, Canada.
33. Farmer, C.J., L.J. Goodrich, E.R. Inzunza & J.P. Smith. 2008. Conservation status of North America's birds of prey. In *State of North America's Birds of Prey. Series in Ornithology 3.* K.L. Bildstein, J.P. Smith, E.R. Inzunza, & R.R. Veit Eds.: 303–419. Nuttall Ornithological Club, Cambridge, MA, and American Ornithologists' Union, Washington DC.
34. Díaz, D., M. Cuesta, T. Abreu & E. Mujica. 2000. *Estrategia de conservaci'on para el condor andino* (Vultur gryphus). World Wildlife Fund and Fundacion BioAndina. Caracas, Venezuela.
35. Cuthbert, R., R.E. Green, S. Ranade, *et al.* 2006. Rapid population declines of Egyptian Vulture (*Neophron percnopterus*) and red-headed vulture (*Sarcogyps calvus*) in India. *Anim. Conserv.* **9:** 349–354.
36. Liberatori, F. & V. Penteriani. 2001. A long-term analysis of the declining population of the Egyptian vulture in the Italian peninsula: distribution, habitat preference, productivity and conservation implications. *Biol. Conserv.* **101:** 381–389.
37. Heredia, B. 1996. Action plan for the Cinereous Vulture (Aegypius monachus) in Europe. Report to BirdLife International. 22p.
38. Ogada, D.L. & R. Buij. 2011. Decline of the Hooded Vulture *Necrosyrtes monachus* across its African range. *Ostrich* **82:** 101–113.
39. Anderson, M.D. 2004b. African white-backed vulture *Gyps africanus.* In *The Vultures of Southern Africa– Quo Vadis?* A. Monadjem, M.D. Anderson, S.E. Piper & A.F. Boshoff, Eds.: 15–27. Birds of Prey Working Group. Johannesburg, South Africa.
40. Rondeau G. & J.M. Thiollay. 2004. West African vulture decline. *Vulture News* **51:** 13–33.

41. Thiollay, J.M. 2006. The decline of raptors in West Africa: long-term assessment and the role of protected areas. *Ibis* **148:** 240–254.
42. Virani, M.Z., C. Kendall, P. Njoroge & S. Thomsett. 2011. Major declines in the abundance of vultures and other scavenging raptors in and around the Masai Mara ecosystem, Kenya. *Biol. Conserv.* **144:** 746–752.
43. Brown, C.J. & S.E. Piper. 1988. Status of Cape Vultures in the Natal Drakensberg and their cliff site selection. *Ostrich* **59:** 126–136.
44. Brown, C.J. 1985. The status and conservation of the Cape Vulture in SWA/Namibia. *Vulture News* **14:** 4–15.
45. Piper, S.E. 2004. Cape vulture *Gyps coprotheres*. In *Vultures in the Vultures of Southern Africa—Quo Vadis?* A. Monadjem, M.D. Anderson, S.E. Piper & A.F. Boshoff Eds.: 5–11. Proceedings of a workshop on vulture research and conservation in southern Africa. Birds of Prey Working Group, Johannesburg.
46. Gilbert, M., M.Z. Virani, R.T. Watson, *et al.* 2002. Breeding and mortality of Oriental White-backed Vulture *Gyps bengalensis* in Punjab Province, Pakistan. *Bird Conserv. Int.* **12:** 311–326.
47. Gilbert, M., J.L. Oaks, M.Z. Virani, *et al.* 2004. The status and decline of vultures in the provinces of Punjab and Sind, Pakistan: a 2003 update. In *Raptors Worldwide*. R.C. Chancellor & B.-U. Meyburg, Eds.: 221–234. Proceedings of the 6th world conference on birds of prey and owls.WWGBP and MME/Birdlife Hungary. Berlin and Budapest.
48. Prakash, V. 1999. Status of vultures in Keoladeo National Park, Bharatpur, Rajasthan, with special reference to population crash in *Gyps* species. *J. Bombay Nat. Hist. Soc.* **96:** 365–378.
49. Prakash, V., R.E. Green, D.J. Pain, *et al.* 2007. Recent changes in populations of resident *Gyps* vultures in India. *J. Bombay Nat. Hist. Soc.* **104:** 129–135.
50. Gilbert, M., R.T. Watson, M.Z. Virani, *et al.* 2006. Rapid population declines and mortality clusters in three Oriental white-backed vulture *Gyps bengalensis* colonies in Pakistan due to diclofenac poisoning. *Oryx* **40:** 388–399.
51. Cunningham, P.L. 2002. Vultures declining in the United Arab Emirates. *Vulture News.* **46:** 8–10.
52. Mendelssohn, H. & Y Leshem. 1983. The status and conservation of vultures in Israel. In *Vulture Biology and Management*. S.R. Wilbur & J.A. Jackson Eds.: 86–98. University of California Press. Berkeley, CA.
53. Ríos-Uzeda, B. & R.B. Wallace. 2007. Estimating the size of the Andean Condor population in the Apolobamba Mountains of Bolivia. *J. Field Ornithol.* **78:** 170–175.
54. Temple, S.A. & M.P. Wallace. 1989. Survivorship patterns in a population of Andean Condors (Vultur gryphus). In *Raptors in the Modern World.* B.U. Meyburg & R.D. Chancellor, Eds.: 247–249.World Working Group on Birds of Prey. Berlin, Germany.
55. Wallace, M.P. & S.A. Temple. 1988. Impacts of the 1982–1983 El nino on population dynamics of Andean condors in Peru. *Biotropica.* **20:**144–150.
56. Baral, H.S., J.B. Giri & M.Z. Virani. 2004. On the decline of Oriental Whitebacked Vultures Gyps bengalensis in lowland Nepal. In *Raptors Worldwide Proceedings of the 6th World Conference on Birds of Prey and Owls.* R.D. Chancellor & B.-U. Meyburg, Eds.: 215–219. WWGBP and MME/Birdlife Hungary. Berlin and Budapest.
57. Acharya, R., R. Cuthbert, H.S. Baral & K.B. Shah. 2009. Rapid population declines of Himalayan Griffon *Gyps himalayensis* in Upper Mustang, Nepal. *Bird. Conserv. Int.* **19:** 99–107.
58. Thiollay J.M. 2007. Raptor population decline in West Africa. *Ostrich.* **78:** 405–413.
59. Ogada, D.L. & F. Keesing. 2010. Decline of raptors over a three-year period in Laikipia, central Kenya. *J. Raptor Res.* **44:** 43–49.
60. Kendall C. & M.Z. Virani. A comparison of two methods—wing tagging and gsm-gps transmitters – for understanding habitat use and mortality in african vultures. *J. Raptor Res.* In press.
61. Mundy, P.J. 2000. The status of vultures in Africa during the 1990s. In *Raptors at Risk.* R.D. Chancellor & Meyburg B.-U. Eds.: 151–164. WWGBP/Hancock House. Berlin.
62. Anderson, M.D. 2000. Egyptian vulture. In *The Eskom Red Data Book of birds of South Africa, Lesotho and Swaziland.* Barnes, K.N. Ed.: 20. BirdLife South Africa. Johannesburg.
63. Houston, D.C. 1996. The effect of altered environments on vultures. In *Raptors in Human Landscapes: Adaptations to Built and Altered Landscapes.* D.M. Bird, D.E. Varland & J.J. Negro, Eds.: 327–335. Academic Press Ltd. London.
64. Gangoso, L., P. Alvarez-Lloret, A.A.B. Rodriguez-Navarro, *et al.* 2009. Long-term effects of lead poisoning on bone mineralization in vultures exposed to ammunition sources. *Env. Pollution.* **157:** 569–574.
65. Bridgeford, P. 2001. More vulture deaths in Namibia. *Vulture News.* **44:** 22–26.
66. Bridgeford, P. 2002. Recent vulture mortalities in Namibia. *Vulture News.* **46:** 38.
67. Anderson, M.D. 1994. Mass African Whitebacked Vulture poisoning in the northern Cape. *Vulture News.* **29:** 31–32.
68. Anderson, M.D. 1995. Mortality of African White-backed Vultures in the Northwest Province, South Africa. *Vulture News* **33:** 10–13.
69. Hernández, M. & A. Margalida. 2008. Pesticide abuse in Europe: effects on the Cinereous vulture (*Aegypius monachus*) population in Spain. *Ecotoxicology.* **17:** 264–272.
70. Hernández, M. & A. Margalida. 2009. Poison-related mortality effects in the endangered Egyptian vulture (*Neophron percnopterus*) population in Spain. *Eur. J. Wildl. Res.* **55:** 415–423. doi: 10.1007/s10344–009-0255–6.
71. Allan, D.G. 1989. Strychnine poison and the conservation of avian scavengers in the Karoo, South Africa. *S. Afr. J. Wildl. Res.* **19:** 102–106.
72. Verdoorn G.H., N. van Zijl, T.V. Snow, *et al.* 2004. Vulture poisoning in southern Africa. In *Vultures in the Vultures of Southern Africa—Quo Vadis?* A. Monadjem, M.D. Anderson, S.E. Piper, & A.F. Boshoff, Eds.: 195–201. Proceedings of a workshop on vulture research and conservation in southern Africa.Birds of Prey Working Group. Johannesburg.
73. Borello, W.D. 1985. Poisoned vultures in Botswana: known facts. *Babbler* **9:** 22–23.

74. Simmons, R.E. 1995. Mass poisoning of Lappet-faced vultures in the Namib Desert. *J. Afr. Raptor Biol.* **10:** 3.
75. Mijele, D. 2009. *Incidences of Poisoning of Vultures and Lions in the Masai Mara National Reserve*. Kenya Wildlife Service Masai Mara Veterinary Report. Nairobi, Kenya.
76. Fajardo, I., A. Ruiz, I. Zorrilla, *et al.* 2011. Use of specialised canine units to detect poisoned baits and recover forensic evidence in Andalucía (Southern Spain). In *Carbofuran and Wildlife Poisoning: Global Perspectives and Forensic Approaches*, N. L. Richards, Ed.: 147–155. Wiley. UK.
77. Taylor, M.J. 2011. Monitoring carbofuran abuse in Scotland. In *Carbofuran and Wildlife Poisoning: Global Perspectives and Forensic Approaches*. N. L. Richards, Ed.: 181–186. Wiley. UK.
78. Garcia, A.R., A.V. Garruta, E.S. Menzano & F.M.M. Garrido. 2011. Capítulo 7: La unidad canina especializada en la detección de venenos. In *Manual para la protección legal de la biodiversidad para agentes de la autoridad ambiental en Andalucía.* I. Fajardo & J. Martín, Eds.: 181–199. Consejería de Medio Ambiente, Junta de Andalucía. Seville, Spain.
79. Oaks, J.L, M. Gilbert, M.Z. Virani, *et al.* 2004. Diclofenac residues as the cause of vulture population decline in Pakistan. *Nature* **427:** 630–633.
80. Green, R.E., I. Newton, S. Shultz, *et al.* 2004. Diclofenac poisoning as a cause of vulture population declines across the Indian subcontinent. *J. Appl. Ecol.* **41:** 793–800.
81. Shultz, S., H.S. Baral, S. Charman, *et al.* 2004. Diclofenac poisoning is widespread in declining vulture populations across the Indian subcontinent. *Proc. Roy. Soc. Lond. B* **271**(Suppl 6): S458–S460. doi:10.1098/rsbl.2004.0223.
82. Pain D.J., C.G.R. Bowden, A.A. Cunningham, *et al.* 2008. The race to prevent the extinction of South Asian vultures. *Bird Conserv. Int.* **18:** S30–S48.
83. Cuthbert, R., M.A. Taggart, V. Prakash, *et al.* 2011. Effectiveness of action in India to reduce exposure of *Gyps* vultures to the toxic veterinary drug Diclofenac. *Plos One* **6:** 1–11.
84. Parish, C.N., W.R. Heinrich & W.G. Hunt. 2007. Lead exposure, diagnosis, and treatment in California Condors released in Arizona. In *California Condors in the 21st Century*. Series in Ornithology, no. 2. A. Mee, L. S. Hall & J. Grantham , Eds.: 97–108. American Ornithologists Union, Washington, DC, and Nuttall Ornithological Club, Cambridge, MA.
85. Wiemeyer S.N., J.M. Scott, M.P. Anderson, *et al.* 1988. Environmental contaminants in California condors. *J. Wildl. Manage.* **52:** 238–247.
86. Snyder, N.F.R. & H.A. Snyder. 1989. Biology and conservation of the California Condor. *Curr. Ornithol.* **6:** 175–267.
87. Parish, C.N., W.G. Hunt, E. Feltes, *et al.* 2009. Lead exposure among a reintroduced population of California Condors in northern Arizona and southern Utah. In *Ingestion of Lead from Spent Ammunition: Implications for Wildlife and Humans.* R.T. Watson, M. Fuller, M. Pokras & W.G. Hunt Eds.: 259–264. The Peregrine Fund. Boise, Idaho. doi:10.4080/ilsa.2009.0217.
88. Campbell, M. 2009. Factors for the presence of avian scavengers in Accra and Kumasi, Ghana. *Area* **41:** 341–349.
89. Hancock P. 2009. Poisons devastate vultures throughout Africa. *Afr. Raptors Newsletter* No. 2. November.
90. Hancock P. 2010. Vulture poisoning at Khutse. *Birds and People* **27:** 7.
91. McKean, S. 2004. Traditional use of vultures: some perspectives. In *Vultures in The Vultures of Southern Africa—Quo Vadis?* A. Monadjem, M.D. Anderson, S.E. Piper & A.F. Boshoff, Eds.: 214–219. Proceedings of a workshop on vulture research and conservation in southern Africa. Johannesburg: Birds of Prey Working Group.
92. Nikolaus G. 2001. Bird exploitation for traditional medicine in Nigeria. *Malimbus* **23:** 45–55.
93. Sodeinde S.O. & D.A. Soewu. 1999. Pilot study of the traditional medicine trade in Nigeria. *Traffic Bulletin* **18:** 35–40.
94. Mander, M., N. Diederichs, L. Ntuli, *et al.* 2007. Survey of the trade in vultures for the traditional health industry in South Africa. Unpublished Report. p.54.
95. van Rooyen, C.S. 2000. An overview of vulture electrocutions in South Africa. *Vulture News* **43:** 5–22.
96. Anderson, M.D. & R. Kruger. 1995. Powerline electrocution of eighteen White-backed vultures. *Vulture News* **32:** 16–18.
97. Markus, M.B. 1972. Mortality of vultures caused by electrocution. *Nature* **238:** 228.
98. Gangoso, L. & C.J. Palacios. 2002. Endangered Egyptian vulture (Neophron percnopterus) entangled in a power line ground-wire stabilizer. *J. Raptor Res.* **36:** 239–240.
99. Janss, G.F.E. 2000. Avian mortality from power lines: a morphologic approach of a species-specific mortality. *Biol. Conserv.* **95:** 353–359.
100. Barrios, L. & A. Rodríguez. 2004. Behavioural and environmental correlates of soaring-bird mortality at on-shore wind turbines. *J. Appl. Ecol.* **41:** 72–81.
101. de Lucas, M., G.F.E. Janss, D.P. Whitfield & M. Ferrer. 2008. Collison fatality of raptors in wind farms does not depend on raptor abundance. *J. Appl. Ecol.* **45:** 1695–1703.
102. Anderson, M.D., A.W.A. Maritz & E. Oosthuysen. 1999. Raptors drowning in farm reservoirs in South Africa. *Ostrich* **70:** 139–144.
103. Donázar, J.A., A. Margalida, M. Carrete, & J.A. Sánchez-Zapata. 2009a. Too sanitary for vultures. *Science* **326:** 664.
104. Donázar, J.A., A. Margalida & D. Campión, Eds. 2009b. *Vultures, Feeding Stations and Sanitary Legislation: A Conflict and Its Consequences from the Perspective of Conservation Biology* Munibe 29 (suppl.), Sociedad de Ciencias Aranzadi, Donostia, Spain.
105. Sekercioglu, C.H., G.C. Daily & P.R. Ehrlich. 2004. Ecosystem consequences of bird declines. *Proc. Natl. Acad. Sci. USA* **101:** 18042–18047.
106. Sekercioglu, C.H. 2006. Increasing awareness of avian ecological function. *Trends Ecol. Evol.* **21:** 464–471.
107. Wenny, D.G., T.L. DeVault, M.D. Johnson, *et al.* 2011. On the need to quantify ecosystem services provided by birds. *Auk* **128:** 1–14.
108. Selva, N. & M.A. Fortuna. 2007. The nested structure of a scavenger community. *P. Roy. Soc. B* **274:** 1101–1108.
109. Butler, J.R.A., & J.T. du Toit. 2002. Diet of free-ranging domestic dogs (Canis familiaris) in rural Zimbabwe:

implications for wild scavengers on the periphery of wildlife reserves. *Anim. Cons.* **5:** 29–37.
110. Mudur, G. 2001. Human anthrax in India may be linked to vulture decline. *Brit. Med. J.* **322:** 320.
111. Markandya, A., T. Taylor, A. Longo, *et al.* 2008. Counting the cost of vulture decline—an appraisal of the human health and other benefits of vultures in India. *Ecol. Econ.* **67:** 194–204.
112. Subramanian, M. 2008. Towering silence. *Science & Spirit* May/June.
113. van Dooren, T. 2010. Vultures and their people in India: equity and entanglement in a time of extinctions. *Manoa.* **22:** 130–146.
114. Baral, N., R. Gautam, N. Timilsina & M.G. Bhat. 2007. Conservation implications of contingent valuation of critically endangered white-rumped vulture Gyps bengalensis in South Asia. *Int. J. Bio. Sci. Manage.* **3:** 145–156.
115. Pomeroy, D.E. 1975. Birds as scavengers of refuse in Uganda. *Ibis* **117:** 69–81.
116. Koening, R. 2006. Vulture research soars as the scavengers' numbers decline. *Science* **312:** 1591–1592.
117. Virani, M. Z. & M. Muchai. 2004. Vulture conservation in the Masai Mara National Reserve, Kenya: proceedings and recommendations of a seminar and workshop held at the Masai Mara National Reserve, 23 June 2004. *Ornithol. Res. Rep.* 57: 2–19.
118. Chaudhry, M.J.I., D.L. Ogada, R.N. Malik & M.Z. Virani. In review. Assessment of productivity, population changes, and mortality in long-billed vultures Gyps indicus in Pakistan since the onset of the Asian vulture crisis. *Bird Conserv. Int.*
119. The Peregrine Fund. 2010. Annual Report. 34.
120. The Peregrine Fund. 2003. Annual Report. Pp 28–29.
121. Watson, R.T., M. Gilbert, J.L. Oaks, & M. Virani. 2004. The collapse of vulture populations in South Asia. *Biodiversity* **5:** 3–7.
122. Anderson, M.D. 2004a. Vulture crises in South Asia and West Africa. . . . and monitoring, or the lack thereof, in Africa. *Vulture News* **52:** 3–4.
123. Monadjem, A., Anderson, M.D., Piper, S.E. & Boshoff, A.F. (Eds). 2004. *Vultures in the Vultures of Southern Africa—Quo Vadis?* Proceedings of a workshop on vulture research and conservation in southern Africa. Birds of Prey Working Group, Johannesburg.
124. Ssemmanda, R. & D. Pomeroy. 2010. Scavenging birds of Kampala: 1973–2009. *Scopus* **30:** 26–31.
125. Otieno P.O., J.O. Lalah, M. Virani, *et al.* 2010. Carbofuran and its toxic metabolites provide forensic evidence for Furadan exposure in vultures (*Gyps africanus*) in Kenya. *B. Environ. Contam. Tox.* **84:** 536–544. doi:10.1007/s00128-010-9956–5.
126. Frank, L. 2010. Hey presto! We made the lions disappear. *Swara* **4:** 17–21.
127. Lalah, J.O. & P.O. Otieno. 2011. A chronicling of long-standing carbofuran use and its menace to wildlife in Kenya: Background on pesticide use and environmental monitoring in Kenya. In *Carbofuran and Wildlife Poisoning: Global Perspectives and Forensic Approaches.* N. L. Richards, Ed.: 43–52. John Wiley & Sons. Chichester, UK.
128. Odino, M. & D.L. Ogada. 2008a. Furadan use in Kenya and its impacts on birds and other wildlife: a survey of the regulatory agency, distributors, and end-users of this highly toxic pesticide. Report to the Bird Committee of Nature Kenya. p. 17.
129. Odino, M. & D.L. Ogada. 2008b. Furadan use in Kenya: a survey of the distributors and end-users of toxic Carbofuran (Furadan) in pastoralist and rice growing areas. Report to Kenya Wildlife Trust. 19 pp.
130. Odino, M. 2011. A chronicling of long-standing carbofuran use and its menace to wildlife in Kenya: Measuring the conservation threat that deliberate poisoning causes to birds in Kenya: The case of pesticide hunting with Furadan in the Bunyala Rice Irrigation Scheme. In *Carbofuran and Wildlife Poisoning: Global Perspectives and Forensic Approaches.* N. L. Richards, Ed.: 53–70. John Wiley & Sons. Chichester, UK.
131. Kahumbu, P. 2010. Evidence for revoking carbofuran registration in Kenya. Report to the Ministry of Agriculture Taskforce. p. 39.
132. Anonymous. 2011. Man jailed for using poison to catch fish. *Daily Nation*, May 11. Nairobi, Kenya.
133. Houston, D.C. 2006. Reintroduction programmes for vulture species. D.C. Houston & S.E. Piper (eds.). 2006. In *Proceedings of the International Conference on Conservation and Management of Vulture Populations.* 14–16 November 2005,Thessaloniki, Greece. Natural History Museum of Crete & WWF Greece. 87–97.
134. Sullivan, K., R. Sieg, & C. Parish. 2007. Arizona's efforts to reduce lead exposure in California Condors.In *California Condors in the 21st Century*. A. Mee & L.S. Hall Eds.: 109–121. The Nuttall Ornithological Club and the American Ornithologists' Union. Lancaster, PA.
135. Terrasse, M. 2006. Long-term reintroduction projects of Griffon Gyps fulvus and Black vultures Aegypius monachus in France. D.C. Houston & S.E. Piper Eds.: 98–107. *Proceedings of the International Conference on Conservation and Management of Vulture Populations.* 14–16 November 2005, Thessaloniki, Greece. Natural History Museum of Crete & WWF Greece. 176 pages.
136. Jurek, R.M. 1990. An historical review of California condor recovery programmes. *Vulture News* **23:** 3–7.
137. Avery, M.L. 2004. Trends in North American vulture populations. In *Proceedings of the 2lst Vertebrate Pest Conference.* R M. Timm & W. P. Gorenzel, Eds.: 116–121. University of California, Davis, CA.
138. Mineau P., M.R. Fletcher, L.C. Glaser, *et al.* 1999. Poisoning of raptors with organophosphorus and carbamate pesticides with emphasis on Canada, the United States and the United Kingdom. *J. Raptor Res.* **33:** 1–37.
139. Fisher, I.J., D.J. Pain & V.G. Thomas. 2006. A review of lead poisoning from ammunition sources in terrestrial birds. *Biol. Conserv.* **131:** 421–432.
140. Rattner, B.A., M.A. Whitehead, G. Gasper, *et al.* 2008. Apparent tolerance of turkey vultures (*Cathartes aura*) to the non-steroidal anti-inflammatory drug diclofenac. *Environ. Toxicol. Chem.* **27:** 2341–2345.
141. Apanius, V., S.A. Temple & M. Bale. 1983. Serum proteins of wild turkey vultures (*Cathartes aura*). *Comp. Biochem. Phys. B.* **76:** 907–913.

Ann. N.Y. Acad. Sci. ISSN 0077-8923

ANNALS OF THE NEW YORK ACADEMY OF SCIENCES
Issue: *The Year in Ecology and Conservation Biology*

Modeling population dynamics, landscape structure, and management decisions for controlling the spread of invasive plants

Paul Caplat,[1] Shaun Coutts,[2] and Yvonne M. Buckley[1,2]

[1]CSIRO Ecosystem Sciences, Brisbane, Queensland, Australia [2]The University of Queensland, School of Biological Sciences, Queensland, Australia

Address for correspondence: Dr. Paul Caplat, CSIRO Ecosystem Sciences, GPO Box 2583, Brisbane, QLD 4001, Australia. paul.caplat@gmail.com

Invasive plants cause substantial economic and environmental damage throughout the world. However, eradication of most invasive species is impossible and, in some cases, undesirable. An alternative is to slow the spread of an invasive species, which can delay impacts or reduce their extent. We identify three main areas where models are used extensively in the study of plant spread and its management: (i) identifying the key drivers of spread to better target management, (ii) determining the role spatial structure of landscapes plays in plant invasions, and (iii) integrating management structures and limitations to guide the implementation of control measures. We show how these three components have been approached in the ecological literature as well as their potential for improving management practices. Particularly, we argue that scientists can help managers of invasive species by providing information about plant invasion on which managers can base their decisions (i and ii) and by modeling the decision process through optimization and agent-based models (iii). Finally, we show how these approaches can be articulated for integrative studies.

Keywords: landscape; seed dispersal; exotic species; integrodifference equation; network theory

Introduction

Invasive plant species cause major environmental and economic damage worldwide through harmful impacts of local populations and extensive ranges of some invaders.[1,2] The dispersal and spread of invasive plants is important throughout all phases of the invasion including establishment, expansion, and reinvasion during and after management. Management of spread can therefore contribute to reductions in invasion extent and impact.[3,4] For the majority of invasive species, eradication is either not achievable, or even desirable, for example, if invaders are of economic benefit to some stakeholders. An alternative to eradication is containment of a potentially invasive population, which reduces the rate of spread in order to delay or reduce impacts across the whole landscape. Depending on the management goal, prevention or delay of spread into areas of high value for production or biodiversity could be prioritized. An understanding of the spread process and management models that take dispersal into account are therefore essential for a variety of management goals. Many managers acknowledge the importance of understanding spread processes for informing management strategies; however, until recently, the modeling tools to match managers' ambitions have been limited.

New models for conceptualizing spread have been developed that, when combined with data, enable us to gain a better understanding of the ecological drivers of spread rates[5–7] and to test a diverse array of management strategies.[3,6] However, there is an increasing need to clarify how spread models can be best used to target management strategies effectively. To do so, it is crucial to understand the way invasive plants establish and spread, how management can impact spread, and how management

doi: 10.1111/j.1749-6632.2011.06313.x

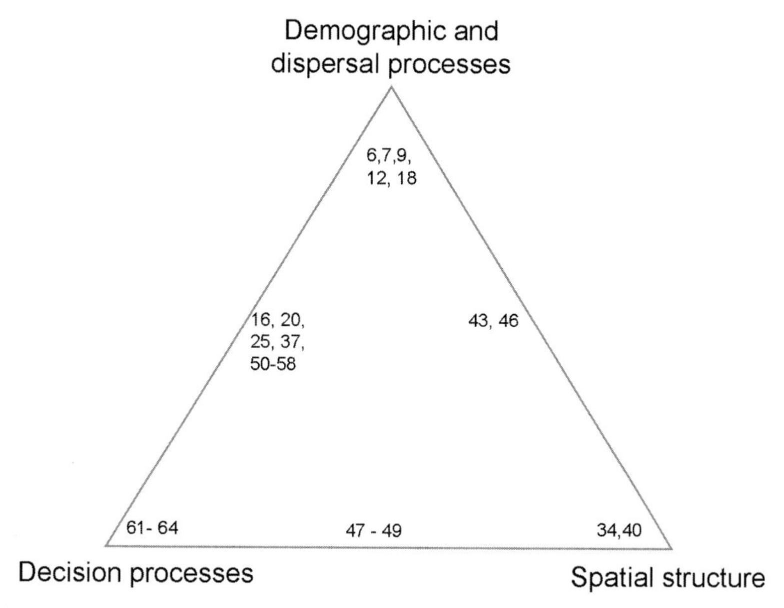

Figure 1. Articulation between the three modeling approaches outlined in this review. Numbers refer to the cited studies that illustrate a specific angle or, if they are on the side of the triangle, integration between the two adjacent angles. Studies that achieve integration of the three components should be represented at the center of the triangle; we are not aware of any.

constraints affect implementation and cost effectiveness. Different modeling approaches have been developed that allow both analysis and prediction of plant spread and have been well summarized recently.[5] Models of management implementation targeting spread are more rarely reported (but see Ref. 3). Here, we aim to outline and clarify the methods available to predict the effect of management on invasive plant spread.

Among the studies that model spread of invasive plants with the general goal of enhancing management practices, we identified three main areas of research, which seek to determine: (1) identifying the key drivers of spread to better target management, (2) determining the role spatial structure of landscapes plays in plant invasions, and (3) integrating management structures and limitations to guide the implementation of control measures. Although these three approaches are characterized by different sets of objectives, concepts, and methods, they are complementary and necessary for the complex decision making involved in determining effective management of invader spread over whole landscapes (Fig. 1). We detail each approach and show how they are, or could be, used to address management of invasive plant spread at the landscape scale. We use examples from the literature throughout to focus attention on the important methods and concepts in the field.

Identification of drivers of population spread to provide management targets

At the scale of a plant population, spread can be modeled as a combination of two sets of processes: demography, driving population growth, and dispersal, driving the spatial component of spread. If management aims at slowing down (or stopping) the spread of an invasive plant, it may be necessary to identify both the demographic and dispersal drivers of spread and which of these drivers are most likely to be impacted by management actions. Although literature identifying drivers of spread is abundant, identification of those parameters or processes most likely to be impacted by management is rare yet is crucial for management success. If the most important drivers of spread cannot be managed effectively, then it is essential to assess whether spread management targets are achievable by focusing management on less important drivers of spread. Here, we discuss methods that enable both identification of drivers of spread and assessment of management feasibility by targeting those drivers. Different modeling frameworks approach population spread from very different perspectives. Typically, two types of models are used to inform population management: analytical models that are usually based on differential or difference equations and describe entire populations, and individual-based models that describe the fate of individuals or groups of individuals.

In their simplest form, analytical models view a population as a homogenous entity and consider implicitly that movement of offspring (dispersal) occurs randomly within the lifetime of an individual.[5] They are usually expressed in the form of a diffusion equation. Their flexibility and generality makes them a useful tool for testing hypotheses such as effect of the shape of dispersal kernels,[8] density dependence and Allee effects,[9] or the role of stochasticity.[10] However, because they offer a highly simplified view of population dynamics, they are difficult to relate directly to specific management actions (although see Ref. 9).

Stage-structured models (usually matrix models, see Ref. 11 and Box 1) detail demographic processes by describing the complete life cycle of organisms and allow identification of which stage transitions and/or underlying parameters are most likely to impact population growth. These transitions or parameters can then be targeted for management with maximum impact on the population.[12,13] Matrix models are frequently used in management-oriented modeling, and there are a number of recently developed additional tools that improve the utility of matrix models for management purposes. Integral projection models,[14] based on continuous functions of vital rates on size, can improve estimates of population growth rate particularly for low sample sizes and in populations where categorical life stages may not be appropriate.[15] Economic sensitivity analysis allows the most cost-effective management strategies to be identified,[16] rather than just the parameters with the largest impact on population growth rate (see also our third section). However, until recently, the great level of detail in describing population dynamics meant that stage-structured models focused on population growth and ignored the spatial component of spread.

The introduction of the stage-based integrodifference equation (IDE) model of spread[17] has allowed a shift in focus from population growth rate to spread rate by combining demographic matrix models with stage-specific dispersal kernels. The IDE model enables the integration of dispersal into matrix models, which can then be exposed to analytical perturbation analysis similar to that used for demographic models (i.e., calculation of sensitivities and elasticities). Studies using structured matrix models[7,18–20] or integral projection models[21] have identified which population processes determine how fast a population can spread, providing initial management targets. The integration of dispersal can significantly change the relative importance of a model's parameters,[20] meaning that results from a demographic matrix model to inform management of population growth rate cannot necessarily be transferred to the spread management problem. An important issue for the IDE approach, however, is the level of description of dispersal.[22] Although demographic transitions are well parameterized from field studies, dispersal is usually described using simple, phenomenological dispersal kernels.[23] Indeed, to be included in the IDE model, a dispersal kernel needs to have a moment-generating function, which limits the choice of candidate functional forms. For wind-dispersed species, however, a semi-mechanistic wind dispersal model was recently simplified into an inverse-Gaussian function (WALD model[24]), which has a moment-generating function. The WALD model can be parameterized from field data, and Caplat *et al.*[7] developed a method that integrates WALD wind dispersal parameters fully into the stage-structured IDE spread model. This semimechanistic method has a major advantage in enabling the characterization of dispersal across the landscape where wind statistics vary between locations (e.g., due to differences in topography). Incorporation of a semimechanistic dispersal model also allowed identification of the mechanisms that drive spread rate, of which a few can be targeted by management (Box 1).

Analytical models have been used extensively because of their simplicity and their tractability, which enables the use of optimization techniques to design management strategies.[10,25] However, a controversial assumption of these models is that populations are homogeneous and that individual variability does not impact on model outcomes significantly.[26] Simulation approaches, such as individual-based modeling (see Ref. 27 and also our third section below), offer an interesting alternative to that assumption by modeling the fate of individuals or small groups of individuals, thus accounting for variability in population dynamics[6] and dispersal, as dispersal can be affected by plant height at inter- and intraspecific levels.[7,28] Individual-based models also have flexibility in describing abiotic and biotic processes in space and time.[29–31] Because individual-based models are often difficult to analyze,[32] they have not been used as frequently as analytical models when it comes to modeling invasive plant management. However, newly applied analysis techniques, such as boosted regression trees, are enabling generalizations about key drivers of spread to be drawn from complex individual-based models (e.g., see Ref. 6).

There are many papers that have sought to identify key drivers of spread from matrix models, IDEs, other analytical approaches and analysis of individual-based models. However, there is a need to emphasize what kinds of management actions are actually achievable by identifying the demographic or dispersal mechanisms that can be acted upon—e.g., reducing fecundity can be achieved using biocontrol agents, while reducing wind speed is far more difficult to achieve! New methods that address cost-effectiveness of actions[3,16] are highly promising for improving real-world applications of modeling outcomes. To effectively inform management, it is also necessary to put population models in a broader context, by accounting better for environmental heterogeneity and spatial dynamics as well as human decision making. The two following sections offer a review of methods that address these two points.

Box 1. Integrating dispersal mechanisms in the integrodifference model

Matrix models describe the transitions between different stages of a life cycle. Let's consider a perennial plant whose life cycle can be described in three stages (seed, juvenile, adult), as in Fig. 2, with establishment occurring at rate e, growth into adults at rate g, and individuals die at a constant rate m. Adults produce f seeds per year.

This can be expressed in form of a matrix A as follows:

$$A = \begin{bmatrix} (1-e)(1-m) & 0 & f \\ e & (1-g)(1-m) & 0 \\ 0 & g(1-m) & 1-m \end{bmatrix}.$$

The long-term average growth rate of the population can be calculated using a mathematical property of matrices, the dominant eigenvalue.[11] Similarly it is possible to calculate the spread rate from an IDE if dispersal is described with a kernel that has a moment-generating function (another mathematical property).[17]

Perturbation analysis calculates the absolute or relative importance of each model parameter for population growth or spread rate using the mathematical properties of the IDE model. The usefulness of the method for management depends on the parameters that are used to define the model.

In the simple case described by A, for instance, the four demographic parameters (e, m, g, f) can have differing importance for spread. From this knowledge, it is easy to derive management priorities (if fecundity f is the most important parameter for spread, then a biocontrol agent targeting seeds should be efficient to control the population's expansion; if mortality m ranks high, then killing individuals can be recommended).

Most dispersal functions are based on generic parameters (e.g., scale and shape of a dispersal kernel) that cannot be easily used to inform management. However, if the dispersal kernel is a function of physical or biological variables (e.g., plant height, seed properties, or wind statistics), perturbation analysis of dispersal parameters can be used to derive management actions, although implementation of management on dispersal parameters is often not as straightforward as for the demographic parameters.[7]

Understanding the role of landscape structure to guide management

Dispersal interacts with other landscape-scale processes such as fragmentation and arrangement of suitable habitat to determine spread rates and population extent. A landscape can be visualized as composed of patches of differing composition. Patches may vary in how well an invader will establish, the population growth rate achievable, the carrying capacity, and the potential for dispersal out of that patch. Parameterizing patch suitability at the landscape scale can be achieved using both population ecology and habitat suitability models. For a population to spread across a landscape, propagules have to disperse out of the natal patch, reach a suitable patch, and establish. Some models allow

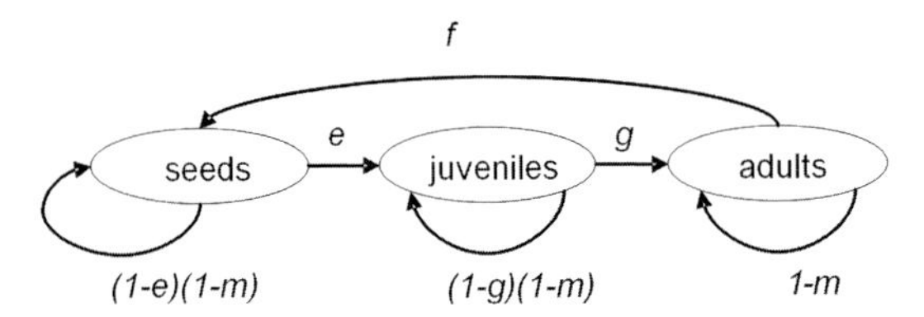

Figure 2. Life cycle of a theoretical perennial plant with a population structured in three stages (in Box 1).

characterization of spread in relation to the spatial configuration of populations and have been useful for analyzing the spatial aspects of management.[3,33,34] The simplest representation of a landscape might consist of sites occupied and unoccupied by an invasive plant. This can be further abstracted to a proportion of the total landscape occupied by the invader. These simple models may not be spatially explicit (i.e., the arrangement of occupied patches is not specified).[2,35–37] Here, we focus on spatially explicit landscape models that specify the arrangement and sometimes the size and shape of patches.

While graph theory has seldom been used in invasion ecology (Refs. 38 and 39 are exceptions), it has a much longer history in the study of both disease and information invasion into social networks (e.g., Refs. 40 and 41). Percolation models are the most commonly used graph theory approach applied to invasive plant spread. Early work used percolation models to analyze how the proportion and arrangement of suitable patches in a landscape affect spread speed and patterns.[38] An invader can spread to a new cell if that cell is within the local dispersal neighborhood. Under these simple rules, the amount of suitable habitat required for a binary landscape (one composed of only suitable and unsuitable habitat) to become traversable was related to both how clumped the suitable habitat was and the dispersal ability (size of the dispersal neighborhood) of the organism.[38] Wider-dispersing organisms achieved higher occupancy in more fragmented landscapes[38] than organisms that dispersed more locally. In this framework, wide and narrowly dispersing organisms differed in the number of neighboring cells they could move to in a single turn. As a result, when dispersal ability increased, so did the total reproductive output, making it impossible to say whether changes in reproduction or dispersal distances were responsible for changes in spread rates. Models that decouple reproductive output from dispersal processes are necessary in order to tease out the independent effects of each.

Simple percolation models give valuable insight into the potential effects of habitat fragmentation; however, they do not take into account rare long-distance, gap-crossing dispersal. One feature of fragmented landscapes that acts to limit dispersal is gap size.[39,42] Rare long-distance dispersal negates some of the effects of fragmented landscapes by allowing propagules to jump gaps.[42] It remains an open question as to how much long-distance dispersal is enough to overcome the effects of fragmentation in two-dimensional landscapes and how the pattern of fragmentation affects this.

Propagules may be able to establish in habitat that they cannot disperse through and disperse through habitat that they cannot establish in. For example, water-dispersed propagules may not be able to establish in a water body but can readily spread along those routes, establishing in suitable habitat that intersects with the dispersal route.[43] Simple percolation models do not take these complexities into account. Network models that decouple suitable habitat from dispersal pathways may, in some cases, better represent invasive plant spread dynamics.

Network models express the dynamics of a plant invasion in terms of a connected graph by defining landscape patches as cells (also known as vertices) and dispersal events as connections (also known as edges), which link the cells (see Box 2). Methods from the field of epidemiology are useful here, as they enable the prediction of speed of spread when one knows the spatial structure of the landscape/network.[44] Knowledge about the rules that link spread and landscape organization has the potential to significantly improve decision making for invasive species management.

In theory, management could be focused on manipulating patch suitability in order to decrease the proportion of suitable patches or to manipulate the pattern of suitable habitat in order to slow spread. These kinds of approaches are more commonly explored in the area-wide management and pest control literature[45] but are still rare. Another potential application is in identifying areas of the landscape where spread will be faster or slower and using this information to prioritize management actions to particular areas of the landscape to achieve the most cost-effective reductions in spread rate or prevention of impact. The application of management optimization models to spatial spread models is an emerging field of research.

Chades *et al.*,[46] for instance, have recently shown the importance of accounting for different patterns of connectivity in a network of patches to prioritize management (Box 2). Chades *et al.* used small network motifs to derive rules of thumb for where to start managing in a network to minimize reinvasion of the network after control. These rules of thumb consistently out-performed simulated "outside-in" management approaches, where the periphery of an invading population was managed before the core infestation. A major advantage of Chades *et al.*'s method is that partial detectability of patch occupancy can be incorporated into the management prioritization process. Invasive plants are often hard to detect, especially in the early stages of invasion when management may be most cost effective. The presence of a soil seed bank can also mean that invasions remain undetected when above ground plant parts are not present.[47] Chades *et al.*'s approach provides guidance on how long we should manage or survey networks for hard-to-detect organisms. By including uncertainty into the model, an optimal sequence of actions can be determined in order to minimize costs of too much or too little management when a population may not be detected perfectly.

While some useful landscape structure methods exist in theory, they have yet to be widely applied to model management options explicitly or optimize management prioritization for controlling invasive plant spread (but see Refs. 46 and 47). The incorporation of decision theoretic methods with spatially structured landscape-scale spread models has the potential to contribute strongly to the invasive plant management field.

Box 2. Network management models

A spatially structured invasive species population can be represented as a network of infested and uninfested patches linked by dispersal (Fig. 3). Commonly, the whole network cannot be managed simultaneously due to economic or logistical constraints. We are therefore faced with the problem of where in the network to start managing and how to proceed through time. Chades *et al.*'s spatial optimization model incorporates management success, dispersal, economic cost and imperfect detection together with network structure to determine management priorities (Fig. 4).[46] This is an example of integrating decision processes together with spatial configuration (Fig. 1), with simple dispersal assumptions feeding into the network model. Dispersal could be parameterized from data or from a lower-level model; this latter modeling approach would shift the study into the center of the triangle (Fig. 1), integrating all three methods.

Network motifs are small subgraphs that capture specific patterns of interactions.

The optimal order for managing within a network motif can be applied to multiple motifs connected together in larger networks (Fig. 5). This avoids the problem of optimizing management over large networks, which is computationally expensive and limits the size of the problem that can be tackled.

The role of human management structures for effecting landscape-scale management

Modeling management at landscape scales requires the description of ecological processes, the impact of management actions, and the structure and constraints of management. Here, we discuss two powerful constraints that can determine the success or failure of spread management programs: resource constraints and multiple managers. Many of the above models of spread management assume a single manager with unlimited funds and who has knowledge of, and access to, all of the invasive population. This implies that managers can apply control uniformly and synchronously across the landscape, which is unrealistic. In most cases, managers of invasive populations do not have enough resources to control the entire invasive population and also differ in their knowledge and choice of actions.

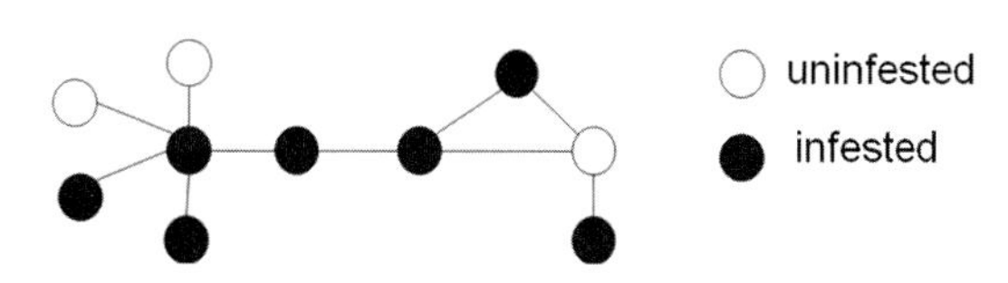

Figure 3. Sites or populations are connected by dispersal routes, which in this case are bidirectional (in Box 2).

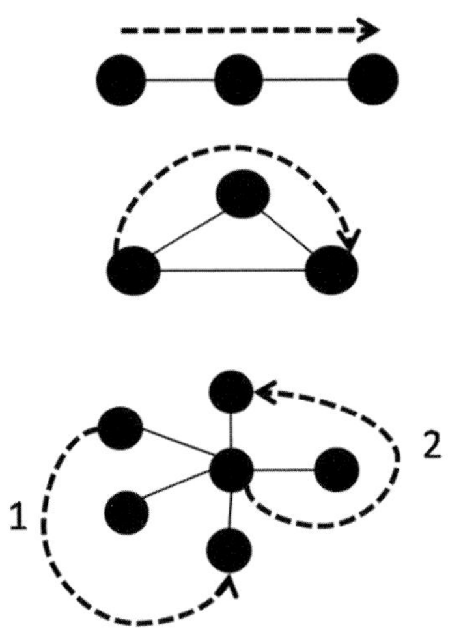

Figure 4. Optimal management for network motifs (in Box 2).

Management with limited resources and/or knowledge

When managers cannot survey or control all populations, resources must be prioritized in some way. Prioritization can be made solely on ecological grounds[33] or admit a limited budget, in which case the goal is to control as much of the invasive population as possible with the available resources. Where the number of management schemes is small, each can be modeled. This is commonly done by simulating ecological processes, the effect various management strategies have on those processes, and management costs.[48,49] The management scheme that reduced the weeds population the most for a given budget is considered the best strategy.

However, if there are many possible management schemes, simulating them all becomes impossible. The number of management schemes can escalate quickly. For instance, allocating one million dollars between two management options leads to 1,000 different prioritization schemes if the budget is allocated in blocks of $1,000. If more management options are available, or the budget is allocated dynamically, the number of possible schemes increases dramatically. Tools such as stochastic dynamic programming[25] can optimize control in both space and time, so that the greatest possible control for a given budget is achieved. This is achieved by backward iteration. Here, one starts at the final time step (normally the end of a planning time horizon) and works out the best decision assuming any future decisions don't matter (because they are beyond the planning horizon). Then, one time step back is taken and the best decision at this time step is calculated, assuming that in the future time step the optimal decision will be made (i.e., the decision calculated in the previous step). This process is repeated until the present time step is reached. The assumption that all future decisions are optimal is crucial, because it means one does not have to calculate the optimal decision for every possible future, only for the future where the optimal decisions are made at each time step.

Although formulating, writing, and parameterizing optimization models are nontrivial challenges,[50] there are important benefits to doing so. First, it forces management objectives to be explicitly and concisely (often mathematically) stated.[51] This is crucial to determine the success or failure of a program and is useful even if the optimization is never done.[52] Optimization models can also suggest rules of thumb.[51] For instance, optimal control strategies often advocate that the majority of money be spent at the start of a control program,[53,54] because control done in the present makes control in the future cheaper by reducing the size of future populations.

Several studies have also used optimization to find the value of information versus direct action. Should one look before they leap, or shoot first and ask questions later? The answer seems dependant on the model used and situation described. Baxter and Possingham[55] suggest one should look then leap. They find that, early in an eradication program for fire ants (*Solenopsis invicta*), it is better to spend resources accurately mapping the population, even if this diverts resources from destroying nests and allows the population to expand

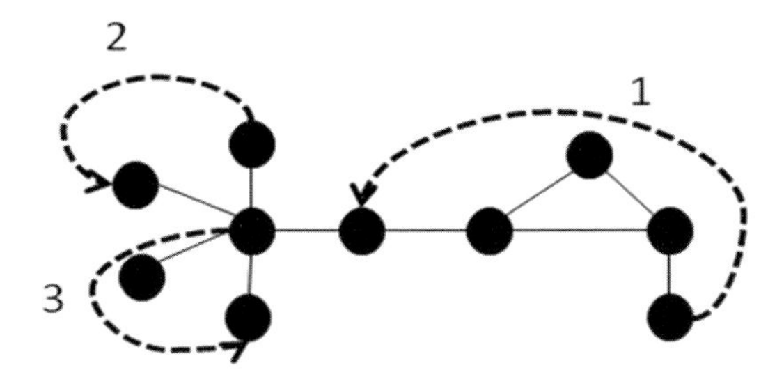

Figure 5. Optimal management of a network composed of connected motifs (in Box 2).

in its early stages. Offering qualified support for monitoring, Haight and Polasky[56] show that it is worth monitoring at the expense of direct action, but only if monitoring is cheap and yields high-quality data relevant for directing management action. In contrast, Carrasco *et al.*[53] show that optimal management strategies under perfect certainty and optimal risk averse strategies under severe uncertainty can be very similar. This suggests that reducing uncertainty would not change management decisions on the ground, although it could still alter how effective management actions are.

Optimization can also be used to compare management under different constraints (e.g., different budgets or available management options). A problem when comparing different constraints is that managers might apply control options in a different way under, for example, different budgets. Optimization avoids this problem by assuming that decision makers act optimally for each available budget. This technique has been used to show how current management budgets compare to optimal budgets and how much more control could be achieved under optimal budgets.[54,57] In a similar fashion, optimization studies have shown that using a single method optimally applied is far less effective than optimized integrated pest management.[58]

Multiple managers

At large scales, invasive populations often have multiple managers. For example, the management of most agricultural weeds is decided and implemented by individual farmers, even if the weeds' ranges extend over multiple properties. Modeling multiple managers of invasive species requires that the ecological system be modeled along with the behavior of managers, interactions and feedbacks between managers, and feedback between manages and the ecological system.[59,60]

An insightful tool to model interactions between managers is game theory. Weed control has been described as a weakest-link public-goods game.[61] The public good provided by each player is a reduction in invasion risk to all other players if they choose to take some sort of control action (this could be direct control or quarantine). The game is referred to as weakest link because if any one player decides not to control, or lets quarantine fail (the weak link), the weed can (re)infest everyone else.

A good example of the logic behind this approach can be found in the model developed by Hennessy,[62] which considers a lake and its users. Each lake user can reduce their risk of introducing a pest to the lake by washing down their boat. Each individual pays the associated cost, but the benefit all users get from the lake is reduced if just one person introduces a pest. An important concept in game theory is the Nash equilibrium, a state where no individuals can improve their personal outcome by unilaterally changing their behavior. In the lake game, there are two Nash equilibriums: either everybody washes down their boat or no one does. A switch from one equilibrium to the other can depend on the actions of a single lake user. When one user makes the effort and washes their boat, the others are encouraged to do so because there efforts are less likely to be in vain. On the other hand, seeing someone who does not wash their boat is an incentive to be equally careless. It is easy to see the crucial role communication has in such a model, and a consistent finding from weakest-link games is that communication between players greatly increases the chances of socially optimal outcomes.[62–64] The logic of weakest-link models also suggests that the whole group can benefit by subsidizing the control efforts of the players with the least incentive to control. Such subsidies increase the confidence of the whole group that the invasion will be prevented or slowed, and this becomes a self-fulfilling belief.[62,63]

While game theory provides powerful tools to model interactions between decision makers, spatially explicit ecological processes, such as plant invasions, can only be included indirectly. Agent-based modeling (ABM) is a more direct and flexible approach to modeling plant population dynamics and multiple managers. ABM simulates population dynamics and managers (or agents in ABM jargon) and can include a wide range submodels, including game theory models.[65]

A common finding of ABM is that coordination between managers is crucial to controlling invasive species.[66] Even if there are only two managers, differences in priorities mean that transfer payments between them are required to coordinate management and achieve eradication.[67] A similar message emerges from ABM based on insect pests in landscapes with both commercial and noncommercial farmers.[68] This study found that if noncommercial farmers considered the impacts of their pest control decisions on the wider area, then commercial farmers could spend far less on pest control and total landscape income was higher. However, if noncommercial farmers considered only themselves, then they seldom controlled, leading to a much lower income produced over the entire landscape. ABM is also used to show how common policy tools, such as subsidies, can have unintended side effects. Albers *et al.*[69] used ABM to show how fertilizer subsidies to land managers in tropical forests can cause an increase in the spread of the invasive grass *Imperata cylindrica*. Subsidies encourage short-term increases in production, which degraded land more quickly and led to much higher prevalence of *I. cylindrica*.

At very large scales, managers are almost always organized in hierarchies; for example, the local government that sets the policy on particular invasive species and on-ground managers who decide how best to implement that policy. Relatively little work has addressed the effects of hierarchical decision making on biological invasions. An exception is a model of zebra mussels (*Dreissena polymorpha*) invading a lake in the midwest of the United States.[59] A key finding is that if higher-level managers mistake how lower-level managers will react to the invasion, they will misallocate resources, making control efforts both less effective and more costly.

Integrating models for management

The three components of spread management (population dynamics, landscape processes, and human management) all need to be addressed in most instances of plant invasions. Because they are complementary and usually occur at the same time, integrative approaches are needed to solve management problems.

The complexity inherent to each component is a barrier to high-level integration. A nested approach, where local population models are incorporated in spatial models through dispersal and spatial models feed into management optimization approaches, is a way around that barrier. By making explicit how the three approaches are articulated (Fig. 1), it is possible to simplify the outputs of population models to use them as inputs in spatial models, and similarly the outputs of spatial models can be incorporated into large-scale, highly integrated management models. The success of integration relies on three considerations: (i) a powerful decision framework will only be as good as the worst of its submodels; (ii) the outputs of the ecological models must be in a form that can be used as an input from the higher level models; and (iii) the risk of propagating error being higher in a complex model makes it crucial for the higher-level models to account for uncertainty.

Because of (i), only when the outputs of the best available ecological models of populations and spread are fed into optimization approaches can the latter be really useful for management. Inversely, using a wrong decision framework or resource allocation method can annihilate the achievements of the ecological submodels.[4] Points (ii) and (iii) are highly dependent on the power of the decision tools. This has been, and still is, limiting the complexity that can be addressed in integrated studies. For instance, Hauser and McCarthy[57] developed an optimal surveillance–eradication model where ecological parameters were based on a dispersal-constrained habitat suitability model parameterized in the field. Despite the relative sophistication of the ecological model, its results needed to be reduced to a simple, static, probability of presence for the weed so that the optimization remained tractable. In many cases, the complexity arising from the interactions between ecological processes at different scales and human decisions makes integrative approaches very challenging. However, the advancement of computer science in parallel with new modeling tools is promising. Approaches such as partially observable decision processes (POMDP, see for instance Ref. 47) provide ways to account for possible gaps in the integration process when modeling management actions.

Conclusion

We described three components of spread management (population dynamics, landscape processes, and human management). They all need to be addressed in most instances of plant invasions. There is a rapidly growing literature that uses some aspects of the three approaches in modeling spread and management of invasive plants.[3,6,7,20,21,46,55] It is important that research efforts should continue in that direction, and interdisciplinary research that combines population biology, landscape ecology, economics, and social sciences must be encouraged.[70] Management optimization in the absence of a good understanding of the underlying ecological, economic, and social contexts of management is not likely to improve our solutions to the urgent invasion management problems that face us.

We stress the need for more management- and decision making–oriented research.

To inform management, research must be read by (or communicated to) managers and policy makers—at the moment, it is still a challenge to extract usable information from the literature, and as scientists we can only plead guilty. A major problem is the sheer volume and diversity of the literature on management. Indeed, a simple keyword search is often not enough to select the studies relevant for a particular management issue. Literature reviews (like this one) are a good way of highlighting the useful studies or approaches. Journal forums (e.g., Ref. 71) and information sessions are important to help bridge the gap between scientists and managers.[72] In addition, throughout the world, governmental and nongovernmental organizations are making a point of publishing reports that have both management and scientific value,[73,74] an important output for applied research.

Acknowledgments

The authors acknowledge funding provided by the Australian Research Council Australian Research Fellowship (YMB), CSIRO OCE postdoctoral fellowship (PC), and RIRDC National Weeds and Productivity Research program (SC). The authors thank Rieks van Klinken, Iadine Chadès, and an anonymous reviewer for their comments on the manuscript.

Conflicts of interest

The authors declare no conflicts of interest.

References

1. Drake, J.A. *et al.* 1989. Biological invasions: a global perspective. In *SCOPE* 37. John Wiley and Sons. Chichester, UK.
2. Hastings, A., R.J. Hall & C.M. Taylor. 2006. A simple approach to optimal control of invasive species. *Theor. Populat. Biol.* **70:** 431–435.
3. Epanchin-Niell, R.S. & A. Hastings. 2010. Controlling established invaders: integrating economics and spread dynamics to determine optimal management. *Ecol. Lett.* **13:** 528–541.
4. Yokomizo, H. *et al.* 2009. Managing the impact of invasive species: the value of knowing the density-impact curve. *Ecol. Appl.* **19:** 376–386.
5. Hastings, A. *et al.* 2005. The spatial spread of invasions: new developments in theory and evidence. *Ecol. Lett.* **8:** 91–101.
6. Coutts, S.R. *et al.* 2011. What are the key drivers of spread in invasive plants: dispersal, demography or landscape: and how can we use this knowledge to aid management? *Biol. Invasions.* **13:** 1649–1661.
7. Caplat, P., R. Nathan & Y.M. Buckley. in press. Seed terminal velocity, wind turbulence and demography drive the spread of an invasive tree in an analytical model. *Ecology*.
8. Kot, M., M.A. Lewis & P. van den Driessche. 1996. Dispersal data and the spread of invading organisms. *Ecology*. **77:** 2027–2042.
9. Tobin, P.C., L. Berec & A.M. Liebhold. 2011. Exploiting Allee effects for managing biological invasions. *Ecol. Lett.* **14:** 615–624.
10. Taylor, C.M. & A. Hastings. 2004. Finding optimal control strategies for invasive species: a density-structured model for Spartina alterniflora. *J. Appl. Ecol.* **41:** 1049–1057.
11. Caswell, H. 2001. *Matrix Population Models: Construction, Analysis, and Interpretation.* Sinauer Associates Inc. Sunderland, MA.
12. Ramula, S. *et al.* 2008. General guidelines for invasive plant management based on comparative demography of invasive and native plant populations. *J. Appl. Ecol.* **45:** 1124–1133.
13. Shea, K. & D. Kelly. 1998. Estimating biocontrol agent impact with matrix models: Carduus nutans in New Zealand. *Ecol. Appl.* **8:** 824–832.
14. Easterling, M.R., S.P. Ellner & P.M. Dixon. 2000. Size-specific sensitivity: applying a new structured population model. *Ecology.* **81:** 694–708.
15. Ramula, S., M. Rees & Y.M. Buckley. 2009. Integral projection models perform better for small demographic data sets than matrix population models: a case study of two perennial herbs. *J. Appl. Ecol.* **46:** 1048–1053.
16. Baxter, P.W.J., M.A. McCarthy & P.W. Menkhorst. 2006. Accounting for managment costs in sensitivity analyses of matrix population models. *Conserv. Biol.* **20:** 893–905.
17. Neubert, M.G. & H. Caswell. 2000. Demography and dispersal: calculation and sensitivity analysis of invasion speed for structured populations. *Ecology.* **81:** 1613–1628.
18. Buckley, Y.M. *et al.* 2005. Slowing down a pine invasion despite uncertainty in demography and dispersal. *J. Appl. Ecol.* **42:** 1020–1030.
19. Neubert, M.G. & I.M. Parker. 2004. Projecting rates of spread for invasive species. *Risk Anal.* **24:** 817–831.
20. Shea, K. *et al.* 2010. Optimal management strategies to control local population growth or population spread may not be the same. *Ecol. Appl.* **20:** 1148–1161.

21. Jongejans, E. *et al.* 2011. Importance of individual and environmental variation for invasive species spread: a spatial integral projection model. *Ecology.* **92:** 86–97.
22. Nathan, R. *et al.* 2008. Mechanisms of long-distance seed dispersal. *Trends Ecol. Evol.* **23:** 638–647.
23. Jongejans, E., O. Skarpaas & K. Shea. 2008. Dispersal, demography and spatial population models for conservation and control management. *Perspect. Plant Ecol., Evol. System.* **9:** 153–170.
24. Katul, G.G. *et al.* 2005. Mechanistic analytical models for long-distance seed dispersal by wind. *Am. Nat.* **166:** 368–381.
25. Shea, K. & H.P. Possingham. 2000. Optimal release strategies for biological control agents: an application of stochastic dynamic programming to population management. *J. Appl. Ecol.* **37:** 77–86.
26. Kendall, B.E. & G.A. Fox. 2003. Unstructured individual variation and demographic stochasticity. *Conserv. Biol.* **17:** 1170–1172.
27. Grimm, V. *et al.* 2005. Pattern-oriented modeling of agent-based complex systems: lessons from ecology. *Science.* **310:** 987–991.
28. Thomson, F.J. *et al.* 2011. Seed dispersal distance is more strongly correlated with plant height than with seed mass. *J. Ecol.* **99:** 1299–1307.
29. Breckling, B., U. Middelhoff & H. Reuter. 2006. Individual-based models as tools for ecological theory and application: Understanding the emergence of organisational properties in ecological systems. *Ecol. Model.* **194:** 102–113.
30. Nehrbass, N. *et al.* 2007. A simulation model of plant invasion: long-distance dispersal determines the pattern of spread. *Biol. Invasions.* **9:** 383–395.
31. Caplat, P., M. Anand & C. Bauch. 2010. Modelling invasibility in endogenously oscillating tree populations: timing of invasion matters. *Biol. Invasions.* **12:** 219–231.
32. Jongejans, E., O. Skarpaas & K. Shea. 2008. Dispersal, demography and spatial population models for conservation and control management. *Perspect. Plant Ecol. Evol. System.* **9:** 153–170.
33. Moody, M.E. & R.N. Mack. 1988. Controlling the spread of plant invasions—the importance of nascent foci. *J. Appl. Ecol.* **25:** 1009–1021.
34. Travis, J.M.J. & K.J. Park. 2004. Spatial structure and the control of invasive alien species. *Anim. Conserv.* **7:** 321–330.
35. Rees, M. & Q. Paynter. 1997. Biological control of Scotch broom: modelling the determinants of abundance and the potential impact of introduced insect herbivores. *J. Appl. Ecol.* **34:** 1203–1221.
36. Buckley, Y.M. *et al.* 2004. Modelling integrated weed management of an invasive shrub in tropical Australia. *J. Appl. Ecol.* **41:** 547–560.
37. Firn, J. *et al.* 2008. Managing beyond the invader: manipulating disturbance of natives simplifies control efforts. *J. Appl. Ecol.* **45:** 1143–1151.
38. With, K.A. 2002. The landscape ecology of invasive species. *Conserv. Biol.* **16:** 1192–1203.
39. With, K.A. & A.W. King. 1999. Dispersal success on fractal landscapes: a consequence of lacunarity thresholds. *Landsc. Ecol.* **14:** 73–82.
40. Keeling, M.J. & K.T.D. Eames. 2005. Networks and epidemic models. *J. Roy. Soc. Interface.* **2:** 295–307.
41. Watts, D.J. 2004. The "new" science of networks. *Ann. Rev. Sociol.* **30:** 243–270.
42. Malanson, G.P. & D.M. Cairns. 1997. Effects of dispersal, population delays, and forest fragmentation on tree migration rates. *Plant Ecol.* **131:** 67–79.
43. Nilsson, C. *et al.* 2010. The role of hydrochory in structuring riparian and wetland vegetation. *Biol. Rev.* **85:** 837–858.
44. May, R.M. 2006. Network structure and the biology of populations. *Trends Ecol. Evol.* **21:** 394–399.
45. Bianchi, F.J.J.A. *et al.* 2010. Spatial variability in ecosystem services: simple rules for predator-mediated pest suppression. *Ecol. Appl.* **20:** 2322–2333.
46. Chades, I. *et al.* 2011. General rules for managing and surveying networks of pests, diseases, and endangered species. *Proc. Natl. Acad. Sci. USA* **108:** 8323–8328.
47. Regan, T.J., I. Chadès & H.P. Possingham. 2011. Optimally managing under imperfect detection: a method for plant invasions. *J. Appl. Ecol.* **48:** 76–85.
48. Cacho, O.J. *et al.* 2010. Allocating surveillance effort in the management of invasive species: a spatially-explicit model. *Environ. Modell. Software.* **25:** 444–454.
49. Higgins, S.I., D.M. Richardson & R.M. Cowling. 2000. Using a dynamic landscape model for planning the management of alien plant invasions. *Ecol. Appl.* **10:** 1833–1848.
50. Eiswerth, M.E. & W.S. Johnson. 2002. Managing nonindigenous invasive species: Insights from dynamic analysis. *Environ. Resour. Econ.* **23:** 319–342.
51. Shea, K. &N. W. G. o. P. Management. 1998. Management of populations in conservation, harvesting and control. *Trends Ecol. Evol.* **13:** 371–375.
52. Possingham, H. *et al.* 2001. Making smart conservation decisions. In *Conservation Biology: Research Priorities for the Next Decade.* M.E. Soule & G.H. Orians, Eds.: 225–244. Island Press. Washington.
53. Carrasco, L.R. *et al.* 2010. Optimal and robust control of invasive alien species spreading in homogeneous landscapes. *J. Roy. Soc. Interface.* **7:** 529–540.
54. Lee, D.J., D.C. Adams & C.S. Kim. 2009. Managing invasive plants on public conservation forestlands: application of a bio-economic model. *Forest Policy Econ.* **11:** 237–243.
55. Baxter, P.W.J. & H.P. Possingham. 2011. Optimizing search strategies for invasive pests: learn before you leap. *J. Appl. Ecol.* **48:** 86–95.
56. Haight, R.G. & S. Polasky. 2010. Optimal control of an invasive species with imperfect information about the level of infestation. *Resour. Ener. Econ.* **32:** 519–533.
57. Hauser, C.E. & M.A. McCarthy. 2009. Streamlining 'search and destroy': cost-effective surveillance for invasive species management. *Ecol. Lett.* **12:** 683–692.
58. Hyder, A., B. Leung & Z.W. Miao. 2008. Integrating data, biology, and decision models for invasive species management: application to leafy spurge (*Euphorbia esula*). *Ecol. Soc.* **13**(2): 12 Available at http://www.ecologyandsociety.org/vol13/iss2/art12/ Accessed 24 November 2011.
59. Finnoff, D. *et al.* 2005. The importance of bioeconomic feedback in invasive species management. *Ecol. Econ.* **52:** 367–381.

60. Settle, C., T.D. Crocker & J.F. Shogren. 2002. On the joint determination of biological and economic systems. *Ecol. Econ.* **42:** 301–311.
61. Perrings, C. *et al.* 2002. Biological invasion risks and the public good: an economic perspective. *Conserv. Ecol.* **6:**(1): 1. Available at: http://www.consecol.org/vol6/iss1/art1/ Accessed 24 November 2011.
62. Hennessy, D.A. 2008. Biosecurity incentives, network effects, and entry of a rapidly spreading pest. *Ecol. Econ.* **68:** 230–239.
63. Burnett, K.M. 2006. Introductions of invasive species: failure of the weaker link. *Agric. Resour. Econ. Rev.* **35:** 21–28.
64. Devetag, G. & A. Ortmann. 2007. When and why? A critical survey on coordination failure in the laboratory. *Exp. Econ.* **10:** 331–344.
65. McAllister, R.R.J. *et al.* 2006. Pastoralists' responses to variation of rangeland resources in time and space. *Ecol. Appl.* **16:** 572–583.
66. McKee, G.J. 2011. Coordinated pest management decisions in the presence of management externalities: the case of greenhouse whitefly in California-grown strawberries. *Agric. Syst.* **104:** 94–103.
67. Grimsrud, K.M. *et al.* 2008. A two-agent dynamic model with an invasive weed diffusion externality: An application to Yellow Starthistle (Centaurea solstitialis L.) in New Mexico. *J. Environ. Manag.* **89:** 322–335.
68. Ceddia, M.G., J. Heikkila & J. Peltola. 2009. Managing invasive alien species with professional and hobby farmers: insights from ecological-economic modelling. *Ecol. Econ.* **68:** 1366–1374.
69. Albers, H.J., M.J. Goldbach & D.T. Kaffine. 2006. Implications of agricultural policy for species invasion in shifting cultivation systems. *Environ. Develop. Econ.* **11:** 429–452.
70. Caplat, P. & S. Coutts. 2011. Integrating ecological knowledge, public perception and urgency of action into invasive species management. *Environ. Manag* **48:** 878–881.
71. Gibbons, D.W., J.D. Wilson & R.E. Green. 2010. Using conservation science to solve conservation problems. *J. Appl. Ecol.* **48:** 505–508.
72. Shaw, J., J. Wilson & D. Richardson. 2010. Initiating dialogue between scientists and managers of biological invasions. *Biol. Invasions.* **12:** 4077–4083.
73. Krueger-Mangold, J.M., R.L. Sheley & T.J. Svejcar. 2006. Toward ecologically-based invasive plant management on rangeland. *Weed Sci.* **54:** 597–605.
74. Poon, E., D.A. Westcott, D. Burrows and A. Webb. 2007. Assessment of research needs for the management of iInvasive species in the terrestrial and aquatic ecosystems of the Wet Tropics. Report to the Marine and Tropical Sciences Research Facility. Reef and Rainforest Research Centre Limited. Cairns.

Ann. N.Y. Acad. Sci. ISSN 0077-8923

ANNALS OF THE NEW YORK ACADEMY OF SCIENCES
Issue: *The Year in Ecology and Conservation Biology*

Sustainable seaweed cutting? The rockweed (*Ascophyllum nodosum*) industry of Maine and the Maritime Provinces

Robin Hadlock Seeley[1] and William H. Schlesinger[2]

[1]Shoals Marine Laboratory and Department of Ecology and Evolutionary Biology, Cornell University, Ithaca, New York. [2]Cary Institute of Ecosystem Studies, Millbrook, New York

Address for correspondence: Robin Hadlock Seeley, Shoals Marine Laboratory, Cornell University, 106A Kennedy Hall, Ithaca, NY 14853. rhs4@cornell.edu

Burgeoning global demand for products derived from seaweeds is driving the increased removal of wild coastal seaweed biomass, an emerging low trophic level industry. These products are marketed as organic and "sustainable." Brown macroalgae, such as kelps (Laminariales) and rockweeds (Fucales), are foundational species that form underwater forests and thus support a diverse vertebrate, invertebrate, and algal community—including important commercial species—and deliver organic matter to coastal ecosystems. The measure of sustainability used by the rockweed (*Ascophyllum nodosum* (L.) LeJolis) industry, maximum sustainable yield, accounts for neither rockweed's role as habitat for 150+ species, including species of commercial or conservation significance, nor its role in coastal and estuarine ecosystems. To determine whether rockweed cutting is "sustainable" will require data on the long-term and ecosystem-wide impacts of cutting rockweed. Once a sustainable level of cutting is determined, strict regulation by resource managers will be required to protect rockweed habitat. Until sustainable levels of cutting and appropriate regulations are identified, commercial-scale rockweed cutting presents a risk to coastal ecosystems and the human communities that depend on those ecosystems.

Keywords: *Ascophyllum nodosum*; rockweed; sustainability; marine macroalgae; seaweed harvest

Introduction

Benthic marine macroalgae, or seaweeds, are critical components of coastal ecosystems. As primary producers they contribute to nutrient cycling, sequester nitrogen and carbon,[1] contribute energy to food webs,[2–4] and provide three-dimensional structure and habitat for fish,[5–7] waterfowl,[8,9] algae,[10] shellfish,[11] shellfish spat,[10,12] and other invertebrates.[11,13]

The rate of wild seaweed extraction is increasing globally[14] as human populations find more and more uses for marine plant-based products in agriculture, animal production, pet foods, cosmetics, and human food.[15] The intensity of seaweed extraction ranges from small operations for individual use, to artisanal collection of seaweeds as food, to commercial harvests of hundreds to thousands of tons daily.[16] The volume of seaweed processed for human uses across the globe is outpacing the ability of wild seaweed populations to supply the demand.[14] Yet the removal of wild seaweed species at the commercial level, marketed as organic and sustainable, continues and intensifies, leading to conservation concerns: should wild seaweeds be harvested at all?[17] Management of commercial seaweed operations globally has not been sufficient to prevent overharvest of some of these species.[18] Is the cutting of wild seaweeds on a commercial scale ecologically sustainable?

The brown macroalgal species in the families Laminariales (kelps) and Fucales (fucoids or rockweeds) are foundational species that form underwater forests that provide numerous ecological services to coastal ecosystems: food, habitat, and shelter for a diverse invertebrate and vertebrate community, as well as essential nutrients to productive coastal ecosystems.[13,19] *Ascophyllum nodosum*, a brown fucoid seaweed (also known as rockweed, bladderwrack, Norwegian kelp, or simply "kelp"), inhabits coastal shorelines from the Canadian Arctic, Greenland, Iceland, and Norway to Portugal and

doi: 10.1111/j.1749-6632.2012.06443.x

Figure 1. The rockweed zone exposed at midtide, Friendship Long Island, Maine.

the east coast of the United States south to New Jersey. *Ascophyllum*'s dominance in the intertidal zone creates an eponymous band in the intertidal zone, the "rockweed zone."[20]

At low tide, dense layers of *Ascophyllum* fronds lie prostrate on the substratum (rocky shore; Fig. 1), creating a protective blanket for other species fixed to the rocks below, living in sediments underneath, or residing in its canopy. The prostrate fronds rise on the incoming tide and form a rockweed forest[21] (Fig. 2). *Ascophyllum* provides habitat to 34 species of fish[22–25] (Table 1) and 100+ invertebrate taxa,[11] which are a food source for fish[26] and birds.[8,27,28] A complete rockweed zone food web was documented by Golléty *et al.*[29]

Ascophyllum is characterized by a slow growth rate[30] and long life span (holdfasts, which attach the alga to the substrate, may be up to 400 years old[31]). The great height (length[32–35]), high fecundity, high within-population genetic diversity, and weak large-scale population phenotypic differentiation[30] mean that *Ascophyllum* plants are analogous to trees,[30,36] and rockweed beds are the underwater equivalent of old-growth forests.[30] Because of the diversity and significance of its ecological services, rockweed is valued as a "high-priority" species (ranking fourth out of 161 monitored marine species) for protection in the United States and Canada,[37] a "priority species" in Northern Ireland,[38] and a "high-sensitivity species" in the UK.[39] Rockweed harvesting has been ranked one of the top human impacts across coastal habitat types in the Gulf of Maine.[40]

Ecological services

Biomass and net primary production

The best ecosystem-level study of rockweed is found in the work by Vadas *et al.*,[3] who studied its biomass and productivity. They reported wet-weight biomass of 11.4–28.9 kg/m^2 at five sites in Cobscook Bay. Separately, Trott and Larsen[41] reported mean wet-weight biomass of 10–17 kg/m^2 in Cobscook Bay, Sutherland[42] reports mean biomass ranging from 6.7 to 12.9 kg/m^2 in Southern New Brunswick, and Keser *et al.*[43] reported a range of 5.0–17.5 kg/m^2 in Nova Scotia.

As seen in ecosystem studies of grasslands, estimates of net primary productivity (NPP) for rockweed are complicated by its multiple points of growth, continuous loss of biomass to herbivory and breakage, and high local, spatial variation in its occurrence. Vadas *et al.*[3] estimated NPP of rockweed in Cobscook Bay at 8.15–11.65 kg wet weight/m^2/yr, giving an annual turnover of 40–70%/yr for rockweed biomass (equivalent to a mean-residence time of 1.4–2.5 years). Schmidt *et al.*[13] estimated primary production at 1.3 kg wet weight/m^2/yr in Nova Scotia. For nine sites in Nova Scotia, Cousens[44] reported mean net primary production of 0.820–2.120 kg/m^2/yr (dry weight) and a mean turnover of 44%/yr. The highest values for net primary production in Cobscook Bay equate to 0.894 kg C/m^2/yr,[4] which is equivalent to net primary production in the world's most productive terrestrial forests.[45]

Most of the carbon in rockweed biomass enters detrital pools. In Great Bay, New Hampshire, 4.62 mt of nitrogen (N) and 0.31 mt of phosphorus (P)

Figure 2. Stickleback fish in a rockweed forest at high tide, Hallowell Island, Moosehorn National Wildlife Refuge, Edmunds, Maine.

are estimated as available to the detrital pool from shed *Ascophyllum* receptacles.[46] After benthic diatoms and phytoplankton, rockweeds are the third most productive component of the Cobscook Bay ecosystem.[4] The productivity of fucoids, primarily *Ascophyllum*, may underlie the productivity of scallops and soft-shell clams and the high diversity of the filter-feeding invertebrate community in Cobscook Bay.[3] Sea scallops (*Placopecten magellanicus*) feed on brown algal detritus when phytoplankton populations are reduced,[47] and Cobscook Bay harbors the most productive scallop beds in Maine[48] (worth $1.0 million in 2004–2005[49]). The work of Vadas *et al.*[3] suggests that *Ascophyllum* plays a significant role in the productivity of those beds as well as the productivity of other commercially important shellfish species. Up to 82% of macrophyte (primarily *Ascophyllum*) production leaves the intertidal zone in Passamaquoddy Bay (NB Canada) during the summer months and is assumed available for decomposition into the bay.[50] Decaying macroalgae in the strand line of Passamaquoddy Bay is released to the sea during spring tides and represents ~7.3% of the primary productivity of the bay.[50]

Habitat

In Maine, coastal wetlands (including the rockweed-covered shore) and significant wildlife habitat are protected under the Natural Resources Protection Act (NRPA).[51] "Seaweed communities" were included as part of Significant Wildlife Habitat for tidal waterfowl/wading birds.[52]

Ascophyllum creates complex habitat in three forms: "wrack" (fronds that are unattached and stranded by tidal action in the high intertidal zone), unattached macroalgal mats drifting near the coast, and plants attached by a holdfast to hard substratum in the intertidal zone.

Macroalgae generally, and *Ascophyllum* wrack in Maine and the Maritimes, decomposes and creates habitat for small invertebrates, especially insects and amphipods[50] that are prey for migrating shorebirds.[53,54] Rockweed rafts floating at sea increase habitat complexity there,[55] providing habitat for isopods, amphipods, and fishes.[56,57]

Over 100 taxa of invertebrates,[11] including lobster, clams, and snails; 34 species of fish, including pollock, herring, flounder, and cod (Table 1); shorebirds, seabirds, and waterfowl (Table 1); and several other algal species[10] use attached *Ascophyllum* habitat. Epifaunal density is correlated with rockweed biomass.[58] "Some species are attracted to rockweed to graze upon it or its associated epiphytes,[59] while others benefit from the physical structure it provides. At low tide, rockweed thalli protect against

Table 1. Conservation status in the United States and Canada of fish and bird species that use rockweed habitat

Species	Common name	Conservation status United States	Conservation status Canada
Fish			
Alosa pseudoharengus[22,25]	Alewife	2011 candidate for ESA listing;[69] USFWS priority species[70]	COSEWIC candidate Group 2[71]
Ammodytes americanus[22]	American sand lance		
Anguilla rostrata[22,25]	American eel	USFWS priority species;[70] MCWCS: SGCN 1;[72] IFW special concern[73]	
Apeltes quadracus[22]	Fourspine stickleback		
Clupea harengus[22,24]	Atlantic herring		
Cyclopterus lumpus[22,24]	Lumpfish		
Fundulus heteroclitis[22,25]	Mummichog		
Gadus morhua[22]	Atlantic cod		
Gasterosteus wheatlandi[22]	Blackspotted stickleback		
Gasterosteus aculeatus[22,24]	Threespine stickleback		
Hemitripterus americanus[22,24,25]	Sea raven		
Liparis atlanticus[22,23]	Atlantic seasnail		
Macrozoarces americanus[22]	Ocean pout		
Menidia menidia[22]	Atlantic silverside		
Merluccius bilinearis[22,24]	Silver hake		
Microgadus tomcod[22,25]	Atlantic tomcod		
Myoxocephalus aenaeus[22,23,25]	Grubby		
Myoxocephalus octodecem-spinosus[22,24,25]	Longhorn sculpin		
Myoxocephalus scorpius[22–25]	Shorthorn sculpin		
Osmerus mordax[22,24,25]	American smelt	MCWCS: SGCN 2;[72] NMFS species of concern[74]	
Pholis fasciata[24]	Rock gunnel		
Pholis gunnellus[22,23]	Rock gunnel		
Pollachius virens[22,24,25]	Pollock		COSEWIC candidate Group 2[71]
Pleuronectes (=Pseudopleuronectes) americanus[22,24,25]	Winter flounder	USFWS priority species[70]	
Pungitius pungitius[22]	Ninespine stickleback		
Raja radiata[22]	Thorny skate		
Salmo salar[22,24]	Atlantic salmon	Endangered species status;[75] USFWS priority species;[70] MCWCS: SGCN 1[72]	Endangered (COSEWIC and SARA)[76]
Scomber scombrus[22,25]	Atlantic mackeral		
Squalus acanthias[24]	Dogfish		
Syngnathus fuscus[22]	Northern pipefish		

Continued

Table 1. *Continued*

Species	Common name	Conservation status United States	Conservation status Canada
Tautogolabrus adspersus[22,25]	Cunner		
Tautoga onitus[25]	Tautog		
Ulvaria subbifurcata[22]	Radiated shanny		
Urophycis tenuis[22,25]	White hake		
Birds[27]			
Somateria mollissima	Common eider	MCWCS: SGCN 2[72]	
Anas rubripes	American black duck	USFWS priority species;[70] MCWCS: SGCN 2[72]	
Anas platyrhynchos	Mallard duck		
Calidris maritima	Purple sandpiper	USFWS priority species;[70] MCWCS: SGCN 2;[72] U.S. Shorebird Plan Cat. 2 [77]	Canadian Shorebird Plan, Cat. 2[78]
Phalacrocorax auritus	Double-crested cormorant		
Pandion haliaetus	Osprey	USFWS priority species[70]	
Gavia immer	Common loon	USFWS priority species;[70] MCWCS: SGCN 2[72]	
Ardea herodias	Great blue heron	IFW: special concern;[73] MCWCS: SGCN 2[72]	
Larus philadephia	Bonaparte's gull	IFW special concern (breeding pop only)[73]	
Sterna hirundo	Terns, including common tern	IFW special concern;[73] USFWS priority species;[70] MCWCS: SGCN 2[72]	
Bucephala albeola	Bufflehead		
Melanitta spp.	Scoters		
Scolopacidae, including *Calidris pusilla*	Including semipalmated sandpiper	(*Calidris pusilla*) IFW special concern;[73] USFWS priority species;[70] MCWCS: SGCN 2;[72] U.S. Shorebird Plan cat. 3[77]	Canadian Shorebird Plan cat. 3[78]
Charadriidae, including *Charadrius semipalmatus*	Including semipalmated plover	(*C. semipalmatus*) U.S. Shorebird Plan cat. 2[77]	U.S./Canadian Shorebird Plan cat. 2[78]
	Kingfishers		
	Mergansers		
	Grebes		

heat, light, desiccation, and predation, allowing for enhanced intertidal survival of invertebrate fauna, especially at the higher tidal levels."[60] At high tide, the expanded floating plant canopy serves as a predation refuge for juvenile fishes and a feeding site for birds.[6,8] Furthermore, the reduced water velocity within rockweed beds facilitates the settlement and attachment of pelagic larvae of species such as barnacles and mussels.[11] Commercially important invertebrates such as lobster (*Homarus americanus*) (Cowan, 2011, personal communication to R.H. Seeley),[13,24,61,62] common periwinkles,[11] blue mussels,[11] soft-shell clams,[11] and bivalve spat[12] all use the shelter of rockweed stands. Noncommercial, but ecologically important, members of the rocky intertidal community include amphipods (food for shorebirds, ducks; and fish, such as pollock and herring[26]), mysid shrimp (food for commercially important fish, such as herring and cod[26] and longhorn sculpin[63]), native periwinkles (*Littorina obtusata* [grazer]), and dog winkles (*Nucella lapillus* [predator]).[11] *L. obtusata* also deposits egg masses on *Ascophyllum* fronds.

Lobsters (*H. americanus*) have been found in rockweed in the Maritimes[13,24,61] and in Maine (Cowan, 2011, personal communication to R.H. Seeley).[62] Lobsters in rockweed are noted in a Gulf of Maine Council on the Marine Environment report.[64] Although the reported density of lobsters found in rockweed is low,[13,24,61] the commercial importance of lobster to Maine's coastal economy,[65] and the observation that much of the habitat for juvenile lobsters is *Ascophyllum*-covered rocks (Cowan, 2011, personal communication to R.H. Seeley),[62] suggest that more research attention should be paid to day-and-night use of *Ascophyllum* beds by lobsters.

Few invertebrates in the NW Atlantic rocky intertidal zone consume live rockweed, because it is chemically well defended by polyphenols,[66] but in the NE Atlantic, *Ascophyllum* subsidizes growth of a key intertidal grazer, the limpet *Patella vulgata*.[67]

Thirty-four species of fish are found in *Ascophyllum* habitat (Table 1). Pollock are a particularly good example of a fish species using rockweed habitat. Pollock moved up to 200 m inshore with the tide for feeding and likely predator avoidance.[6,7] Juveniles appeared in the intertidal zone in early May.[7] Juvenile pollock prefer open intertidal habitat, where they school on falling tides, but prefer dense algal intertidal habitat, where they are solitary or in small shoals, on high-rising tides.[7] Juvenile pollock also feed in rockweed habitat, preying on invertebrates associated with rockweed: crustaceans (copepods, isopods, and amphipods) and small gastropods (*L. obtusata*) are the most widely consumed prey types.[6]

Many bird species, including those of conservation concern, use the rockweed zone as a habitat for feeding, reproduction, or sheltering (Table 1). The rockweed canopy is particularly important for Common Eider ducklings, who lack the ability to dive for food for about three weeks after hatching, and thus forage for amphipods in the canopy floating at the surface.[8,9,27] The *Ascophyllum* zone food web is stable, and it is likely that its complexity, with multiple trophic pathways, is responsible for this stability.[29] The rockweed canopy supports numerous trophic pathways and thus the stable functioning of the mid intertidal zone.[29] After simulated species removals (25%) from the rockweed community in Maritime Canada, food webs collapsed.[13] The rockweed food web was less robust than the food web in eelgrass.[13] In Iceland, Sarà *et al.* found that organic matter from *Ascophyllum* flows through the food web to top predators via predation or scavenging.[68]

Biogeochemistry

Most N enters Cobscook Bay during tidal exchange with the Gulf of Maine. Runoff from fertilizer, sewage disposal, and salmon aquaculture add a small amount of available N to the ambient concentrations.[79] Unfortunately, in recent years, red tides have resulted in closure of shellfish fisheries during the summer months, their occurrence largely blamed on excessive N concentrations in the estuary.[80] If its tissue N content averages about 1% (0.8%;[13] 1.15%[81]), the total production of rockweed in Cobscook Bay (6.3×109 gC/yr)[2] sequesters about 60 t/yr of N from the estuarine waters. Alternatively, Garside and Garside estimate the total tidal inflow of N at 70 t/day in Cobscook Bay.[79] Thus, the growth of rockweed can be expected to remove $<1\%$ of the N delivered to Cobscook Bay each year, converting it to organic forms. Nevertheless, the N content in rockweed beds is an important resource for higher trophic levels.[13]

Commercial-scale cutting of *Ascophyllum*

Despite its critical ecological role in coastal ecosystems, *Ascophyllum* is a target of a growing rockweed-processing industry. Rockweed beds are cut with hand-held cutter rakes, sickles, or mechanical cutter devices on boats and processed into fertilizers and agricultural growth stimulants (also dog food, cosmetics, and meat preservative) in Norway, Ireland, Scotland, France, Canada, and the United States (Maine). As a result of global demand for rockweed products, which are marketed as organic and "sustainable," the intensity of *Ascophyllum* (rockweed) cutting in the NW Atlantic (United States and Canada) has risen dramatically since 1986 (Maine Department of Marine Resources, personal communication to R.H. Seeley).[82] In Maine, rockweed landings have nearly doubled in the past five years, from 7.1 (2006) (Maine Department of Marine Resources, personal communication to R.H. Seeley) to 12.7 (2010) million pounds.[83] In the Maritimes, landings of rockweed have increased from ~9.9 (1984–1985, Ref. 82) to ~88.1 million pounds in 2010.[82] Rockweed cut in NB is capped by the Canadian government at ~26 million pounds (11,000 mt; Table 2).

The industrial-scale removal and processing of low trophic level species such as coastal seaweeds is an illustration of "fishing down the food web."[87] In fact, seaweed cutting is often conducted by fishers who are switching from other higher trophic level fisheries that have been depleted, such as groundfish, sea urchins, clams, and scallops.[85] Targeting low trophic level species, such as marine plants, risks impacting high trophic level species and could lead not only to widespread changes in ecosystems, but also to the collapse of traditional fisheries.[88]

History of rockweed harvesting

Ascophyllum has been used for fertilizer and other agricultural uses by coastal farmers and gardeners in New England (United States) at least since the 18th century.[89] *Ascophyllum* was so valuable to farmers that rights to rockweed on the shore were specifically mentioned in legal documents.[90]

The modern era of industrial-scale rockweed cutting began in 1959 in Nova Scotia[91] when the provincial government began area management. In the 1950s, *Ascophyllum* was used primarily as raw material for sodium alginate and "kelp" meal. After introduction of the highly efficient suction mechanical seaweed harvester in 1985,[92,93] rockweed landings jumped from 9,448 t in 1985 to 29,598 t in 1989.[92] Exploitation rates reached 95% of biomass,[94] resulting in overharvested areas.[95] Starting in 1988, areas of overharvest in SW Nova Scotia (Annapolis Basin, St Mary's Bay, Lobster Bay)[95] were closed. There was a mix of open areas (no limitations), exclusively licensed areas with true area management, and other areas that were not monitored.[96] Mechanical harvesting of rockweed was discontinued in Nova Scotia after the overharvest but is currently (2011) returning to Nova Scotia in the form of an experimental mechanical harvesting lease (Nova Scotia Department Fisheries and Agriculture, 2011, personal communication to R.H. Seeley). As recently as 2006, some areas of SW Nova Scotia were still being overharvested.[95]

In the early 1990s, even more rockweed was needed to meet demand, and in New Brunswick there were virgin stands of *Ascophyllum*.[86] New Brunswick was opened to *Ascophyllum* cutting for the first time in 1995,[96] but not without protest from fishing and conservation communities.[97] As the pilot harvest was starting in the island community of Grand Manan, the Grand Manan Fisherman's Association, Chamber of Commerce, Municipal Council, and representatives of traditional dulse and periwinkle harvesters formed the Island Rockweed Committee to protect Grand Manan's rockweed, asserting that "rockweed is as essential to their fisheries as topsoil is to farmers."[98] The citizens of Grand Manan viewed the rockweed cutting as a threat to their economic sustainability.[97] As a result of conservation concerns, a strict set of regulations was enacted in New Brunswick, including a cap on landings (originally 10,000 mt, but after industry reassessed total biomass, the cap was raised to 11,800 mt).[42] Mechanical harvesting is not currently permitted in NB (Table 2). According to the rockweed industry, NS harvest areas have now reached "maximum annual sustainable production."[82]

In Maine, commercial rockweed harvesting apparently started in 1950, when fishermen in Harpswell cut and shipped rockweed from Bailey Island.[99] A large Canadian company, Acadian Seaplants Ltd. (ASL), which is responsible for ~92% of the harvesting in the Maritimes,[82] expanded its cutting area from Canada into eastern Maine in the 1990s. ASL announced its intention to expand the rockweed harvest from eastern Maine to the rest of the Maine coast.[100] A 2009 Maine state law established the first and only rockweed management area in Maine (Cobscook Bay Rockweed Management Area).[84]

Regulation of rockweed harvesting

Rockweed cutting in New Brunswick and Nova Scotia (Canada) is managed more strictly than in Maine (Table 2). The most important points of difference are area management (leases and sectors), limits on take, closed conservation areas, and reporting (pre- and postharvest plan).

Outside the Cobscook Bay Rockweed Management Area, there is no area management (any person

Table 2. Rockweed harvest management in Maine (United States) and the Maritime Provinces (Canada)

	United States[84] ME outside CBRMA	United States [84] ME within CBRMA	Canada (Conservation Council of New Brunswick, Fundy Baykeeper and K. Watson, personal communication to R.H. Seeley) NB	Canada[85] NS
Harvest				
Management authority	State	State	Province and federal	Province and federal
Harvest method	Any	Any	Hand only	Hand only (one experimental mechanical harvest license)
Limit (tons)	None	Annual limit: 17% of available tons per sector	11,000 mt[42]	None
Area management in place?	No	Yes	Yes, lease areas	Yes, lease areas
Area management scheme	None	Sectors assigned to companies	Yes	Exclusive areas and open areas
Cutting height mandated	16″	16″	5″	5" in lease areas and nonleased areas
Required holdfast incidence in landings, by weight	Cutting height is 16" above holdfast	Cutting height is 16" above holdfast	<10%[86]	≤15%
Data and reporting				
Biomass study required prior to cut	No	Yes	Yes	Yes
Required biomass study conducted by?	Industry	Industry	Industry	
Conservation areas closed to cutting	Federal properties only: NPS, USFWS	Yes	Yes	None (the minister has the authority to designate conservation areas)
Formal opportunity for objections to cutting in a sector/lease area	No	No	Yes	Yes
Licenses and fees				
Harvester license required	Yes	Yes	Yes	Yes
Harvester license fee	US $58	US $58	$50 Can. personal fishers reg, $50 Can. for vessel license number	Federal (DFO); $100 Can.
Nonresident harvester license	Yes, allowed	Yes, allowed	No	No
Nonresident license fee	US $230	US $230	N/A	N/A
Lease fee required			No	Yes
Lease fee amount			NA	$608.24 Can.
Seaweed buyer fee	Yes	Yes	N/A	Yes
Nonresident buyers	Yes	Yes	No	No
Nonresident buyer fee	US $500	US $500	N/A	N/A
Royalty fees required	Yes	Yes	Yes	Yes
Royalty fees	US $1.50/wet ton	US $1.50/wet ton	Confidential	$2.25 Can/wet ton
Harvester reporting				
Management plan required prior to harvest	No	Yes	Yes	Yes
Postharvest report required	No	No	Yes	No
Postharvest business report required	No	No	No	Yes

Continued

Table 2. *Continued*

	United States[84] ME outside CBRMA	United States [84] ME within CBRMA	Canada (Conservation Council of New Brunswick, Fundy Baykeeper and K. Watson, personal communication to R.H. Seeley) NB	Canada[85] NS
Third-party verification of landings?	No	Required	Dockside monitoring program: tonnage. Third party certified by DFO monitors 10% of landings	No
Landings data reported by harvester	Yes	Yes	Yes	No
Landings data reported by company/leaseholder	No	No	Yes	Yes
Deadline for landings reporting	December of following year (e.g., 2010 take data must be reported by December 2011)	December of following year (e.g., 2010 take data must be reported by December 2011)	Feb. 2012 for 2011 season	January 15 of following year
Penalties for noncompliance	Summons and fine	Summons and fine	Ranges from verbal warning to company license revocation	License revocation

CBRMA, Cobscook Bay Rockweed Management Area.

licensed by the state can cut anywhere), non-Maine residents and non-U.S. citizens may obtain licenses, and a cutting height of 16″ is mandated but not effectively enforced (e.g., holdfasts in Maine landings[32] represent a 0″, not a 16″, cut height). Closed conservation or research areas are limited to federal property (USFWS and USNPS lands in Maine have been closed to commercial rockweed cutting since ~2000). Within the Cobscook Bay Rockweed Management Area,[84] there is an annual limit (17%) on take (based on the available rockweed biomass, which is estimated by industry and reported verbally to Department of Marine Resources staff [Maine Department of Marine Resources, personal communication to R.H. Seeley]), area management through assigned sectors, and state and private conservation areas closed to rockweed cutting. Unlike Canada, Maine allows mechanical harvesting anywhere in the state. Complicating rockweed management in Maine is private ownership of intertidal areas. This has led to a debate over whether or not there is a public trust right to remove seaweed from private shores. The Maine Department of Marine Resources issues permits for seaweed harvesting while not taking a position on the question of rockweed ownership.[83]

Impacts of rockweed cutting

Impacts of rockweed harvests on the rockweed community are indirect (e.g., through changing habitat architecture, food, or cover availability) or direct (removal of species in bycatch). Rockweed harvest also impacts the rockweed itself (Table 3).

Indirect impacts of rockweed harvesting on common eider (*Somateria mollissima*) ducklings stem from duckling feeding behavior. For about three weeks, ducklings are unable to dive for mussels and must forage in the floating rockweed canopy (for small prey such as amphipods[8]). During this time, ducklings are vulnerable to predation by bald eagles and great black backed gulls.[118,119] Rockweed cutting activity can disturb eider crèches at a critical time and removes rockweed canopy where the ducklings feed,[8] as a result of harvesters targeting the largest and heaviest clumps with most rockweed biomass in the upper canopy.[111,112] For this reason, researchers in New Brunswick recommended that the start of rockweed harvesting be delayed until early to mid-July.[8,27,120] Hamilton[8] also warned that rockweed cutting should not be conducted in a way that changes height and structure of the rockweed canopy, but both eider ducklings and harvesters target the canopy. The longest *Ascophyllum* plants, with

Table 3. Documented impacts after rockweed harvest to vertebrate, invertebrate, and algal species and to the rockweed community

Group 1	Group 2	Level of impact	Common name	Species	Impact
Vertebrate	Bird	Species	Common eider	*Somateria mollissima*	Duckling numbers decreased,[9] reduced feeding of ducks[28]
Vertebrate	Fish	Species	Cunner	*Tautogolabrus adspersus*	Reduced feeding[25]
Vertebrate	Fish	Community	Fishes: Atlantic tomcod, cunner, pollock, grubby, short-horned sculpin, long-horned sculpin, Atlantic sea raven, winter flounder	*Microgadus tomcod, Tautogolabrus adspersus, Pollachius virens, Myoxocephalus aeneus, Myoxocephalus scorpius, Myoxocephalus octodecemspinosus, Hemitripterus americanus, Pseudopleuronectes americanus*	Biomass reduced[25]
Invertebrate	Coelenterate	Species	Hydroid	*Dynamena pumila*	Reduced abundance[10]
Invertebrate	Crustacean	Species	Isopods	*Jaera marina, Idotea balthica*	Removed in bycatch[101,102]
Invertebrate	Crustacean	Species	Amphipods	*Gammarus, Marinogammarus*	Removed in bycatch[101,102]
Invertebrate	Crustacean	Species	Green crab	*Carcinus maenas*	Reduced abundance,[10] removed in bycatch[101,102]
Invertebrate	Crustacean	Species	Barnacle	*barnacle (larvae)*	Removed in bycatch[101,102]
Invertebrate	Crustacean	Species	Barnacle	*Balanus*	Reduced abundance[103]
Invertebrate	Crustacean	Taxonomic group	Crustaceans	*Planktonic crustaceans*	Site-specific: reduced abundance[25]
Invertebrate	Crustacean	Taxonomic group	Sediment meiofauna	*crustaceans*	Reduced abundance[103]
Invertebrate	Mollusk	Species	Blue mussel	*Mytilus edulis*	Removed in bycatch[101,102]
Invertebrate	Mollusk	Species	Bivalves	*Mya arenaria, Hiatella arctica*	Removed in bycatch[101,102]
Invertebrate	Mollusk	Species	Macoma clams	*Macoma*	Removed in bycatch[101,102]
Invertebrate	Mollusk	Species	Snails	*Littorina saxatilis, Lacuna vincta, Nucella lapillus*	Removed in bycatch[101,102]
Invertebrate	Mollusk	Species	Rough periwinkle	*Littorina saxatilis*[a]	Removed in bycatch[32]
Invertebrate	Mollusk	Species	Smooth periwinkle	*Littorina obtusata*	Removed in bycatch;[101,102] reduced abundance in winter[104]
Invertebrate	Mollusk	Species	Common periwinkle, wrinkle	*Littorina littorea*	Removed in bycatch[32,101,102,105]
Invertebrate	Mollusk	Species	Bivalve spat		Removed in bycatch[12,106]
Invertebrate	Sponge	Species	Sponge	*Hymeniacodon/Hymeniacidon*	Reduced abundance[103]
Invertebrate	Sponge	Species	Sponge	*Halichondria*	Reduced abundance[103]
Invertebrate	Worm	Taxonomic group	Bristle worms	*polychaetes*	Removed in bycatch[101,102]
Invertebrate	Worm	Taxonomic group	Aquatic earthworms	*oligochaetes*	Removed in bycatch[101,102]
Alga	Alga	Species	Encrusting algae	*Phymatolithon lenormandii*	Reduced abundance (% cover) or death[10]
Alga	Alga	Species	Encrusting algae	*Hildenbrundia rubra*	Reduced abundance[10]
Alga	Alga	Species	Wrack	*Fucus vesiculosus*	Removed in bycatch;[107] reduced abundance[10,103]
Alga	Alga	Species	Ephemeral algae	*Enteromorpha, Ulva,* unspecified	Increased abundance[108,103,104]
Alga	Alga	Species	Green alga	*Cladophora*	Increased abundance[103]
Alga	Alga	Species	Rockweed plant morphology	*Ascophyllum nodosum*	Shorter plants[10,42,109,28]
Alga	Alga	Species	Rockweed plant morphology	*Ascophyllum nodosum*	Increased number of laterals on stipe[10,34,103,110]
Alga	Alga	Species	Rockweed plant morphology	*Ascophyllum nodosum*	Decrease in proportion of reproductive laterals[44]

Continued

Table 3. *Continued*

Group 1	Group 2	Level of impact	Common name	Species	Impact
Alga	Alga	Species	Rockweed plant morphology	*Ascophyllum nodosum*	Holdfast loss[32,93,111,112,102,113,114]
Alga	Alga	Species	Rockweed	*Ascophyllum nodosum*	Reduced abundance (% cover)[10,103,104]
Alga	Alga	Population	Rockweed	*Ascophyllum nodosum*	Reduced biomass, length;[109] reduced % cover[10]
Alga	Alga	Population	Rockweed plant morphology	*Ascophyllum nodosum*	"Long-term changes" noted in harvesting areas[115]
Alga	Alga	Population	Rockweed beds	*Ascophyllum nodosum*	Decrease in biomass[116]
Alga	Alga	Population	Rockweed beds	*Ascophyllum nodosum*	Reduced structural complexity[10]
Alga	Alga	Population	Rockweed population structure	*Ascophyllum nodosum*	Population structure altered by long-term harvesting[115]
Alga	Alga	Population	Rockweed population structure	*Ascophyllum nodosum*	Population structure altered by mechanical harvesting[93]
Community	Community	Community	Rockweed community	*sessile invertebrates*	Site-specific: reduced abundance[25]
Community	Community	Community	Rockweed community	*invertebrates and algae*	Boulders: animal cover decreased 66%;[103] mean number of species reduced by 33%[103]
Community	Community	Community	Rockweed community	*invertebrates and algae*	Species richness reduced[10]
Physical environment			Sediment size		Sediment coarser[103]

[a]Questionable species identification based on reported numbers and sizes of *L. saxatilis*. High-spired *L. obtusata*[117] may have been misidentified as *L. saxatilis*. No voucher specimens from the study were retained.

a high canopy available to ducklings, are also the largest (most of the biomass in plants over 80 cm in length [height] are distal[111,112]) and thus more attractive to rockweed harvesters paid by the pound. Rockweed clumps over 130 cm that are cut are typically reduced up to 55% of length (height) and 78% of biomass.[109] Rockweed areas where common eiders breed are protected in some areas of New Brunswick and on some conservation lands in Maine (within the Cobscook Bay Rockweed Management Area, federal wildlife refuges, and Acadia National Park).

Allen notes that several factors, including rockweed harvesting, are likely acting in concert to limit eider populations in the Gulf of Maine.[118] Allen *et al.* note that Quahog Bay, Maine, which has suitable eider habitat, has no eiders.[119] Rockweed has been harvested in Quahog Bay for many years.[34]

Other indirect impacts on fish that use the rockweed zone have been detected. Black and Miller[25] found that a fish species abundant in rockweed, cunner (*Tautogolabrus adspersus*), consumed less food in intertidal areas where *Ascophyllum* was cut than in areas where *Ascophyllum* was intact. Total abundance of fish (numbers) in cut and intact rockweed areas did not differ, but fish biomass in areas from which rockweed was removed was significantly reduced.[25]

Impacts on the rockweed community as a result of simulated rockweed cutting were investigated in Maine by Fegley:[10] overall species richness declined and did not recover during the two-year study. In light of these results, it is noteworthy that the rockweed community in Brittany, France, a region with a long rockweed harvesting history, has reduced species diversity (50 taxa)[121] relative to the high species diversity reported for rockweed communities in Maine (100+ taxa).[11]

Rockweed cutting also significantly affected abundance of common intertidal species: green crabs (*Carcinus meanas*), common periwinkles (*L. littorea*), blue mussels (*Mytilus edulis*), limpets (*Tectura testudinalis*), a colonial hydroid (*Dynamena pumila*), blue mussel spat (*M. edulis* recruits), and barnacles (*Semibalanus balanoides*).[10]

Data on direct impacts of rockweed harvesting to the rockweed community are found in studies of bycatch conducted by government[101,102] and industry.[32] Species directly impacted through bycatch removal include *L. littorea, L. obtusata, L. saxatilis,*

C. maenas, M. edulis, D. pumila, Gammarus spp., *Marinogammarus* sp., *oligochaetes, polychaetes, Lacuna vincta, N. lapillus, Macoma, Mya arenaria, Hiatella arctica, Jaera marina, Idotea balthica,* barnacle larvae, and *Fucus vesiculosus* (Table 3). Bycatch of algal species in rockweed harvest has not been addressed[120] by Maine DMR resource managers, but reports of unwanted *Fucus* in the harvest landings have been published by industry,[107,122] indicating there is algal bycatch.

Rockweed harvesting has direct impacts on plant morphology (individual), rockweed clump, and rockweed bed structure (population). Before cutting, rockweed plants form an underwater forest (Fig. 2) up to 200 cm tall;[32,33] after cutting, plants are shorter (e.g., Maine's 16″ [40 cm] regulation; Table 2) and produce numerous lateral branches,[34,35] creating a much shorter rockweed "bush." Many have noted short- to medium-term effects on the plant itself of cutting rockweed, including reductions in rockweed length, an increase in the number of lateral branches, and the proportion of laterals that are reproductive (Table 3). Fegley noted that full recovery of *Ascophyllum* had not occurred even two years after cutting: cut rockweed was significantly shorter.[10] Rockweed harvesting can also result in a loss of habitat complexity, since the most complex part of the clump in the canopy is removed. Holdfast loss occurs when the cutter rake has a dull blade[93] and when harvesters cut in areas where rockweed is easily dislodged.[120,123] Long-term effects noted in Canada include an altered population structure[115] and effects on habitat (Table 3 and references therein).

The "17% of biomass" guideline has been adopted in rockweed harvest management in NB as a "precautionary" level of harvest.[124] In the Cobscook Bay Rockweed Management Area (ME), 17% of available biomass in each bay sector may be removed.[84] What is the source of this number, and should it continue to be used as a "precautionary" management tool? The 17% figure is derived from a Canadian estimate that 50% of biomass could be removed every three years, although the state of Maine reports that rockweed biomass recovers from harvest in 3–11 years.[83] Rockweed is cut annually, and assuming 50% regrowth in three years, ~17% could be removed per year.[125] In the late 1990s, government managers of the rockweed harvest in New Brunswick questioned whether the biomass removed should be 17% everywhere in Canada,[12] recognizing that 17% may not be the right amount for particular areas, "including beds that, because of the substrate, would be vulnerable to holdfast removal during harvest. . .[and] beds known to be used by young eider ducklings, and beds adjacent to designated protected areas. . .These areas. . .might be harvested at a different exploitation rate, following a schedule which takes into consideration use by various species, or might be excluded from the management plan entirely."[12] We agree, and note in addition to these concerns, that 17% (which is ~34% of net primary production [NPP]) is too high to sustain NPP levels (this paper; see below).

To evaluate sustainability, many variables need to be considered. Data on rates of regrowth of rockweed are one. The list of factors affecting rockweed growth rates is long, and includes plant age,[126] size,[31] wave exposure,[43,44] solar irradiance,[44,127] temperature,[127,128] position in the intertidal zone,[127,129] season of the year (related to seasonal variation in temperature and light[127,130]), and site.[129] The large number of factors affecting regrowth rates is one reason that rockweed management on wide geographic scale (e.g., the state of Maine) is problematic.

Holdfasts are the primary source of frond recruitment to the population each year,[111] since recruitment via sexual reproduction is limited.[131] The estimate of "% holdfasts" (% of wet weight of clumps with holdfasts attached[112]) has been assumed to represent rockweed plant mortality.[112] Mortality has been estimated at 4–15% for cutter rake harvests[111] and 20–36% for mechanical harvests.[111] Recent work indicates that hand cutter rakes actually remove 17% of the holdfast tissue area when rockweed plants with attached holdfasts are removed,[112] suggesting that rather than clump mortality, "% holdfast material" represents plant injury, since some holdfast tissue remains. However, the ability of the holdfast to regenerate lost tissue is unknown.[112] Industry points to the observation that naturally storm-cast rockweed in the strand line also contains holdfasts as indicative that the impact of rockweed harvests is small compared to natural events. However, naturally storm-cast rockweed collected on the shore had significantly smaller area of holdfast material attached compared to the harvested clumps.[112] Furthermore, the naturally cast clumps with holdfasts decompose in the wrack and provide a source of nutrients to the ecosystem and food for crustaceans and insects[50] and birds,[54] whereas harvested rockweed clumps are lost from the marine ecosystem.

It is not simple to compare the impact of hand cutter rake and mechanical harvesting. Holdfast mortality from cutter rake harvests depends on individual harvester maintenance of the rake.[132] Mechanical harvesters shift the structure of the rockweed population from bimodal to unimodal, an effect that appears to last for approximately three years.[93] In general, mechanical harvesting removes more holdfasts than hand harvesting,[111] and mechanical harvesters are capable of taking more rockweed biomass per tide (hand: 2–3 t per boat; mechanical: 50 t/day).[133]

Is rockweed cutting "sustainable?"

The challenge of determining sustainable rockweed harvests has been taken up by industry representatives, resource managers, and conservationists in countries along the North Atlantic for at least the past decade (United States[125] and Canada,[97,134] Scotland,[15] Northern Ireland,[135,136] Ireland,[16,116] and Iceland[137]). Governments grapple with the question of how much rockweed can be "sustainably" removed in a harvest, yet they lack the information needed to determine these limits. For example, as the rockweed cutting industry moved into Grand Manan Island, NB, the Canadian government set the goal of ensuring that "harvesting and processing are undertaken in an environmentally acceptable manner;"[97] however, it did not establish criteria for determining the acceptability or sustainability of harvests. Consequently, it had no way to assess whether harvests were sustainable or not.[97]

What variables should be included in a definition of sustainable rockweed harvests? We can start to answer this question by looking at the relationship between NPP and the consumption of biomass by native herbivores in terrestrial ecosystems. McNaughton *et al.* found that about 15% of aboveground primary productivity is typically consumed by herbivores in grasslands.[138] Similarly, Cebrian reported that about 10% of NPP is consumed in marine macroalgal communities, drawing on data from 28 studies, one of which examined fucoid algae in Canada.[139] Given these estimates, a commercial harvest in which 17% of rockweed biomass is removed (e.g., NB and Cobscook Bay Rockweed Management Area, ME) is excessive. This is because removing 17% of biomass removes 34% of NPP, given that the biomass turnover rate is typically about 0.5. This points out the serious shortcomings of using biomass recovery at a 17%/yr removal rate or using maximum sustainable yield (MSY) as a measure of sustainable harvests.[82,140–142]

Besides setting the removal rate at the right level, other critical parameters that need to be considered in defining an ecologically sustainable harvest include recovery of preharvest rockweed morphology, rockweed bed structure, rockweed community structure and function, and ecosystem function. There is agreement in the scientific community that identifying ecologically sustainable harvests will require impact studies that include cumulative impacts of annual harvests and impacts on a landscape scale.[88] Fegley hinted that the effects of repeated cutting on a large scale are likely to be larger than the effects documented in her smaller-scale experimental study and recommended landscape-scale studies to understand the full impact of commercial rockweed harvesting.[10] A cumulative impact assessment is necessary to achieve ecologically sustainable harvests,[143] but cumulative impacts are still unknown.[83,88]

In addition to problems defining a ecologically sustainable level of rockweed harvest and the challenge of understanding cumulative impacts on a landscape scale, there are three further difficulties in assessing an ecologically sustainable level of harvest: a complex web of interactions in a diverse rockweed community[29] creates challenges for direct harvest impact assessments;[22] the rockweed community (vertebrates, invertebrates, algae) varies in space and time,[10,22,144] creating site- and time-specific impacts,[10,25] and high-power statistical tests of impact will require large sample sizes or long-term studies,[22] or both.

The challenges described above result in a lack of adequate measures of the full impact of rockweed harvesting. The dearth of impact assessment, combined with the tendency of state and provincial governments to allow rockweed harvests unless and until negative impact information is available, has produced a situation in which rockweed harvests are intensifying even though critical scientific information is lacking. In contrast, participants at the 1999 rockweed workshop in St. Andrews, New Brunswick, suggested that independent studies demonstrating no long-term impacts should be required prior to the start of rockweed harvesting, rather the procedure usually in place in which scientists bear the burden of demonstrating negative impacts before large-scale harvests are curtailed.[27]

Sustainability concerns in the NE Atlantic

The Scottish Environmental Protection Agency (SEPA) has stated that "SEPA's primary concern would be to see that any harvesting is undertaken in a manner that is sustainable and does not therefore harm the ecosystem" (Baird).[15] Scottish National Heritage concluded that assessments of sustainable rockweed harvest show "ecosystem as well as *Ascophyllum* recovery. . .such studies would have needed a much longer term study."[15] In Northern Ireland, the Environment and Heritage Service (EHS, currently the Department of Environment) refused to support mechanical harvesting "unless it can be demonstrated that it will not have an adverse impact on the environment."[135] The EHS proposed an Environmentally Sustainable Seaweed Harvesting Code of Conduct,[135] including consent from landowners, preharvesting plan, rotation cycles, harvesting methods, environmental protection measures, and harvest records. The Code of Conduct was proposed to industry but never adopted. With respect to sustainability, the final EHS report concluded that "there is a lack of specific information on the carrying capacity of marine ecosystems to support seaweed harvesting."[136] In Ireland, there is concern about the impact of repeat harvesting and a recommendation that mechanical harvesting be prohibited.[16]

Once a sustainable level of harvest is determined for an area, strict enforcement of regulations is necessary, because the cutter with the rake will be biased toward immediate benefit rather than long-term health of the rockweed.[113,145]

Is *Ascophyllum* a species under threat?

The old-growth rockweed forest is already at risk from warming ocean temperatures[122] and pollution, which has led to the loss of *Ascophyllum* in the Baltic.[146,108] There is evidence that *Ascophyllum* is slowly being replaced by *Fucus*[107] in the NW Atlantic, as *Chondrus* was replaced by *Furcellaria* in the Maritime provinces.[147] *Chondrus* raking was also once a thriving industry in Maritime Canada and the Gulf of Maine, but as a percentage of total seaweed landed, *Chondrus* is now near zero.[147] *Ascophyllum* stands may, in fact, be relicts, established prior to the arrival of the common periwinkle, *L. littorea*.[148]

Landed Maine seaweed, primarily rockweed (∼92–96% of landings since 2006 [Maine Department of Marine Resources, personal communication to R.H. Seeley]), is valued at two cents a pound, ranked last (by value) in the list of commercial harvests listed by the ME DMR.[149] Millions of pounds of rockweed are removed annually from the Maine coast. Without ecologically appropriate sustainability data, the level of rockweed harvests placing at risk species much higher on Maine's commercial value ranking—including lobster, scallop, clam, periwinkle, and groundfish (pollock, herring, flounder)—is unknown. Maine's management of rockweed harvesting, outside of the Cobscook Bay Rockweed Management Area, seems particularly weak compared to Canada's much stricter requirements. The rockweed industry in most areas of Maine is left to manage itself.

There is huge potential for mismanagement of rockweed harvesting, with significant ecosystem consequences. We are pessimistic about any improvement in Maine's management of rockweed harvests as they intensify, given declining state resources. For this reason, it would be prudent to curtail rockweed harvests until appropriate resources are found to devote to rockweed management, including landscape-scaled and long-term studies of community impacts of cutting rockweed and corresponding design and enforcement of regulations that protect rockweed habitat. Despite its protected status under Maine's NRPA, "Significant Wildlife Habitat" (including "seaweed communities") of special value to wading birds, shorebirds, and ducks is harvested for rockweed, as is areas of "High Value Habitat for Priority Trust Species" (e.g., the Jonesport/Beals area[52]).

There is no evidence that rockweed cutting in Maine and the Maritime provinces is ecologically sustainable, but there is a long history in fisheries management of allowing resource extraction when scientific information is missing until a crisis of overharvest occurs.[150] Should resource extractors be required to show evidence of ecological sustainability before commercial extraction of low trophic level, habitat-forming seaweeds is permitted? Or is it acceptable and wise to permit the extraction despite missing scientific information?[150]

Conclusions

(1) Rockweed has critical value as habitat, as food, and as a nutrient source supporting a community of over 150 other organisms in Maine and the Maritime Provinces (Ref. 11; Table 1).

(2) Cutting rockweed has documented impacts on the alga itself and on the rockweed community as a whole (Table 3).

(3) The current metric for "sustainable" harvests—MSY—is inappropriately narrow. A metric for an ecologically sustainable harvest must be based on the data from large-scale, long-term studies of postharvest recovery of rockweed morphology, of rockweed community structure and function, and of ecosystem impacts. Until this metric is developed and enforceable regulations based on it are developed, commercial-scale rockweed cutting should not be permitted.

As Smith *et al.*[151] have recently reported in a study of global fisheries, removing low-trophic level species that constitute a high proportion of the biomass in the ecosystem or are highly connected in the food web at standard levels of MSY can have large impacts on the rest of the ecosystem, including commercially valuable fish. *Ascophyllum* fits all three criteria (low-trophic level, high proportion of biomass,[3] highly connected in the food web[13,29]). Therefore, Smith *et al.*'s[151] findings serve as a warning about the dangers of failing to manage rockweed and other wild seaweeds sustainably.

Acknowledgments

R. H. S. gratefully acknowledges support from a TogetherGreen Conservation Leadership Fellowship and from W. E. Bemis (Shoals Marine Laboratory) during the writing of this paper. We appreciate the comments of an anonymous reviewer and K. Ross, critical bibliographic research conducted by V. Constant, and N and P analyses run by A. Townsend.

Conflicts of interest

The authors declare no conflicts of interest.

References

1. Chung, I.K., J. Beardall, S. Mehta, *et al.* 2011. Using marine macroalgae for carbon sequestration: a critical appraisal. *J. Appl. Phycol.* **23:** 877–886.
2. Gollety, C., A. Migné & D. Davoult. 2008. Benthic metabolism on a sheltered rocky shore: role of the canopy in the carbon budget. *J. Phycol.* **44:** 1146–1153.
3. Vadas, R.L., W.A. Wright & B.F. Beal. 2004. Ecosystem modelling in Cobscook Bay, Maine: a boreal, macrotidal estuary. *Northeastern Naturalist* **11:** 123–142.
4. Campbell, D. E. 2004. Ecosystem modelling in Cobscook Bay, Maine: a boreal, macrotidal estuary. *Northeastern Naturalist* **11:** 355–424.
5. Levin, P.S. & M.E. Hay. 1996. Responses of temperate reef fishes to alterations in algal structure and species composition. *Mar. Ecol. Prog. Ser.* **134:** 37–47.
6. Rangeley, R.W. & D.L. Kramer. 1995. Tidal effects on habitat selection and aggregation by juvenile pollock *Pollachius virens* in the rocky intertidal zone. *Mar. Ecol. Prog. Ser.* **126:** 19–29.
7. Rangeley, R.W. & D.L. Kramer. 1995. Use of rocky intertidal habitats by juvenile pollock *Pollachius virens. Mar. Ecol. Prog. Ser.* **126:** 9–17.
8. Hamilton, D.J. 2001. Feeding behavior of common eider ducklings in relation to availability of rockweed habitat and duckling age. *Waterbirds* **24:** 233–241.
9. Blinn, B.M., A.W. Diamond & D.J. Hamilton. 2008. Factors affecting selection of brood-rearing habitat by Common Eiders (*Somateria mollissima*) in the Bay of Fundy, New Brunswick, Canada. *Waterbirds* **31:** 520–529.
10. Fegley, J. 2001. Ecological Implications of Rockweed, *Ascophyllum nodosum*, (L.) Le Jolis, Harvesting. PhD Thesis, University of Maine (USA).
11. Larsen, P.F. 2010. The macroinvertebrate fauna of rockweed (*Ascophyllum nodosum*)–dominated low-energy rocky shores of the northern Gulf of Maine. *J. Coast. Res. online.* doi: 10.2112/JCOASTRES-D-10-00004.1
12. DFO. 1999. The impact of the rockweed harvest on the habitat of southwest New Brunswick. DFO Maritimes Regional Habitat Status Report 99/2E.
13. Schmidt, A.L., M. Coll, T.N. Romanuk & H.K. Lotze. 2011. Ecosystem structure and services in eelgrass *Zostera marina* and rockweed *Ascophyllum nodosum* habitats. *Mar. Ecol. Prog. Ser.* **437:** 51–68.
14. McHugh, D.J. 2003. A guide to the seaweed industry. FAO Fisheries Technical Paper, No. 441.
15. Burrows, M.T., M. Macleod & K. Orr. 2010. Mapping the intertidal seaweed resources of the Outer Hebrides. Scottish Association for Marine Science Internal Report No. 269.
16. McLaughlin, E., J. Kelly, D. Birkett, *et al.* 2006. Assessment of the effects of commercial seaweed harvesting on intertidal and subtidal ecology in Northern Ireland. Environment and Heritage Service Research and Development Series. No. 06.
17. Burt, M. 2000. The challenge: why should we care about rockweed? Gulf of Maine rockweed: management in the face of scientific uncertainty. In *Proceedings of the Global Programme of Action Coalition for the Gulf of Maine (GPAC) workshop*, Huntsman Marine Science Centre, St. Andrews, New Brunswick, December 5–7, 1999. R.W. Rangeley & J. Davies, Eds. Huntsman Marine Science Centre Occasional Report No. 00/1: 6–9.
18. Ugarte, R. 2000. Harvest and management strategies of seaweed: a global and local perspective. Gulf of Maine rockweed: management in the face of scientific uncertainty. *Proceedings of the Global Programme of Action Coalition for the Gulf of Maine (GPAC) Workshop*, Huntsman Marine

Science Centre, St. Andrews, New Brunswick, December 5–7, 1999. R.W. Rangeley & J. Davies, Eds. Huntsman Marine Science Centre Occasional Report No. 00/1: 38.
19. Bates, C.R., G.W. Saunders & T. Chopin. 2009. Historical versus contemporary measures of seaweed biodiversity in the Bay of Fundy. *Botany* **87:** 1066–1076.
20. Mann, K.H. 2000. *Ecology of Coastal Waters with Implications for Management*. Blackwell Science. Malden, MA.
21. Carson, R. 1955. *The Edge of the Sea*. Houghton Mifflin. Boston, MA.
22. Rangeley, R.W. 2000. Aquatic macrophytes as foraging and refuging habitats for fishes. Gulf of Maine rockweed: management in the face of scientific uncertainty. *Proceedings of the Global Programme of St. Andrews*, New Brunswick, December 5–7, 1999. R.W. Rangeley & J. Davies, Eds. Huntsman Marine Science Centre Occasional Report No. 00/1: 18–24.
23. Gullo, A. 2002. The value of rockweed (*Ascophyllum nodosum*) as habitat for tidepool fishes. MS Thesis, University of Maine (USA).
24. Corrigan, S.E. & R.A. Curry. 2002. Fish community of rockweed beds in the Bay of Fundy. New Brunswick Cooperative Fish and Wildlife Research Unit Report 02-01.
25. Black, R. & R.J. Miller. 1991. Use of the intertidal zone by fish in Nova Scotia. *Env. Biol. Fishes* **31:** 109–121.
26. Bowman, R.E., C.E. Stillwill, W.L. Michaels & M.D. Grosslein. 2000. Food of Northwest Atlantic fishes and two common species of squid. US Department of Commerce, NOAA Tech Memo NMFS NE 155.
27. Hamilton, D.J. 2000. Community-level interactions between birds and aquatic macrophytes: lessons for a rockweed harvest? Gulf of Maine rockweed: management in the face of scientific uncertainty. *Proceedings of the Global Programme of Action Coalition for the Gulf of Maine (GPAC) Workshop*, Huntsman Marine Science Centre, St. Andrews, New Brunswick, December 5–7, 1999. R.W. Rangeley & J. Davies, Eds. Huntsman Marine Science Centre Occasional Report No. 00/1: 25–30.
28. Hamilton, D.J. 2003. Effects of predation by common eiders (*Somateria mollissima*) in an intertidal rockweed bed relative to an adjacent mussel bed. *Mar. Biol.* **142:** 1–12.
29. Golléty, C., P. Riera & D. Davoult. 2010. Complexity of the food web structure of the *Ascophyllum nodosum* zone evidenced by a δ13C and δ15N study. *J. Sea Res.* **64:** 304–312.
30. Olsen, J.L., F.W. Zechman, G. Hoarau, *et al.* 2010. The phylogeographic architecture of the fucoid seaweed *Ascophyllum nodosum*: an intertidal "marine tree" and survivor of more than one glacial-interglacial cycle. *J. Biogeogr.* **37:** 842–856.
31. Aberg, P. 1992. Size based demography of the seaweed *Ascophyllum nodosum* in stochastic environments. *Ecology* **73:** 1488–1501.
32. Ugarte, R., C. Bartlett & L. Perry. 2010. A preliminary study to monitor periwinkle by-catch and incidence of holdfasts in harvested rockweed, *Ascophyllum nodosum*, from Cobscook Bay, Maine. Unpublished report.
33. Crawford, S. 1999. Results of a rockweed biomass inventory of Cobscook Bay, conducted by Quoddy Spill Prevention Group. Unpublished.
34. Fegley, J.C. 2006. Morphological, population and biomass studies of rockweed (*Ascophyllum nodosum*) in Quahog Bay and Taunton Bay. Unpublished report.
35. Fegley, J.C. & R.L. Vadas. 2001. A quantitative assessment of the rockweed (*Ascophyllum nodosum*) resource at selected sites along the coast of Maine. Final report. Unpublished.
36. Petit, R.J. & A. Hampe. 2006. Some evolutionary consequences of being a tree. *Ann. Rev. Ecol. Evol. Syst.* **37:** 187–214.
37. Wippelhauser, G. 1996. Ecology and management of Maine's eel grass, rockweeds and kelps. Maine Natural Areas Program, Department of Conservation.
38. Morton, O. 2011. Northern Ireland priority species. *Ascophyllum nodosum*. Retrieved February 1, 2012, from http://www.habitas.org.uk/priority/species.asp?item=660
39. Hill, J. & N. White. 2008. *Ascophyllum nodosum*. Knotted wrack. Marine Life Information Network: Biology and Sensitivity Key Information Sub-programme [online]. Plymouth: Marine Biological Association of the United Kingdom. Retrieved February 1, 2012, from http://www.marlin.ac.uk/speciessensitivity.php?speciesID=2632.
40. Gulf of Maine Council (GOMC) on the Marine Environment. 2005. Marine habitats in the Gulf of Maine: assessing human impacts and developing management strategies.
41. Trott, T. & P.F. Larsen. Evaluation of short-term changes in rockweed (*Ascophyllum nodosum*) and associated epifaunal communities following cutter rake harvesting in Maine. Unpublished report to ME Department of Marine Resources.
42. Sutherland, B. 2005. *An Independent Study and Review of the New Brunswick Rockweed Harvest - Phase 2*. Eastern Charlotte Waterways Inc. Black Harbour, NB.
43. Keser, M., R.L. Vadas & B.R. Larson. 1981. Regrowth of *Ascophyllum nodosum* and *Fucus vesiculosus* under various harvesting regimes in Maine USA. *Bot. Mar.* **24:** 29–38.
44. Cousens, R. 1985. Frond size distributions and the effects of the algal canopy on the behavior of *Ascophyllum nodosum* (L.) LeJolis. *J. Exp. Mar. Biol. Ecol.* **92:** 231–249.
45. Schlesinger, W.H. 1997. *Biogeochemistry: An Analysis of Global Change*. Academic Press/Elsevier. San Diego, CA.
46. Josselyn, M.N. & A.C. Mathieson. 1978. Contribution of receptacles from the fucoid *Ascophyllum nodosum* to the detrital pool of a north temperate estuary. *Estuaries* **1:** 258–261.
47. Cranford, P.J. & J. Grant. 1990. Particle clearance and absorption of phytoplankton and detritus by the sea scallop *Placopecten magellanicus* (Gmelin). *J. Exp. Mar. Biol. Ecol.* **137:** 105–121.
48. Northeast Fisheries Science Center. Kelly, K. 2010. Appendix B5: Results from Maine sea scallop surveys, 2002-2008. B. Atlantic sea scallop stock assessment for 2010. From Northeast Fisheries Science Center. 2010. 50th Northeast Regional Stock Assessment Workshop (50th SAW) Assessment Report. US Dept Commerce, Northeast Fisheries Science Center Reference Document 10–17.
49. Ahearn, K. 2005. Cobscook bay sea scallops: the fishery and markets. Unpublished report for Cobscook Bay Resource Center, Eastport, ME.

50. Bradford, B. 1989. A demonstration of possible links for a detrital pathway from intertidal macro-algae in the Bay of Fundy. MSc Thesis, Acadia University, Wolfville, NS.
51. Maine Revised Statutes. Title 38. Chapter 3. Subchapter 1. Article 5-A: Natural Resource Protection Act. § 480-B. Retrieved February 1, 2012, from http://www.mainelegislature.org/legis/statutes/38/title38sec480-B.html
52. Beginning with habitat: high value plant & animal habitats. Primary map 2. Town of Jonesport. April 8, 2010. Retrieved February 1, 2012, from www.beginningwithhabitat.org/the_maps/pdfs/Jonesport/Map2.pdf.
53. Dugan, J.E., D.M. Hubbard, M.D. McCrary & M.O. Pierson. 2003. The response of macrofauna communities and shorebirds to macrophyte wrack subsidies on exposed sandy beaches of southern California. *Estuar. Coast. Shelf Sci.* **58S:** 25–40.
54. Maine Department of Inland Fisheries and Wildlife. Maine Inland Fisheries and Wildlife Outdoors Report, March 21, 2009. Retrieved February 2, 2012, from http://www.dextermaine.com/dexterlakes/news/ifwweekly.html.
55. Kingsford, M.J. 1995. Drift algae: a contribution to near-shore habitat complexity in the pelagic environment and an attractant for fish. *Mar. Ecol. Prog. Ser.* **116:** 297–301.
56. Locke, A. & S. Corey. 1989. Amphipods, isopods and surface currents: a case for passive dispersal in the Bay of Fundy, Canada. *J. Plank. Res.* **11:** 419–430.
57. Ingolfsson, A. 1995. Floating clumps of seaweed around Iceland: natural microcosms and a means of dispersal for shore fauna. *Mar. Biol.* **122:** 13–21.
58. Johnson, S.C. & R.E. Scheibling. 1987. Structure and dynamics of epifaunal assemblages on the intertidal macroalga *Ascophyllum nodosum* and *Fucus vesiculosus* in Nova Scotia, Canada. *Mar. Ecol. Prog. Ser.* **37:** 209–227.
59. Vadas, R.L. 2000. Impacts of grazers on rockweed and macroalgae. Gulf of Maine rockweed: management in the face of scientific uncertainty. *Proceedings of the Global Programme of Action Coalition for the Gulf of Maine (GPAC) Workshop*, Huntsman Marine Science Centre, St. Andrews, New Brunswick, December 5–7, 1999. R.W. Rangeley & J. Davies, Eds. Huntsman Marine Science Centre Occasional Report No. 00/1: 12.
60. Bertness, M.D., G.H. Leonard, J.M. Levine, *et al.* 1999. Testing the relative contribution of positive and negative interactions in rocky intertidal communities. *Ecology* **80:** 2711–2726.
61. Sharp, G.J., I. Barkhouse & R. Semple. 1999. Abundance of canopy invertebrates in harvested and unharvested *Ascophyllum nodosum* clumps. RAP Working Paper.
62. Cowan, D.F. 1999. Method for assessing relative abundance, size distribution, and growth of recently settled and early juvenile lobsters (*Homarus americanus*) in the lower intertidal zone. *J. Crust. Biol.* **19:** 738–751.
63. Hacunda, J.S. 1981. Trophic relationships among demersal fishes in a coastal area of the Gulf of Maine. *Fish. Bull.* **79:** 775–788.
64. Gulf of Maine Council Habitat Restoration Subcommittee. 2004. The Gulf of Maine Habitat Restoration Strategy. Gulf of Maine Council on the Marine Environment.
65. Steneck,R.S., T.P. Hughes, J.E. Cinner, *et al.* 2011. Creation of a gilded trap by the high economic value of the Maine lobster fishery. *Cons. Biol.* **25:** 904–912.
66. Pavia, H. & G.B. Toth. 2000. Inducible chemical resistance to herbivory in the brown seaweed *Ascophyllum nodosum*. *Ecology* **81:** 3212–3225.
67. Davies, A.J., M.P. Johnson & C.A. Maggs. 2008. Subsidy by *Ascophyllum nodosum* increases growth rate and survivorship of *Patella vulgata*. *Mar. Ecol. Prog. Ser.* **366:** 43–48.
68. Sarà, G., M. DePirro, C. Romano, *et al.* 2007. Sources of organic matter for intertidal consumers on *Ascophyllum* shores (SW Iceland): a multi-stable isotope approach. *Helg. Mar. Res.* **61:** 297–302.
69. NOAA, NMFS. Alewife (*Alosa pseudoharengus*). Status. Retrieved February 1, 2012, from http://www.nmfs.noaa.gov/pr/species/fish/alewife.htm.
70. USFWS. Gulf of Maine Coastal Program. Identifying, ranking and mapping habitat for priority trust species in the Gulf of Maine watershed. Retrieved February 1, 2012, from http://www.fws.gov/GOMCP/pdfs/gomanalysis08.pdf
71. COSEWIC. Candidate Wildlife Species. Retrieved February 1, 2012, from http://www.cosewic.gc.ca/eng/sct3/index_e.cfm.
72. Maine's Comprehensive Wildlife Conservation Strategy, Chapter 4: Key habitats and natural communities. Retrieved February 1, 2012, from www.maine.gov/ifw/wildlife/groupsprogram_s/comprehensive_strategy/pdfs/chapter4.pdf.
73. State of Maine, Department of Inland Fisheries and Wildlife. Species of Special Concern. March 17, 2011. Retrieved February 2, 2012, from http://www.maine.gov/ifw/wildlife/species/endangered_species/specialconcern.htm.
74. NOAA, NMFS. Proactive Conservation Program: species of concern. Retrieved February 1, 2012, from http://www.nmfs.noaa.gov/pr/species/concern/#list.
75. NOAA, NMFS. Atlantic salmon (*Salmo salar*). Status. Retrieved February 1, 2012, from http://www.nmfs.noaa.gov/pr/species/fish/atlanticsalmon.htm.
76. Environment Canada. Species at risk. A to Z index. Retrieved February 1, 2012, from http://www.sararegistry.gc.ca/sar/index/default_e.cfm.
77. US shorebird plan: Brown, S., C. Hickey, B. Harrington & R. Gill, Eds. 2001. *The U.S. Shorebird Conservation Plan*, 2nd ed. Manomet Center for Conservation Sciences. Manomet, MA. Retrieved February 1, 2012, from http://www.fws.gov/shorebirdplan/USShorebird/PlanDocuments.htm.
78. Canadian shorebird plan: Donaldson, G.M., C. Hyslop, R.I.G. Morrison, *et al.* 2000. *Canadian shorebird conservation plan.* Special Publication, Canadian Wildlife Service, Environment Canada, Ottawa. Retrieved February 1, 2012, from http://www.ec.gc.ca/Publications/default.asp?lang=En&xml=4A90A2;A1-1260-41CC-B4F2-4E736D6F6E0E.
79. Garside, C. & J.C. Garside. 2004. Nutrient sources and distributions in Cobscook Bay. *Northeastern Naturalist* **11:** 75–86.
80. Townsend, D.W., N.R. Pettigrew & A.C. Thomas. 2001. Offshore blooms of the red tide dinoflagellate, *Alexandrium* sp. in the Gulf of Maine. *Cont. Shelf Res.* **21:** 347–369.
81. Schlesinger, W.H. 2011. Unpublished data.

82. Ugarte, R. & G.J. Sharp. 2011. Management and production of the brown algae *Ascophyllum nodosum* in the Canadian maritimes. *J. Appl. Phycol.* doi: 10.1007/s10811-011-9753-5.
83. Maine Department of Marine Resources. 2011. Rockweed: ecology, industry, management. Retrieved February 1, 2012, from www.maine.gov/dmr/rm/rockweed/2011facts.pdf.
84. State of Maine regulations governing the harvesting of rockweed. Retrieved February 1, 2012, from www.maine.gov/dmr/lawsandregs/regs/29.pdf. State of Maine law on harvesting of rockweed in the Cobscook Bay Management Area (LD 345 as amended). Retrieved February 1, 2012, from http://www.mainelegislature.org/legis/bills/bills_124th/billtexts/SP010902.asp.
85. Lotze, H.K, I. Milewski & B. Worm. 2004. Two hundred years of ecosystem change in the outer Bay of Fundy. Part I. Changes in species and the food web. In *Health of the Bay of Fundy: Assessing Key Issues*. Environment Canada. Atlantic Region, Occasional report. **21**. P.G. Wells, G.R. Daborn, J.A. Percy, J. Harvey & S.J. Rolston, Eds.: 320–326.
86. Ugarte, R. 2007. Review of the management review of rockweed (*Ascophyllum nodosum*) harvesting in New Brunswick after a decade of its initiation. In Challenges in environmental management in the Bay of Fundy-Gulf of Maine. *Proceedings of the 7th Bay of Fundy Science Workshop*, St. Andrews, New Brunswick, 24–27 October 2006. Bay of Fundy Ecosystem Partnership Technical Report No. 3. G.W. Pohle, P.G. Wells & S.J. Rolston, Eds.: 108–116. Bay of Fundy Ecosystem Partnership, Wolfville, NS.
87. Pauly, D., V. Christensen, J. Dalsgaard, *et al.* 1998. Fishing down marine food webs. *Science* **279:** 860–863.
88. Lotze, H.K. & I. Milewski. 2004. Two centuries of multiple human impacts and successive changes in a North Atlantic food web. *Ecol. Appl.* **14:** 1428–1447.
89. Deane, S. 1790. *The New England Farmer; or Georgical Dictionary, Containing a Compendious Account of the Ways and Methods in Which the Most Important Art of Husbandry in All Its Branches Is, or May Be, Practised to the Greatest Advantage in This Country*. Wells and Lilly. Boston, MA.
90. Mason, G.C. 1876. Nicholas Easton v. The City of Newport. *Proceedings of the Rhode Island Historical Society*. 1876–1877: 15–17.
91. Anon. 1959. Common form of seaweed now big business. *Montreal Gazette*. June 23.
92. Milewski, I. 1996. Remarkable rockweed. *Island J.* **12:** 74–79.
93. Ang, P.O., G.J. Sharp & R.E. Semple. 1993. Change in the population structure of *Ascophyllum nodosum* (L.) Le Jolis due to mechanical harvesting. *Hydrobiologia* **260/261:** 321–326.
94. Sharp, G.J., P. Ang, Jr. & D. MacKinnon. 1994. Rockweed (*Ascophyllum nodosum* (L.) Le Jolis) harvesting in Nova Scotia: its socioeconomic and biological implications for coastal zone management. *Proceedings of the Coastal Zone Canada* **'94**: 1632–1644.
95. Anderson, S. 2006. How sustainable are emerging low-trophic level fisheries on the Scotian Shelf? Honours Bachelor of Science Thesis, Dalhousie University, Halifax NS Canada.
96. Ugarte, R. & G.J. Sharp. 2001. A new approach to seaweed management in Eastern Canada: the case of *Ascophyllum nodosum*. *Cah. Biol. Mar.* **42:** 63–70.
97. Marshall, J. 1999. Bitter harvest. *Altern. J.* **25:** 10–15.
98. Coon, D. 2000. In Welcome and introduction comments. Gulf of Maine rockweed: management in the face of scientific uncertainty. *Proceedings of the Global Programme of Action Coalition for the Gulf of Maine (GPAC) Workshop*, Huntsman Marine Science Centre, St. Andrews, New Brunswick, December 5–7, 1999. R.W. Rangeley & J. Davies, Eds.: 4. Huntsman Marine Science Centre Occasional Report No. 00/1.
99. Anon. 1950. Harpswell lobstermen to rake rockweed. *Portland Press Herald*; 15 April, p. 4.
100. Anon. 2009. Seaweed harvest dispute swirling. *Bangor Daily News*, 14 September 2009.
101. McEachreon, T. 1999. Compliance monitoring in the New Brunswick Rockweed Fishery (1996-1998). Stock Status report for Regional Advisory Process meeting.
102. McEachreon, T. 2000. In Compliance monitoring in the New Brunswick rockweed fishery (1996-1998). Gulf of Maine rockweed: management in the face of scientific uncertainty. *Proceedings of the Global Programme of Action Coalition for the Gulf of Maine (GPAC) Workshop*, Huntsman Marine Science Centre, St. Andrews, New Brunswick, December 5–7, 1999. R.W. Rangeley & J. Davies, Eds.: 64—68. Huntsman Marine Science Centre Occasional Report No. 00/1.
103. Boaden, P.J.S. & M.T. Dring. 1980. A quantitative evaluation of the effects of *Ascophyllum* harvesting on the littoral ecosystem. *Helgo. Meeres.* **33:** 700–710.
104. Kelly, L., L. Collier, M.J. Costello, *et al.* 2001. Impact assessment of hand and mechanical harvesting of *Ascophyllum nodosum* on regeneration and biodiversity. Marine Fisheries Services Division, Marine Institute, Abbotstown, Dublin 15.
105. Sharp, G.J., R.Semple & T. MacEachreon. 1998. Rockweed and periwinkle harvests: conflict or complement? In Coastal monitoring and the Bay of Fundy. *Proceedings of the Maritime Atlantic Ecozine Science Workshop* held in St. Andrews NB, Nov 11–15, 1997. M.D.B. Burt & P.G. Wells, Eds. Huntsman Marine Science Center. St Andrews, NB.
106. DFO. 1998. Periwinkle (*Littorina littorea*). DFO stock status report C3-46.
107. Ugarte, R., A. Critchley, A.R. Serdynska & J. P. Deveau. 2009. Changes in composition of rockweed (*Ascophyllum nodosum*) beds due to possible recent increase in sea temperature in eastern Canada. *J. Appl. Phycol.* **21:** 591–598.
108. Worm, B. & H.K. Lotze. 2000. In Nutrient pollution, low-trophic level harvesting and cumulative human impact on coastal ecosystems. Gulf of Maine rockweed: management in the face of scientific uncertainty. *Proceedings of the Global Programme of Action Coalition for the Gulf of Maine (GPAC) Workshop*, Huntsman Marine Science Centre, St. Andrews, New Brunswick, December 5–7, 1999. R.W. Rangeley & J. Davies, Eds.: Huntsman Marine Science Centre Occasional Report No. 00/1: 40–41.
109. Ugarte, R., G.J. Sharp & B. Moore. 2006. Changes in the brown seaweed *Ascophyllum nodosum* (Le Jolis) plant

structure and biomass produced by cutter rake harvests in southern New Brunswick. *J. Appl. Phycol.* **18**: 351–359.
110. Kerin, B.K. 1995. Impact of harvesting on the nitrogen, phosphorus and carbon contents of the brown alga, *Ascophyllum nodosum* (L.) Le Jolis (rockweed). MS Thesis, University of New Brunswick, St John.
111. Sharp, G.J., R. Ugarte & R.E. Semple. 2006. The ecological impact of marine plant harvesting in the Canadian maritimes, implications for coastal zone management. *ScienceAsia 32 Suppl.* **1**: 77–86.
112. Ugarte, R. 2011. An evaluation of the mortality of the brown seaweed *Ascophyllum nodosum* (L.) Le Jol. produced by cutter rake harvests in southern New Brunswick, Canada. *J. Appl. Phycol.* **23**: 401–407.
113. Ang, P.O., G.J. Sharp & R.E. Semple. 1996. Comparison of the structure of populations of *Ascophyllum nodosum* (Fucales, Phaeophyta) with different harvest histories. *Hydrobiologia* **326/327:** 179–184.
114. Sharp, G.J. 1981. An assessment of *Ascophyllum nodosum* harvesting methods in southwestern Nova Scotia. *Can. Tech. Rep. Fish. Aquat. Sci.* No. 1012
115. Sharp, G.J. & J.D. Pringle. 1990. Ecological impact of marine plant harvesting in the northwest Atlantic: a review. *Hydrobiologia* **204/205:** 17–24.
116. Eschmann, C. & D.B. Stengel 2011. Recovery of *Ascophyllum nodosum* after harvesting in Ireland and potential interactions with climate change. *Proceedings of the Fourth Congress of the International Society for Applied Phycology*, June 19–24, 2011, Halifax, Canada.
117. Seeley, R.H. 1986. Intense natural selection caused a rapid morphological transition in a living marine snail. *Proc. Natl. Acad. Sci. USA* **83:** 6897–6901.
118. Allen, B. 2007. Status of the Common Eider in Maine. Unpublished report, Maine Department of Inland Fisheries and Wildlife, Bangor, Maine.
119. Allen, B., D.G. McAuley & A. Tur. 2008. Survival, nest success and productivity of female common eiders (*Somateria mollissima dresseri*) on Flag Island, Harspwell, Casco Bay, Maine. Maine Department of Inland Fisheries and Wildlife files.
120. DFO. 1999. *Proceedings of the Maritimes Advisory Process of Rockweed Stocks Canadian Stock Assessment Proceedings Series* 99/36.
121. Gollety, C., E. Thiebaut & D. Davoult. 2011. Characteristics of the *Ascophyllum nodosum* stands and their associated diversity along the coast of Brittany, France. *J. Mar. Biol. Assoc. UK* **91:** 569–577.
122. Ugarte, R., J.S. Craigie & A.T. Critchley. 2010. Fucoid flora of the rocky intertidal of the canadian maritimes: implications for the future with rapid climate change. In *Seaweeds and Their Role in Globally Changing Environments Cellular Origin, Life in Extreme Habitats and Astrobiology*, Vol. 15, Part 2. A. Israel, R. Einav & J. Seckbach, Eds.: 73–90. Springer. Dordrecht, Heidelberg, London, New York.
123. Norton, T.A. 1986. The ecology of macroalgae in the Firth of Clyde. *Proc. Roy. Soc. Edinburgh* **90B:** 255–269.
124. Sharp, G.J. & C. Bodiguel. 2001. Introducing integrated management, ecosystem and precautionary approaches in seaweed management: the *Ascophyllum nodosum* (rockweed) harvest in New Brunswick Canada and implications for industry. In *Proceedings of the 17th International Seaweed Symposium*. A.R.O. Chapman, Ed.: 107–114. Oxford University Press. Oxford.
125. Maine DMR. 2010. Maine Department of Marine Resources; coastal fishery research priorities: rockweed (*Ascophyllum nodosum*). Minutes. Retrieved February 1, 2012, from www.maine.gov/dmr/rm/rockweed/symposium2010/minutes.pdf.
126. Lazo, L. & A.R.O. Chapman. 1996. Effects of harvesting on *Ascophyllum nodosum* (L.) Le Jol. (Fucales, Phaeophyta): a demographic approach. *J. Appl. Phycol.* **8:** 87–103.
127. Stengel, D.B. & M.J. Dring. 1997. Morphology and in situ growth rates of plants of *Ascophyllum nodosum* (Phaeophyta) from different shore levels and responses of plants to vertical transplantation. *Eur. J. Phycol.* **32:** 193–202.
128. Strömgren, T. 1977. Short-term effect of temperature upon the growth of intertidal Fucales. *J. Exp. Mar. Biol. Ecol.* **29:** 181–195.
129. Keser, M. & B.R. Larson. 1984. Colonization and growth of *Ascophyllum nodosum* (Phaeophyta) in Maine. *J. Phycol.* **20:** 83–87.
130. Mathieson, A.C., J.W. Shipman, J.R.O'Shea & R.C. Hasevlat. 1976. Seasonal growth and reproduction of estuarine fucoid algae in New England. *J. Exp. Mar. Biol. Ecol.* **25:** 273–284.
131. Dudgeon, S.R. & P.S. Petraitis. 2005. First year demography of a foundation species, *Ascophyllum nodosum*, and its community implications. *Oikos* **109:** 405–415.
132. DFO. 1998. Rockweed in the Maritimes. DFO stock status report C3-57.
133. Sharp, G.J. 1987. *Ascophyllum nodosum* and its harvesting in Eastern Canada. Case studies of seven commercial seaweeds resources. FAO Technical Report 281: 3–46.
134. Davies, J. 2000. In Gulf of Maine rockweed: management in the face of scientific uncertainty. *Proceedings of the Global Programme of Action Coalition for the Gulf of Maine (GPAC) Workshop*, Huntsman Marine Science Centre, St. Andrews, New Brunswick, December 5–7, 1999. R.W. Rangeley & J. Davies, Eds. Huntsman Marine Science Centre Occasional Report No. 00/1: 76.
135. McAdam, J. 2006. Comments for FIDC on the Environment and Heritage Service's Sustainable Seaweed Harvesting Workshop. Retrieved February 1, 2012, from www.ukfit.org/reports/Sustainable%20Seaweed%20Harvesting.pdf.
136. Environment and Heritage Service, Northern Ireland. 2007. Environmentally sustainable seaweed harvesting in Northern Ireland. Retrieved February 1, 2012, from http://www.doeni.gov.uk/niea/index/seaweedharvestingniehspositionstatement.pdf.
137. Ingolfsson, A. 2010. The conservation value of the Icelandic intertidal, and major concerns. *Natturuverndargildi Islensku Fjorunnar Og Adstedjandi Haettur* **79:** 19–28.
138. McNaughton, S.J., M. Osterheld, D.A. Frank & K.J. Williams. 1989. Ecosystem-level patterns of primary productivity and herbivory in terrestrial habitats. *Nature* **341:** 142–144.
139. Cebrian, J. 2002. Variability and control of carbon consumption, export, and accumulation in marine communities. *Limnol. Oceanogr.* **47:** 11–22.

140. Hession, C., M.D. Guiry, S. McGarvey & D. Joyce. 1998. Mapping and assessment of the seaweed resources (*Ascophyllum nodosum*, *Laminaria* spp.) off the West Coast of Ireland. Marine Resource Series 5, Marine Institute, 1998.
141. Chambers, P., R.E. DeWreede, E.A. Iriandi & H. Vandermeulen. 1999. Management issues in aquatic macrophyte ecology: a Canadian perspective. *Can. J. Bot.* **77:** 471–487.
142. Vea, J. & E. Ask. 2011. Creating a sustainable commercial harvest of *Laminaria hyperborea*, in Norway. *J. Appl. Phycol.* **23:** 489–494.
143. Ernst, M. 2004. A survey of coastal managers' science and technology needs prompts a retrospective look at science-based management in the Gulf of Maine. NOAA National Ocean Service.
144. Rangeley, R.W. 1998. Variability in the use of rockweed habitats by fishes: implications for detecting environmental impacts. In: Coastal monitoring and the Bay of Fundy. *Proceedings of the Maritime Atlantic Ecozone Science Workshop.* M.D.B. Burt & P.G. Wells, Eds.: 28–29.
145. Hardin, G. 1968. The tragedy of the commons. *Science* **162:** 1243–1248.
146. Hinrichsen, D. 1999. *Coastal Waters of the World: Trends, Threats, and Strategies.* Island Press. Washington, DC.
147. Mory, C. 2005. *Chondrus crispus* fact sheet. History and status of Irish Moss (*Chondrus crispus*) stocks in the Gulf region. Retrieved February 1, 2012, from http://www.glf.dfo-mpo.gc.ca/e0006845.
148. Petraitis, P.S., E.T. Methratta, E.C. Rhile, *et al.* 2009. Experimental confirmation of multiple community states in a marine ecosystem. *Oecologia* **16:** 139–148.
149. Maine Department of Marine Resources. Commercial landings. Retrieved February 1, 2012, from http://www.maine.gov/dmr/commercialfishing/recentlandings.htm.
150. Wilson, J. 2004. Learning to govern at many scales: lessons from Maine's Fisheries. Retrieved February 1, 2012, from muskie.usm.maine.edu/changingmaine/lectures/JWilsonLec.pdf.
151. Smith, A.D.M, C.J. Brown, C.M. Bulman, *et al.* 2011. Impacts of fishing low–trophic level species on marine ecosystems. *Science* **333:** 1147–1150.

Ann. N.Y. Acad. Sci. ISSN 0077-8923

ANNALS OF THE NEW YORK ACADEMY OF SCIENCES
Issue: *The Year in Ecology and Conservation Biology*

Artificial persons against nature: environmental governmentality, economic corporations, and ecological ethics

Michael S. Northcott
School of Divinity, University of Edinburgh, Edinburgh, United Kingdom

Address for correspondence: M. S. Northcott New College, Edinburgh EH1 2LX, UK. m.northcott@ed.ac.uk

Despite the 194 nation-state signatories to the global Convention on Biological Diversity, the conservation effort is failing to halt an ongoing spiral of decline in most habitats and ecological communities on land and ocean. Environmental ethicists argue that the failure to halt the unsustainable predation on the ecosystems that sustain industrial civilization is indicative of a moral as well as a scientific crisis. Principal ethical interventions in ecology include the ascription of value to species and ecosystems, wilderness ethics, and ecological virtue. Ecological virtue ethics identifies agency, character, institutions, and practices as crucial to moral formation and outcomes. However, the dominant role of the economic corporation in ecological destruction subverts a virtues approach. Corporations as fictive persons will not learn ecological virtue absent of legal and regulatory reform and the ecological education of business leaders and owners.

Keywords: environmental; corporations; value; species; economic

The year 2010 was a bad year for the health of habitats and species on the home planet. The failure of the Copenhagen Climate Conference in December 2009 to produce a draft treaty governing international greenhouse gas emissions cast a pall over the prospects for the mitigation of anthropogenic climate change (ACC). ACC represents the most significant global-scale threat to biodiversity; 2010 saw the publication of research indicating that global climate change is producing growing droughts and heavier rainfall in shorter time periods. Together these have resulted in an unexpected net decline of plant productivity of 1% in the exceptionally warm decade of 2000 and 2009.[1] This contrasts with the previous decade in which the same scientists recorded a net global gain of 6%. Climate change above 2 C° on preindustrial levels threatens the survival of tropical forests, which contain approximately 80% of global biodiversity.[2] Rising global emissions of carbon dioxide are also causing increased ocean acidification, which is reducing marine fertility and biodiversity.[3]

The other major global treaty designed to protect the planet's biodiversity, the Convention on Biological Diversity (CBD), is also failing to reduce biodiversity losses. The year 2010 saw the publication of a global audit of plant biodiversity in the light of the approaching tenth anniversary of the CBD, to which 193 nations have acceded. The review found that one in five of the plants studied in the sample was at risk of extinction, while ongoing threats to amphibians and mammals were even higher.[4] Hence the goal of the CBD "to achieve by 2010 a significant reduction of the current rate of biodiversity loss at the global, regional and national level as a contribution to poverty alleviation and to the benefit of all life on Earth" has not been realized.[5]

The biggest cause of biodiversity loss is continuing destruction of moist tropical forest prior to land-use change toward commercial-scale monocrop plantations such as soya and oil palm. This destruction is driven in large part by economic corporations rather than by indigenous populations whose traditional harvesting practices depend on natural

doi: 10.1111/j.1749-6632.2011.06294.x

biodiversity, not on a single cultivated species.[6] If maximizing the income of indigenous forest dwellers were the priority in the use of tropical forests, then community-based harvesting of timber and nontimber forest products would always win over forest burning and replacement with monocrops. Community-based timber projects in Costa Rica give considerable economic benefit to local people, while the small-scale and labor-intensive nature of this approach results in much less damage to the forest than large-scale commercial logging.[7,8] Conservationists are also exploring the potential of local community forest harvesting of nontimber forest products, such as nuts, fruits, resins, bamboo, and herbs for medicine and cooking. One case study of arecanut palm harvesting in Southeast Asia shows that this approach retains 90% of local bird species, which is far better than the outcomes for conversion from tropical forest to oil palm monocrop.[9] Markets in these products need to be balanced with conservation concerns if they are not to unsustainably diminish supplies.[10] But local harvesting of these products has the potential to sustain common property regimes in forest management while also feeding local supply chains and hence further maximizing regional employment and income potential from the forests.[11] On the other hand, the principal markets for tropical monocrop products are in North America, Europe, Japan, and Australasia, where human populations are relatively stable but consumption continues to grow. Soya is used primarily as animal feed, and palm oil is primarily used in processed foods, cosmetics and detergents, and increasingly as biofuel. Action in developed countries to limit imports of products sourced unsustainably from tropical regions is insufficient, though the European Union made a start with a ban on illegally sourced Indonesian timber in 2011.[12] Without such action, the CBD will not drive reduced degradation of biodiversity hotspots in tropical regions.

The second biggest cause of habitat destruction is pelagic ocean fishing.[13] Again the principal agents are not peasant fishers but large technologically sophisticated fishing fleets owned and managed by commercial corporations. The global catch capacity of commercial fishing fleets far exceeds the capacity of the oceans to supply them.[14] Overharvesting threatens turning large areas of the marine environment into marine deserts by midcentury at present rates of destruction.[15] It also undermines the ability of traditional fisher communities to fish locally and sustainably. Here, the principal markets are in both rich developed nations in North America, Europe, and Japan, and increasingly in China, and other developing nations as they emulate excessive and unsustainable consumption patterns in developed countries. While consumer awareness of the unsustainable sourcing of fish is growing, voluntary action—without legislation—is insufficient to stem the destruction of biodiversity from unsustainable fishing.

Industrial corporations, often licensed or subsidized by national governments, are presiding over what a growing number of scientists are calling the "sixth extinction." Though it will be the sixth major wave of species loss in the earth's history, it is the first caused by one species, and the rate of species loss is already faster than any previous event.[16,17,18] The failure of the first global treaty on biodiversity to stem the tide of extinction reflects an enduring conflict between the political influence and power of the primary agents of economic development—economic agencies and corporations owned by governments and private shareholders—and the aspirations and projects of conservationists in government, in nongovernmental organizations (NGOs), and in the scientific community.[19]

Aspirations to conserve the environment are not confined to conservation experts. A 2008 survey found that 95% of individual citizens in Europe believe that environmental protection is very important, though most think action is more important at the governmental level and through global agreements rather than at the level of the individual consumer and householder.[20] Cross-national studies that include data from North America also indicate that there is a sustained level of concern for a healthy environment in all developed countries.[21] However, there are few occasions on which an environmental issue has played a major role in voting patterns in either Europe or North America, although the election of the Rudd government in Australia in 2007 was partly due to its stance on climate change.[22]

Conflicts between citizen aspirations and conservation strategies and the commercial interests of private corporations are evident in the failure of global scale treaties to address ecological problems such as climate change and biodiversity loss. Such conflict is also clearly implicated in national-scale failures. These conflicts reflect an enduring and structural problem of power in relation to conservation. Conservation biologists seek to restrain the ecological impact of economic development on wildlife by strategies that involve a focus on particular endangered species or habitats. Thus a range of national laws, international treaties, and scientific conservation projects involve the identification, conservation, and, where possible, reintroduction of individual endangered species that are often also "charismatic species" such as orangutans or polar bears.[23] Another array of legal, scientific, and nongovernmental activities involve attending to and saving "biodiversity hotspots," "sites of special scientific interest," or "scenic" places.

The construction of conservationist spaces involves the reconfiguring of human–nature relations in these spaces, which often privileges external actors over indigenous ones. Consequently, indigenous people—residents, tribal peoples, local communities, traditional users of lands appropriated for conservation—find themselves drawn into a new mode of environmental governmentality.[24,25] In many cases, this new governmentality is added to existing disciplining and restraining invasions by government and corporate agents that are the principal cause of biodiversity losses. This is evident, for example, in the contrast between indigenous management of tropical forests in the Amazon, Borneo, and the Congo with the effects of governmental and corporate uses of these forests. Traditional users—whose uses may have endured for thousands of years without significant biodiversity reductions—lose their agential power to manage environments in ways that enable them to feed and house themselves while not diminishing ecosystemic richness.[26,27,28] As access to their traditional lands is denied, and in most cases land-use rights are conferred instead on corporations, they find themselves turned into poachers or even refugees. Hence the outcome of ecological governmentality is that local populations are subject to increased surveillance and control by the state while their use rights to local forest and land-derived produce are handed over to private corporations whose record of environmental destruction is infinitely greater than that of indigenous communities. Community-based conservation was introduced to address the problem of environmental exclusion of local communities from national parks and nature reserves, but this approach is criticized as having failed to promote species protection as effectively as the earlier approach.[29]

There is a contradiction at the heart of the modern conservation movement, modern environmentalism, and ecological science. Economic growth sustains growth in conservation and ecology journals, conferences, books, organizations, projects, scientists, and volunteers. But this activity depends upon the same growth engine that is the root of the problem, which is growth in corporately driven and sustained consumption and production in the developed, and increasingly in the developing, world. Wendell Berry argues that the problem is ultimately one of a failure of moral character.

> "While conservationists are exploring, enjoying, and protecting the nation's resources, they are also using them. They are drawing their lives from the nation's resources, scenic and unscenic. If the resolve to explore, enjoy, and protect does not create a moral energy that will define and enforce responsible use, then organized conservation will prove ultimately futile. And this, again, will be a failure of character."[30]

Berry is surely correct that the ecological crisis is a failure of character, but whose character? In the developed world, many citizens are no longer actively engaged in food growing or manufacturing, as corporations have outsourced production to other countries. But citizens have to eat and have to find clothing and housing. Social scientific studies reveal that citizens of post-industrial societies increasingly value nonmaterial goods, including environmental quality.[31] But the condition of being a consumer without land and without the tools of a trade is one of extreme dependence on an economy increasingly dominated by corporate actors in which preferences for goods those corporations do not value are hard to express through the thin medium of consumer choice. To blame the moral character of citizens for the ecological crisis leaves out of the picture the most influential agents of this crisis, which are not citizens but rather economic corporations that have the rights of citizens, and far greater powers than citizens, but lack citizens' capabilities for moral discernment.

Gus Speth is also critical of environmentalists, but not because they lack good intentions. Instead, he suggests environmentalists are guilty of incrementalism: they have focused too much on the effects of ecological destruction and not nearly enough on the underlying causes. For Speth, the causes are the governmental quest for unrestrained economic growth and the quest of large private corporations to maximize shareholder value regardless of ecological or even social costs to the communities and habitats in which corporations operate.[32] Focusing as Berry does on the moral failings of citizens—as though they are the principal agents of environmental destruction—runs the risk of colluding with the disempowering effects of ecological governmentality on citizens and local communities while leaving the main agents and promoters of ecologically destructive production and consumption—large economic corporations—free to continue with business as usual. Furthermore, it neglects the way in which the political powers of citizens and local communities have been transferred by the state to economic corporations in the last 200 years.

The rights of corporations emanate from the chartering and licensing of corporations by the nation state. This process began in the fifteenth century when the monarchs and emergent nation states of modern Europe began to derogate their own powers of incorporation to other corporate entities and so granted to corporations a share of the sovereign powers, which the nation state drew to itself in the post-Reformation period over land and natural resources contained within the boundaried spaces claimed by the state.[33,34] In post-revolutionary America, the status of joint stockholding corporations was ambiguous. Some argued that since they were chartered by the King of England they should, after the Revolution, be made subject to the will of the people in the towns in which they were situated. This claim was soon tested in the courts in a conflict between the Trustees of Dartmouth College and the governor of New Hampshire, which went all the way to the Supreme Court. The governor had declared the College a public body, removed its charter and seal of office, and countermanded decisions of the college trustees in an effort to turn it into a public university. The trustees and their lawyers argued that the Royal Charter of George II, which had established the college, was a contract that remained in force even after the Revolution. In a precedent-setting case, the Supreme Court found for the corporation of Dartmouth College over the State of New Hampshire.[35] A subsequent judgment in the United Kingdom—Salomon versus Salomon—had a similar effect in British company law.[36] The result was the creation of a fictive legal personality—or artificial person—to which was ascribed in law the same rights as were ascribed to actually existing persons.

The corporation began as a means for monarchs and governments to charter and commission corporate entities such as the East India Company and the Hudson Bay Company in the exploration and exploitation of territories beyond national jurisdictions. But this device to extend the agency of the state beyond its territory became a means to affirm the independent sovereign power of such royally chartered companies in the post-revolutionary and republican space of the United States and then in the British homeland as well. Thus was born the modern Anglo-Saxon business corporation, independent of the sovereign power of government and answerable only to its stockholders. Today, economic corporations possess more wealth than most of the national jurisdictions that birthed them. Through this wealth they are able to lobby parliaments and fund political parties and the election expenses and offices of politicians who claim to represent the will of really existing people, not artificial people. Similarly corporations, such as News Corporation, can purchase media across different formats and create outlets whose purpose is not to inform the public in a balanced way over environmental and other public concerns, nor to seek the truth, but to advocate a "corporate-friendly" worldview that focuses on the sexual and other misdemeanors of celebrities or politicians while neglecting corporate crime and distorting public perceptions of major issues that might otherwise provoke more corporate regulation—and most notably ACC.[37,38] In all three domains, where News Corporation holds a major share of print and broadcast media—the United States, Australia, and Britain—skepticism about the science of climate change is widespread. Content analysis of four media outlets in these domains in the six months between January and June 2010 by the author of this paper revealed that *Fox News*, the *Wall Street Journal*, the *London Times*, and the *Australian*—all Murdoch-owned outlets—carried a disproportionate number of stories suggesting that arguments for ACC are based on bad science, that there is considerable disagreement among scientists as to the causes of climate change, or that the climate is not changing beyond historic norms. News Corporation does not act alone. The corporate management of public perceptions of climate change science was first proposed by the American Petroleum Institute in 1998, which aimed "to inform the American public that science does not support the precipitous actions (the Kyoto climate treaty) would dictate, thereby providing a climate for the right policy decisions to be made."[39]

If they were real human beings, corporate artificial persons would be regarded as sociopaths because the governing interest of the corporation is the bottomline of profit, and hence the value of its stock to its owners, and not conventional moral concerns for other people or species.[40] If shareholder value rises by burning other peoples' forests, fishing out oceans, or polluting the atmosphere with climate-changing quantities of CO_2, this is of no concern to the company since these burdens do not show up on company accounts. It would, however, be an oversimplification to say that such activities happen without the consent or collusion of governments. In many cases of corporate environmental destruction, governments are also involved: through tax-funded corporate subsidies or corporate tax breaks or through the issue of government licenses for forest conversion, fossil fuel, or fishing extraction within national jurisdictions.[41] Governments also subsidize or underwrite—as most notably in the 2008 financial crash—the activities of large commercial banks that are the source of investment funds for fossil fuel exploration and monoculture conversion of forests. The largest failed bank in Europe—the Royal Bank of Scotland—which was rescued from bankruptcy by an 85% British government buyout, is also the largest single global funder of highly polluting or ecologically risky oil and gas projects, including oil exploration in the Arctic Ocean and tar sands oil extraction in Canada.[42]

Governments fund conservation agencies and issue legislation designed to control the worst harms arising from the activities of private corporations and citizens. But public funds devoted to conservation activities and to the policing of environmental regulation represent a fraction of funds devoted to corporate subsidies. U.S. federal subsidies to the coal, oil, and gas industries in the period 2002–2008 amounted to just over $12 billion annually. Federal government support of Wall Street banks, and spending on wars in oil-rich regions in the same period, came to more than $6 trillion.[43] By contrast, the budget for the Environmental Protection Agency (EPA) in the same period was approximately $55 billion.[44]

The principal approaches to environmental ethics developed by philosophers in the United States and Europe neglect corporate behaviors and the role of governments in authorizing and subsidizing them. The most influential approach involves the ascription of moral value to ecosystems, land, and species. For Aldo Leopold, ecological science reveals the interconnections that sustain the life community on which human life depends. Ecological and evolutionary science indicates that humans are members of the "land community" and that instead of its conquerors they must act as its citizens. Ecology generates a new sense of right and wrong in the earth community, which Leopold called the "land ethic:" "a thing is right," he argued, "when it tends to preserve the integrity, stability, and beauty of the biotic community. It is wrong when it tends otherwise."[45] In a similar vein, Holmes Rolston proposed that species and communities of life have "intrinsic value" to the extent that they display goal-directed behaviors. Species rely on other species—and on natural processes such as photosynthesis—for the pursuit of their own purposes in enduring and reproducing. In this way, Rolston argues, values in nature have a prior or given existence independent of human culture.[46] The land ethic and intrinsic value provide an extracultural referent for human value, a measure of right and wrong that is more ecologically sensitive than culturally originated value systems, such as pricing mechanisms or aesthetic judgments. This approach is also said to overcome the mind–body, subject–object, and nature–culture divides that environmental philosophers argue have distorted human perceptions of the natural world since the Enlightenment and hence challenge the ideational roots of modern environmental destruction.[47,48]

The concept of intrinsic value is reflected in the U.S. Endangered Species Act (1973). The purpose of the act states that certain species have been rendered extinct because of economic growth and development "untempered by concern and conservation" and others have been depleted in numbers and are therefore "threatened with extinction." The Act therefore commits the United States as a sovereign state to "conserve to the extent practicable" other threatened species of fish, wildlife, and plants that "are of esthetic, ecological, educational, historical, recreational, and scientific value to the Nation and its people."[49] New Zealand's more recent Resource Management Act (RMA) uses the phrase "intrinsic value," not only with respect to endangered species but also to ecosystems. It states that the act is designed to promote the conservation of "aspects of ecosystems and their constituent parts which have value in their own right, including (a) their biological and genetic diversity and (b) the essential characteristics that determine an ecosystem's integrity, form, functioning, and resilience."[50] However, in both domains, the rise of the anti-statist and growth-at-any-cost rhetoric of neoliberalism is diminishing democratic preparedness to challenge industrial developments that threaten biodiversity. Thus while the preamble of the New Zealand RMA recognizes the intrinsic value of ecosystems, the 788 pages of the RMA contain many clauses that diminish the legal capacities of citizens or local authorities to appeal or halt corporate behaviors or practices that threaten ecosystem integrity in particular places.[51,52] In the United States, the annual budget of the EPA was diminished during the Bush administration (2001–2009) and again since 2011. So too did the ability of the EPA to decide on scientific matters, such as whether climate change is caused by fossil fuels or threatens endangered species within U.S. territory. Changes in federal law during the Bush–Cheney administrations also reduced the ability of citizens democratically to resist corporate environmental pollution at county or state level.[53,54] The Bush–Cheney administration also strengthened efforts to curb the ability of citizens to take out tort actions against corporate environmental harms in state and federal courts.[55]

Intrinsic value is an important philosophical idea, and in the U.S. Endangered Species Act and New Zealand RMA, it has signally entered political and legal discourse, providing a new cultural warrant for the executive powers of government to restrain corporate environmental destruction. In practice, however, corporations continue to use their greater wealth, and their lobbying and monetary influence over parliamentary and government executive processes, to pursue environmentally destructive developments in terrains that have formally recognized the intrinsic value of species and ecosystems. There is also an unintended consequence in the adoption of intrinsic value as a political value by the state. In the absence of a formal mechanism internal to the accounts and practices of private corporate entities and citizens, attempts to enforce interpretations of intrinsic or natural value inevitably involve growth in the powers of the state relative to corporations and the citizen: in other words, they increase environmental governmentality of citizens and local communities while neglecting the corporate root of most environmental destruction. In libertarian discourse, this increase in environmental governmentality is represented as growth in the powers of the state over the individual. Hence Sunstein represents precautionary environmental regulation in relation to air and water pollution as being based on a climate of fear and as over-reaction to "worst case scenarios."[56]

Attempts to legislate for the recognition of intrinsic value also have another disadvantage, for they suggest that the nation state is the principal agency responsible for protecting or redeeming nature from ecological destruction. This is an ambiguous claim when in many cases corporate activities are licensed, or even subsidized, by the same governments that claim to "protect" nature.[57] The promotion of the nation-state as guardian of the environment is also problematic because in the great majority of cases environmental destruction is place specific and hence local in character. But the nation-state in its history has had a tendency to gather to itself and its executive agencies the local powers of people and communities that reside in particular places.

The second main approach to environmental ethics may be best described as "wilderness ethics." This approach is variously attributed to Ruskin and Wordsworth in England, Muir and Thoreau in America, and Goethe and Hölderlin in Germany, though it has ancient antecedents. Perhaps the best known antecedent in Western culture is the wilderness poetry of the Hebrew Psalms and injunctions in Israelite law codes to leave space from agrarian uses for wild animals to inhabit, as well as the command in Leviticus to give the land a "sabbath" every seventh year for the benefit of the land itself and for wild animals.[58] Modern wilderness ethics are encapsulated in John Muir's account of the wilderness of the Californian Sierra as "godful" and as "full of tricks and plans to draw us up into God's light."

Wilderness ethics are also encapsulated in his claim that there is more wonder, beauty, strength, and gracefulness in animals "cared for by Nature alone" in mountain pastures and ancient forests than in domestic animals in cities and plains.[59] Thoreau argued that humans have a cultural and spiritual need to experience natural beauty in relatively unmodified natural areas and that "wildness is the preservation of the world."[60] Wordsworth, Ruskin, and Muir argued that industrial civilization must place limits on its use of wild lands and set apart a proportion of wild lands for conservation. They pressed for the establishment of National Parks in the Sierra Mountains in California and in the Lake District in England. For the Romantics, the choice of which lands to set aside was determined principally by aesthetics: certain kinds of landscape—mountains, waterfalls, lakes, and wide rivers running through lowland hills—were identified with the Romantic ideal of the natural sublime.[61] For conservation scientists, the choice of land to be set aside is determined more by ecological values such as species richness or the presence of endangered and endemic species in particular areas.[62] Recent iterations of this approach include "rewilding," which involves efforts to reintroduce species to restored habitats and hence recover a species mix characteristic of preindustrial or prehistoric eras, and the ascription of rights to nature as well as humans, as in the constitution of Ecuador and in efforts to grant rights to primates.[63–66]

While the wilderness ethic has played a valuable role in the setting aside of certain species-rich or aesthetically pleasing areas of land from industrial development, it has not proved sufficient to global-scale problems, such as ozone depletion, pelagic ocean fishing, or ACC. Further, the wilderness ethic may even have the perverse effect of giving an ecological gloss to industrial civilization while permitting citizens and corporations to continue destructive behaviors in those parts of the earth not specifically designated as wildlife reserves or national parks. Creating wilderness areas will have little ultimate ecological effect if corporate extraction of fossil fuels, or corporate extraction or pollution of watersheds, destroy the climatic conditions or fresh water on which species in these areas depend for their survival. The wilderness approach has also had an analogous effect on the intrinsic value approach when it valorized the nation-state as the "owner" and maintainer of wilderness areas while excluding or restraining local residents from traditional use rights to areas set aside from normal uses. The unintended consequence of the wilderness or set-aside approach, as noted above, is often to alienate local communities from central government-led or corporately managed conservation projects.

Given the failures of earlier approaches to ecological ethics to stem global and local reductions in biodiversity and destruction of habitats, a third approach has emerged in the last 20 years in an effort to measure the monetary value of the services ecosystems and species render to human beings. This involves a significant intervention in the modern history of financial accounting. The origins of modern accounting may be traced to the invention of double-entry bookkeeping by the Franciscan monk Luca Pacioli in the fourteenth century.[67] The great advantage of Pacioli's approach over previous accounting methods was its simplicity. Pacioli, in effect, changed the nature of accounting practice from a set of narratives to a set of numbers. With the requirement to keep two columns of numbers, it became possible to arrive at a reliable measure of the outcome of the activities of a monastery, or a modern corporation, without reference to detailed and context-specific narratives about the factors of production. The practice also involved a shift in accounting practice from a broad focus on what economists sometimes call "use values" to a narrower focus on exchange values, either as realized in corporate activities or remaining as potential in the assets of the corporation. This shift meant that descriptions of the lives of workers or domestic animals, or the conditions of plants, rocks, soils, or water sources used during the accounting year, could be left out of the accounts.[68,69]

Double-entry bookkeeping is the historical root of modern economics. Under its aegis, modern economists hold that measures of exchange value are the appropriate means for determining the monetary value of a corporation's annual activities or the annual product of the natural and human resources of a nation state. Where scarcity causes the market price of a nonrenewable resource—such as copper—to rise, classical economists argue that alternative substitutable resources will be found or developed through human ingenuity and technical innovation.[70] But there are a range of cultural and natural goods—endangered species, a stable climate, potable water, clean air, human dignity, and quality of life—that are not easily reducible to market mechanisms or measures of exchange value. Many of these goods are not conventionally bought or sold in markets, and they therefore do not show up on corporate or national accounting mechanisms. Ecological or environmental economists have therefore begun to develop measures of the economic value of the benefits or services ecosystems and species confer upon humans. They propose that such measures should be factored into the profit and loss or cost and benefit accounting procedures that guide economic planning and development, and hence corporate behaviors, in modern nation-states. In a global study, in which a range of studies of the value of particular ecosystem services were synthesized, it was estimated that the annual value contributed to economic activities by the ecosystem services provided by the planet was $33 trillion. The total value of the human trading economy, at the time of the study, was estimated at $18 trillion.[71] In this light, the human economy is clearly less valuable than the economy of the earth in which it is set and furthermore is depleting natural wealth at a rate faster than it is being replaced by human wealth.

Ecological economists propose that Aristotle's classical distinction between exchange and use values—encapsulated in his distinction between *oikonomia* and *chrematistics*—remains valid. Hence citizens and parliaments ought to be capable of deliberating on environmental use values in such a way as to resist the modern default subjugation of use values to exchange values. They propose that such deliberation ought to inform environmental policy making and hence require market actors to value use and not only exchange values.[72]

In practice, however, efforts to require market actors to recognize and account for use values lack cultural or political purchase. To give just one recent example, in the United States and Europe, growing numbers of bees are suffering from the recent phenomenon of colony collapse disorder (CCD). Two thirds of the plants eaten by humans are pollinated by bees. Entomologists identified a likely link between CCD and the introduction in the last 10 years of systemic pesticides—and in particular neonicotinoids—which are expressed through plants coated in such substances as seeds instead of being topically applied.[73] This discovery, combined with protests by beekeepers, led to the banning of neonicotinoid pesticides in France and Germany. However, in the United States, where such seeds were first introduced and where CCD was first identified, the use of neonicotinoids continues. The UK government has similarly refused to ban neonicotinoids. The annual contribution of domesticated bee pollination services to U.S. agriculture is estimated at $20 billion and the value of wild insect services to U.S. agriculture is estimated at $57 billion.[74,75] The combined annual profit of Monsanto and Cargill—the two largest global seed companies—were just $2.9 billion in 2009. But in a contest between exchange and use values, exchange values win in the economically liberal domains of the United States and the United Kingdom.

Biodiversity and a stable climate are goods of inestimable value when their use to future generations is taken into account. But the moral ethos sustained by the dominance in economically liberal domains of the interests of private corporations gives priority to exchange over use values and to the values realized by present over future persons. Artificial persons, though treated as equivalent to human beings in law, lack the capacity to become the kinds of people who express the moral character of those who love nature and who use nature carefully and with aforethought for future generations and other users both human and nonhuman. This is because the exclusive guiding goal of an economic corporation is the maximization of shareholder value, or profit. Efforts by conservation-minded citizens, NGOs, and conservation scientists to restrain corporate economic activities therefore require resort to the nation-state and the law courts as the original source of the sovereign power of corporations. But there is a political, and hence a legal, reluctance to put environmental goods, and their use values to present and future generations, above corporate interests. This reluctance does not just reflect the political influence, and legal status and powers, of corporations. It is also informed by the preference of modern economists and a growing number of politicians for market allocation of the social and environmental costs of natural resource use, since markets are said to be more "efficient" than more deliberative forms of cost allocation by rational choice theorists.[76]

The refusal of corporate actors, and classical economists, to properly account for the inestimable gifts that the earth and the sun confer on humankind leads to a deep moral and philosophical conundrum that is not resolvable by conservation projects that sequester parts of the environment—diversity hotspots, scenic places, endangered species—from the ravaging effects of an economy devoted to exchange over use values. Moral and legal appeals to intrinsic values have proven similarly insufficient. The "conservation contradiction" identified above is thus revealed to have a corporate origin: the power and wealth of modern economic corporations, and their control over ever larger areas of the planet's surface, including subterranean minerals and fuels, grows apace. Neither scientific conservation nor environmental ethics are capable of preventing the resultant and persistent depletion of the health of ecosystems and destabilization of the earth's atmosphere. This points to a deeper contradiction between the guiding logic of Western and, now, global civilization and the health of ecosystems and the creatures that constitute and inhabit them.

On the face of it, the project to measure the monetary value of ecosystem services and to incorporate these as "natural capital" into corporate and national accounting procedures, presents the best device for reforming the advancing pace of ecological destruction. Unlike intrinsic value or wilderness ethics, it presents the ecological problem in a numerical and monetary form that ought to be capable of incorporation into the accounting procedures and decision-making processes of economic corporations and government agencies. However, efforts to incorporate natural capital into corporate and government accounts have proven ineffective. The best known of these efforts concerns the creation of markets in carbon emissions. But these markets are expensive to create, they encourage fraud, and they have proven ineffective.[77] In essence, carbon credits are financial derivatives. As with other derivatives markets, trading in them has provided lucrative business for banks and investment houses but has made no impact on the global rate of fossil fuel extraction.[78] In Europe, carbon trading was introduced in response to the Kyoto Protocol, but it has proven ineffective in persuading corporations, government agencies, and private citizens to reduce their carbon-emitting activities. Trading in pollution permits began in the U.S. power industry as a way to reduce sulfur pollutants from coal-fired generators. Trade in pollution is designed to change the economic status of pollution from a social cost external to the market to an external good analogous to profit and loss measured on company accounts. Advocates argue that this approach is more efficient than government regulation. But this claim only holds true where corporate compliance with environmental regulation is weak, as it is in the United States.

There is a fourth major approach to environmental ethics that I have not yet touched on, and this arises from the Aristotelian-influenced style of moral reasoning known as virtue ethics. In this approach, the focus moves from commands, duties, and values regarding nature to the virtues that characterize those individuals and communities who act rightly toward nature. English philosophers argued after the Second World War that the duties and values that philosophers since the Enlightenment have made the principal language of ethics are reductive because they fail to describe what makes some individuals behave well, and others badly, in similar circumstances.[79] Aristotle's emphasis on virtue requires attention to those institutions, practices, relationships, and roles that shape the character of good people and direct their lives toward the good.[80] For Alasdair MacIntyre, a key figure in the revival of virtue ethics, family, school, and communities of residence, work, and worship are institutions where individuals are shaped either to live well or badly. For MacIntyre, living well indicates an ability to realize the classical and Christian moral excellences, which include justice, humility, love, and care.[81] The music teacher not only teaches what is and is not a good way to play Mozart, but also shapes the pupil to express those goods that are internal to being a good musician, such as delight in good music and patience in the learning of good technique and performance.

The virtues approach has been taken up by a number of environmental philosophers. Northcott describes the ecological potential of the traditional virtues as follows: love for nonhuman beings and wild places, temperance in balancing human needs and the carrying capacities of ecosystems, justice in the use of the environments of other people and other animals, prudence in attending to the needs of future generations for a habitable planet that remains aesthetically enjoyable, and courage in preparedness to challenge ecologically destructive behaviors and make the case for radical reform of industrial and corporate practices.[82] Van Houtan argues that the virtues arise from the practices of conservation science when understood as a social tradition as well as empirical science.[83] Exemplary practices that inculcate ecological virtues include learning to identify species by name and spotting them in trees or undergrowth; walking, cycling, or canoeing in wild places; clearing trash or invasive species from foreshores, lakes, rivers, or woodlands; recycling used materials; and reducing waste and energy and material consumption in households and the workplace. Institutions where individuals might learn such skills include the family, schools, clubs, higher education and conservation organizations devoted to such activities, and workplaces where sustainability is encouraged. Ecosystem communities in the natural world also play a crucial role. The biographies of environmental and ecological pioneers, such as Rachel Carson and E. O. Wilson, reveal that they were shaped in their love of nature by childhood encounters with wild places such as ponds and woodland.[84]

How though might a corporation be shaped and formed virtuously, so as to love nature? For Aristotle, business is devoted to exchange values, or *chrematistics*, and this promotes the vices of avarice and usury, or lending money at interest. Only households can properly be said to be virtuous in their pursuit of *oeconomics*, since in the household, economic activity has the purpose of sustaining family and community and in this personal context is more likely to be characterized by the virtues of justice and temperance.[85] However, the claim of modern economists since Adam Smith is that the modern business is capable of promoting the good of others regardless of whether it is a family business or an economic corporation, provided those who work in it devote themselves to making well the products they purvey. On this account, individuals who work in business learn the virtues from the internal goods and practices that business promotes such as accountability to others, skill in manufacture, or personal service and diligence.

For Solomon, a leading philosopher in business ethics, Aristotle's claim that businesses cannot be ethical neglects that they, and the individuals who work in business, are part of the *polis* or the wider community that sustains citizens and forms them toward the good. Citizens work in businesses, and if they are rightly formed by the wider community in which they and the businesses they work for dwell, they will behave responsibly in business. Businesses also foster many of the practices through which other moral communities—such as families or schools—foster virtuous individuals. Businesses are communities that foster loyalty and inspire or require teamwork and cooperation. Businesses inculcate excellence in relevant skills such as bargaining and negotiating. Businesses foster interpersonal roles between seniors and juniors. All of these features make businesses analogous to other moral communities. There is therefore no ultimate antagonism, according to Solomon, between business and the common good.[86]

Have we then misidentified the root of the ecological crisis with reckless corporate profit seeking? On Solomon's account, businesses are no different from other communities of persons. They may love or hate nature no more or less. But there is a crucial problem that Solomon does not admit, and this is that the guiding profit motive requires that businesses devote themselves to that good above other goods. The history of modern business has therefore required the elaboration of a great body of corporate law designed to restrain businesses from exploiting their workers unjustly or unduly polluting the air and water around their premises or further afield. Moore argues that businesses may still be virtuous if they exercise a balancing judgment between the bottom line—or external good—of profit and other internal goods, such as care for their employees and diligence in maintaining the quality of their products, though he neglects ecological concerns in his account.[87]

In the absence of changes in accounting rules and corporate law, few corporations have yet included the full social or environmental costs of their activities for species and for present and future generations on their balance sheets. It cannot be said of a music teacher that she instills the good of excellence in performance or appreciation of great music in her pupils without exploiting them only because the law restrains her. A music teacher who needed such legal restraint would not be considered a good or a virtuous teacher. But a corporation that is not restrained by law will coerce its workers, as for example did the Chinese company Foxconn, which drove its workers so hard that they began to commit suicide.[88] Foxconn makes iPads and iPhones for Apple Corporation—the world's second richest company. What of Apple Corporation's duties to the "global community" (Solomon's term) from which it derives mammoth shareholder value? It designs its products in California but exploits Chinese workers to make them because to employ Californians it would have to employ them, and run their manufacturing processes, according to North American environmental and labor laws, which would, marginally, reduce its gargantuan profits.

Virtue ethics has the considerable advantage over other approaches to environmental ethics of highlighting the importance of agency, character, and power relations. In particular, virtue ethics can explain in a way other ethical frameworks cannot how it is that corporations continue to neglect the moral responsibilities to people and planet that are increasingly enshrined in international and national laws to protect biodiversity. Corporations are the most powerful agents on the planet. Their devotion to profit and wealth accumulation generates a vicious character in their behavior that outweighs the virtues of the individuals that work for them or own their shares. Corporate agents in Indonesia, the most biodiverse nation on the planet, continue to extract fossil fuels, mine ores, and replace rainforests with oil palms despite rising global knowledge of the unique biodiversity the archipelago contains. When Indonesian environmentalists and scholars gathered at a conference to discuss environmental destruction on the island of Sumbawa, they were violently set upon by an armed gang acting on behalf of mining interests.[89] Corporations who behave like this may seek to shroud their activities in the discourse of "corporate social responsibility (CSR)." But CSR does not prevent the ongoing corporate destruction of species-rich habitats or of the traditional use rights of local fishers and farmers.

Like the other approaches to ecological ethics reviewed here, virtue ethics still leaves us with a problem. How might citizens and communities teach environmentally vicious corporations environmental virtues? The conventional answer is government regulation. But this is still prone to the libertarian critique of the growth of what Sunstein calls "laws of fear."[90] Government regulation is also prone to money politics and corporate lobbying. Where corporations can buy political influence by funding politicians' election expenses, it is unlikely that politicians will promote effective environmental regulation of their activities. Is there another way? Some argue that the answer is shareholder democracy. In this approach, individual environmentally aware consumers buy shares in corporations and use their votes at shareholder meetings to persuade the companies to adopt more ecologically sustainable practices.[91] However, the votes of individual shareholders rarely change corporate policies since the majority of shares are owned by other corporations. Another approach is consumer pressure. Corporate environmental campaigns—such as the "Green My Apple" campaign—have shown some success.[92] Before the campaign began in 2006, Apple computers were loaded with fire retardants, PVC, lead, mercury, and other toxic elements, while Apple's 2011 products are actively promoted by Apple as free of such elements and hence "environmentally friendly." But against such gains, it must still be acknowledged that if Apple made its products in the United States, its factories would have to meet much higher environmental and social criteria than they do in China.

To sum up, the conservation contradiction identified in this review reveals a mismatch between the growth of environmental governmentality and the continuing devotion of private corporations to growth in shareholder value regardless of environmental destruction. On this account, the devolution by the nation-state to economic corporations of its monopoly powers over the ecosystems and environmental spaces of its citizens and local communities is an important but largely unacknowledged root cause of the ecological crisis. Major legal reform of accounting practices designed to identify and bring into company balance sheets the full social and ecological costs of business practices is essential if the guiding principle of shareholder value is not to continue the advancing pace of species and habitat destruction.

If, however, special interests and short-sighted law makers continue to resist social and ecological accounting reform and regulation, there is one other possible approach that might be pursued. As Solomon argues, companies are controlled by CEOs and board chairs and directors. Ecological education of this group of people—and many directors and chairs serve on more than one board—might offer another way forward. A major funded program of environmental education for business leaders in the United States, Europe, Japan, Beijing, India, and Brazil—the locations of the majority of the world's largest company headquarters—might offer significant rewards. This program would involve funded visits to wild and endangered habitats, during which scientific information about threats to natural capital represented by declining species and habitats would be explained by leading scientists. The program would also involve in-company education in the business and social benefits, as well as the ecological advantages, of corporations' adopting the ecological virtues as enumerated above. Perhaps major donors such as the large philanthropic foundations in the United States might be approached by the AAAS and NAS to fund such an educational program. Ecologically virtuous corporations will ultimately only be shaped by changes in legal regulation of their responsibilities and of their accounting practices. But such corporations will also need to be led and managed by business leaders who have a fuller understanding of the mid- and long-term threats to human as well as species flourishing from continuing with "business as usual."

Conflicts of interest

The author declares no conflicts of interest.

References

1. Zhao, M. & S. Running. 2010. Drought-induced reduction in global terrestrial net primary production from 2000 through 2009. *Science* **329:** 940–943.
2. Asner, G.P., *et al.* 2010. Combined effects of climate and land-use change on the future of humid tropical forests. *Conserv. Lett.* **3:** 495–403. doi:10.1111/j.1755-263X.2010.00133
3. Raven, J., *et al.* 2005. *Ocean Acidification Due to Increasing Atmospheric Carbon Dioxide*, Royal Society. London.
4. Sampled Red List Index for Plants 2010 at http://threatenedplants.myspecies.info/ visited on October 2, 2010.
5. Butchart, S.H.M., *et al.* 2010. Global biodiversity: indicators of recent declines. *Science* **328:** 1164–1168
6. Hecht, S.B. 1985.Environment, development and politics: capital accumulation and the Eastern Amazonia. *World Develop.* **13.6:** 663–684.
7. Hartshorn, G.S. 1995. Ecological basis for sustainable development in tropical forests. *Annu. Rev. Ecol. Syst.* **26:** 155–175.
8. Myers, N. 1988. Tropical forests: much more than stocks of wood. *J. Trop. Ecol.* **4:** 209–221.
9. Ranganathan, J., *et al.* 2008. Sustaining biodiversity in ancient tropical countryside. *Proc. Natl. Acad. Sci.* **105:** 17852–17854.
10. Arnold, J.E.M. & M.R. Perez. 2001. Can non-timber forest products match tropical forest conservation and development objectives? *Ecol. Econ.* **39:** 437–447.
11. Vadjunec, J.M. 2011. Extracting a livelihood: institutional and social dimensions of deforestation in the Chico Mendes Extractive Reserve, Acre, Brazil. *J. Lat. Am. Geogr.* **10:** 151–174.
12. Black, R. 2011. EU and Indonesia sign deal on illegal timber. *BBC News* 4 May: http://www.bbc.co.uk/news/science-environment-13272393
13. Harwood, J. 2001. Marine mammals and their environment in the twenty-first century. *J. Mammal.* **82:** 630–640.
14. Pauly, D. 2009. Beyond duplicity and ignorance in global fisheries. *Scientia Marina* **73:** 215–224.
15. Worm, B., *et al.* 2006. Impacts of biodiversity loss on ocean ecosystem services. *Science* **314:** 787–90.
16. Leakey, R., *et al.* 1995. *The Sixth Extinction: Patterns of Life and the Future of Humankind.* Doubleday. New York.
17. Pimm, S.L., *et al.* 2000. The sixth extinction: how large, where and when? In *Nature and Human Society: the Quest for a Sustaiable World.* P.H. Raven, Ed.: 46–62. National Academy Press. Washington, DC.
18. Wake, D.B., *et al.* 2008. Are we in the midst of the sixth mass extinction? A view from the world of amphibians. *Proc. Natl. Acad. Sci.* **105**(Suppl. 1): 11466–11473.
19. Chan, K.M., *et al.* 2007. When agendas collide: human welfare and biological conservation. *Conserv. Biol.* **21:** 59–68.
20. Attitudes of European Citizens Towards the Environment. 2008. Directorate General Environment, Brussells.
21. Franzen, A., *et al.* 2010. Environmental attitudes in cross-national perspective: a multilevel analysis of the ISSP 1993 and 2000. *Eur. Sociol. Rev.* **26:** 219–234.
22. Economist. 2009. Cap, trade and block. *The Economist* **393:** 51–52.
23. Rolston, H. 1994. *Conserving Natural Value.* Columbia University Press. New York.
24. Luke, T.W. 1999. Environmentality as green governmentality. In *Discourses of the Environment.* E. Darier, Ed.: 121–151. Blackwell. Oxford.
25. Foucault, M. Governmentality. In *The Foucault Effect: Studies in Governmentality.* G. Burchell, *et al.*, Eds.: 87–104.Harvester Wheatsheaf. Hemel Hempstead .
26. Hardin, R. 2011. Concessionary politics: property, patronage and political rivalry in Central African forest management. *Curr. Anthropol.* **52:** S113–S125.
27. Brosius, J.P. 1999. Green dots, pink hearts: displacing politics from the Malaysian Rain Forest. *Am. Anthropol.* **101:** 36–57.
28. Thuk-Poe, L. 2004. *Changing Pathways: Forest Degradation and the Batek of Pahang.* Lexington Books. Oxford.
29. Fletcher, R. 2010. Neoliberal environmentality: towards a poststructuralist political ecology of the conservation debate. *Conserv. Soc.* **8:** 171–81.
30. Berry, W. 1989. *The Unsettling of America*, Sierra Club Books. San Francisco, CA. 23.
31. Franzen, A. 2004. Environmental attitudes in international comparison: an analysis of the ISSP Surveys 1993 and 2000. *Soc. Sci. Quart.* **84:** 297–308.
32. Speth, G. 2008. *The Bridge at the End of the World: Capitalism, the Environment, and Crossing from Crisis to Sustainability.* Yale University Press. New Haven, CT.
33. Maitland, F.W. 1905. Moral personality and legal personality. *J. Soc. Compar. Legisl.* **6:** 192–200.
34. Ferguson, J. & A. Gupta 2002. Spatializing states: towards an ethnography of neoliberal governmentality. *Am. Ethnol.* **29:** 981–1002.

35. United States Supreme Court. 1819. *Trustees of Dartmouth College V. Woodward*, **17:** 518-715.
36. House of Lords. 1897. *Salomon V. Salomon and Co. Ltd.* London: House of Lords, A.C. 22.
37. Hughes, P. 1997. Can governments weather the storm in the new communications climate? *Aus. J. Pub. Administration* **56:** 78–86
38. Domke, D. 2004. *God Willing? Political Fundamentalism in the White House, the 'War on Terror' and the Echoing Press.* Pluto Press. London.
39. Holmes T. 2009. Balancing acts: PR, 'impartiality' and power in mass media coverage of climate change. In *Climate Change and the Media.* T. Boyce & J. Lewis, Eds.: 92–100. Peter Lang. New York.
40. Bakan, J. *The Corporation: The Pathological Pursuit of Profit and Power*. Free Press: New York.
41. Myers, N., *et al.* 2001. *Perverse Subsidies: How Tax Dollars Can Undercut the Environment and the Economy.* Island Press. Washington DC.
42. Willams, N. 2010. Tar troubles. *Curr. Biol.* **20:** R260.
43. Environmental Law Institute. 2009. *Estimating U. S. Government Subsidies to Energy Sources: 2002–2008.* Environmental Law Institute. Washington DC.
44. Environmental Protection Agency. 2011. *EPA's Budget and Spending.* Available at http://www.epa.gov/planandbudget/budget.html (Accessed on May 16, 2011).
45. Leopold, A. 1949. *A Sand County Almanac and Sketches Here and There.* Oxford University Press. New York: 262.
46. Rolston, H. 1987. *Environmental Ethics: Duties to and Values in Natural World.* Temple University Press. Philadelphia: 186–187.
47. Callicott, J.B. 1986. On the intrinsic value of nonhuman species. In *The Preservation of Species: The Value of Biological Diversity.* B. Norton Ed.: 138–72. Princeton University Press. Princeton NJ.
48. Zimmerman, M.J. 2001. *The Nature of Intrinsic Value.* Rowman and Littlefield. Lanham, VA.
49. US Senate. 1973. *Endangered Species Act of 1973.* House of Congress. Washington, DC. Sec. 2. 1–5.
50. New Zealand Legislation: Acts. 1991. *Resource Management Act 1991.* Parliament of New Zealand. Wellington, NZ. 2.1 (1).
51. Arnoux, R., *et al.* 1993. The logic of death and sacrifice in the resource management law reforms of Aotearoa/New Zealand. *J. Econ. Issues* **27:** 1059–1096.
52. Grundy, K.J. *et. al.* 1996. Sustainable management and the market: The politics of planning reform in New Zealand. *Land Use Pol.* **13:** 197–211.
53. Austin, A. & Phoenix L. 2005. The neoconservative assault on the Earth: the environmental imperialism of the Bush administration. *Capitalism Nat. Socialism* **16:** 25–44.
54. Northcott, M.S. 2009. The dominion lie: how millennial theology erodes creation care. In *Diversity and Dominion: Dialogues in Ecology, Ethics, and Theology*. K.S. Vanhouten & M.S. Northcott, Eds.: 89–108. Wipf and Stock. Portland, ON.
55. Gavin, S.F. 2008. Stealth tort reform. *Valparaiso Univ. Law Rev.* **42:** 431–9.
56. Sunstein, C.R. 2007. *Worst-Case Scenarios.* Harvard University Press. Cambridge, MA.
57. Vanhouten, K.S. & M.S. Northcott 2009. Nature and the nation-state: ambivalence, evil and American environmentalism. In *Diversity and Dominion: Dialogues in Ecology, Ethics, and Theology*. K. Vanhouten & M. Northcott, Eds.: 138–156. Wipf and Stock. Portland, ON.
58. Viswanath, P.V. & M. Szenberg. 2003. Examining the biblical perspective on the environment in a costly contracting framework. In *Economics of Judaism.* C. Chiswick & T. Lehrer Eds.: 122–141. Bar Ilan University Press. Ramat Gan.
59. Muir, J. 1911. *My First Summer in the Sierra.* Houghton Mifflin. Boston: **189:** 331
60. Thoreau, H.D. 1862. Walking. *Atlantic Monthly* **3:** 12–16.
61. Nash, R. 1967. *Wilderness and the American Mind.* Yale University Press. New Haven: 44–49
62. Margules, C. & Usher M.B. 1981. Criteria used in assessing wildlife conservation potential: a review. *Biol.l Conserv.* **21:** 79–109
63. Cullinan, C. 2003. *Wild Law: A Manifesto for Earth Justice* 2003. Green Books. Totnes, Devon.
64. Donlan, J. 2005. Re-wilding North America. *Nature* **436:** 913–914.
65. Charman, K. 2008. Ecuador first to grant nature constitutional rights. *Capitalism Nat. Socialism* **19:** 131–132.
66. de Waal, F. 2006. *Primates and Philosophers: How Morality Evolved.* Princeton University Press. Princeton, NJ.
67. Carruthers, B.G. & W.N. Espeland. 1991. Accounting for rationality: double-entry bookkeeping and the rhetoric of economic rationality. *Am. J. Sociol.* **97:** 31–69
68. Geijsbeek, J.B. 1972. *Ancient Double-Entry Bookkeeping: Luca Pacioli's Treatise.* Scholar's Book Co. Houston, TX.
69. Crosby, A.W. 1997. *The Measure of Reality: Quantification and Western Society 1250–1600.* Cambridge University Press. Cambridge.
70. Krautkraemer, J.B. 1998. Nonrenewable resource scarcity. *J. Econ. Literature* **36:** 2065–2107.
71. Costanza, R., *et al.* 1987. The value of the world's ecosystem services and natural capital. *Nature* **387:** 253–60.
72. Farber, S.C., *et al.* 2002. Economic and ecological concepts for valuing ecosystem services. *Ecolog. Econ.* **41:** 375–92.
73. Girolami, V., *et al.* 2009. Translocation of neonicotinoid insecticides from coated seeds to seedling guttation drops: a novel way of intoxication for bees. *J. Econ. Entomol.* **102:** 1808–1815.
74. Morse, R.A., *et al.* 2000. The value of honey bees as pollinators of US crops. *Bee Culture* **128:** 1–15.
75. Losey, J.E., *et al.* 2006. Ecological services provided by insects. *Bioscience* **56:** 311–323.
76. Coase, R. 1960. The problem of social cost. *J. Law Econ.* **3:** 1–44.
77. Bachram, H. 2004. Climate fraud and carbon colonialism: the new trade in greenhouse gases. *Capitalism Nat. Socialism* **15:** 1–16.
78. Northcott, M.S. 2010. The concealments of carbon markets and the publicity of love in a time of climate change. *Int. J. Pub. Theol.* **4:** 294–313.
79. Anscombe, G.E. 1958. Modern moral philosophy. *Philosophy* **33:** 1–16.
80. Murdoch, I. 1970. *The Sovereignty of Good.* Routledge. London

81. MacIntyre, A. 1981. *After Virtue: A Study in Moral Theory*. Duckworth. London.
82. Northcott, M.S. 1996. *The Environment and Christian Ethics*. Cambridge University Press. Cambridge, 314–316.
83. Van Houtan, K.S. 2006. Conservation as virtue: a scientific and social process for conservation ethics. *Conserv. Biol.* **20:** 1367–1372.
84. Carson, R. 1988. From the sense of wonder. *Landscape J.* **7:** 1–17.
85. Daly, H. & J. Cobb. 1989. *For the Common Good: Redirecting the Economy Toward Community, the Environment and a Sustainable Future*. Beacon Press. Boston.
86. Solomon, R.C. 1992. Corporate roles, personal virtues: an Aristotelean approach to business ethics. *Bus. Ethics Quart.* **2:** 317–339.
87. Moore, G. 2005. Corporate character: modern virtue ethics and the virtuous corporation. *Bus. Ethics Quart.* **15:** 659–685.
88. Barboz, D. 2010. String of suicides continues at electronics supplier in China. *New York Times*. May 25, 2010.
89. Welker, M.A. 2009. "Corporate security begins in the community:" mining, the corporate social responsibility industry, and environmental advocacy in Indonesia. *Cultural Anthropol.* **24:** 142– 179.
90. Sunstein, C.R. 2005. *Laws of Fear: Beyond the Precautionary Principle*. Cambridge University Press. Cambridge.
91. O'Rourke, A. 2003. A new politics of engagement: shareholder activism for corporate social responsibility. *Bus. Strat. Environ.* **12:** 227–239.
92. *Green My Apple Archive* 2007. Available at http://www.greenpeace.org/apple/ Accessed on July 20, 2011.
93. Lohmann, L. 2006. Carbon trading: a critical conversation on climate change, privatisation and power. *Develop. Dialog.* **48:** 3–328.
94. Wilson, E.O. 1995. *Naturalist*. Time Warner: Washington, DC.

Ann. N.Y. Acad. Sci. ISSN 0077-8923

ANNALS OF THE NEW YORK ACADEMY OF SCIENCES
Issue: *The Year in Ecology and Conservation Biology*

The impacts of nature experience on human cognitive function and mental health

Gregory N. Bratman,[1] J. Paul Hamilton,[2] and Gretchen C. Daily[3]

[1]Emmett Interdisciplinary Program in Environment and Resources, Stanford University, Stanford, California. [2]Department of Psychology, Stanford University, Stanford, California. [3]Department of Biology, Stanford University, Stanford, California

Address for correspondence: Gregory N. Bratman, Emmett Interdisciplinary Program in Environment and Resources, 473 Via Ortega, Suite 226, Stanford University, Stanford, CA 94305. gbratman@stanford.edu

Scholars spanning a variety of disciplines have studied the ways in which contact with natural environments may impact human well-being. We review the effects of such nature experience on human cognitive function and mental health, synthesizing work from environmental psychology, urban planning, the medical literature, and landscape aesthetics. We provide an overview of the prevailing explanatory theories of these effects, the ways in which exposure to nature has been considered, and the role that individuals' preferences for nature may play in the impact of the environment on psychological functioning. Drawing from the highly productive but disparate programs of research in this area, we conclude by proposing a system of categorization for different types of nature experience. We also outline key questions for future work, including further inquiry into which elements of the natural environment may have impacts on cognitive function and mental health; what the most effective type, duration, and frequency of contact may be; and what the possible neural mechanisms are that could be responsible for the documented effects.

Keywords: ecosystem services; nature experiences; psychology; cognitive function; mental health

Introduction

For hundreds of years and across many cultures of the world, influential traditions in science, philosophy, poetry, and religion have emphasized the role that nature plays in providing feelings of well-being. In the modern era of scientific enterprise, a large body of work has demonstrated the importance of nature to human physical health, characterizing the numerous ways in which people depend on the natural environment for security in the supply of food, water, energy, climate stability, and other material ingredients of well-being. And now, in the face of intensifying human impacts on the natural environment—perhaps most visible in the form of land conversion, urban sprawl, and pollution of air and water—researchers have begun to document the importance of nature for mental functioning as well. For example, recent work has shown, though not yet explained causally, the disadvantage that individuals from urban environments have in processing stress when compared to their rural counterparts.[1]

Beliefs about the role of nature experience in mental health have played a role in the civic and political discussions surrounding conservation for a long time. In the United States, for example, writers such as John Muir and the originators of the Wilderness Act discussed nature's contributions to mental health specifically, albeit qualitatively.[2] This discourse extends well beyond "wilderness." In their work on the history of healing gardens in hospital settings, Marcus and Barnes trace the incorporation of restorative gardens and natural areas in infirmaries back to the Middle Ages, referring to the nearly thousand-year old writings of St. Bernard that support the healing effects of these natural spaces.[3] The authors follow these "courtyard traditions" in hospitals through the English, German, and French designs of the 1600s–1800s. The benefits of natural areas were

doi: 10.1111/j.1749-6632.2011.06400.x

thought to span physiological and mental aspects of well-being. Remnants of these traditions can still be found in the inclusion of *Kur* ("course of treatment" involving nature walks, herbal remedies, and mud baths) in mainstream German healthcare.[4,5]

The incorporation of nature into the estates of the rich is another example of the extent to which people have been willing to invest resources in aesthetically pleasing landscapes throughout history. The reasons for this may vary from a display of power and control over nature (as in the gardens of Versailles) to a sense of peace and enlightenment that these landscapes create in the mind of the landowner.[6] Modern environmental economics addresses the ways in which people are willing to pay for access to natural landscapes, using travel cost methods, contingent valuation, and hedonic studies of property values that embody a preference for nature in higher prices for places nearer to it.[7–13] But a central question remains: *why* are some people willing to pay more for contact with (or views of) nature?

Today, most people are experiencing significantly lower levels of daily contact with nature as compared to their parents' generation. One study estimates that the typical American now spends nearly 90% of his or her life within buildings.[14] This trend permeates most areas of the world. Many cultures with strong traditional ties to their surrounding natural environs have found themselves under the assault of modernization, development, and environmental degradation, which have been tied conclusively to an increase in feelings of isolation and depression within these communities.[15–18] As we move into cities and indoors at an unprecedented rate, we are faced with a rapid disconnection from the natural world, and this opens a suite of critical questions about repercussions for psychological well-being.

Approach to the review

Here, we review the effects of nature experience on human cognitive function and mental health, synthesizing work from environmental psychology, urban planning, medicine, and landscape aesthetics. We provide an overview of the prevailing explanatory theories of these effects, the ways in which exposure to nature itself has been considered, and the role that individuals' preferences for nature may play in its impact on psychological functioning. Specifically, we consider three possible explanations for the effects of nature experience on cognitive function and mental health. The first two, attention restoration theory and stress reduction theory, stem from effects that may remain unrecognized to the individual, while the third, an idea that has its roots in the traditions of social psychology, relates to the mediating effects of explicitly held preferences about nature.

We include studies that employ a particular set of tools and approaches (traditional psychology tests, surveys, and questionnaires) to quantify impacts of nature experience on specific aspects of cognitive function and/or mental health (attention, concentration, memory, impulse inhibition, stress, and mood). Using a "snowball" method, we began with the work of Stephen and Rachel Kaplan that played a crucial role in establishing modern environmental psychology[19] as well as the work of Roger Ulrich on the measurement of stress in individuals as they respond to different environments.[20] From these groundbreaking and foundational studies, we worked forward by compiling the literature that builds on them. Our search methods included mining the references of these subsequent studies and using computer search engines. We restrict the focus of our review to the benefits that fit under the theories developed from these two strands of thought, along with the additional exploration of the ways in which preferences for nature may or may not influence these particular benefits. Thus, our search brought us through much of the environmental psychology literature, touching occasionally on studies that fall within the bounds of urban planning, medical research, and landscape aesthetics. There is currently exciting, interdisciplinary work underway on broader aspects of "cultural ecosystem services" and relevant decision-making challenges.[21,22]

Nature

Our analysis must begin with a clear notion of nature. In their biophilia hypothesis, Wilson and Kellert claim that we, as human beings, have an innate love for the natural world, universally felt by all, and resulting at least in part from our genetic make-up and evolutionary history.[23] But what do we mean when we speak of the natural world, or nature? These are clearly subjective terms. Studies have shown that most individuals consider the term "wilderness" to consistently and generally apply to areas without discernible human influence.[19,24,25]

But "wilderness" is only one category from a broad spectrum of gradients, and the degree or amount of "nature" that a landscape contains can be culturally or personally defined. Additionally, cultures and individuals differ with respect to what are considered to be the attractive and natural components of landscapes.[26]

The definition of what makes an environment "natural" changes across time, space, and the individual engaged in the defining. Debates span the humanities and natural sciences over whether nature is a social construction or if it exists on its own in an independent and constant form.[27,28] We cannot look to science for an impartial or consistent answer to this question. "Objective" classifications from satellite data have been shown to differ from individuals' assessments of environmental qualities and descriptions of areas in the same place.[29] Philosophical debates over the human definition and representation of nature are numerous and complicated, and a summary of them is beyond the scope of this paper, but we briefly discuss some key issues below.

Most studies included in this review use comparative approaches in which the experience of individuals within one environment is contrasted with that of individuals within another, where one environment is clearly more "natural," within the context of the study (e.g., tree-lined city streets vs. trails through a nature preserve). The ranking of sites along an urban–natural gradient is therefore clear (without specific definition) within the context of each study. The impact of environments natural to different degrees is captured within these studies, though descriptions of these degrees are not categorized in a consistent way. Interestingly, all of the natural environments provided quieter atmospheres and were almost always accompanied by a comparatively larger field of view than the urban environments (e.g., there is no documented instance in which subjects were placed within a cave or another such natural but confined space).

For the purposes of this review, we developed a definition of nature that is applicable to all of the environments considered within these studies: areas containing elements of living systems that include plants and nonhuman animals across a range of scales and degrees of human management, from a small urban park through to relatively "pristine wilderness." A definition of this breadth is necessary, given the large range of landscapes included in the aggregate of these studies and the lack of pertinent ecological details.

This work addresses crucial but relatively unexplored questions about the particular elements of nature that impact the human psyche. At a minimum, it would be most informative were the research to specify the types of environments used in experiments in some detail, using modern quantitative methods at multiple scales. Ideally, further research would seek to understand and define what the "natural" components of these landscapes are that act as input for psychological mechanisms. This would lead to a more coherent and thorough set of postulates about which particular aspects of nature may have impacts on cognitive function and mental health—and ultimately, what the causal pathways are for these effects.

The nexus of nature experience and cognitive function and mental health

We examine studies that have attempted to document the psychological impacts of nature experience in a scientifically rigorous way. Many of us have experienced an emotional fulfillment from viewing—or being physically present within—natural environments. And on an instinctual level, many of us can also relate to Kellert and Wilson's hypothesis that human beings have a universal, innate connection to nature.[23,30] Theory from social psychology emphasizes the importance to the individual of belonging to a group, and Wilson argues that we have a similar need to feel connected to natural environments.[31]

Studies in environmental psychology focus particularly on the questions that follow from this connection: what happens to our cognitive abilities, emotional states, and mental health (all defined below) if we are deprived of experience in nature? Does the human psyche suffer in a measurable way—and across cultures, ages, and genders? If so, and if an increasing proportion of the global human population is experiencing the impacts of a withdrawal from nature, it may be helpful to define and investigate a new type of ecosystem service. This service would encapsulate the ways in which nature benefits our minds; thus, we might call it a *psychological ecosystem service*.

If psychological benefits from nature experience exist, they must come from the *interaction* between the individual and the environment—that is, they come as a result of our biology and cognitive processes within the context of a place, landscape, or seascape.[32] Thus, the service takes place within the mind and body of a person as he or she experiences an environment. These types of phenomena may seem hard to define, but a growing body of research has attempted to identify the consistent benefits that experiences of nature may provide.

We examine the aspects of these benefits that are relevant to *cognitive capacities* (including attention, memory, and impulse inhibition), *emotional states* (mood), and *stress*. This is not to imply that *negative* psychological effects from nature experience are not possible as well. Fear of being attacked by wildlife or struck by certain types of disease, which are particularly possible in natural environments, can cause mental distress. Hurricanes, earthquakes, and other natural disasters bring with them high levels of emotional anguish for those affected.[33,34] Interestingly, there have been few investigations of such potential negative effects; we focus on the growing literature on positive effects.

Nature and evolution of the human psyche

In attempts to tease out consistent contributing forces across many of these studies, some authors theorize that evolutionary influences are at work in our preference for particular natural environments. A popular hypothesis explains the value of grasslands and savannas to human well-being in terms of the sightlines and room for flight that such landscapes would have provided our early ancestors, when most forms of protection or flight available today did not yet exist.[20,23,35–39]

The data supporting this hypothesis are equivocal.[40] Nonetheless, the positive feelings we may experience from viewing these spaces are in stark contrast to the immediate, impulsive, and possibly instinctive repulsive reaction we have toward snakes and spiders—animals that may not have served us well in our evolutionary past. Interestingly, the strength of these aversions has been shown to be significantly greater than that elicited by the far more damaging, modern threat of guns.[23,41]

Prevailing theories about the attraction to (and possible restorative effects of) viewing or having physical contact with natural landscapes most often stem from the supposition that human beings are not fully adapted to urban environments and that something may be missing when we are deprived of contact with nature, however we choose to define it.[19] Details of the arguments vary, but most are based upon the postulate that the overwhelming evolutionary experience of human beings as a species involves natural environments, and we are therefore predisposed to resonate with these surroundings, consciously or not. We consequently come away from them with an increase in our positive affect and decrease in our negative feelings or stress—particularly when we have interacted with those environments that were favorable for our survival as a species.[42,43]

Box 1: Terminology

Definitions used in this article and typical metrics for assessment

Directed attention: The effortful, conscious process of bringing cognitive resources to bear in order to focus on selected stimuli, while avoiding distraction from unrelated perceptual inputs. Assessment of this ability involves tests used to measure concentration, impulse inhibition and memory.

Concentration: Directed attention applied over a relatively long time interval. While a variety of tests access this construct, measures of mental vigilance (e.g., Necker Cube Pattern Control, proofreading, etc.) do so most directly and without reliance on short-term mental storage systems.

Impulse inhibition: The capacity to stop execution of an overlearned or prepotent response. Response conflict tasks such as stop signal and Stroop color-word can be used to measure this construct.

Short-term and working memory: The ability to keep information in mind over short delays has been measured in the literature with simple span tasks such as the forward digit span test. The capacity to manipulate and transform information in memory is typically measured through more complex span tasks, such as the backward digit span and operation span tests.

Mood: A sustained positive or negative affective state that can influence emotions occurring over a shorter time-span—e.g., a bad mood can increase the frequency with which one feels the emotion of anger. Mood can be measured through self-assessment surveys, such as the PANAS (Positive and Negative Affect Schedule).

Nature/natural: Areas containing elements of living systems that include plants and non-human animals across a range of scales and degrees of human management—from a small urban park to "pristine wilderness."

Nature experience: Time spent being physically present within, or viewing from afar, landscapes (or images of these landscapes) that contain elements from the above category. The distinction between physical and visual contact with nature may be important.

Stress: The psychophysiological phenomenon caused when environmental demands reach or exceed an organism's capacity to address those demands.

Introduction to attention restoration and stress reduction theories (theories I and II)

There are two major explanatory theories within the environmental psychology literature that account for the restorative power of nature, and they both draw heavily on the theory of evolution. One of these frameworks, stress reduction theory (SRT), posits a healing power of nature that lies in an unconscious, autonomic response to natural elements that can occur without recognition and most noticeably in individuals who have been stressed before the experience.[44] Certain natural places (especially those along watersides and with visible horizons) may be seen as safe havens—areas in which our species tended to have greater rates of survival. The positive affective response that we feel in these spaces is due to this common evolutionary history. In other words, merely seeing or being present within nature can reduce stress through the automatic generation of physiological and psychological responses, the qualities of which will be explained more below.

The other explanatory theory, attention restoration theory (ART), centers on the power of nature to replenish certain types of attention through unconscious, cognitive processes in response to natural landscapes. Its supporters claim that directed attention is the mechanism most closely related to focus and concentration, and our urban life taxes this capability more consistently than the situations with which human beings have had to deal in our collective past.[19] The experience of interacting with natural environments allows this capability to replenish itself through a process of restoration, also to be described more below.

These theories have much in common, with the major points of departure involving a focus on cognitive versus autonomic processes. Both support the idea that changes in attention and stress load can come from interaction with natural environments, but they differ in their claims about the primary mechanisms at work. There may also be a "blurring" of effects between the two theories. Does a reduction of stress allow an individual to concentrate better, or does a replenishment of directed attention make a person feel less stressed, as an additional benefit?[45,46]

Assertions are controversial regarding the causal mechanisms of nature's impact on an individual's mental and physical state. Nevertheless, a great deal of research in environmental psychology can be seen as falling under one, or both, of these theoretical camps.[19,35,45]

The explanations of changes in measurements of mood fall somewhere between the two theories of ART and SRT and may also follow from a third theory we will discuss below: effects that are tied to conscious preferences. ART and SRT both assert that contact with nature should induce positive affect, either through the replenishment of directed attention (and the relief and relaxation that this brings) or through the benefits of reduced stress. Thus, measurements of mood appear in studies that work within either of the theories' constructs.

SRT

Ulrich suggests that landscapes with views of water and/or vegetation and that contain modest depth, complexity, and curvilinearity would have been most beneficial for survival (allowing for the spotting of food sources, predators, etc.).[42,46] These landscapes, according to SRT, help to moderate and diminish states of arousal and negative thoughts within minutes, through psychophysiological pathways.[24,35]

Ulrich appeals to work in affective psychology from the 1970s and 1980s in stating that emotions occur innately and in some state of constancy across cultures (i.e., nearly all people are born with the ability to feel "sad" or "happy").[37] Many cross-cultural regularities exist in the way these emotions are expressed facially.[47–49] Additionally, these feelings may occur before an individual is consciously aware of them.[50] Building from this work, Ulrich claims that conscious processes are not necessary or required to produce emotion. Thus, affective reactions to environments may happen at a preconscious level and may subsequently impact cognitive processes without an individual's conscious knowledge.[42] Although urban dwellers may *think* they have habituated themselves to factors that cause stress, they may still be having stress-related reactions in their bodies and brains about which they are unaware.

Stress studies. Ulrich put his hypothesis to the test in a series of exploratory studies.[46,51] He instructed a group of mildly stressed participants to view sets of color slides: one group saw nature scenes with vegetation and trees predominating the visual field, while another group viewed city landscapes with little to no vegetation. Self-ratings of positive affect, including elation and affection, were greater in those subjects that viewed the natural, vegetative scenery. Negative feelings such as fear were lower in the nature group as well. Additionally, urban viewers experienced increases in aggravation, anxiety, and feelings of sadness.

More recent work confirms these results, showing decreases in self-reported stress and increases in positive mood after prolonged experience in wilderness areas.[2,52] Ulrich found similar results when students (stressed because of a final exam) reported higher levels of positive affect and lower levels of fear after viewing slides of natural scenes than those who viewed urban ones.[46] Further, Honeyman found the same trend when subjects were presented with urban images containing vegetation versus urban images without vegetation.[53]

To further test this theory, Ulrich *et al.* ran an experiment in which 120 subjects watched a stressful movie for 10 min and then viewed scenes (and sounds) of six different types of settings, ranging from most urban to most natural for another 10 minutes.[37] During this, subjects were monitored for levels of physiological stress through the measures of heart rate, skin conductance, muscles tension, and systolic blood pressure. Subjects were also asked to self-rate their affective states. All measures indicated significantly higher speed of recovery from stress when subjects were viewing nature scenes than when they were viewing urban scenes. In a related area, recent work using functional magnetic resonance imaging (fMRI) has shown that urbanization may tax the neural mechanisms involved in dealing with stress.[1]

Other studies add supporting evidence for SRT. The impact of forests versus urban landscapes on stress relief was explored by transporting 12 subjects between forest and city settings in Japan—and measuring salivary cortisol concentration, diastolic blood pressure, and pulse rate while the subjects were physically present within each. All of these measures indicated significantly decreased stress for the participants after being present in the forests for only 15 min—a result that was not found when they were placed in urban landscapes.[54,55] Using survey techniques, Ottosson and Grahn found that those individuals who were currently dealing with a greater crisis—and the increased level of stress that accompanies this—experienced the stress relief from nature experience to a greater degree than others.[39]

Future work in this area should explore the possibility that a change in context itself plays a role in these observed reductions in stress. Removing oneself from habitual patterns of normal experience may have a psychophysiological effect that is unrelated to the natural elements of the newer, less familiar context. The aesthetics (i.e., degree of "pleasantness") of these settings may have an effect as well. One possibility for a future study might be to examine the relative impacts of "ugly" nature scenes versus "beautiful" urban scenes. We discuss aesthetics more in a section below.

ART

Kaplan and Kaplan formulated a theory that examines the ways in which exposure to nature can have a restorative effect on the brain's ability to focus. These researchers contend that a replenishment of our direct attentional capacities is the primary mechanism underlying effects of exposure to nature.[19] ART uses theoretical constructs dating back to William James, resting upon the proposal that attention can be separated into two distinct components voluntary (directed) and involuntary attention.[56]

The theory posits that directed attention requires the use of cognitive control—individuals must consciously use their faculties to focus on a stimulus that may or may not otherwise have attracted their attention. In order to do this, an individual must inhibit or suppress the urge to pay attention to distractions. After prolonged use, this capability can become fatigued, and this fatigue may reveal itself through difficulties in concentrating and higher rates of irritability.[57] Traditional psychological constructs of working memory, impulse inhibition, and the less specifically defined concept of "concentration" are all capabilities that supposedly require directed attention—and we can therefore measure levels of this type of attention through established, valid psychological testing techniques.

In contrast, involuntary attention is utilized when individuals are presented with stimuli that are "inherently intriguing." ART claims that interaction with natural environments employs faculties of concentration not normally used—involuntary ones—thus allowing the neural mechanisms underlying directed attention a chance to rest and replenish. The experience that comes from viewing or being present within natural landscapes allows attentional reserves to replenish, which in turn can benefit performance on other tasks, delay of gratification, and perhaps even levels of depression and stress. Kaplan appeals to the words of a famous champion of urban parks, Frederick Law Olmsted: "[natural scenery] employs the mind without fatigue and yet exercises it; tranquilizes it and yet enlivens it; and thus, through the influence of the mind over the body, gives the effect of refreshing rest and reinvigoration to the whole system."[45,58]

Situations in which directed attention is rendered unnecessary for a period of time may allow for its restoration.[45] The Kaplans postulate that there are four essential components that a landscape must contain for it to most efficiently provide restorative effects on direct attentional capacity, and these are most often found in natural environments: *extent* (the scope of experience, including the possibility of feeling immersed within it); *being away* (an escape from the habitual activities and concerns of daily life, ranging from "micro"-experiences, such as gazing out a window, to day-long backpacking trips); *fascination* (aspects of an environment that innately capture attention, effortlessly and without directed effort); and *compatibility* (a "match" between an individual's intentions, inclinations, or purposes and the environment).[45,60–62]

Clearly, other settings may satisfy all or some of these conditions, but natural environments most consistently contain all of them simultaneously. Urban stimuli are postulated not to have these four qualities, typically, and therefore do not restore our direct attentional capacities. When removed from contact with nature, human beings are missing out on a critical type of rest. And through cognitive testing, we can measure whether replenishment has occurred after nature experience. There are a variety of behavioral studies that have been conducted in an attempt to test ART; we review these below.

Attention studies. Berman *et al.* tested subjects with a backward digit span task—a test that measures working memory and therefore serves as a proxy in environmental psychology for directed attention capacity.[63] After this test, the experimenters then induced mental fatigue in the subjects with a 35-min test that taxed memory and randomly sorted the participants into two groups: one group that walked through an urban setting, and another that walked through an arboretum—both walks were 2.8 miles and 50–55 minutes. Following this, participants performed the digit span backward task again. The "arboretum" group performed significantly better on the memory/directed attention task than did the "urban" group. The authors also showed increases in positive affect (as measured through the PANAS) in the arboretum-walk group. Significant improvements in working memory were also noted in a second study in which groups viewed pictures of natural versus urban scenery.

Tennessen and Cimprich used the digit span backward tests and the Necker Cube Pattern Test—a task in which the subjects must use their concentration to prohibit a stimulus (an ambiguously drawn cube) from "flipping orientations" as they view it—to test increased capacity for attention in students. The participants lived in dormitories with similar-sized windows that offered views ranging from "all natural" (trees and a lake) to "all built" (city streets, other buildings, or a brick wall).[64] Those students who had the most natural views showed a greater ability to direct attention. The authors consider these results to support the hypothesis that window views provide the opportunity for "micro-restorative activity." Digit span forward and backward tests were also given, but no significant differences were observed.

In a natural experiment, Taylor *et al.* compared children from the same population in a housing complex in Chicago, whose living conditions and demographic characteristics differed only by their views from home: a small pocket of urban park or a barren concrete area.[65] The authors examined the relations between near-home nature and concentration, impulse inhibition, and delay of gratification in inner-city children who had been randomly assigned to live in one of 12 architecturally identical high-rise buildings with varying levels of nearby nature. On average, the more natural a view from home, the better the performance on digit span backward, alphabet backward, matching familiar figures, and the Stroop color-word test (assesses the ability to override the tendency to read a word when it is printed in an incongruent color—e.g., the word "white" written in blue—while the instructions require the naming of the color, instead of reading), as well as a delay of gratification test (in which subjects had to avoid the temptation to eat a bag of candy when the tester leaves the room). Interpreting their data slightly differently from previous studies, the authors consider the aggregate performance on these tests to correspond to a form of "self-discipline." This has been shown to act as a mediating factor for lower levels of aggression and violence, as well as higher levels of scholastic and career success.[66]

In testing the potential usefulness of natural images, Berto induced mental fatigue in subjects through the sustained attention to response test (SART), a five-minute response–control test that requires subjects to press a button when a rarely occurring target digit appears on a computer screen, but not when other digits appear.[67] The experimenter then exposed participants to pictures of natural scenery ("restorative environments") or urban scenery ("nonrestorative environments"). Those exposed to natural pictures performed significantly better on the second administration of the SART than did their counterparts, after exposure to the images. Additionally, these results held when the subjects were exposed to natural versus geometric figures, supporting the assertion of ART that natural scenes in particular have this type of restorative potential.

Table 1. Types of environment

Urban green	Speldewinde *et al.*,[17] Mayer *et al.*[31] de Vries *et al.*,[40] Abkar *et al.*,[43] Hartig *et al.*,[52] Wells,[59] Berman *et al.*,[63] Tennessen and Cimprich,[64] Taylor *et al.*,[65] Kuo and Sullivan,[66] Nisbet and Zelenski,[76] Ulrich,[91] Kaplan,[95] Wells and Evans,[96] Pretty *et al.*,[100] Fuller,[105] Verderber,[107] Leather *et al.*,[108] Evans,[109] Grahn and Stigsdotter,[110] Groenewegen *et al.*,[111] Richardson *et al.*,[112] Coley *et al.*,[115] Kuo *et al.*,[116] Takano *et al.*,[117] Maas *et al.*,[118] Mitchell and Popham,[119] Van den Berg *et al.*[120]
Water bodies	Mayer *et al.*,[31] Ulrich,[37] de Vries *et al.*,[40] Ulrich,[51] Laumann *et al.*,[68] Chang *et al.*[113]
Forest/woodland	Hartig *et al.*,[35] Park *et al.*,[54] Lee *et al.*,[55] Chang *et al.*[113]
Countryside/farmland	Mayer *et al.*,[31] Hartig *et al.*,[35] Ulrich,[37] Ulrich[51]
Wilderness	Cole and Hall,[2] Hartig *et al.*,[52] Paxton and McAvoy[114]

Categories of natural environments in which corresponding studies were conducted. Color-coded categories represent each study by its type in Figure 1. This only includes examples in which exposure is claimed to have made a direct psychological or behavioral impact—it does not include studies in which preferences were examined based upon presentation of images, etc. (e.g., Anderson[85]).

Studies that address attention and stress simultaneously

The Kaplan and Ulrich theories are related but differ in important ways. As Ulrich stresses the importance of the evolutionary aspects of response to environment, he tends to emphasize affective and stress-related components of the individual's relationship with landscapes. The Kaplans' theory is centered more on effects on cognition. Thus, Ulrich emphasizes the importance of a reduction in arousal, with physiological evidence showing decreased stress levels in subjects when viewing natural versus urban images.[37,46] This contrasts with ART, which is more concerned with a replenishment of attentional capacities.[52]

There are some studies that have attempted to address both stress- and attention-related factors at once. Hartig *et al.* used ambulatory blood pressure measurements to assess psychophysiological stress differences in groups of individuals with varying levels of attentional fatigue who viewed, or were present within, urban versus natural environments.[35] Both nature experience groups (natural views and

Table 2. Duration of time in nature

Minutes to hours	Mayer *et al.*,[31] Hartig *et al.*,[35] Ulrich,[37] Abkar *et al.*,[43] Ulrich,[46] Ulrich,[51] Hartig *et al.*,[52] Park *et al.*, Lee *et al.*,[55] Berman *et al.*,[63] Berto,[67] Laumann *et al.*,[68] Nisbet and Zelenski,[76] Pretty *et al.*,[100] Fuller,[104] Chang *et al.*[113]
Days	Cole and Hall,[2] Hartig *et al.*,[52] Ulrich,[91] Verderber,[107] Paxton and McAvoy[114]
Years/longitudinal studies	Speldewinde *et al.*,[17] de Vries *et al.*,[40] Wells,[59] Tennessen and Cimprich,[64] Taylor *et al.*,[65] Kuo and Sullivan,[66] Kaplan,[95] Wells and Evans,[96] Leather *et al.*,[108] Evans,[109] Grahn and Stiggsdotter,[110] Groenewegen *et al.*,[111] Richardson *et al.*,[112] Coley *et al.*,[115] Kuo *et al.*,[116] Takano *et al.*,[117] Mitchell and Popham,[119] Maas *et al.*,[118] Mitchell and Popham,[119] Van den Berg *et al.*[120]

Studies categorized by the duration of time in which subjects were exposed to a particular natural environment. This only includes examples in which exposure is claimed to have made a direct psychological or behavioral impact—it does not include studies in which preferences were examined based upon presentation of images, etc. (e.g., Anderson[85]).

presence within natural landscapes) showed decreased stress, improved mood, and better performances on attention tests (the Necker Cube Pattern Test and proofreading task). Because the authors were able to measure both blood pressure and attentional capacities at various times throughout the walk (instead of just before and after), they were able to conclude that stress and attention impacts happened at different times and were not significantly related, providing evidence for the possibility of different causal pathways for the positive impact of both types of measures.

Both attention- and stress-related theories were studied simultaneously in another example in which subjects viewed videos of urban versus natural (waterside or forest) scenes.[68] In this example, participants had increased attentional load induced by a proofreading task. This was followed by a test of their attentional capacity with Posner's attention-orienting task, a test that allows for the experimenter to distinguish between involuntary and voluntary attention performance through examining the individual's processing of different visual stimuli (peripherally or centrally located) and the ability to shift between the two types of attention demands.[69–71] Laumann *et al.* simultaneously gathered stress reduction data by continuously recording heart rates throughout the experiment using electrocardiogram (EKG) equipment.[68] The "nature group" had significantly lower heart rates than the "urban group" while watching their respective videos, but did not show an increased ability to shift between voluntary and involuntary attention. Thus, these results support SRT, but do not provide compelling evidence for ART.

Future work

As we summarize in Table 4, these studies have used a set of valid, traditional tasks and technologies to measure attention and stress in individuals. There is room for even more finely grained analysis, however. Through the use of other tests, such as filtering tasks, complex operation span tasks, or distraction tasks, we may be able to further isolate effects of prepotent response inhibition and resistance to distractions.[72] As mentioned previously, future studies should work toward pinpointing causal effects and mechanisms, as well as determining whether the removal of subjects from the situation to which they are accustomed and placing them in a novel environment might be responsible for some of the effects that have been attributed specifically to nature in these experiments.

Box 2: Theories of restorative benefits of nature

Theory I

Stress reduction theory (Ulrich)—reduction in stress during experience of natural stimuli. Measured through physiological response.

Theory II

Attention restoration theory (Kaplan and Kaplan)—recovery from directed attention fatigue through experience of natural stimuli. Focused on cognitive processes and responses.

Theory III

Mediating effect of opinions about nature—our conscious opinions about nature relate to the impacts of nature experience on mood and other aspects of cognitive function.

Table 3. Types of exposure

Images	Mayer *et al.*,[31] Ulrich *et al.*,[37] Ulrich,[46] Ulrich,[51] Honeyman,[53] Berman *et al.*,[63] Berto,[67] Laumann *et al.*,[68] Pretty *et al.*,[100] Chang *et al.*[113]
Window views	Abkar *et al.*,[43] Tennessen and Cimprich,[64] Ulrich,[91] Moore,[93] Kaplan,[95] Verderber,[107] Leather *et al.*,[108] Takano *et al.*[117]
Physically present	Cole and Hall,[2] Speldewinde *et al.*,[17] Mayer *et al.*,[31] Hartig *et al.*,[35] de Vries *et al.*,[40] Abkar *et al.*,[43] Hartig *et al.*,[52] Park *et al.*,[54] Lee *et al.*,[55] Wells,[59] Berman *et al.*,[63] Taylor *et al.*,[65] Kuo and Sullivan,[66] Nisbet and Zelenski,[76] Wells and Evans,[96] Fuller,[104] Evans,[109] Grahn and Stigsdotter,[110] Groenewegen *et al.*,[111] Richardson *et al.*,[112] Paxton and McAvoy,[114] Coley *et al.*,[115] Kuo *et al.*,[116] Takano *et al.*,[117] Maas *et al.*,[118] Mitchell and Popham,[119] Van den Berg *et al.*[120]

Studies categorized by three different degrees of exposure: passive viewing of representations ("images"); views of natural landscapes themselves ("window views"); and presence within landscape or environment ("physically present"). This only includes examples in which exposure is claimed to have made a direct psychological or behavioral impact—it does not include studies in which preferences were examined based upon presentation of images, etc. (e.g., Anderson[85]).

Preferences for nature (theory III)

We now return to the issue of opinions about the environment. Conscious preferences for landscape aesthetics may relate to the restorative benefits of nature in a complicated manner. There are a number of ways to think about this issue. Do attitudes about nature directly impact the cognitive and/or mood benefits that interactions with these landscapes might provide? Or do human beings have an innate and even universal preference for the very aspects of nature that are restorative, thus eliminating the degree to which preferences themselves might influence benefits as a mediating factor?

In support of the latter view, one study has shown that the more mentally fatigued the subject, the greater the likelihood that he or she would choose a restorative walk in a natural environment over an urban one.[73] Additionally, Korpela *et al.* examined stated preferences for four of the components that Kaplan and Kaplan claim to be essential restorative qualities of natural environments—"being away, fascination, extent, and compatibility."[74] They found that these qualities correlate very closely with the aspects of an environment that *independently* make it a "favorite place" for subjects. Thus, it may be that people seek out these types of characteristics in the places to which they feel most attached, a postulate that fits well with theories of self-regulation.[75] They may make these choices without being aware of the fact that these qualifications might be most consistently fulfilled by natural settings. Indeed, a recent study demonstrated that people tend to underestimate the degree to which even short exposure to natural environments can increase positive mood.[76]

We will now explore the details of the former view—that an individual's opinions about nature may impact the way in which natural environments ultimately impact his or her mood and cognitive function. Exploration of this concept typically involves a measurement of "connection to nature"[18] and draws on theory from social psychology that the sense of belonging to something greater than oneself, and a resultant decrease in negative rumination, has an effect on feelings of well-being.[77–82]

Explicitly stated connection to nature

Much of the literature in "landscape aesthetics" involves efforts to analyze the ways in which people come to explicitly judge the scenic beauty of an environment through stated preferences and willingness to pay, typically for levels and shapes of openness, obstruction, scale, and depth of views.[83,84] Interestingly, the manner with which a landscape is described may play a crucial role in the way it is rated. Anderson showed that subjects' preconceived notions of landscape descriptions or designations impacted their perceived degree of natural beauty.[85] When shown identical photographs with varying descriptions ("commercial timber stand, leased grazing range, recreation area, national park, or wilderness area"), participants rated (the exact same) scenes higher as the degree of natural qualities increased in the descriptions. Although these studies are interesting in and of themselves, they do not explicitly address cognitive benefits, nor do they

Table 4. Psychological functions and measurements. Techniques of assessment for various functions, and examples in the literature of studies that employed these techniques

Psychological function	Techniques for assessment	Techniques developed by	Examples in the literature
Concentration/ memory	Symbol digit modalities Digit span forward and backward Alphabet backward Necker cube pattern control tests Proofreading Search and memory test Posner's attention-orienting task	Symbol digit modalities (Smith[121]) Digit and alphabet span (Wechsler[122]) Necker cube (Orbach *et al.*[123]) Search and memory test (Smith and Miles[124]) Posner's attention-orienting task (Posner[125])	Kaplan and Kaplan,[19] Mayer *et al.*,[31] Hartig *et al.*,[35] Ottosson and Grahn,[39] Kaplan,[45] Hartig *et al.*,[52] Wells,[59] Berman *et al.*,[63] Tennessen and Cimprich,[64] Taylor *et al.*,[65] Kuo and Sullivan,[66] Berto,[67] Laumann *et al.*[68]
Impulse inhibition/delay of gratification	Matching familiar figures test Stroop color-word test Bag of candy test	Matching familiar figures test (Kagan[126]) Stroop test (Stroop[127]) Bag of candy test (Rodriguez *et al.*[128])	Hartig *et al.*,[35] Taylor *et al.*[65]
Aggression	Conflicts Tactic Scale (CTS) State anger section of the Zuckerman's Inventory of Personal Reactions (ZIPERS)	CTS (Straus[129]) (Zuckerman[130])	Hartig *et al.*,[35] Hartig *et al.*,[52] Kuo and Sullivan[66]
Stress relief	Physiological measurements of heart rate (EKG), heart rate, muscle tension, systolic and diastolic blood pressure (SBP, DBP). Brain electrical activity Skin conductance Survey of Perceived Restorativeness Scale (PRS) fMRI	PRS (Hartig *et al.*[131])	Lederbogen *et al.*,[1] Ulrich,[20] Hartig *et al.*,[35] Ulrich *et al.*,[37] Korpela *et al.*,[61] Laumann *et al.*,[68] Wells and Evans,[96] Grahn and Stigsdotter,[110] Van Den Berg *et al.*[120]
Mood	Profile of Mood States (POMS). Positive and Negative Affect Schedule (PANAS) Zuckerman Inventory of Personal Reactions (ZIPERS). Overall Happiness Scale (OHS)	POMS (McNair *et al.*[132]) PANAS (Watson *et al.*[133]) ZIPERS (Zuckerman[130]) OHS (Campbell *et al.*[134])	Mayer *et al.*,[31] Hartig *et al.*,[35] Ottosson and Grahn,[39] Berman *et al.*,[63] Nisbet and Zelenski,[76] Barton and Pretty[101]
Self-esteem	Rosenberg Self-Esteem Scale (RSE). The Global Self-Worth subscale of the Harter Competency Scale	RSE (Rosenberg[135]) Self-worth subscale of Harter Competency Scale (Harter[136])	Wells and Evans,[96] Barton and Pretty[101]

Continued

Table 4. *Continued*

Psychological function	Techniques for assessment	Techniques developed by	Examples in the literature
"Child development"	Inference drawn from increased social interactions; attention capacity (using mothers' ratings of children's cognitive abilities with attention deficit disorders); Global Self-Worth subscale of the Harter Competency Scale; ADHD symptoms		Wells,[59] Taylor *et al.*,[65] Wells and Evans,[96] Coley *et al.*[117]

attempt to isolate the reasons behind willingness to pay for the aesthetic beauty of landscapes.[86,87]

Mayer and Frantz developed a survey consisting of 14 questions that identify an individual's conscious, stated level of emotional connection to nature.[18] This "connectedness to nature scale" (CNS) is closely related to its predecessors: the New Environmental Paradigm (NEP) Scale and the inclusion of Nature in the Self (INS) scale.[88,89] CNS pays special attention to the ways in which people do or do not feel that they are a part of their surrounding, natural world (based upon Leopold's assertion that this is a necessary precursor for pro-environmental behavior).[90] In Mayer and Frantz's estimation, environmental behavior can often be predicted by the degree to which a person identifies himself or herself with the natural world—the higher the sense of "belonging," the greater the likelihood of sustainable actions in "lifestyle patterns, ecological behavior, and curriculum decisions among students."

The authors also claim that an individual's CNS score is correlated with life satisfaction, overall happiness, and perspective-taking ability.[18] In work on emotion, a decrease in self-awareness—or rumination on aspects of one's conception of self—has been shown to be associated with positive mood.[79–81] This agrees with recent work that shows increased depression can occur with increased rumination.[77,78] The tendency to engage in negative rumination might be linked to CNS scores—specifically, the notion that one is part of a force "greater than oneself" may lift an individual out of rumination on a negative sense of self. In essence, the sense of belonging to nature can provide a benefit in and of itself, in the same way that previous work in social psychology has shown that feelings of belonging to a group can provide a sense of purpose and positive impact for individuals.[31]

Thus, Mayer *et al.* argue that the positive effects of nature on mood are actually mediated by an increase in an individual's feeling of connection to nature through experience.[31] In this paradigm, connection to nature is a causal mechanism for the generation of psychological benefits because of the power of the feelings associated with belonging to a community or something "greater than oneself." A related question is whether or not there may be a relation between an individual's CNS score and the psychological benefits he or she receives from nature experience. If an individual feels more connected to nature, will there be a corresponding difference in the benefits that do or do not accrue to them with exposure to nature? Further research is needed to thoroughly investigate this question.

Urban nature

We have discussed studies that employ a variety of different types of exposure to natural environments for the subjects involved: from views of images and out of windows to being physically present within the landscape (Tables 1–3). Accessible natural areas are posited to be an important part of the mental health of urban citizens, whether viewed from inside a building or experienced while present in accessible city parks.

Windows

The power of scenic views from a window has been shown in several famous studies, including the compelling study by Ulrich in which patients recovering from gallbladder surgery who had a view from their hospital window had shorter postoperative stays, and less potent pain medication requirements, than those who looked out onto a brick wall.[91] Although most clearly related to physical health, this study may fit within the framework of SRT, assuming that decreased stress improves physical recovery. (This provides some affirmation of the investment in gardens throughout several centuries as a critical element in European hospitals.)[3]

Studies have also shown decreased stress levels to be associated with greater job satisfaction in workers with a view through their office windows that included natural elements, as well as greater life satisfaction and attentional capacity in residents (or even prisoners) who have natural views.[93,94] Proponents of ART claim that repeated viewings of natural surroundings may have incremental effects by allowing indirect attention a chance to come to the forefront of mental mechanisms and providing brief periods of replenishment for directed attention. The aggregate of these short exposures could lead to a restoration effect—so-called "potential micro-restorative opportunities."[95]

Urban greenspace

Several observational studies have shown a strong positive correlation between urban greenspace exposure (including gardens) and physical and emotional health.[40,59] Wells showed an increase in cognitive functioning capacity in children who had recently moved to more natural surroundings versus those who had moved to more urban environments.[59] And Wells and Evans demonstrated a correlation between "nearby nature" and parent-reported stress levels and self-reported measures of self-worth in children grades 3–5, leading the authors to postulate that nature may function as a "buffer" for children against stressful life events and threats to their self-esteem.[96]

Although exercise itself has been shown in numerous studies to impact mood, some believe that natural environments benefit the individual over and above the exercise itself, leading to support for "green exercise."[97–99] Pretty *et al.* compared groups of individuals all engaged in identical exercise while viewing either green landscapes or barren urban environments.[100] The former group showed significantly greater reductions in blood pressure as well as increases in positive mood and self-esteem relative to the control group.

In a meta-analysis involving 25 studies, Bowler *et al.* examined the impacts of short-term forms of "green exercise" (nearly all were one hour or less).[60] Subjects in each of the 25 studies engaged in identical physical activities within natural (urban and university greenspace, gardens, woodlands, wilderness parks) versus "synthetic" (gyms, laboratory, or urban space) environments. Results showed that the most significant difference was an increase in positive self-reported emotion for those subjects engaged in "green exercise." The authors did not find significant differences in attention capacity or measures of stress that could be attributed to varying exposures to natural versus synthetic environments, but their analysis provided evidence for additional positive benefits of exercise, specifically in the context of natural spaces.

Duration

The majority of these studies have relied upon either longitudinal designs that examine the impacts of different living conditions' access to nature or cross-sectional designs that examine responses to natural environments for periods of between 10 min and an hour; with urban parks, terrestrial nature preserves, and arboretums being used as the predominant landscape of choice (see Fig. 1).

More recently, researchers have also asked questions about the "ideal" duration of time spent in nature for mental health benefits. Is there duration with which we may find the greatest marginal return for our time—an ideal "dose" that packs the most mental health benefit in the shortest amount of time? In a meta-analysis, Barton and Pretty analyzed previous studies in an attempt to determine the most effective dose of green exercise for mood and self-esteem benefits.[101] Their study showed both men and women improving in measurements of self-esteem after green exercise (decreasing for both sexes with age), but only men exhibited significant improvement in mood. Their dose–response charts are somewhat puzzling: with the greatest benefit (measured by total change in mood [TMD] and self-esteem) coming from only five minutes of activity, followed by an entire day, and lowest improvements coming from 10 min to half-day doses. Further research is required to arrive at satisfactory

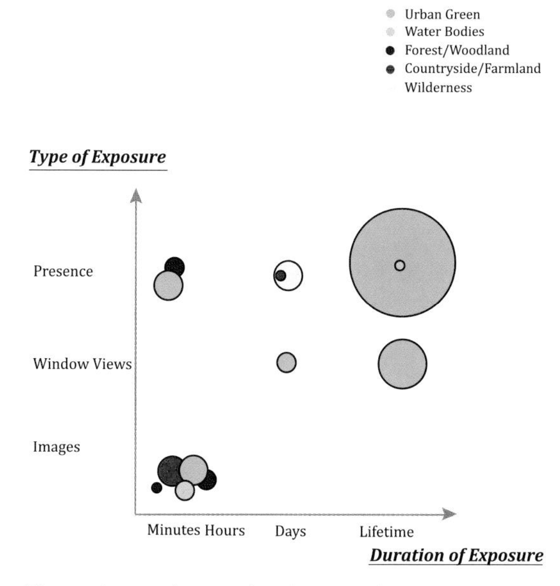

Figure 1. Distribution of studies considered in this review. Color and size of the bubble indicates the nature "type" and number of studies, respectively. Location on the *y*-axis indicates the type of nature exposure that was addressed in the work. Location on the *x*-axis illustrates the duration of exposure to which the subjects were exposed.

explanations for these differences, if they are shown to be reliable.

Ecosystem services

We have examined numerous ways in which nature experience has been shown to impact cognitive and emotional capacities for individuals. If these benefits do exist, discussions about an "ideal" length of exposure to specific types of environment might be framed in a way that has become increasingly familiar to conservation scientists and environmental economists. We may attempt to determine the benefit that comes from a particular type of interaction (e.g., images, window views, physical presence), for a particular amount of time (for example, minutes, days, years), with a particular form of nature (for example, urban parks, forests, water bodies). With further research, we may eventually be able to quantify the marginal benefit that comes with the addition of something as "small" as a single tree. This type of thinking has a role for policy makers, urban planners, and even architects as they ponder the value of spending tax dollars on urban parks, putting gardens in hospitals, or providing outdoor learning experiences for children.

With this in mind, we can begin to envision a way in which environmental psychology fits into the ecosystem services paradigm. Organizations such as the Natural Capital Project have developed tools that present easy ways to visualize the value of particular parcels of land (or pixels in GIS) as a means for carbon sequestration, water purification, flood mitigation, pollination, and other ecosystem services (including a suite for seascapes).[102] Policy makers can then be presented with an analysis of explicit tradeoffs that would result under alternative choices or scenarios. Until now, most of this work has been tied to biophysical processes. If further research is able to attribute "psychological benefit values" reliably to these areas (the impact that comes from specific elements contained within the area), similar results might be presentable. As stated in the Introduction, however, these benefits exist only through the interaction of an individual with the landscape, raising the importance of duration of exposure in a different way than it has been incorporated within previous ecosystem service studies.

The time is ripe for such development within the scientific community that is advancing ecosystem service tools. Policy makers are becoming increasingly engaged in the issue of the health benefits of nature, with a national study commission by the Dutch government in 2004[103] and China investing in a new reserve system to span 25% of the nation for the provision of vital ecosystem services (Personal Communication, Professor Zhiyun Ouyang, Chinese Academy of Sciences, 1 September 2011). Many of these efforts focus on the benefits from urban and suburban greenspace, as well as more extensive natural areas.

Conclusion and future directions

We have reviewed many studies that demonstrate impacts of nature experience on human cognitive function and mental health. These effects have been shown to occur in measures of memory, attention, concentration, impulse inhibition, and mood. The studies considered here span many of the major areas of examination within contemporary psychology and, taken together, constitute a strong foundation for an emerging field of inquiry. We now point to ways of building upon this foundation, both to make different lines of existing work more intercomparable, so that individual experiments add maximum value in the context of advancing the broader field, and also to open promising new directions for exploration.

First, formal, quantifiable, and consistent metrics could be used to compare "urban" and "natural" environments. In existing literature, they are often loosely and vaguely described. Second, a large percentage of these studies has been conducted within a small range of landscape types, and many involve similar durations and frequencies of time spent in nature (see Fig. 1). More extensive work remains to be done to cover the full range of variety, duration, frequency, and spatial scale. Third, there is need to incorporate systematically the possible repercussions of opinions about nature on the individual psychological benefits that a landscape may provide. Fourth, there remains great scope for considering more detail in measured behavioral effects. For example, filtering tasks, complex operation span tasks, distraction tests, and other modern psychology tests may allow for examination of more detailed capabilities in the broad category of "attention." And last, following the precedent of countless intervention studies within psychology, there is need to assess the degree to which the measured impacts persist once subjects have returned to their normal environments.

Explicitly categorizing and quantifying the elements of nature considered within a study will do much to encourage the development of consistency across the field. To challenge or replicate a claim about particular effects, scholars must ensure that the variables of manipulation are duplicated accurately. It may be helpful to develop a common language of "geographical features" that allows for a mutual understanding and shared terminology with respect to the description of various landscapes. This may lead to an agreement upon what aspects of an environment combine to result in its classification as, for example, "forest," "field," or "urban greenspace." Furthermore, when placing a landscape into one of a variety of categories, it would likely prove most fruitful to incorporate basic principles of ecology and integrate considerations of scale, diversity (biological and geographical), topography, and vegetation density. Remotely sensed landscape imagery, coupled with on-site verification, can help formulate rigorous classifications within and across studies. Additionally, the landscape aesthetics literature provides a precedent for a systematic consideration of the "visually appealing" aspects of an environment (consistent to various degrees across individuals).

With the exception of one study, biodiversity has not been shown to correspond independently to psychological benefits.[104] As with all preferences, however, this can change. With increased appreciation for biodiversity might come an increased tendency to appreciate landscapes containing qualities of native habitats and species, thereby eliciting more positive responses from the viewer.[105]

Finally, as we have mentioned throughout, much more research is needed to determine possible causal mechanisms for the observed effects that have been demonstrated in the studies contained within this review. Functional magnetic resonance imaging (fMRI) and other tools in cognitive neuroscience offer exciting possibilities for examinations of brain activity before and after nature experience—and may offer insights into the neural activity changes that are responsible for the compelling observations we have found in the literature.

Interdisciplinary efforts have much to offer in this exploration. For example, combining mapping, tracking, and testing approaches from ecology, psychology, epidemiology, and computer science may allow researchers to study the real-time effects on mood and cognitive function within individuals as they move through a variety of landscape types.[106] There are many other exciting possibilities for synthesis across disciplines in pursuit of this subject matter. Isolating effects that are attributable to environmental change has been of utmost importance in recent decades for the study of population biology, climate change, and other high-profile areas in the sciences. Similar demands for rigor and specificity on the ways in which natural landscapes impact the mind may lead to exciting, compelling, and even completely unanticipated results. These are essential considerations for humanity as we move away from the surroundings with which we have dealt for millennia.

Acknowledgments

We thank the following for the influential conversations and extremely thoughtful feedback: P. Ehrlich, L. Frishkoff, R. Gould, J. Gross, D. Karp, E. Katnelson, B. Levy, C. Mendenhall, H. Mooney, A. Wagner, as well as members of the NCEAS working group on cultural ecosystem services. We are grateful for financial support from Peter and Helen Bing, the

Winslow Foundation, and for the David and Lucile Packard Foundation Stanford Graduate Fellowship to G. Bratman.

Conflicts of interest

The authors are not aware of any biases that might be perceived as affecting the objectivity of this review.

References

1. Lederbogen, F., P. Kirsch, L. Haddad, *et al.* 2011. City living and urban upbringing affect neural social stress processing in humans. *Nature* **474:** 498–501.
2. Cole, D.N. & T.E. Hall. 2010. Experiencing the restorative components of wilderness environments: does congestion interfere and does length of exposure matter? *Environ. Behav.* **42:** 806–823.
3. Marcus, C.C. & M. Barnes. 1999. *Healing Gardens: Therapeutic Benefits and Design Recommendations.* John Wiley & Sons Inc. New York.
4. Maretzki, T.W. 1987. The Kur in West Germany as an interface between naturopathic and allopathic ideologies. *Soc. Sci. Med.* **24:** 1061–1068.
5. Payer, L. 1996. *Medicine and Culture: Varieties of Treatment in the United States, England, West Germany, and France.* Holt Paperbacks. New York.
6. Mukerji, C. 1994. The political mobilization of nature in seventeenth-century French formal gardens. *Theor. Soc.* **23:** 651–677.
7. Bartik, T.J. 1988. Measuring the benefits of amenity improvements in hedonic price models. *Land Econ.* **64:** 172–183.
8. Bastian, C.T., D.M. McLeod, M.J. Germino, *et al.* 2002. Environmental amenities and agricultural land values: a hedonic model using geographic information systems data. *Ecol. Econ.* **40:** 337–349.
9. Cho, S., J.M. Bowker & W.M. Park. 2006. Measuring the contribution of water and green space amenities to housing values: an application and comparison of spatially weighted hedonic models. *J. Agric. Res. Econ.* **31:** 485.
10. Garrod, G.D. & K.G. Willis. 1992. Valuing goods' characteristics: an application of the hedonic price method to environmental attributes. *J. Environ. Manage.* **34:** 59–76.
11. Garrod, G. & K. Willis. 1992. The environmental economic impact of woodland: a two-stage hedonic price model of the amenity value of forestry in Britain. *Appl. Econ.* **24:** 715–728.
12. Ready, R.C., M.C. Berger & G.C. Blomquist. 1997. Measuring amenity benefits from farmland: hedonic pricing vs. contingent valuation. *Growth Change* **28:** 438–458.
13. Shultz, S.D. & D.A. King. 2001. The use of census data for hedonic price estimates of open-space amenities and land use. *J. Real Estate Finance Econ.* **22:** 239–252.
14. Evans, G.W. & J.M. McCoy. 1998. When buildings don't work: the role of architecture in human health. *J. Environ. Psychol.* **18:** 85–94.
15. Van Haaften, E.H. & F.J.R. Van De Vijver. 1996. Psychological consequences of environmental degradation. *J. Health Psychol.* **1:** 411–429.
16. Speldewinde, P.C., A. Cook, P. Davies & P. Weinstein. 2009. A relationship between environmental degradation and mental health in rural Western Australia. *Health Place* **15:** 865–872.
17. Speldewinde, P.C., A. Cook, P. Davies & P. Weinstein. 2011. The hidden health burden of environmental degradation: disease comorbidities and dryland salinity. *EcoHealth* **8:** 82–92.
18. Mayer, F.S. & C.M.P. Frantz. 2004. The connectedness to nature scale: a measure of individuals' feeling in community with nature. *J. Environ. Psychol.* **24:** 503–515.
19. Kaplan, R. & S. Kaplan. 1989. *The Experience of Nature: A Psychological Perspective. Manuscript, University of Nevada, Reno.* Cambridge University Press. Cambridge.
20. Ulrich, R.S. 1986. Human responses to vegetation and landscapes. *Landscape Urban Plan.* **13:** 29–44.
21. Chan, K.M.A., T. Satterfield & J. Goldstein. 2011. Rethinking ecosystem services to better address and navigate cultural values. *Ecol. Econ.* (In press).
22. Chan, K.M.A., A. Guerry, P. Balvanera, *et al.* Where are cultural and social in ecosystem services? A framework for constructive engagement. *Bioscience.* (Accepted).
23. Kellert, S.R. & E.O. Wilson. 1993. *The Biophilia Hypothesis.* Island Press. Washington, DC.
24. Wohlwill, J.F. 1983. The concept of nature: a psychologist's view. *Hum. Behav. Environ. Adv. Theor. Res.* **6:** 5–37.
25. Van den Berg, A.E. & S.L. Koole. 2006. New wilderness in the Netherlands: an investigation of visual preferences for nature development landscapes. *Landscape Urban Plan.* **78:** 362–372.
26. Buijs, A.E., B.H.M. Elands & F. Langers. 2009. No wilderness for immigrants: cultural differences in images of nature and landscape preferences. *Landscape Urban Plan.* **91:** 113–123.
27. Soulé, M.E. & G. Lease. 1995. *Reinventing Nature? Responses to Postmodern Deconstruction.* Island Press. Washington, DC.
28. Callicott, J.B. 1989. *In Defense of the Land Ethic: Essays in Environmental Philosophy.* SUNY Press. Albany, NY.
29. de Jong, K., M. Albin, E. Skärbäck. 2011. Area-aggregated assessments of perceived environmental attributes may overcome single-source bias in studies of green environments and health: results from a cross-sectional survey in southern Sweden. *Environ. Health* **10:** 4.
30. Wilson, E.O. 1984. *Biophilia.* Harvard University Press. Harvard.
31. Mayer, F.S., C.M.P. Frantz, E. Bruehlman-Senecal & K. Dolliver. 2009. Why is nature beneficial? *Environ. Behav.* **41:** 607–643.
32. Lindholst, A.C. 2009. Contracting-out in urban green-space management: instruments, approaches and arrangements. *Urban Forest. Urban Green.* **8:** 257–268.
33. Benight, C.C., G. Ironson, K. Klebe, *et al.* 1999. Conservation of resources and coping self-efficacy predicting distress following a natural disaster: a causal model analysis where the environment meets the mind. *Anxiety, Stress Coping* 107–126.
34. Freedy, J.R., M.E. Saladin, D.G. Kilpatrick, *et al.* 1994. Understanding acute psychological distress following natural disaster. *J. Traumatic Stress* **7:** 257–273.

35. Hartig, T., G. Evans, L. Jamner, *et al.* 2003. Tracking restoration in natural and urban field settings. *J. Environ. Psychol* **23:** 109–123.
36. Ulrich, R.S. 1977. Visual landscape preference: a model and application. *Man Environ. Syst.* **7:** 279–293.
37. Ulrich, R.S., R. Simons, B. Losito, *et al.* 1991. Stress recovery during exposure to natural and urban environments. *J. Environ. Psychol.* **11:** 201–230.
38. Daniel, T.C. & R.S. Boster. 1976. Measuring landscape aesthetics: the scenic beauty estimation method. US Department of Agriculture.
39. Ottosson, J. & P. Grahn. 2008. The role of natural settings in crisis rehabilitation: how does the level of crisis influence the response to experiences of nature with regard to measures of rehabilitation? *Landscape Res.* **33:** 51–70.
40. De Vries, S., R.A. Verheij, P.P. Groenewegen & P. Spreeuwenberg. 2003. Natural environments-healthy environments? An exploratory analysis of the relationship between greenspace and health. *Environ. Planning A* **35:** 1717–1732.
41. Seligman, M.E.P. 1971. Phobias and preparedness. *Behav. Ther.* **2:** 307–320.
42. Ulrich, R.S. 1983. Aesthetic and affective response to natural environment. *Hum. Behav. Environ. Adv. Theor. Res.* **6:** 85–125.
43. Abkar, M., M. Kamal, S. Maulan & M. Mariapan. 2010. Influences of viewing nature through windows. *Aus. J. Basic Appl. Sci.* **4:** 5346–5351.
44. Ulrich, R.S. 1993. Biophilia, biophobia and natural landscapes. In *The Biophilia Hypothesis*. S. R. Kellert & E. O. Wilson, Eds.: 73–137. Island Press. Washington, DC
45. Kaplan, S. 1995. The restorative benefits of nature: toward an integrative framework. *J. Environ. Psychol.* **15:** 169–182.
46. Ulrich, R.S. 1979. Visual landscapes and psychological well-being. *Landscape Res.* **4:** 17–23.
47. Ekman, P. & W.V. Friesen. 1971. Constants across cultures in the face and emotion. *J. Personal. Soc. Psychol.* **17:** 124–129.
48. Ekman, P. 1971. *Universals and Cultural Differences in Facial Expressions of Emotion.* University of Nebraska Press. Lincoln.
49. Ekman, P., W.V. Friesen & P. Ellsworth. 1982. *Emotion in the Human Face.* Pergamon Press. Oxford.
50. Zajonc, R.B. 1980. Feeling and thinking: preferences need no inferences. *Am. Psychologist* **35:** 151–175.
51. Ulrich, R.S. 1981. Natural versus urban scenes. *Environ. Behav.* **13:** 523–556.
52. Hartig, T., M. Mang & G.W. Evans. 1991. Restorative effects of natural environment experiences. *Environ. Behav.* **23:** 3–26.
53. Honeyman, M.K. 1992. Vegetation and stress: a comparison study of varying amounts of vegetation in countryside and urban scenes. In: *The Role of Horticulture in Human Well-Being and Social Development*. D. Relf, Ed: 143–145. Timber Press. Portland, OR.
54. Park, B, Y. Tsunetsugu, T. Kasetani, *et al.* 2007. Physiological effects of Shinrin-yoku (taking in the atmosphere of the forest)—using salivary cortisol and cerebral activity as indicators. *J. Physiol. Anthropol.* **26:** 123–128.
55. Lee, J., B.J. Park, Y. Tsunetsugu, *et al.* 2009. Restorative effects of viewing real forest landscapes, based on a comparison with urban landscapes. *Scandinav. J. Forest Res.* **24:** 227–234.
56. James, W. 1892. *Psychology: The Briefer Course*. Holt. New York.
57. Kaplan, S. 1983. A model of person-environment compatibility. *Environ. Behav.* **15:** 311–332.
58. Olmsted, F. & R. Nash. 1865. The value and care of parks. Report to the Congress of the State of California. Reprinted in: *The American Environment*. R. Nash, Ed.:18–24. Addison-Wesley. Hillsdale, NJ.
59. Wells, N.M. 2000. At home with nature. *Environ. Behav.* **32:** 775–795.
60. Bowler, D.E. *et al.* 2010. A systematic review of evidence for the added benefits to health of exposure to natural environments. *BMC Public Health 2010.* **10:** 456.
61. Korpela, K., M. Kyttä & T. Hartig. 2002. Restorative experience, self-regulation, and children's place preferences. *J. Environ. Psychol.* **22:** 387–398.
62. Han, K. T. 2009. An exploration of relationships among the responses to natural scenes: scenic beauty, preference, and restoration. *Environ. Behav.* **42:** 243–270.
63. Berman, M.G., J. Jonides & S. Kaplan. 2008. The cognitive benefits of interacting with nature. *Psychol. Sci.* **19:** 1207–1212.
64. Tennessen, C.M. & B. Cimprich. 1995. Views to nature: effects on attention. *J. Environ. Psychol.* **15:** 77–85.
65. Taylor, A.F., F.E. Kuo & W.C. Sullivan. 2002. Views of nature and self-discipline: evidence from inner city children. *J. Environ. Psychol.* **22:** 49–63.
66. Kuo, F.E. & W.C. Sullivan. 2001. Aggression and violence in the inner city. *Environ. Behav.* **33:** 543–571.
67. Berto, R. 2005. Exposure to restorative environments helps restore attentional capacity. *J. Environ. Psychol.* **25:** 249–259.
68. Laumann, K., T. Gärling & K.M. Stormark. 2003. Selective attention and heart rate responses to natural and urban environments. *J. Environ. Psychol.* **23:** 125–134.
69. Posner, M.I. 1980. Orienting of attention. *Quar. J. Exp. Psychol.* **32:** 3–25.
70. Jonides, J. 1981. Voluntary versus automatic control over the mind's eye's movement. *Attention and Performance IX* **9:** 187–203.
71. Yantis, S. & J. Jonides. 1984. Abrupt visual onsets and selective attention: evidence from visual search. *J. Exp. Psychol. Hum. Percept. Perform.* **10:** 601–621.
72. Friedman, N.P. & A. Miyake. 2004. The relations among inhibition and interference control functions: a latent-variable analysis. *J. Exp. Psychol. Gen.* **133:** 101.
73. Hartig, T. & H. Staats. 2006. The need for psychological restoration as a determinant of environmental preferences. *J. Environ. Psychol.* **26:** 215–226.
74. Korpela, K.M., T. Hartig, F.G. Kaiser & U. Fuhrer. 2001. Restorative experience and self-regulation in favorite places. *Environ. Behav.* **33:** 572–589.
75. Epstein, S. 1991. Cognitive-experiential self-theory: An integrative theory of personality. In *The relational self:*

Theoretical convergences in psychoanalysis and social psychology. R.C. Curtis, Ed.: 111–137. Guilford. New York.
76. Nisbet, E.K. & J.M. Zelenski. 2011. Underestimating nearby nature. *Psychol. Sci.* **22:** 1101–1106.
77. Nolen-Hoeksema, S. 1991. Responses to depression and their effects on the duration of depressive episodes. *J. Abnor. Psychol.* **100:** 569–582.
78. Siegle, G.J., S.R. Steinhauer, M.E. Thase, V.A. Stenger & C.S. Carter. 2002. Can't shake that feeling: event-related fMRI assessment of sustained amygdala activity in response to emotional information in depressed individuals. *Biol. Psychiatr.* **51:** 693–707.
79. Wicklund, R.A. 1975. Objective self-awareness1. *Adv. Exp. Soc. Psychol.* **8:** 233–275.
80. Carver, C.S. & M.F. Scheier. 1981. *Attention and Self-Regulation: A Control-Theory Approach to Human Behavior*. Springer-Verlag. New York.
81. Gibbons, F.X. 1990. Self-attention and behavior: a review and theoretical update. *Adv. Exp. Soc. Psychol.* **23:** 249–303.
82. Nolen-Hoeksema, S. & J. Morrow. 1993. Effects of rumination and distraction on naturally occurring depressed mood. *Cogn. Emotion* **7:** 561–570.
83. Skřivanová, Z. & O. Kalivoda. 2010. Perception and assessment of landscape aesthetic values in the Czech Republic—a literature review. *J. Landscape Stud.* **3:** 211–220.
84. Ode, A., M.S. Tveit & G. Fry. 2010. Advantages of using different data sources in assessment of landscape change and its effect on visual scale. *Ecological Indicators* **10:** 24–31.
85. Anderson, L.M. 1981. Land use designations affect perception of scenic beauty in forest landscapes. *Forest Sci.* **27:** 392–400.
86. Grêt-Regamey, A., P. Bebi, I.D. Bishop & W.A. Schmid. 2008. Linking GIS-based models to value ecosystem services in an Alpine region. *J. Environ. Manage.* **89:** 197–208.
87. Grêt-Regamey, A., A. Walz & P. Bebi. 2008. Valuing ecosystem services for sustainable landscape planning in alpine regions. *Mount. Res. Develop.* **28:** 156–165.
88. Dunlap, R.E., K.D. Van Liere, A.G. Mertig & R.E. Jones., 2000. New trends in measuring environmental attitudes: measuring endorsement of the new ecological paradigm: a revised NEP scale. *J. Soc. Issues* **56:** 425–442.
89. Schultz, P.W. 2002. Inclusion with nature: the psychology of human-nature relations. *Psychol. Sustain. Develop.* 61–78.
90. Leopold, A. 1949. *A Sand County Almanac*. Oxford University Press. New York.
91. Ulrich, R.S. 1984. View through a window may influence recovery from surgery. *Science* **224:** 420–421.
92. Gerlach-Spriggs, N., R.E. Kaufman & S.B. Warner Jr. 2004. *Restorative Gardens: The Healing Landscape*. Yale University Press. New Haven, CT.
93. Moore, E.O. 1981. A prison environment's effect on health care service demands. *J. Environ. Syst.* **11:** 17–34.
94. Kaplan, S. 1993. The role of natural environment aesthetics in the restorative experience. In: *Managing Urban and High-Use Recreation Settings, General Technical Report NC-163, Forest Service*. P.H. Gobster, Ed.: 46–49. USDA. St. Paul, MN.
95. Kaplan, R. 2001. The nature of the view from home. *Environ. Behav.* **33:** 507–542.
96. Wells, N.M. & G.W. Evans. 2003. Nearby nature. *Environ. Behav.* **35:** 311–330.
97. Berger, B.G. & R.W. Motl. 2000. Exercise and mood: a selective review and synthesis of research employing the profile of mood states. *J. Appl. Sport Psychol.* **12:** 2000.
98. Rethorst, C.D., B.M. Wipfli & D.M. Landers. 2009. The antidepressive effects of exercise: a meta-analysis of randomized trials. *Sports Med.* **39:** 491–511.
99. Kawachi, I., B.P. Kennedy, K. Lochner & D. Prothrow-Stith. 1997. *Social capital, income equality, and mortality. Am. J. Public Health* **87:** 1491–1498.
100. Pretty, J., J. Peacock, M. Sellens & M. Griffin. 2005. The mental and physical health outcomes of green exercise. *Int. J. Environ. Health Res.* **15:** 319–337.
101. Barton, J. & J. Pretty. 2010. What is the best dose of nature and green exercise for improving mental health? A multi-study analysis. *Environ. Sci. Technol.* **44:** 3947–3955.
102. Kareiva, P.K., H. Tallis, T.H. Ricketts, G.C. Daily, S. Polasky, Eds. 2011. *Natural Capital: Theory & Practice of Mapping Ecosystem Services*. Oxford University Press. Oxford.
103. Health Council of the Netherlands and Dutch Advisory Council for Research on Spatial Planning, N. and the E. Health Council of the Netherlands and Dutch Advisory Council for Research on Spatial Planning, Nature and the Environment. 2004. Nature and health. The influence of nature on social, psychological and physical well-being: The Hague. Health Council of the Netherlands. *J. Adv. Nursing* **65:** 1527–1538.
104. Fuller, R.A., K.N. Irvine, P. Devine-Wright, P.H. Warren, *et al.* 2007. Psychological benefits of greenspace increase with biodiversity. *Biol. Lett.* **3:** 390–394.
105. Gobster, P., J. Nassauer, T. Daniel & G. Fry. 2007. The shared landscape: What does aesthetics have to do with ecology? *Landscape Ecol.* 22: 959–972.
106. Killingsworth, M. a. & D.T. Gilbert. 2010. A wandering mind is an unhappy mind. *Science* **330:** 932.
107. Verderber, S. 1986. Dimensions of person–window transactions in the hospital environment. *Environ. Behav.* **18:** 450–466.
108. Leather, P., M. Pyrgas, D. Beale & C. Lawrence. 1998. Windows in the workplace. *Environ. Behav.* **30:** 739–762.
109. Evans, G.W. 2003. The built environment and mental health. *J. Urban Health* **80:** 536–555.
110. Grahn, P. & U.A. Stigsdotter. 2003. Landscape planning and stress. *Urban Forest. Urban Green.* **2:** 1–18.
111. Groenewegen, P., A. van den Berg, S. de Vries & R. Verheij. 2006. Vitamin G: effects of green space on health, well-being, and social safety. *BMC Public Health* **6:** 149.
112. Richardson, E., J. Pearce, R. Mitchell, *et al.* 2010. The association between green space and cause-specific mortality in urban New Zealand: An ecological analysis of green space utility. *BMC Public Health* **10:** 240.
113. Chang, C., W. Hammitt, P. Chen, *et al.* 2008. Psychophysiological responses and restorative values of natural environments in Taiwan. *Landscape Urban Plan.* **85:** 79–84.
114. Paxton, T. & L. McAvoy. 2000. Social psychological benefits of a wilderness adventure program. In: *Wilderness science*

in a time of change conference—Volume 3: Wilderness as a place for scientific inquiry; 1999 May 23–27; Missoula, MT. Proceedings RMRS-P-15-VOL-3. S.F. McCool, D.N. Cole, W.T. Borrie, and J. O'Loughlin, comps. U.S. Department of Agriculture, Forest Service, Rocky Mountain Research Station. Ogden, UT.

115. Coley, R.L., W.C. Sullivan & F.E. Kuo. 1997. Where does community grow? *Environ. Behav.* **29:** 468.
116. Kuo, F.E., M. Bacaicoa & W.C. Sullivan. 1998. Transforming inner-city landscapes. *Environ. Behav.* **30:** 28–59.
117. Takano, T., K. Nakamura & M. Watanabe. 2002. Urban residential environments and senior citizens' longevity in megacity areas: the importance of walkable green spaces. *J. Epidemiol. Commun. Health* **56:** 913–918.
118. Maas, J., R. Verheij, P. Groenewegen, *et al.* 2006. Green space, urbanity, and health: how strong is the relation? *J. Epidemiol. Commun. Health* **60:** 587–592.
119. Mitchell, R. & F. Popham. 2008. Effect of exposure to natural environment on health inequalities: An observational population study. *The Lancet* **372:** 1655–1660.
120. Van Den Berg, A.E., *et al.* 2010. Green space as a buffer between stressful life events and health. *Soc. Sci. Med.* **70:** 1203–1210.
121. Smith, A. Smith. 1982. *The Symbol Digit Modalities Test Manual.* Western Psychological Services. Los Angeles.
122. Wechsler, D. 1955. *Wechsler Adult Intelligence Scale Manual.* Psychological Corporation. New York.
123. Orbach, J., D. Ehrlich & H.A. Heath. 1963. Reversibility of the Necker cube: An examination of the concept of "satiation of orientation." *Percept. Motor Skills.* **17:** 439–458.
124. Smith, A.P. & C. Miles. 1987. The combined effects of occupational health hazards: an experimental investigation of the effects of noise, nightwork, and meals. *Int. Arch. Occup. Environ. Health,* **59:** 83–89.
125. Posner, M.I. 1980. Orienting of attention. *Quart. J. Exp. Psychol.* **32:** 3–25.
126. Kagan, J. 1965. Individual differences in the resolution of response uncertainty. *J. Personal. Soc. Psychol.* **2:** 154–160.
127. Stroop, J.R. 1935. Studies of interference in serial verbal reactions. *J. Exp. Psychol.* **18:** 643–662.
128. Rodriguez, M.L., W. Mischel & Y. Shoda. 1989. Cognitive person variables in the delay of gratification of older children at risk. *J. Personal. Soc. Psychol.* **57:** 358–367.
129. Straus, M.A. 1979. Measuring intrafamily conflict and violence: the conflict tactics (CT) scales. *J. Marriage Family* **41:** 75–88.
130. Zuckerman, M. 1977. Development of a situation-specific trait-state test for the prediction and measurement of affective responses. *J. Consult. Clin. Psychol.* **45:** 513.
131. Hartig, T., K. Korpela, G. Evans & T. Garling. 1997. A measure of restorative quality in environments. *Scand. Hous. Plan. Res.* **14:** 175–194.
132. McNair, D., M. Lorr & L. Droppleman. 1971. *Manual for the Profile of Mood States (POMS).* Educational and Industrial Testing Service. San Diego, CA.
133. Watson, D., L.A. Clark & A. Tellegen. 1988. Development and validation of brief measures of positive and negative affect: the PANAS scales. *J. Personal. Soc. Psychol.* **54:** 1063–1070. Available at: http://psycnet.apa.org/psycinfo/1988-31508-001 [Accessed September 7, 2011].
134. Campbell, A., P.E. Converse & W.L. Rodgers. 1976. *The Quality of American Life: Perceptions, Evaluations, and Satisfactions.* Russell Sage Foundation. New York.
135. Rosenberg, M. 1965. *Rosenberg Self-Esteem Scale.* Princeton University Press. Princeton, NJ.
136. Harter, S. 1985. *Self-Perception Profile for Children.* University of Denver. Denver, CO.

Ann. N.Y. Acad. Sci. ISSN 0077-8923

ANNALS OF THE NEW YORK ACADEMY OF SCIENCES
Issue: *The Year in Ecology and Conservation Biology*

Reducing emissions from deforestation and forest degradation (REDD+): game changer or just another quick fix?

Oscar Venter[1] and Lian Pin Koh[2,3]

[1]Terrestrial Ecology and Sustainability Science and the School of Marine and Tropical Biology, James Cook University, Cairns, Australia. [2]Department of Environmental Sciences, ETH Zurich, CHN G 73.2, Universitatstrasse 16, Switzerland. [3]Department of Biological Sciences, National University of Singapore, Singapore

Address for correspondence: Oscar Venter, School of Marine and Tropical Biology, James Cook University, Smithfield QLD 4878, Australia. oscar.venter1@jcu.edu.au

Reducing emissions from deforestation and forest degradation (REDD+) provides financial compensation to land owners who avoid converting standing forests to other land uses. In this paper, we review the main opportunities and challenges for REDD+ implementation, including expectations for REDD+ to deliver on multiple environmental and societal cobenefits. We also highlight a recent case study, the Norway–Indonesia REDD+ agreement and discuss how it might be a harbinger of outcomes in other forest-rich nations seeking REDD+ funds. Looking forward, we critically examine the fundamental assumptions of REDD+ as a solution for the atmospheric buildup of greenhouse gas emissions and tropical deforestation. We conclude that REDD+ is currently the most promising mechanism driving the conservation of tropical forests. Yet, to emerge as a true game changer, REDD+ must still demonstrate that it can access low transaction cost and high-volume carbon markets or funds, while also providing or complimenting a suite of nonmonetary incentives to encourage a developing nation's transition from forest losing to forest gaining, and align with, not undermine, a globally cohesive attempt to mitigate anthropogenic climate change.

Keywords: carbon payment scheme; biodiversity conservation; climate change; REDD; tropical deforestation

Introduction

The growing world population is driving global demand for food, feed, fuel, and other raw materials. These demands create economic opportunities—mediated by commodity markets and trade policies—that many developing countries respond to by exploiting and exporting their natural resources. As a consequence, vast areas of tropical forests are being cleared for agriculture, timber extraction, oil and gas development, and mining and infrastructure expansion.

Recognizing the economic impetus behind these exploitative activities, conservationists have been developing and implementing "Payments for Ecosystem Service" (PES) schemes, which provide financial compensation to land owners who avoid converting standing forests to other land uses.[1] These schemes typically involve monetization of key environmental services that society values and that forests might provide. The most prominent of such PES schemes is Reducing Emissions from Deforestation and Forest Degradation plus the conservation, sustainable management, and enhancement of forest carbon stocks (REDD+).

What is REDD+?

REDD+ is simply a method for putting a price tag on the carbon storage and sequestration services provided by forests. The basic premise of REDD+ is that, without it, a certain amount of carbon dioxide would be emitted due to the loss or degradation of forests. These expected emissions serve as a reference level to measure efforts to protect forests and reduce carbon emissions. If realized emissions are below the reference level, these reductions are considered "additional," and a corresponding amount

doi: 10.1111/j.1749-6632.2011.06306.x

of carbon credits may be awarded. To be sold on carbon markets, carbon credits must be generated through a rigorous process of measuring, reporting, and verification (or MRV), which might be subjected to certification by recognized standards, such as the Verified Carbon Standard (www.v-c-s.org). Early REDD+ demonstration and early action projects have been largely site based, protecting a particular tract of forests,[2] whereas for full-scale REDD+ implementation, the MRV process and financial compensation will take place at the national scale.[3]

The "+" after REDD comes from more recent discussions that have broadened the mechanism's scope to also recognize the carbon benefits of forest conservation, improved forest management, and the sequestration potential of afforestation and reforestation. At the same time, expectations for REDD+ to deliver on multiple environmental and societal issues have risen. These "cobenefits" might include conservation of biodiversity and improvements to rural livelihoods.

Much of the hype surrounding REDD+ stems from the perceived scale of financing,[4] yet this financing remains surprisingly nebulous. Almost all the demand for carbon credits comes from what are known as compliance carbon markets,[5] which are markets that sell credits for meeting obligations under the United Nations Framework Convention on Climate Change (UNFCCC). The UNFCCC is an international treaty produced at the United Nations Conference on Environment and Development (or Earth Summit) in Rio de Janeiro in 1992. The Kyoto Protocol to the UNFCCC has been ratified by 37 industrialized countries who commit themselves to a reduction of greenhouse gas (GHG) emissions by 5.2% from 1990 baseline levels. To help these "Annex I" countries meet their commitments, they are allowed to offset emissions by purchasing certified emission reduction (CER) credits, each equivalent to the reduction of one metric ton of carbon dioxide. These CER credits can be generated from the implementation of Clean Development Mechanism (CDM) projects in developing countries, which lead to net emission reductions. These CDM projects typically involve production of renewable energy (e.g., hydro-, thermal, or wind power) and improvements in the handling of industrial and agricultural byproducts.

When REDD+ (at the time it was just RED) was first proposed in 2005 by the Coalition for Rainforest Nations (www.rainforestcoalition.org) as a strategy for correcting the market failures that lead to rampant deforestation, it was envisaged that demand for REDD+ credits would be created, and conservation funds generated, from REDD+'s inclusion as a legitimate emissions reduction activity of the CDM. Despite intense lobbying by proponents of REDD+—particularly by the World Bank's Forest Carbon Partnership Facility (http://wbcarbonfinance.org/) and the United Nations–REDD Program (http://un-redd.org)—REDD+ is still not recognized as a legitimate CDM activity. As REDD+ negotiations are now taking place as part of long-term cooperative action under the UNFCCC, rather than the Kyoto Protocol, REDD+ is unlikely to ever be incorporated in the CDM.

Despite this limitation, REDD+ initiatives are being financed and implemented across the developing world. In their review of 64 developing countries, Cerbu *et al.*[6] found that as of October 2009, there were at least 79 REDD+ readiness activities (such as monitoring carbon stocks and identifying sources and drivers of emissions) and 100 REDD+ demonstration activities, largely in Indonesia (21) and Brazil (17). Some of the finance for these early initiatives comes from voluntary carbon markets, such as the Chicago Climate Exchange (www.chicagoclimatex.com), where credits are bought by companies that voluntarily wish to offset their emissions. Most of the funding, however, comes through voluntary carbon funds, many of them linked to the REDD+ Partnership, which has already helped to channel almost US $8 billion for REDD+, and bilateral agreements, such as the Norway–Indonesia REDD+ deal involving $1 billion.

It appears, for now at least, that REDD+ has gained momentum and access to financing independent of the UNFCCC process. The REDD+ partnership (http://reddpluspartnership.org), which includes 72 partner countries, is acting as the interim platform for REDD+ financing, policy development, and knowledge sharing. Yet, the continued growth and success of REDD+ still hangs in the balance. It depends not only on the mechanisms eventual formal inclusion in a post-Kyoto climate agreement, but also on its perceived ability to deliver meaningful climate benefits and surmount ongoing challenges.

In this paper, we review the main opportunities and challenges for REDD+ implementation. We also highlight a recent case study: the Norway–Indonesia REDD+ agreement. Looking forward, we critically examine the fundamental assumptions of REDD+ as a solution for the atmospheric buildup of GHG emissions and tropical deforestation and discuss recent proposals to expand the scope of REDD+ in terms of both the land uses it encompasses and its revenue sources.

Promises and opportunities

Emissions reductions

By reducing negative changes in forest carbon stores and promoting positive changes through forest restoration, reforestation, and afforestation, the first and most important promise of REDD+ lies in its potential to mitigate anthropogenic climate change. While subtropical and temperate forests in developing countries are technically eligible, tropical forests have emerged as the central candidate for REDD+ implementation due to their high carbon content and rates of forest loss, as well as their heightened potential for carbon sequestration.[7,8]

In 2005, tropical humid forests covered roughly 11 million km^2 and tropical dry forests covered a further 7.5 million km[29] combined, storing 248 Gt of biomass carbon in the vegetation,[10] which is roughly equivalent to the total fossil fuel emissions since the industrial revolution.[11] When these forests are cleared or logged, much of this carbon is quickly released into the atmosphere through fires and decomposition.[12]

When REDD+ was first proposed, it was thought that tropical deforestation was responsible for 17–25% of anthropogenic carbon emissions.[11,13,14] Since then, these figures have been revised downward, partly because rates of deforestation are lower than previously estimated,[15–18] and partly because fossil fuel emissions have increased.[19] The most recent estimates indicate that through the early 2000s, net deforestation and forest degradation released 1.22–1.30 PgC per year,[7,20] plus a further 0.30 PgC per year from peat degradation,[20] which is closely associated with deforestation.[21] This makes tropical deforestation and forest degradation responsible for roughly 15% of all anthropogenic carbon emissions, and in turn, this is the theoretical maximum by which REDD+ could reduce carbon emissions.

Global carbon markets traded a staggering $142 billion in 2010.[5] If REDD+ is connected to compliance markets, a portion of these funds is potentially available to help finance the protection of forests and the reduction in forest-based emissions. Even if REDD+ is instead financed through carbon funds, the size of early funds, such as those associated with the REDD+ partnership, indicates that REDD+ will have access to considerable financing to tackle tropical deforestation.

To look at it another way, forests could play a major role in reducing the costs and increasing the effectiveness of climate mitigation efforts. The Eliasch review,[22] the follow-up to the Stern Review,[23] estimates that including REDD+ in carbon markets could provide financing to reduce forest-based emissions by 75% by 2030; taken together with afforestation and reforestation, this would make forests carbon neutral. Because REDD+ appears to be a relatively cost-effective mechanism for reducing emissions, especially when compared with options such as carbon capture and storage,[23] including forests could lower the overall costs of meeting global emissions reductions targets. If REDD+ is included in a post-Kyoto climate deal, the cost of reducing global carbon emissions to half of 1990 levels would be reduced by 50% in 2030 and 40% in 2050.[22] These reduced costs could be taken as a means to make climate mitigation more palatable to the pocket-wise voter or instead as an opportunity to pursue more ambitious targets at no additional cost.

Forest and biodiversity conservation

At least half of all known and yet undiscovered species are found in tropical forests,[24–26] and their widespread loss is one of the major drivers of the current loss of global biological diversity.[27–31] Consequently, 18 of 25 global priorities for the conservation of imperiled biodiversity, or "biodiversity hotspots," are tropical forest regions.[32] Despite the importance of these areas for conservation, there has arguably been little success at protecting tropical forest species from the threats they face.[33–35]

Its tropical forest focus and considerable financial clout have rapidly propelled REDD+ into the role of potential remedy to the current biodiversity crisis.[36–41] The solution seems simple. Because of their vast and vulnerable carbon stocks that could be protected at relatively low cost (though we question this assumption later in the paper), tropical forest regions are poised to receive unprecedented funds to incentivize their protection. As these very areas are also the planet's unquestioned treasure troves of biodiversity, it seems likely that a major collateral benefit of REDD+ will be widespread biodiversity protection. But recent research has revealed that it might not be so straightforward.

First, REDD+ is likely to focus on protecting forests in regions that are most cost-effective for reducing carbon emissions.[37,42] An important determinant of the biodiversity benefits of REDD+ will therefore be the degree to which biodiversity and carbon priorities overlap. At the global scale, carbon and biodiversity are correlated.[39,43] However, if deforestation rates and the opportunity costs of implementing REDD+ are considered when identifying global priorities for carbon, this correlation breaks down, and REDD+ appears likely to perform poorly for biodiversity.[37] Perversely, this lack of coincidence could even lead to biodiversity-negative results, if REDD+ displaces deforestation effort from its priorities into high-biodiversity locations.[42] It is worth noting, however, that early REDD+ activities do show a preference for high-biodiversity locations,[2,6] though this may be due to the fact that early project developers are often organizations working primarily for biodiversity conservation.[2] Additionally, a regional-scale analysis found greater congruence between REDD and biodiversity priorities,[36] but more work at this scale is required to determine if this is a general relationship.

Biodiversity may also miss out if REDD+ is implemented in such a way that only protects a forest's trees, which represent the bulk of its biomass carbon, but not its other species. The persistence of forest-dwelling species can be jeopardized by a range of threats other than the loss of the forest.[44–46] In the tropics, these species are often threatened by climate change,[47] altered fire regimes,[29,48] and hunting.[46,49] Thus, even if REDD+ were to act in the right places, if it does not address the other threats to forest biodiversity, it could deliver minimal outcomes for conservation.

A number of studies have also warned of the danger that REDD+ could promote the expansion of monoculture plantations of nonnative species or "carbon farms," potentially at the expense of native forests.[50–53] This would undoubtedly circumvent any potential biodiversity benefits and, in many cases, probably lead to biodiversity declines.[54,55] A final issue is the risk that REDD+ operates on a shorter timeframe than necessary for protecting biodiversity in the long-term, especially long-lived species.[56]

To avoid these deleterious biodiversity outcomes, conservation scientists have called for the UNFCCC to adopt strict biodiversity safeguards.[40] And this is exactly what happened at the UNFCCC's Conference of the Parties in Cancun, 2010. While the exact mechanism for implementing the decision remains undecided, Annex I paragraph 2 of the Cancun decision (Dec. 1/CP16) states that REDD+ must incentivize "the conservation of natural forests and biological diversity" and "not be used for the conversion of natural forest areas." Moreover, it must be "consistent with [...] relevant international conventions," which includes the Convention on Biological Diversity. In our opinion, these decisions are an important step toward putting to rest the issue of biodiversity negative outcomes from REDD+.

Promoting biodiversity-positive outcomes, such as targeting biodiversity hotspots for protection, would incur little extra cost within an international REDD+ scheme.[37] Two main options exist for capitalizing on this opportunity. The first is to develop biodiversity credits by monitoring, reporting, and verifying biodiversity outcomes from REDD+ actions[57] and linking these credits to biodiversity offsetting markets or funds.[58,59] The second is to harness biodiversity conservation funds, which exceed $1.5 billion annually,[60] to help finance REDD+ implementation in biodiversity priorities. Whatever the mechanism, it seems that achieving the best outcomes for biodiversity from REDD+ will take some extra effort and financing to realize.

What's different about REDD+?

The importance of tropical forests as carbon sinks and biodiversity hotspots has long been recognized,[30,31,61] and the fight to save them was not born with REDD+. Reforestation and afforestation in the tropics have been a sanctioned activity under Kyoto's CDM since 2001, yet a mere 0.54% of CDM carbon credits come from forests.[62] Moreover, while strategies employed to protect biodiversity in the tropics can be effective,[63] they have failed to arrest widespread declines in forest area[17,64] and metrics of biodiversity.[65] Do we have reason to think that REDD+ will be any different?

One of the most compelling reasons for renewed hope is the level at which REDD+ is currently, and likely to continue, generating financing for tropical forest conservation. An early projection estimated that REDD+ could generate $10–60 billion per year in payments to developing countries if deforestation were to be halved.[4] At the time, the prediction seemed outlandish. But today, billions of dollars are already pouring into REDD+ readiness and early-action initiatives. More significantly, Annex I countries agreed, in the Copenhagen accord, to mobilize $30 billion in the period 2010–2012 and $100 billion per year by 2020 for climate mitigation and adaptation in developing countries, much of which could to be used for REDD+.[66] In contrast, data from the early 1990s indicate that the total protected area budget for developing countries was $0.7 billion.[67] Clearly, regardless of the exact figures, REDD+ represents a major financial boost for conserving tropical forests.

But there's more to it than simply bankrolling business-as-usual conservation. When REDD+ reaches full implementation, it is expected that payments for carbon credits will only be made when participants have demonstrated, through an approved MRV process, that emissions have been successfully reduced. Payments being conditional on performance are typical of carbon offsetting schemes, but this represents largely new ground for forest conservation. For instance, since 1989, $5.8 billion has been funneled through Indonesia's reforestation fund to promote forest recovery on degraded lands and forests. However, most of these funds have been lost to fraud or wasted on ineffectual projects, partly because payments were made without holding recipients accountable for outcomes.[68] Another difference is the time horizon of REDD+ projects. Most early REDD+ projects have expected time horizons of 20–30 years,[2] whereas the average World Bank–funded biodiversity project lasts only 6.5 years.[69] Moreover, the Copenhagen accord recognizes the need for financing to be "adequate, predictable and sustained."[66] The conditionality of payments and the long-term nature of projects give reason to believe that the financial incentives provided by REDD+ may catalyze greater change per dollar invested than past initiatives.

Challenges and limitations

Technical and implementation issues

REDD+ is still in its learning-while-doing phase, and many of the design and implementation methods are still being developed and trialed. This is not compromising financing at present, as most initiatives are currently funded through voluntary contributions, without rigorous demands on maximizing and accurately quantifying realized emissions reductions. But when the time comes for REDD+ credits to be traded on international carbon markets, the ability to convince markets that emissions reductions are real and lasting, which largely hinges on technical methods of implementing actions and monitoring outcomes, will determine market demand, and in turn, the price of REDD+ credits.[70]

Determining additionality. One of the founding principles of carbon offsetting is that carbon credits are only awarded for activities that are additional, meaning they reduce emissions to a level lower than would have occurred without the activity. In REDD+, the counterfactual scenario of "what would occur" to forests without intervention is known as the reference level, though this level could also include negotiated adjustments to fine-tune how carbon credits are generated.[71] Reference levels act as the yardstick for measuring REDD+ outcomes and can be set at subnational scales, as is often the required by early REDD+ projects, or national scales, which is the anticipated scale for REDD+ accounting.[3] Because they are the basis for generating carbon credits, reference levels determine of the potential distribution of REDD+ payments and participation as well as the perceived integrity and equity of carbon credits. This makes defining these levels one of the most important and contested aspect of any REDD+ program.

The problem is that reference levels are determined *ex ante*, and predicting the future can be a perilous undertaking. At present, there is no set method for developing reference levels; instead, a range of options have been proposed.[72–74] Most proposals involve extrapolating forward recent past trends in forest loss. At the national scale at least, recent historic emissions do a reasonable job of predicting near-term future emissions.[72] But basing reference levels solely on historic emissions has three major drawbacks. First, national or regional circumstances can change, causing unanticipated deviations from historic trends. For instance, according to national forest statistics,[18] Indonesia doubled its forest carbon emissions in the period 2000–2010 compared to the 1990s. If REDD+ had been implemented in 2000 using a 10-year historic reference level, Indonesia would have been short-changed $18 billion over a 10-year period, assuming it could reduce its forest carbon emissions and receive compensation at a CO_{2e} price of $10. The second drawback is that a historic emissions approach rewards countries pursuing rapid deforestation and punishes countries that have maintained high levels of forest cover and low levels of deforestation, which is ethically questionable. Third, offering high-forest, low-deforestation countries little incentive to participate in REDD+ risks accelerating deforestation in these areas as internationally mobile industries switch their investments into areas unrestricted by REDD+ policy measures.

To circumvent these drawbacks, it has been proposed that reference levels be set based on forest carbon stocks, not fluxes.[75] Under such a scheme, countries that have lots of forest but little forest loss, like Suriname and Gabon, could receive carbon credits for their forests that would otherwise be overlooked under a history-based scheme.[72] While it may be ethically sound to reward countries that have exercised good forest stewardship, it is questionable whether paying to protect forests that are not under threat, at least in the near-term, has any additional climate benefits.[76]

Clearly, how reference levels are set can either lead to short-changing participants, as with our Indonesian example, or wasted funds. Participants whose reference levels are set too high will be able to generate credits without taking additional action to reduce emissions. Because these emissions reductions will incur no or low implementation and opportunity costs, they will be the most price competitive on carbon markets. Such falsely generated credits could dominate carbon markets, especially if markets are fairly small.

At present, the best way forward for setting reference levels appears to be to combine incentives to reduce high levels of deforestation with incentives to maintain low rates of forest loss.[77,78] Modeling shows that a combined approach would maximize participation and minimize the costs of emission reductions by reducing the potential for increased deforestation in nonparticipating countries.[73] Additionally, historic trends could be further modified by modeling land-cover change to anticipate deviations from this trend, though the predictive accuracy of such models leaves much to be desired.[79] Finally, reference levels are unlikely to be determined by data alone, but also through political negotiation between participating countries trying to carve out a decent slice of REDD+ revenues and Annex I countries trying to maximize the integrity of emission reductions. The impact of these negotiations on scientifically justified reference levels should be minimized.

Permanence and leakage. Since its inception, REDD+ proponents have been challenged to convince investors that interventions will lead to lasting climate benefits.[80,81] Skeptics fear that forest carbon protected by REDD+ activities could simply be released into the atmosphere at a later date through land clearing, climatic change, or fire, undermining the intended climate benefits.[48,82] The perceived lack of permanence—the longevity of a carbon pool and the stability of its stocks—in emissions reductions could seriously undermine demand for REDD+ credits.

Permanence has been a particularly important issue for early REDD+ projects, as they typically protect a particular tract of forest,[2] which can be vulnerable to encroachment and stochastic events.[33] The transition of REDD+ from site-based projects to national-scale implementation changed this considerably. When operating at the national scale, the MRV process is based on the realized rate of net national forest-carbon emissions, not whether a particular tract of forest is protected. This makes reducing emissions from forests appear identical to reducing the rate of fossil fuel emissions, where rates of emissions are reduced, not particular fossil fuel reserves. It is therefore argued that, like efforts targeting fossil fuel emissions, REDD+ will lead to permanent benefits, as even if countries opt out in the future, emissions should simply return to the pre-REDD+ rates and not undue previous reductions.[4,83]

However, comparing historic trends in deforestation with those of fossil fuel emissions raises doubts about the similarity of the two sources. For instance, aside from a slight increase each year, fossil fuel

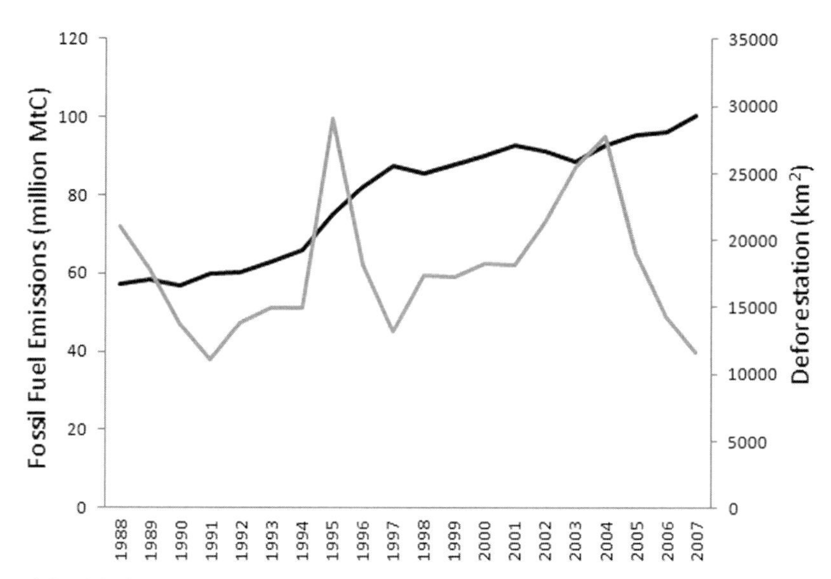

Figure 1. Annual fossil fuel emissions (black line) and deforested area (gray line) for Brazil, 1988–2007. Data are from the Brazilian National Institute of Space Research (http://www.inpe.br/ingles/) and the Carbon Dioxide Information Analysis Centre (http://cdiac.ornl.gov/ftp/trends/emissions/bra.dat).

emissions in Brazil show little interannual variation (Fig. 1). On the other hand, deforestation is marked by a series of peaks and troughs, determined largely by national and international economic conditions and government policies and investments.[84] Given the capacity for deforestation rates to increase over short periods of time, there is a risk that REDD+ achievements are undone by accelerated deforestation following periods of low forest loss.

A number of options exist to guard against the risk of nonpermanence. These include withholding credits from the market to act as a buffer or insurance policy against future emissions, discounting the value of credits, and creating temporary instead of permanent credits. These options have the effect of either paying less per ton of averted emissions or delaying payments. Both options will reduce the cost-competitiveness of REDD+ compared with the drivers of forest loss, and how to strike the right balance between reducing the risks of nonpermanence and sufficiently rewarding participants for emissions reductions remains to be determined.

REDD+ carbon credits could also be undermined by the effects of leakage, which is the shifting or displacement of deforestation from an area under REDD+ restrictions to an area without these restrictions.[85] The clearest empirical evidence for leakage comes from work in the Peruvian Amazon. By observing changes in forest loss rates inside and outside of protected areas both before and after park establishment, Oliveira *et al.*[86] were able to show that while deforestation rates decreased inside parks, these benefits were undermined by increased deforestation in the adjacent areas. Even if leakage is unavoidable, the integrity of REDD+ credits could be maintained if leakage is accurately quantified. However, to quote from the leakage assessment for the Noel Kempff climate action project, "through the process of assessing leakage [...] it has become ever more apparent that quantifying leakage is not a straightforward process."[87]

As with the issue of permanence, the transition from site-based projects to national-scale implementation has considerably changed the potential impacts of leakage on REDD+ activities. The national-scale MRVing of carbon credits implicitly accounts for any within-country leakage of deforestation effort, effectively side-stepping the need to do so explicitly. Still, the drivers of deforestation are increasingly able to shift from one country to another,[88] raising the possibility of leakage occurring across national borders. In fact, economic modeling suggests that when logging is reduced in one country, 42–95% of this reduction may be leaked as increased logging in other countries.[89] International leakage could be minimized by encouraging all developing countries to participate in REDD+,[90] partially through offering attractive reference levels to high-forest, low-deforestation countries.

Monitoring forest carbon. To set reference levels and track progress, REDD+ programs must monitor and report on changes in forest cover and carbon stocks through time. With the widespread availability of remotely sensed data—LANDSAT has been producing data continually since 1972—and ongoing national forest census programs,[91] it would seem as though monitoring forests would not present a major hurdle for REDD+ implementation. However, in 2009, only 3% of countries had the capacity to monitor and report on changes in forest cover for REDD+,[92] and still no global forest monitoring network exists.[93] Even more concerning are the findings by Pelletier *et al.*[94] that, in Panama, deforestation would have to be reduced by half before it could be detected above the levels of statistical uncertainty from using available data sources. Most of the uncertainty in quantifying forest carbon stocks surrounds our ability to map the spatial variation of carbon within a given land-cover type, such as primary forests.[94] The problem is that most optical remote sensing technologies are unable to determine carbon variation above a fairly low carbon value, and ground-based techniques are too slow and expensive to employ across vast areas.[91]

The limitations of current forest monitoring programs has not gone unnoticed, and in 2010–2011 \$6.3 billion was spent on monitoring land carbon,[95] with some areas of progress showing considerable potential. Notable among them are recent advancements in using airplane-based LiDAR (light detection and ranging) technology in conjunction with field surveys and satellite-based optical remote sensing. This hybrid approach has been used to map the carbon content of 4.3 million hectares of the Amazon basin with high accuracy and precision and at a cost of only \$0.08 per hectare.[96] While it may be cost-effective, using LiDAR combined with other remote sensing data is technically demanding, and whether this approach can be scaled up and replicated to deliver wall-to-wall mapping remains to be seen.

Socioeconomic and political issues

A REDD+ mechanism would require decision makers to commit to long-term decisions that might have implications that extend far beyond the immediate REDD+ affected areas and communities. Such impacts might include not only lost financial opportunities from withholding conventional land use activities, but also less-direct and tangible socioeconomic costs associated with REDD+. Therefore, the success of REDD+ implementation requires careful consideration of the full range of economic, social, and political implications of avoiding deforestation.[97]

The most obvious examples of such costs include the loss of employment and revenue generation from raw material processing and other value-adding downstream industries. The Malaysian oil palm industry, which continues to threaten the country's remaining rainforests, particularly in Borneo, contributes about 5–6% of Malaysian GDP and provides direct employment for 570,000 workers. The industry is associated with a further 830,000 workers employed in downstream activities. Oil palm exports also generate foreign exchange earnings amounting to around \$10 billion annually. These funds contribute to rural development including the improved provision of piped water, electricity, communications, roads, schools, and healthcare through land schemes coordinated by the Federal Land Development Authority.[98] Any downscaling of oil palm activities and associated industries could lead to significant socioeconomic impacts, manifest in the form of constrained opportunities for regional economic growth, including reduced tax revenues for regional and local governments. Governments might become less inclined to invest in such regions, particularly if REDD+ forfeits any opportunity for development over a period of 20–30 years. Consequently, compared to a non-REDD+ area, REDD+ regions might have poorer infrastructure and telecommunications, limited mobility and market access, and lower-quality public services. Recognizing this issue, some forest-rich nations are already advocating for REDD+ funds to be allocated to development outside the forestry sector. Notably, Guyana specifically calls for REDD investments in private sector entrepreneurship, electricity, roads, health, and education.[99]

REDD+ might also lead to less tangible but equally important political and socioeconomic costs relating to national and local development. The future livelihood options of local communities might be constrained, as is arguably the case with protected area systems.[100] Perhaps more importantly, forest-using communities with unclear land tenure might be denied access to the lands they manage and depend on, leading to conservation refugees.[101] At the national level, REDD+ implementation might encourage the migration of people from REDD+ affected areas to urban centers. Such demographic changes could raise political and social concerns and further undermine government priorities for investment in depopulated rural areas. By promoting national institutions charged with forest management, there is also the risk that REDD+ will reverse decades of work toward decentralized forest governance.[102,103]

Perhaps the least-considered aspect of REDD+ is its potential implications for international relations. A REDD+ mechanism that limits the supply of key commodities could alter conventional political ties where such relations are strongly based on trade. For example, Japan's historical relationships with Indonesia and the Philippines have largely been based on the trade of natural resources such as timber.[104] A central priority of Japan's foreign development aid has been to secure the development and import into Japan of natural resources.[97] Such consumer–client relationships, as well as other political and business ties that shape political and economic interactions between countries, might limit the widespread implementation of REDD+.

While there are important risks that REDD+ could constrain development options and lead to widespread and possibly unanticipated socioeconomic costs, as detailed above, these risks have been at least partially mitigated by the addition of socioeconomic safeguards into REDD+ negotiations (Cancun Decision 1/CP.16). These safeguards include the requirement that activities are "undertaken in accordance with national development priorities" and "in the context of sustainable development and reducing poverty." What remains is identifying how REDD+ can best be implemented to fulfill both development and climate mitigation objectives.

Norway–Indonesia REDD+ agreement

In May 2010, the governments of Norway and Indonesia signed a Letter of Intent for a REDD+ partnership that would contribute to significant reductions in greenhouse gas emissions from deforestation, forest degradation, and peatland conversion in Indonesia.[105] Phase 2 of this agreement requires Indonesia to implement a two-year moratorium to suspend all new concessions for the conversion of peat and natural forests. The wider goal of this moratorium is to create a baseline on critical elements of forests, peatlands, and "degraded lands" that is strategic to the effective implementation of a nationwide REDD+ strategy in the future. This partnership offers a good opportunity for Indonesia to mitigate its deforestation and carbon emission levels. However, it has also galvanized intense debates within Indonesian society regarding the scope of the moratorium and its consequent environmental and socioeconomic ramifications.

After much political wrangling between the Indonesian Ministry of Forestry and the country's REDD+ Task Force, the moratorium was signed into effect in May 2011 by Indonesian President Dr. H. Susilo Bambang Yudhoyono (Presidential Instruction No. 10/2011). An accompanying first draft of an "indicative map" displays land areas protected by the moratorium. Critics point out that the lack of clarity and transparency about how the indicative map was generated makes it difficult to assess its efficacy for reducing greenhouse gas emissions.[106] Environmentalists are also disappointed with the fact that large swaths of secondary forests, as well as any forest that has already been offered up to plantation companies as concessions, will be exempted from the moratorium.[107] On the other hand, some community and village forestry development projects are now under threat of being abandoned because such activities do fall within the remit of the moratorium. Social nongovernmental organizations argue that the moratorium is biased toward big agribusinesses while severely threatening ongoing efforts at developing sustainable forestry programs at the grassroots level.[108]

Putting aside the controversies about the moratorium, the success of the Norway–Indonesia partnership, and Indonesia's emissions reduction efforts in general, will depend on its ability to overcome its patchy record in forest conservation.[109] A ban on the expansion of oil palm plantations on peatlands in 2007 was repealed only two years later to allow the conversion of 2 million ha of peatlands.[110,111] Furthermore, despite promises by the Indonesian government to reduce forest fires, the number of human-induced fires continues to increase.[112] Most crucially, the Indonesian government has been criticized for rampant and systematic corruption. The country's Reforestation Fund has allegedly been used for politically expedient projects unrelated to forest restoration.[68]

The Norway–Indonesia REDD+ agreement faces tremendous challenges. But the authorities can ill-afford to allow this initiative to fail. Many observers regard the Norway–Indonesia partnership as a harbinger of outcomes in other forest-rich nations seeking REDD+ funds, such as Papua New Guinea, Democratic Republic of Congo, and Colombia. Therefore, the Indonesian case study will likely have profound implications for climate change mitigation and conservation across the developing tropics.

Way forward

Clarifying and examining fundamental assumptions of REDD+

One of the fundamental assumptions of REDD+ is that the lack of financial incentives to protect forests is the primary reason behind deforestation and forest degradation in forest-rich developing nations. However, tropical deforestation is, in fact, often driven by a complex suite of social, political, and economic factors, including corruption, weak governance, human migration, and national and international political pressures. As we discussed above, even if REDD+ is able to fully compensate for the direct financial opportunity costs of avoided deforestation, there still might be other indirect and less tangible factors that would continue to drive forest conversion and undermine REDD+. Therefore, the expectation for a well-funded REDD+ to fully deliver on avoiding deforestation might have to be tempered with the realization it must either be part of a portfolio of solutions for mitigating forest and carbon loss or further expanded to include such solutions.

In the past few decades, without the help of REDD+, a number of countries transitioned from net deforesters to forest-gaining countries, including China, India, Vietnam, the Republic of Korea, and Chile. What drove these transitions can provide lessons for how best to compliment REDD+ financial incentives with additional socioeconomic and political reforms. Most notably, the transition from forest-losing to forest gaining was characterized by rapid economic development, an increase in the perceived value of wood, a strengthening of the rights of indigenous and forest-using peoples, and the establishment of aggressive afforestation and reforestation programs.[113] In addition to putting a price on forest carbon, incentives to stimulate these changes could also include stimulating nonforestry sectors of the economy, increasing the value of wood by increasing in-country downstream processing, and tackling ongoing indigenous land rights disputes.

There has been a tendency in the scientific literature and by political and environmental lobbyists to view REDD+ not strictly as a tool for mitigating climate change, but rather as a silver bullet for tropical forest conservation. Clearly, the opportunities to derive cobenefits resulting from avoided deforestation appear too important to pass up. Nevertheless, there is a real risk of overburdening REDD+ with too many good intentions, and emission reductions must therefore remain the main purpose of REDD+. The challenge ahead is to develop complimentary mechanisms and markets to allow for the bundling of benefits, such as biodiversity conservation, without increasing transaction costs of participating in REDD+ carbon markets.

It is also important to consider the dilemma posed by REDD+ for climate change mitigation. At the UN climate negotiations in Copenhagen in 2010, developed countries, including European Union members, pledged to reduce emissions by up to 20% by 2020. The more successful that REDD+ might eventually become at supplying carbon credits to Annex I countries to meet their emissions targets, the less incentives there would be for domestic actions to cut emissions.[95] Therefore, somewhat paradoxically, to achieve effective emissions reductions globally, either ceilings will have to be imposed on the use of REDD+ credits for offsetting emissions in developed countries, or separate REDD+ targets will need to be set.

Conclusion

In our opinion, REDD+ is currently the most promising mechanism for conserving tropical forests, largely because of its unprecedented ability to harness funds and put them to use in ways that are fundamentally more effective than past efforts. But to the dismay of some, after six years of negotiations, an officially endorsed REDD+ mechanism is yet to be finalized. At the same time, this delay presents an opportunity to further shape REDD+ to address persistent issues and create new opportunities. To ensure that REDD+ acts as a true game changer in the struggle to protect tropical forests, and not just a fleeting quick-fix, improvements should focus on REDD+'s ability to provide or compliment a suite of nonmonetary incentives to encourage developing nations to transition away from net deforestation while also accessing low transaction cost and high-volume carbon markets or funds, and align with, not undermine, a globally cohesive attempt to mitigate climate change.

Acknowledgments

O.V. is supported by the Australian Research Council and L.P.K. is supported by the Swiss National Science Foundation and the North-South Centre of ETH Zurich.

Conflicts of interest

The authors declare no conflicts of interest.

References

1. Engel, S., S. Pagiola & S. Wunder. 2008. Designing payments for environmental services in theory and practice: an overview of the issues. *Ecol. Econ.* **65:** 663–674.
2. Caplow, S., P. Jagger, K. Lawlor & E. Sills. 2011. Evaluating land use and livelihood impacts of early forest carbon projects: lessons for learning about REDD+. *Environ. Sci. Policy.* **14:** 152–167.
3. UNFCCC. 2007. Reducing emissions from deforestation in developing countries: approaches to stimulate action, conference of the parties, Bali, 3 December to 11 December 2007. Available from http://unfccc.int/resource/docs/2007/sbsta/eng/l23a01r01.pdf. United Nations Framework Convention on Climate Change (UNFCCC). Bonn, Germany.
4. Ebeling, J. & M. Yasue. 2008. Generating carbon finance through avoided deforestation and its potential to create climatic, conservation and human development benefits. *Philos. Trans. R. Soc. Lond. B. Biol. Sci.* **363:** 1917–1924.
5. World Bank. 2011. *State and Trends of the Carbon Market 2011.* The World Bank. Washington, DC.
6. Cerbu, G. A., B. M. Swallow & D. Y. Thompson. 2011. Locating REDD: a global survey and analysis of REDD readiness and demonstration activities. *Environ. Sci. Policy* **14:** 168–180.
7. Malhi, Y. 2010. The carbon balance of tropical forest regions, 1990–2005. *Curr. Opin. Environ. Sustain.* **2:** 237–244.
8. Gullison, R. E., P. C. Frumhoff, J. G. Canadell, *et al.* 2007. Tropical forests and climate policy. *Science* **316:** 985–986.
9. Bartholome, E. & A.S. Belward. 2005. GLC2000: a new approach to global land cover mapping from Earth observation data. *Int. J. Remote. Sens.* **26:** 1959–1977.
10. Saatchi, S. S., N. L. Harris, S. Brown, *et al.* 2011. Benchmark map of forest carbon stocks in tropical regions across three continents. *Proc. Natl. Acad. Sci. USA* **108:** 9899–9904.
11. IPCC. 2007. Working group I Report. The physical science basis—contribution to the fourth assessment report of the Intergovernmental Panel on Climate Change. Available from http://www.ipcc.ch/ipccreports/assessments-reports.htm. Intergovernmental Panel on Climate Change. Geneva, Switzerland.
12. Winjum, J. K., S. Brown & B. Schlamadinger. 1998. Forest harvests and wood products: sources and sinks of atmospheric carbon dioxide. *For. Sci.* **44:** 272–284.
13. DeFries, R. S., R. A. Houghton, M. C. Hansen, *et al.* 2002. Carbon emissions from tropical deforestation and regrowth based on satellite observations for the 1980s and 1990s. *Proc. Natl. Acad. Sci. USA* **99:** 14256–14261.
14. Houghton, R. A. 2001. *Revised Estimates of the Annual Net Flux of Carbon to the Atmosphere from Changes in Land Use and Land Management 1850–2000*: 378–390. Blackwell Munksgaard, Sendai, Japan.
15. FAO. 2006. Global forest resource assessment 2005: progress towards sustainable forest management. In Forestry Paper: 147. Food and Agriculture Organization of the United Nations. Rome, available from http://www.fao.org.
16. Achard, F., H. D. Eva, H. J. Stibig, *et al.* 2002. Determination of deforestation rates of the world's humid tropical forests. *Science* **297:** 999–1002.
17. Hansen, M. C., S. V. Stehman, P. V. Potapov, *et al.* 2008. Humid tropical forest clearing from 2000 to 2005 quantified by using multitemporal and multiresolution remotely sensed data. *Proc. Natl. Acad. Sci. USA* **105:** 9439–9444.
18. FAO. 2010. Global forest resources assessment 2010. In *FAO Forestry Paper* 163. Food and Agriculture Organization of the United Nations. Rome.
19. JRCNEA 2011. The Emissions Database for Global Atmospheric Research (EDGAR). Available from http://edgar.jrc.ec.europa.eu/index.php, Inspra, Italy.
20. van der Werf, G. R., D. C. Morton, R. S. DeFries, *et al.* 2009. CO_2 emissions from forest loss. *Nat. Geosci.* **2:** 737–738.
21. Page, S. E., F. Siegert, J. O. Rieley, *et al.* 2002. The amount of carbon released from peat and forest fires in Indonesia during 1997. *Nature* **420:** 61–65.
22. Eliasch, J. 2008. Climate change. Financing global forests: the Eliasch review. Report to the UK government.
23. Stern, N. 2007. *The Economics of Climate Change: The Stern Review.* Cambridge University Press. Cambridge, UK.
24. Groombridge, B. & M. D. Jenkins. 2003. *World Atlas of Biodiversity: Earth's living Resources in the 21st Century.* University of California Press. Berkeley, Los Angeles, London.
25. Wilson, E. O. 1988. *Biodiversity.* National Academy Press. Washington, DC.
26. Joppa, L. N., D. L. Roberts, N. Myers & S. L. Pimm. 2011. Biodiversity hotspots house most undiscovered plant species. *Proc. Natl. Acad. Sci. USA* **108:** 13171–13176.
27. Laurance, W. F. 2007. Have we overstated the tropical biodiversity crisis? *Trends Ecol. Evol.* **22:** 65–70.
28. Pimm, S.L., G.J. Russell, J. L. Gittleman & T. M. Brooks. 1995. The future of biodiversity. *Science* **269:** 347–350.
29. Sodhi, N.S., L.P. Koh, B.W. Brook & P.K.L. Ng. 2004. Southeast Asian biodiversity: an impending disaster. *Trends Ecol. Evol.* **19:** 654–660.

30. Whitmore, T. C. & J. A. Sayer. 1992. *Tropical Deforestation and Species Extinctions*. Chapman and Hall. London.
31. Myers, N. 1984. *The Primary Source: Tropical Forests and Our Future*. W.W. Norton. New York.
32. Myers, N., R. A. Mittermeier, C. G. Mittermeier, *et al.* 2000. Biodiversity hotspots for conservation priorities. *Nature* **403:** 853–858.
33. Curran, L. M., S. N. Trigg, A. K. McDonald, *et al.* 2004. Lowland forest loss in protected areas of Indonesian Borneo. *Science* **303:** 1000–1003.
34. Brooks, T. M., R. A. Mittermeier, C. G. Mittermeier, *et al.* 2002. Habitat loss and extinction in the hotspots of biodiversity. *Conserv. Biol.* **16:** 909–923.
35. Bradshaw, C. J., N. S. Sodhi & B. W. Brook. 2009. Tropical turmoil: a biodiversity tragedy in progress. *Front. Ecol. Environ.* **7:** 79–87.
36. Venter, O., E. Meijaard, H. Possingham, *et al.* 2009. Carbon payments as a safeguard for threatened tropical mammals. *Conserv. Lett.* **2:** 123–129.
37. Venter, O., W. F. Laurance, T. Iwamura, *et al.* 2009. Harnessing carbon payments to protect biodiversity. *Science* **326:** 1368.
38. Laurance, W. F. 2007. A new initiative to use carbon trading for tropical forest conservation. *Biotropica* **39:** 20–24.
39. Strassburg, B. B. N., A. Kelly, A. Balmford, *et al.* 2010. Global congruence of carbon storage and biodiversity in terrestrial ecosystems. *Conserv. Lett.* **3:** 98–105.
40. Grainger, A., D. H. Boucher, P. C. Frumhoff, *et al.* 2009. Biodiversity and REDD at Copenhagen. *Curr. Biol.* **19:** 974–976.
41. Harvey, C. A., B. Dickson & C. Kormos. 2010. Opportunities for achieving biodiversity conservation through REDD. *Conserv. Lett.* **3:** 53–61.
42. Miles, L. & V. Kapos. 2008. Reducing greenhouse gas emissions from deforestation and forest degradation: global land-use implications. *Science* **320:** 1454–1455.
43. UNEP-WCMC. 2008. *Carbon and Biodiversity: a Demonstration Atlas*. Eds. Kapos V., Ravilious C., Campbell A., Diskson B., Gibbs H., Hansen M., Lysenko I., Miles L., Price J., Scharlemann J.P.W., Trumper K. UNEP WCMC. Cambridge, UK.
44. Burgman, M. A., D. Keith, S. D. Hopper, *et al.* 2007. Threat syndromes and conservation of the Australian flora. *Biol. Conserv.* **134:** 73–82.
45. Venter, O., N. N. Brodeur, L. Nemiroff, *et al.* 2006. Threats to endangered species in Canada. *Bioscience* **56:** 903–910.
46. Laurance, W. F. & C. A. Peres. 2006. *Emerging Threats to Topical Forests*. University of Chicago Press. Chicago.
47. Malhi, Y., J. T. Roberts, R. A. Betts, *et al.* 2008. Climate change, deforestation, and the fate of the Amazon. *Science* **319:** 169–172.
48. Aragao, L. E. O. C. & Y. E. Shimabukuro. 2010. The incidence of fire in amazonian forests with implications for REDD. *Science* **328:** 1275–1278.
49. Bodmer, R. E., J. F. Eisenberg & K. H. Redford. 1997. Hunting and the likelihood of extinction of Amazonian mammals. *Conserv. Biol.* **11:** 460–466.
50. Putz, F. E. & K. H. Redford. 2009. Dangers of carbon-based conservation. *Glob. Environ. Change Hum. Pol. Dimensions.* **19:** 400–401.
51. Stickler, C. M., D. C. Nepstad, M. T. Coe, *et al.* 2009. The potential ecological costs and cobenefits of REDD: a critical review and case study from the Amazon region. *Glob. Change Biol.* **15:** 2803–2824.
52. Sasaki, N. & F. E. Putz. 2009. Critical need for new definitions of "forest" and "forest degradation" in global climate change agreements. *Conserv. Lett.* **2:** 226–232.
53. Bekessy, S. A. & B. A. Wintle. 2008. Using carbon investment to grow the biodiversity bank. *Conserv. Biol.* **22:** 510–513.
54. Brockerhoff, E. G., H. Jactel, J. A. Parrotta, *et al.* 2008. Plantation forests and biodiversity: oxymoron or opportunity? *Biodivers. Conserv*. **17:** 925–951.
55. Barlow, J., T. A. Gardner, I. S. Araujo, *et al.* 2007. Quantifying the biodiversity value of tropical primary, secondary, and plantation forests. *Proc. Natl. Acad. Sci. USA* **104:** 18555–18560.
56. Phelps, J., E. L. Webb & L. P. Koh. 2011. Risky business: an uncertain future for biodiversity conservation finance through REDD+. *Conserv. Lett.* **4:** 88–94.
57. Epple, C., E. Dunning, B. Dickson & C. A. Harvey. 2011. *Making Biodiversity Safeguards for REDD+ Work in Practice: Developing Operational Guidelines and Identifying Capacity Requirements*. UNEP-WCMC. Cambridge, UK.
58. Bekessy, S. A., B. A. Wintle, D. B. Lindenmayer, *et al.* 2010. The biodiversity bank cannot be a lending bank. *Conserv. Lett.* **3:** 151–158.
59. Ring, I., M. Drechler, A. J. A. Teeffelen, S. Irawan & O. Venter. 2010. Biodiversity conservation and climate change mitigation: what role can economic instruments play? *Curr. Opin. Environ. Sustain.* **2:** 50–58.
60. Halpern, B. S., C. R. Pyke, H. E. Fox, *et al.* 2006. Gaps and mismatches between global conservation priorities and spending. *Conserv. Biol.* **20:** 56–64.
61. Palm, C. A., R. A. Houghton, J. M. Melillo & D. L. Skole. 1986. Atmospheric carbon-dioxide from deforestation in Southeast Asia. *Biotropica* **18:** 177–188.
62. UNFCCC 2010. CDM statistics. Available from http://cdm.unfccc.int/Statistics/index.html.
63. Brooks, T. M., S. J. Wright & D. Sheil. 2009. Evaluating the success of conservation actions in safeguarding tropical forest biodiversity. *Conserv. Biol.* **23:** 1448–1457.
64. Pimm, S. L. & P. Raven. 2000. Biodiversity—extinction by numbers. *Nature* **403:** 843–845.
65. Butchart, S. H. M., A. J. Stattersfield, J. Baillie, *et al.* 2005. Using red list indices to measure progress towards the 2010 target and beyond. *Philos. Trans. R. Soc. Lond. B. Biol. Sci.* **360:** 255–268.
66. UNFCCC. 2009. Copenhagen accord. FCCC/CP/2009/L.7. Available from http://unfccc.int/resource/docs/2009/cop15/eng/l07.pdf. United Nations Framework Convention on Climate Change (UNFCCC). Bonn, Germany.
67. James, A. N., M. J. B. Green & J. R. Paine. 1999. A global review of protected area budgets and staff. In *WCMC Biodiversity* Series No. 10. WCMC—World Conservation Press. Cambridge, UK.

68. Barr, C., A. Dermawan, H. Purnomo & H. Komarudin. 2010. Financial governance and Indonesia's reforestation fund during the Soeharto and post-Soeharto periods, 1989–2009: a political economic analysis of lessons for REDD+. In *Occasional Paper 52*. CIFOR. Bogor, Indonesia.
69. Kareiva, P., A. Chang & M. Marvier. 2008. Environmental economics—development and conservation goals in World Bank projects. *Science* **321:** 1638–1639.
70. Dargusch, P., K. Lawrence, J. Herbohn & Medrilzam. 2010. A small-scale forestry perspective on constraints to including REDD in international carbon markets. *Small-Scale For.* **9:** 485–499.
71. Angelsen, A. 2008. How do we set the reference levels for REDD payments? In *Moving Ahead with REDD: issues, Options and Implications*. Angelsen, A., Ed. CIFOR. Bogor, Indonesia.
72. Griscom, B., D. Shoch, B. Stanley, *et al.* 2009. Sensitivity of amounts and distribution of tropical forest carbon credits depending on baseline rules. *Environ. Sci. Policy.* **12:** 897–911.
73. Busch, J., B. Strassburg, A. Cattaneo, *et al.* 2009. Comparing climate and cost impacts of reference levels for reducing emissions from deforestation. *Environ. Res. Lett.* **4:** 044006.
74. Leischner, B. & P. Elsasser. 2010. Reference emission levels for REDD: implications of four different approaches applied to past period's forest area development in 84 countries. *Landbauforschung* **60:** 119–130.
75. TCG. 2009. *Estimating Tropical Forest Carbon at Risk of Emission from Deforestation Globally: Applying the Terrestrial Carbon Group Reference Emission Level Approach.* The Terrestrial Carbon Group Project. Australia.
76. Gaveau, D. L. A., S. Wich, J. Epting, *et al.* 2009. The future of forests and orangutans (Pongo abelii) in Sumatra: predicting impacts of oil palm plantations, road construction, and mechanisms for reducing carbon emissions from deforestation. *Environ. Res. Lett.* **4.**
77. Griscom, B. 2011. *Establishing Efficient, Equitable, and Environmentally Sound Reference Emissions Levels for REDD+: A Stock Flow Approach.* The Nature Conservancy. Arlington, VA.
78. Strassburg, B., R. K. Turner, B. Fisher, *et al.* 2009. Reducing emissions from deforestation-The "combined incentives" mechanism and empirical simulations. *Glob. Environ. Change* **19:** 265–278.
79. Pontius, R. G., W. Boersma, J. C. Castella, *et al.* 2008. Comparing the input, output, and validation maps for several models of land change. *Ann. Reg. Sci.* **42:** 11–37.
80. Australia. 2009. Reducing emissions from deforestation and forest degradation in developing countries (REDD). Submissions to the AWG-LCA, AWG-KP and SBSTA.
81. Tomich, T. P., H. Foresta, R. Dennis, *et al.* 2001. Carbon offsets for conservation and development in Indonesia? *Am. J. Alternative Agr.* **17:** 13.
82. Gumpenberger, M., K. Vohland, U. Heyder, *et al.* 2010. Predicting pan-tropical climate change induced forest stock gains and losses-implications for REDD. *Environ. Res. Lett.* **5:** 014013.
83. Skutsch, M. & B. J. De Jong. 2010. The permanence debate. *Science* **327:** 1079–1079.
84. Butler, R. A. 2011. Deforestation in the Amazon. Available from URL http://www.mongabay.com/brazil.html.
85. Ewers, R. M. & A. S. L. Rodrigues. 2008. Estimates of reserve effectiveness are confounded by leakage. *Trends Ecol. Evol.* **23:** 113–116.
86. Oliveira, P. J. C., G. P. Asner, D. E. Knapp, *et al.* 2007. Land-use allocation protects the Peruvian Amazon. *Science* **317:** 1233–1236.
87. Winrock. 2002. *2001 Analysis of Leakage, Baselines, and Carbon Benefits for the Noel Kempff Climate Action Project*. Winrock International. Washington, DC.
88. Butler, R. A. & W. F. Laurance. 2008. New strategies for conserving tropical forests. *Trends Ecol. Evol.* **23:** 469–472.
89. Gan, J. B. & B. A. McCarl. 2007. Measuring transnational leakage of forest conservation. *Ecol. Econ.* **64:** 423–432.
90. Murray, B. C. 2008. Leakage from an avoided deforestation compensation policy: empirical evidence, and corrective policy solutions. Working paper NI-WP 08–02. Nicholas Institute for Environmental Policy Solutions, Duke University. Durham, USA.
91. Gibbs, H. K., S. Brown, J. O. Niles & J. A. Foley. 2007. Monitoring and estimating tropical forest carbon stocks: making REDD a reality. *Environ. Res. Lett.* **2:** 1–13.
92. Herold, M. 2009. *An Assessment of National Forest Monitoring Capabilities in Tropical Non-annex 1 countries: Recommendations for Capacity Building*. Friedrich-Schiller-Universität Jena and GOFC–GOLD Land Cover Project Office. Jena, Germany.
93. Grainger, A. & M. Obersteiner. 2011. A framework for structuring the global forest monitoring landscape in the REDD +era. *Environ. Sci. Policy* **14:** 127–139.
94. Pelletier, J. & *et al.* 2011. Diagnosing the uncertainty and detectability of emission reductions for REDD +under current capabilities: an example for Panama. *Environ. Res. Lett.* **6:** 024005.
95. Maslin, M. & J. Scott. 2011. Carbon trading needs a multi-level approach. *Nature* **475:** 445–447.
96. Asner, G. P., G. V. N. Powell, J. Mascaro, *et al.* 2010. High-resolution forest carbon stocks and emissions in the Amazon. *Proc. Natl. Acad. Sci. USA* **107:** 16738–16742.
97. Ghazoul, J., R. A. Butler, J. Mateo-Vega & L. P. Koh. 2010. REDD: a reckoning of environment and development implications. *Trends Ecol. Evol.* **25:** 396–402.
98. MPOC. 2008. *Fact Sheet on Malaysian Oil Palm.* Available from http://www.mpoc.org.pk/page/fact-sheets-malaysian-palm-oil. Malaysian Palm Oil Council. Selangor Darul Ehsan, Malaysia.
99. Guyana. 2010. Transforming Guyana's Economy While Combating Climate Change. Available from http://www.undp.org.gy/documents/bk/Revised-LCDS-May-20–2010-draft-for-MSSC.pdf. Office of the President. Georgetown, Guyana.
100. Kaimowitz, D. & D. Sheil. 2007. Conserving what and for whom? Why conservation should help meet basic human needs in the tropics. *Biotropica* **39:** 567–574.
101. FoE. 2008. REDD myths: a critical review of proposed mechanisms to reduce emissions from deforestation and degradation in developing countries. Friends of the Earth International Secretariat. Available from www.foei.org/en/publications/pdfs/redd-myths.
102. Phelps, J., E. L. Webb & A. Agrawal. 2010. Does REDD plus threaten to recentralize forest governance? *Science* **328:** 312–313.

103. Anonymous. 2011. Seeing REDD. *Nature* **472:** 390–390.
104. Dauvergne, P. 1997. *Shadows of the Forest*. Massachusetts Institute of Technology Press. Cambridge, MA.
105. Norway-Indonesia. 2010. Letter of Intent between the Government of the Kingdom of Norway and the Government of the Republic of Indonesia on "Cooperation on reducing greenhouse gas emissions from deforestation and forest degradation" Available from www.norway.or.id/PageFiles/404362/Letter_of_Intent_Norway_Indonesia_26_May_2010.pdf.
106. Wells, P. & G. Paoli. 2011. An Analysis of Presidential Instruction No. 10, 2011: Moratorium on Granting of New Licenses and Improvement of Natural Primary Forest and Peatland Governance. Available from URL www.daemeter.org/wp-content/files/Daemeter_Moratorium_Analysis_20110527_FINAL.pdf. Daemeter Consulting. Bogor, Indonesia.
107. Butler, R. A. 2011. Indonesia's moratorium disappoints environmentalists Available from http://news.mongabay.com/2011/0520-indonesia_moratorium_defined.html.
108. Butler, R. A. 2011. Indonesia's moratorium undermines community forestry in favor of industrial interests. Available from http://news.mongabay.com/2011/0621-community˙forests˙moratorium.html.
109. Clements, G. R., J. Sayer, A. K. Boedhihartono, *et al.* 2010. Cautious optimisim over Norway-Indonesia REDD pact. *Conserv. Biol.* **24:** 1437–1438.
110. Koh, L. P. 2009. Calling Indonesia's US$ 13 Billion Bluff. *Conserv. Biol.* **23:** 789–789.
111. Butler, R. A. 2009. Indonesia confirms that peatlands will be converted for plantations Available from http://news.mongabay.com/2009/0219-indonesia.html.
112. Simamora, A. P. 2010. More hotspots detected despite pledge to reduce forest fires. In *Jakarta Post*. Available from www.thejakartapost.com/news/2010/06/11/more-hotspots-detected-despite-pledge-reduce-forest-fires.html, Jakarta.
113. Gregersen, H., H. El-Lakany, L. Bailey & A. White. 2011. The greener side of REDD+: lessons from REDD+ from countries where forest area is increasing. Rights and Resources Initiative. Washington, DC.

Ann. N.Y. Acad. Sci. ISSN 0077-8923

ANNALS OF THE NEW YORK ACADEMY OF SCIENCES
Issue: *The Year in Ecology and Conservation Biology*

The boreal forest as a cultural landscape

Edward A. Johnson[1] and Kiyoko Miyanishi[2]

[1]Biogeoscience Institute and Department of Biological Sciences, University of Calgary, Calgary, Alberta, Canada.
[2]Department of Geography, University of Guelph, Guelph, Ontario, Canada

Address for correspondence: Edward A. Johnson, Department of Biological Sciences, University of Calgary, Calgary, Alberta T2N 1N4. johnsone@ucalgary.ca

Because of its generally low density of humans and few settlements, the circumpolar boreal forest is often viewed as an untouched wilderness. However, archeological evidence indicates that humans have inhabited the region since the continental glaciers disappeared 8,000–12,000 years ago. This paper discusses the ecological impacts that humans have had on the boreal forest ecosystem through their activities in prehistoric, historic, and recent times and argues that the boreal forest has always been a cultural landscape with a gradient of impacts both spatially and temporally. These activities include hunting, trapping, herding, agriculture, forestry, hydroelectric dam projects, oil and natural gas development, and mining. In prehistoric times, human impacts would generally have been more temporary and spatially localized. However, the megafaunal extinctions coincident with arrival of humans were very significant ecological impacts. In historic times, the spread of Europeans and their exploitation of the boreal's natural resources as well as agricultural expansion has altered the composition and continuity of the boreal forest ecosystem in North America, Fennoscandia, and Asia. Particularly over the last century, these impacts have increased significantly (e.g., some hydroelectric dams and tar sands developments that have altered and destroyed vast areas of the boreal forest). Although the atmospheric changes and resulting climatic changes due to human activities are causing the most significant changes to the high-latitude boreal forest ecosystem, any discussion of these impacts are beyond the limits of this paper and therefore are not included.

Keywords: boreal region; ecosystems; forest

Introduction

Although often viewed as a vast untouched wilderness, the circumpolar boreal forest (taiga) region (Fig. 1), which extends roughly between 50° and 70° N latitude, is, in fact, a cultural landscape[1] that exhibits a gradient of human influence. At one end of the gradient are remote areas far from transportation networks such as roads and railways, inhabited largely by indigenous people, and whose ecosystems are still largely unmodified by humans. Figure 2 shows the extent of such "intact" forest landscapes, defined as "essentially undisturbed by human development with an area of at least 50,000 ha,"[2] in Canada and Russia. At the opposite end of the gradient are areas where the natural landscape has been significantly modified or replaced completely by agriculture, cities/towns, and industrial developments. Examples of cities in the boreal include Tromsø, Norway (pop. 63,596); Sudbury, Canada (pop. 157,857); Anchorage, United States (pop. 291,826); and Moscow, Russia (pop. 11,514,300). Between these extremes lie areas that have been partially converted to agricultural fields or whose natural resources are being exploited with varying degrees of effects/impacts on native plant and animal communities through activities such as herding, agricultural fields, strip and open pit mining, oil extraction, tree harvesting, and animal trapping. In all of the boreal regions, there are strong north–south changes in the types of environmental impact (e.g., agriculture being largely confined to the south). In Sweden, only small patches of unlogged forest remain, and in Fennoscandia generally most current forest stands classified as virgin and/or undisturbed have been shown to have had a history

doi: 10.1111/j.1749-6632.2011.06312.x

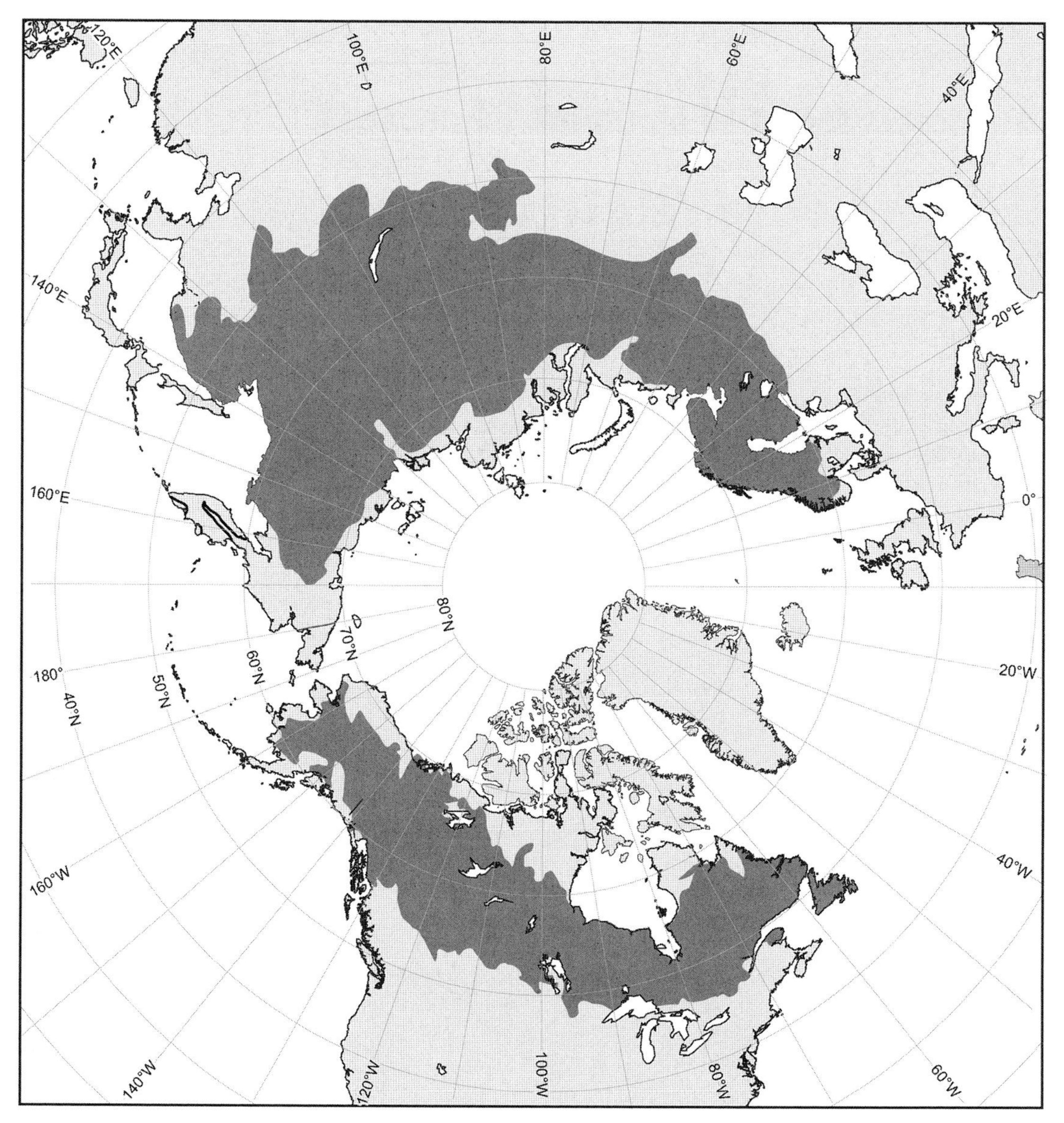

Figure 1. Circumpolar boreal region (adapted from Hare and Ritchie[102]). Reproduced with permission from The American Geographical Society.

of human disturbance and impact.[3,4] Such intensive forest use is not the case in the North American boreal, where much of the forest has not yet been cut for the first time; very little, if any, of the forest has experienced a complete harvest rotation; and large areas are still far from any roads or rail lines. However, even here, the forest has experienced the impacts of exploitation of nontimber forest resources such as fur-bearing animals in the 1700s and, more recently, oil and minerals.

Up to the middle of the last century, ecologists considered the boreal forest as having few species and poorly defined communities because it was thought to be an immature ecosystem that had not had time to reach the maturity of the deciduous forest further south.[5] The traditional ideas of succession and climax did not seem to work in the boreal forest. The boreal ecosystem, particularly in North America, was seen as wilderness. Natural and human disturbances were often not recognized, while

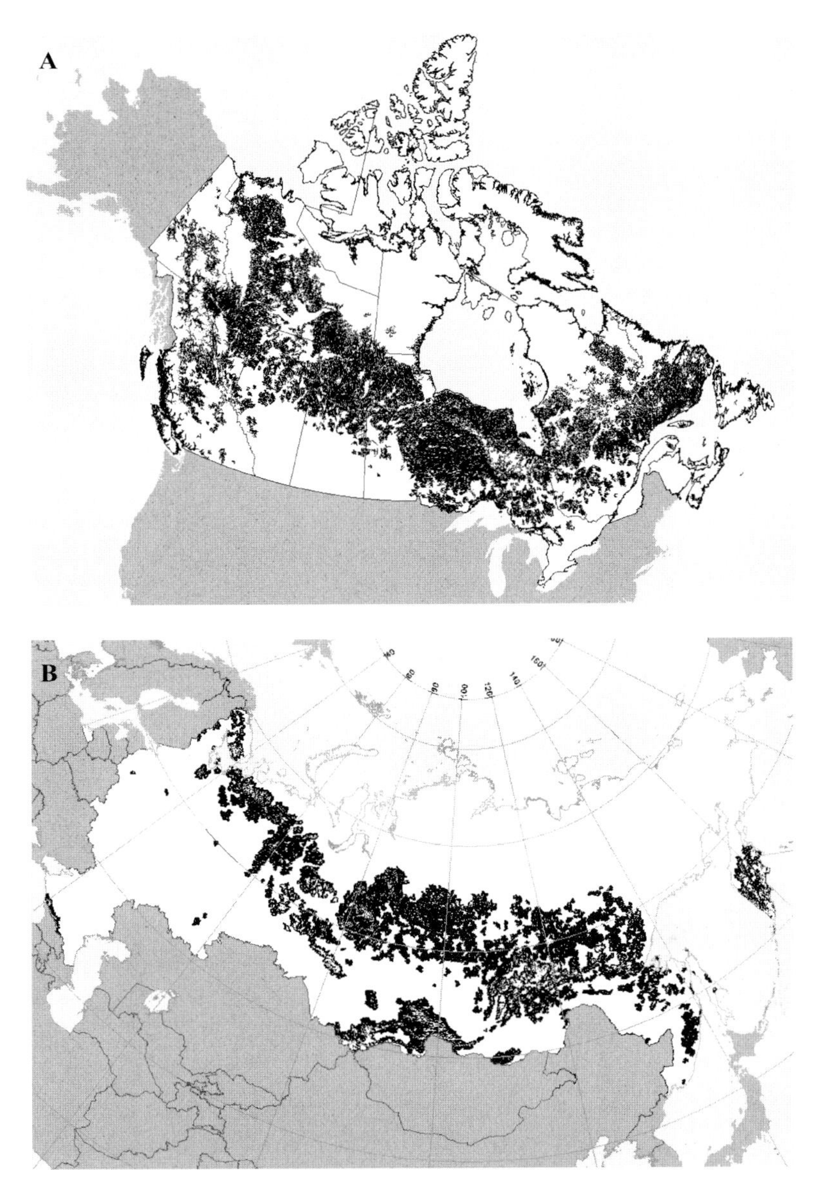

Figure 2. Intact forest of (A) Canada and (B) Russia (adapted from Global Forest Watch).[103–105] Black dots represent areas of at least 50,000 ha essentially undisturbed by human development. Reproduced with permission from Global Forest Watch.

in other situations any disturbance was assumed to be human-caused (e.g., forest fires).[6] Both viewpoints led to myopia of the cultural impact on the landscape.

In this paper, we will discuss the cultural impact on the ecology of the boreal forest. By *cultural*, we mean the effects that both indigenous and nonindigenous people have had on boreal ecosystems. We focus on the ecological impacts and not social and economic impacts that these activities have had.

Prehistoric boreal cultural landscape

Much of the circumpolar boreal region was covered several times by continental glaciers or by tundra during the Pleistocene. Humans thus had to invade, along with the reinvasion of vegetation and wildlife, as the ice retreated. Palynologists and archeologists have found evidence that much, if not all, of the current circumpolar boreal region was inhabited at some time by humans since retreat of the last continental glaciation. Scandinavia was completely ice free approximately 8,500 years BP, and the oldest archeological sites in northern Sweden and Norway are estimated to be approximately 8,000 and 10,000 years old, respectively.[7–11] The North American boreal became ice free between 11,600 and 8,400 BP, and archeological sites attributed to Paleo Indians in the boreal shield of northwestern Ontario date between 10,000 and 7,500 years BP.[12] In the Russian boreal, human occupation came 13,000 to 12,500 BP.[13]

These early boreal inhabitants who lived primarily by fishing, hunting, and trapping had mostly local impacts.[10] In general, the people inhabiting the boreal regions around the world, such as the Saami in Scandinavia and the Athabasca in North America, were nomadic. The seasonal cycle varied by location, but generally movement was for access to seasonally available foods. The freezing of water had a major influence on travel by allowing a change from water-borne transport by boats and canoes to land transport by snowshoe, toboggan, and dog or reindeer sled. Changes in food and other subsistence materials meant changes in residence and numbers of individuals in the group. Large groups only assembled for short periods at sites of concentrated resources—e.g., fish, water birds, and material for tools. Knowledge of habitat and its distribution on the larger landscape was important for knowing the location and seasonal availability of berries; roots and medicinal plants; animals with small ranges, like the bear and beaver; and nomadic animals like the moose, caribou, and reindeer. All of these activities were done at times of least effort and the best time for success with the tools available made of stone, native copper, wood and other fiber, and animal material. Spring and fall were usually critical times when summer and winter food was less available and people were dependent on dried game meat, berries, and smoked fish. Additionally, firewood was required year round and acquired using only stone tools.

The extent of impacts on ecosystems by indigenous peoples in Fennoscandia was limited before the development of agriculture and the fur trade. They would have exploited their boreal environment by fishing, hunting, and trapping for subsistence. In Europe, numerous archeological sites such as those in Norrland, Sweden, provide evidence of a traditional subsistence hunting culture dating from 8,000 BP that used primarily Eurasian elk (*Alces alces*; a close relative to the North American moose, *Alces americanus*), beaver, and fish.[14] Sometime around 4,000 BP, a change occurred from subsistence hunting to surplus production, presumably for trading for metals.[15] Also at this time, reindeer were added to the traditionally hunted Eurasian elk. Eurasian elk had been widespread from Scandinavia through Russia to Siberia but disappeared from much of Europe from the 11th to 19th centuries.

Prior to the arrival of Europeans in North America, exploitation by indigenous peoples was limited by the tools available, such as bows and arrows, snares, spears, hooks, etc. Furthermore, their social units were relatively small, with several nuclear families forming a hunting group and several hunting groups forming a band of 50–100 people.[16] As a result, their impacts on the fish and wildlife would have been limited. However, there is clear evidence that extinction of megafauna occurred whenever humans first entered areas where the fauna had no previous contact with humans.[17,18] It is also clear that indigenous peoples hunted the megafauna. Further, in areas where this megafauna had previous, long-term contact with humans, extinction was less and slower. So it is possible in the boreal forest of North America and parts of Russia that there could have been a major impact of humans on populations of both large herbivores and predators immediately after the retreat of the continental ice sheet.[19] The evidence of effects of this megafaunal extinction on vegetation and other parts of the trophic cascade is still lacking.

In North America, indigenous people used small controlled fires in selected areas and under specific weather conditions to encourage certain plant species (e.g., berry patches, forage for game animals) as well as to maintain campgrounds and trails.[20,21] Thus, they created small-scale cultural landscapes.[22,23] However, Carcaillet *et al.* could find no evidence of human-set fires in northern Sweden that might indicate management of habitat for game.[24]

Historic boreal cultural landscape

Trapping

In the North America boreal region, the arrival of European fur traders and establishment of their trading posts starting in the 1600s introduced steel and copper tools as well as guns to the indigenous people in exchange for furs. Fur trading started earlier in Fennoscandia and as early as the Middle Ages in parts of Russia but appeared to follow in a general manner the experience of North America.[25]

The fur trade resulted in a major change in the indigenous people's life ways and had a significant ecological impact, first directly on the populations of fur-bearing animals and second indirectly on the forests and ungulate populations. In North America, the indigenous people often had been trading for decades before seeing Europeans because of the use of indigenous middle men as traders far from the European trading posts.[26] However, as the fur trade and trading posts became more widespread, the indigenous people became more settled and concentrated around the trading posts, building more permanent homes, and providing the trading post inhabitants with not only furs but also game meat (largely caribou and moose). Thus, by the early 19th century, census estimates of the indigenous populations around two trading posts in northern Ontario, Canada ranged from 218 to 339.[27] Indigenous people in North American were exposed to European contagious diseases for the first time following European contact and then experienced cycles of epidemics at intervals of less than 30 years until the early 1900s.[28] First contact with European communicable diseases such as smallpox usually reduced populations by more than one half. These epidemics specifically affected the very young and old. This was very disruptive to societies in which the elders were responsible for maintaining the cultural traditions and making important group decisions.

The effects of trapping on North American boreal ecosystems are difficult to evaluate since the indigenous populations had already changed to a trapping-for-goods economy when Europeans first contacted them directly. Additionally, a major redistribution of indigenous groups resulted from both epidemic European diseases and the trading economy. Many groups moved into different ecosystems than those they had occupied in pretrading times. For example, many boreal groups moved from the boreal forest into the grassland plains with the arrival of the horse.

However, there is a consensus that trapping-for-trade reduced the populations of the beaver, wolf, bear, and other fur-bearing animals.[26,29–32] There is also some evidence that the moose, deer, and other sources of meat were reduced around trading posts. Carlos and Lewis have presented research that show trapping-for-trade reduced beaver populations in the western boreal of Canada through overharvesting.[33–35] This overharvesting was due to competition between trading companies who raised the prices paid for fur, thus leading to increased harvest by indigenous people. There was little overharvesting in regions where the price of furs stayed lower because of little or no competition between traders. Consequently, if a trading company could control access to the resource commons, it could—if it desired—maintain a sustained yield. But if it had competition for this resource commons from other trading companies, it had to raise the price for furs, resulting in the trappers increasing the harvest to an unsustainable level and causing the fur-bearer populations to decline.

The animals hunted and trapped for trade in the boreal divide roughly into two categories: herbivores (for meat and pelts) and predators (for pelts). The ecological effects of the trapping of these two groups of animals on the rest of the ecosystem are different. Removal of top predators *could* have caused trophic cascades, although evidence from the trapping–trading period is thin and anecdotal.[36] The effects on vegetation in this trophic cascade do not seem to have been commented on by European traders or explorers in their journals or reports. However, the reduction of the beaver would have had significant effects on streams and riparian and adjacent upland vegetation.[37–39] Again, these effects are not seriously commented on in the historic records and not systematically recorded. This is probably a result of the traders not going far from their trading posts and depending on the trappers for information on the state of the trapping region. The state of the ecosystems in the hinterlands was only commented on when large forest fires occurred and when meat was scarce.

Fire seems to have been used by indigenous peoples in specific habitats to improve trapping in areas that did not burn as frequently by wildfires and in small areas to attract game species at certain seasons.[21,40] Records by traders and their employees are often unreliable because of a lack of cultural awareness of the indigenous population. The traders often accused indigenous people of setting fires indiscriminately. In fact, into the 1970s, government agencies still believed that most forest fires in the boreal subarctic were human-caused rather than lightning-caused.[6] At present, we lack systematic information

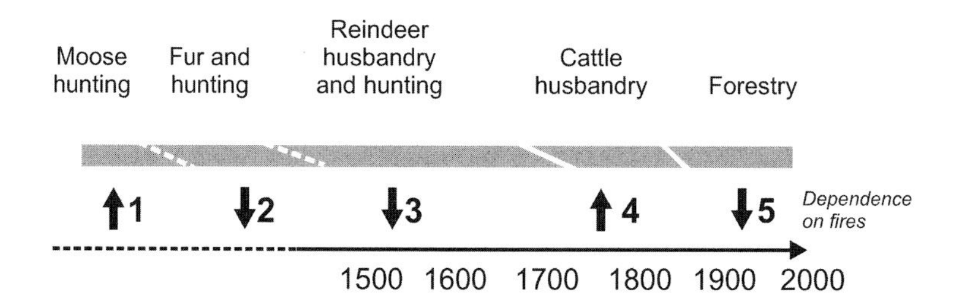

Figure 3. Approximate time frame for the sequence of cultures in the interior of northern Sweden (source: Granström and Niklasson[41]). Reproduced with permission from the authors and Royal Society Publishing.

on most of the ecological effects on boreal ecosystems during the trapping–trading periods in order to draw more general conclusions. Certainly, some effects were widespread if diffuse and thus not apparent to observers interested only in trading and with little environmental understanding or experience in the hinterlands away from the trading posts.

Herding

Herding of indigenous mammals is not common in the boreal forest. Hunters have sometimes used game drives and pit traps when the opportunity presented itself and the animals were gathered into herds, but in general only *Rangifer tarandus* (called caribou in North America and reindeer in Europe) regularly gather into herds and have seasonal migration. Further, only in Fennoscandia and Russia has semidomestication of reindeer occurred. The development of this herding is unknown but seems to have developed from the ease of forming loose attachments to herds for milking and culling for meat. Reindeer and indigenous human populations have occupied the boreal forest and tundra of Fennoscandia and Russia for all of the Holocene. Reindeer are migratory, spending summers in the forest eating herbs, grasses, and deciduous leaves of birch. Starting ∼ 2000 BP, Saami people in Fennoscandia shifted to herding of reindeer (Fig. 3) at the same time that their trade in furs was increasing with an emerging market in Europe.[41] Reindeer herding reached its full development in the 1500s. Before this, reindeer were less important than Eurasian elk hunting and fishing.[42] Today, large areas of the boreal forest in Norway, Sweden, and Finland are grazed by semidomesticated reindeer, herded by the Saami exclusively in Norway and Sweden and by both Saami and other groups in Finland. Reindeer populations increased markedly after the 1940s and have increasingly come into conflict with forestry interests.

Effects on the boreal forest by reindeer grazing depend on their density and seasonal habitat needs. However, in general, when grazing increases in intensity, trampling and changes in composition of lichens, herbs, and deciduous trees occur (den Herder *et al.*[43]). The effects are not always detrimental, but at high reindeer densities and short recurrences of grazing, they can have significant impacts on the ecosystem. The most obvious effect is a more open forest with little intermediate forest canopy. Evidence indicates that fires decreased after reindeer herding began because of the importance of lichens in the diet of reindeer and their slow recovery after fire.[44,45]

Agriculture

Despite the long summer day length, agriculture in the boreal is limited by harsh climate, thin soils, and distance to markets.[46] Consequently, most farming is restricted to mixed crops and animal husbandry. Forestry and trapping are often winter activities for farmers. It seems throughout history the southern edge of the boreal forest was invaded by southern agricultural groups when it was or appeared to be climatically more accommodating or when population and political pressure moved populations.[47] In Fennoscandia, farming began at the end of the 1600s with the spread of farmers from further south into areas traditionally used by Saami. Burning for agricultural clearing increased at this time.[48] During the 1800s, along with emigration to North America, there was increased colonization of the inland boreal areas in Sweden, Norway, and Finland.[46] Unlike in North America, this agricultural settlement included some indigenous Saami who gave up their nomadic reindeer herding to establish permanent farms. However, by the end of the 1800s, many farms were beginning to be abandoned as immigration to America increased.

In the Canadian and Russian boreal, agricultural colonization on the southern edge occurred only in the early 1900s and reached a peak in both regions after the First World War.[47] Again, as in Fennoscandia, by the 1940s farms were being abandoned. In all cases in the southern boreal, it appears that the spread of agriculture into the forest was decided by what farmers thought were favorable forest types that could be cleared for agriculture and favorable locations with respect to roads, railways, and nearby farms for cooperative help. Tchir *et al.* showed that settlers in the Canadian boreal selected higher hill slope positions irrespective of substrate (glaciolacustrine or glacial till).[49] Settlers appear to have used observable attributes such as stoniness, soil texture, and hill slope position rather than soil productivity in making settlement decisions. Thus, the species-richer upper hill slopes of aspen forest (glaciolacustrine) and aspen and white spruce forest (glacial till) were settled first, while the species-poorer lower hill slopes of aspen forest (glaciolacustrine) and white spruce and balsam fir forest (glacial till) were settled later.

Forest clearance for agriculture and colonization had the usual ecological effects in replacing natural or seminatural ecosystems with crops and introduced grasses supported by fertilizer and herbicide and fragmenting the remaining natural landscape. Fencing areas for cattle had the effect seen everywhere on indigenous wildlife populations. In the clearing phase of agricultural settlement, fires spread beyond the agricultural clearings into the intact forests. These settlement fires came at very short intervals, causing tree species composition to change by eliminating species that required longer fire intervals to become sexually reproductive (e.g., white spruce) and increasing species that could regenerate by sprouting (e.g., aspen).[50] These fires often spread tens of kilometers into the adjacent forest. In Canada, Fennoscandia, Russia, and the Baltic countries, agricultural colonization and the resulting increase in fires produced land-use conflicts with the indigenous populations who used the land for trapping, herding, and hunting.

Because these agricultural colonizations were often ill advised for both economic and climate–soil reasons, many farms were abandoned and the areas reverted to forest. The regenerating forests were rarely the same as that before agriculture (notice we did not say *natural* because some areas, particularly in Fennoscandia and Russia, had already been affected by herding and trapping for some time). Often, the regenerating forest's composition depends on the kinds of cropping or grazing that occurred on the field before abandonment.[51–53]

The spread of agriculture into the boreal forest often results in the drainage and mining of peatlands with the resulting changes in hydrology and release of methane and carbon dioxide.[54–56]

Contemporary boreal cultural landscape

Forestry

Throughout the boreal forest, indigenous people would have used wood for fires and structures. This impact was usually local and temporary because of seasonal movement; in Eurasia, forests were cleared in slash and burn agriculture.[57] Cutting and logging are not new to the boreal forest, but only in the 1800s did larger industrial forestry start. The principal limitations to the development of industrial forestry in the boreal forest are the slow growth of trees, the limited distribution of "productive" sites often making up less than 40% of the landscape, and access and transport of the wood to distant markets. In fact, much of the remaining intact boreal forest has very low productivity.[1] Early transport of logs to mills was by water and starting in the mid-1800s by permanent or temporary railroads. River transport of floating logs and rafts of logs and damming and sometimes changing or reversing the course of rivers had significant riparian ecological impacts that in many areas persist today and often go unrecognized.[58]

In boreal North America, Fennoscandia, and Russia, early industrial forestry was along the southern edge of the boreal forests. Early logging involved selective cutting of the largest and most easily accessed and transported trees. This logging was done with little knowledge of or concern for changes in forest composition or sustainability. The logging affected species composition by removing large pine and spruce and leaving smaller species, particularly deciduous trees.[59] In Fennoscandia, logging also affected reindeer grazing by reducing lichen cover.[60] While fires increased during clearance for agriculture, they decreased when forestry became more important.[48,61,62]

By the 1900s, industrial forestry was more widespread in North America, Fennoscandia, and Russia, largely due to improved access to markets, the emerging pulp industry, and the simultaneous spread of settlements into the southern edge of the boreal forest. Industrial forestry expanded rapidly after the Second World War in all regions of the *southern* boreal forest. At about this time in North America and Fennoscandia, sustained yield forestry began to be practiced with resulting increase in silviculture treatments, landscape forest rotation age regulation, and single-species plantations. The Soviet system in Russia also saw a major increase in logging in the postwar years with the then-strong centralized control and subsidy.[63]

The ecological effect of this century and a half of increased industrial forestry across the southern edge of the boreal forest has led to younger forests that increasingly reflect the relatively short rotation ages. The shorter rotation age has reduced the area in older "old-growth" forests. However, in boreal North America, unlike Fennoscandia and Russia, most forest leases are still in their first rotation and in areas that have never been cut. Natural disturbances (wildfire and insect outbreaks) are still "harvesting" an important part of industrial forests compared to commercial operations.[64] The importance of fire and insect outbreaks has led to an increase in salvage logging after these disturbances and often results in disruption of the natural regeneration.[65] Replacement of mixed species stands with single-species plantations was very popular after 1945 and has again been used in recent decades to change the composition of the forest to reduce the risk from future insect epidemics.

Dams and hydroelectric power

In recent decades, the drive to find secure energy sources has led to extensive development of hydropower in the boreal regions. Since the 1950s, the Russians have built 13 hydroelectric dams on the Oh, Yenisei, Lena, and Kolyma Rivers. The James Bay hydroelectric development project in Québec, Canada, begun in 1974, has flooded an area of 177, 000 km^2 (bigger than the state of Florida, United States) and has 11 generating stations. There are other smaller dams in the rest of the Canadian boreal, again mostly built since the 1970s. In Sweden with electric dams, hydroelectric dam construction started in the early part of the 1900s but most occurred after the Second World War.[66] As with some Canadian dams (e.g., Taltson River Dam), some dams in other parts of the boreal were constructed to produce power for mining operations. Along with dam creation, in recent years there have been economic, social, and environmental needs to remove dams and to restore the rivers to their predam flows.[67] Dam removal also has many impacts on ecosystems that had adjusted to conditions with the dam.[68,69]

Ecological effects of dams in the boreal are similar to those in other ecosystems. Dams have not generally changed river discharge, but they have changed the natural discharge patterns with decreased magnitude and timing of spring runoff and peak flows with return times longer than 10 years mostly eliminated.[70–74] Significant effects on the downstream hydrographs have been found in some boreal regions to extend 1,100 km downstream.[75] Dams have resulted in effects on stream chemistry, water temperatures, sediment loads, riparian vegetation, fish populations, and benthic populations.[76–79] Additionally, these dams lead to extensive flooding of forests and are barriers to traditional migration routes of caribou and reindeer. Flooding has created a problem of methylmercury contamination; on flooded lands, biologically unavailable inorganic mercury is converted into a biologically available and toxic form.[80] Dams also require the construction of power lines and road access with the associated ecological effects that such linear disturbances have. All dams have limited life spans before their reservoirs are filled with sediment. This has not always been considered in many dams built into the 1970s.[81]

Dams big and small, natural or human, have effects on the fluvial geomorphology of streams and rivers, both up and down the stream.[82,83] Recent work in geomorphology has shown that changes in flow regime, bank strength, and sediment budgets can have long-term effects on stream morphology and thus ecosystems.[84,85] These effects often extend well upstream and downstream of the dam and pond/reservoir by disrupting the connection of fluvial processes and hillslope processes.[86] In boreal regions, the time of adjustment may be very long, on the order of 1,000 years, although some features, particularly stream width and bar exposure caused by tree and shrub colonization, respond in decades.[84] The interaction with regional water tables and peatlands has indirect effects on biogeochemical cycling, flora and fauna composition, and productivity, both in streams and on adjacent uplands.[87] Since these effects may be slow and subtle, they are often missed or seen as part of the normal variation.

Oil and natural gas development

Oil and gas development are limited to sedimentary basins that can produce oil, natural gas, shale gas, and tar sands. In Alaska, United States, and Newfoundland, Canada, oil and gas are found in the Arctic and Atlantic Oceans, not in the boreal. In Canada's boreal, the sedimentary basins are around the western edge of the Precambrian Shield in Alberta, Saskatchewan, and British Columbia. Sweden has very small deposits, mostly as shale gas. Norway's oil and gas deposits are all offshore. Russia has major deposits of oil and gas in Siberia. In all of these areas, there has been extensive exploration, drilling, and production, particularly since the Second World War.

The environmental impacts of oil and gas development occur during exploration seismic, drilling, production and pipeline construction, operation, and closure. The impact of these activities has been reduced in Canada and Alaska in the recent decades because of increased regulatory requirements and technological advances. Traditional 2D seismic required the clearing of forest by bulldozers of sets of ~eight meter-wide linear corridors at ~one-half kilometer intervals and then holes drilled at intervals along the corridors in which dynamite charges were set. The charges are used to create seismic waves to map the subsurface features. In recent years, the use of low-impact seismic requires smaller vehicles and reduced forest clearing in construction of seismic lines; additionally, GPS instruments allow surveying with less linear corridors. However, with the need for smaller scale mapping in 3D, a finer grid is often required. Most of the seismic before the 1980s was by traditional methods, so that many areas of the boreal are crisscrossed by seismic lines (e.g., see Fig. 4). In areas with forestry operations, the area cut by seismic lines is often greater than that harvested for trees.[88] Regeneration in seismic lines is slow, primarily because boreal trees are not good at filling gaps of this small size and tree replanting was not done.[89] Seismic lines fragment the landscape; provide access for exotic species; provide corridors for easy predator movement and hunter and recreational access; contribute to soil compaction and topsoil loss from traffic used to construct the corridors; and have impacts on peatlands, wetlands, and streams.

Drilling technology has changed in the last decades, resulting in a reduced footprint of the well site during drilling and afterwards in production. The actual drilling process uses a large number of chemicals, many of which are toxic and carcinogenic. These materials are used as lubricants in the bore hole and are injected into the surrounding rock in order to increase flow through the sediment pores. Most fields required additional wells drilled to allow enhanced recovery. In the past, this led to multiple well platforms and roads, but today directional drilling allows many wells from a common platform, thus reducing the number of well sites and supporting infrastructure. Producing wells must then be connected by pipelines that require further access routes and support roads and clearing for aerial view of the pipeline route. Natural gas requires processing plants near the well fields. This discussion is simplified; oil and gas differ in diameter of their pipelines, regulations, and, in the case of natural gas, if it is shallow or deep deposits.

The environmental impacts of oil and natural gas well sites, pipelines, and roads are often very high and often persist for long periods on the boreal landscape.[90] The drilling of oil and natural gas well sites, although now more carefully managed, still uses many hazardous chemicals that have real possibility of leaking above and below ground, producing soil and water contamination. In areas in which forestry; oil and gas development; and other industrial, municipal, and recreational land-use occur, the synergistic ecological impacts are spatially extensive and disruptive of the ecosystem processes of population dynamics of animals, biogeochemical cycles, and biotic community structure and organization.

Tar sands development is more of a mining operation at present than traditional oil and gas drilling. Tar sands mining consists of removing the surface material to get at the sediment layer in which the bitumen is found.[91] The surface material is then replaced. There are also large artificial ponds in which the toxic waste from the extraction process is stored. Large amounts of water are withdrawn from rivers, reducing the annual magnitude and pattern of flow. Ecological restoration of the landscape is

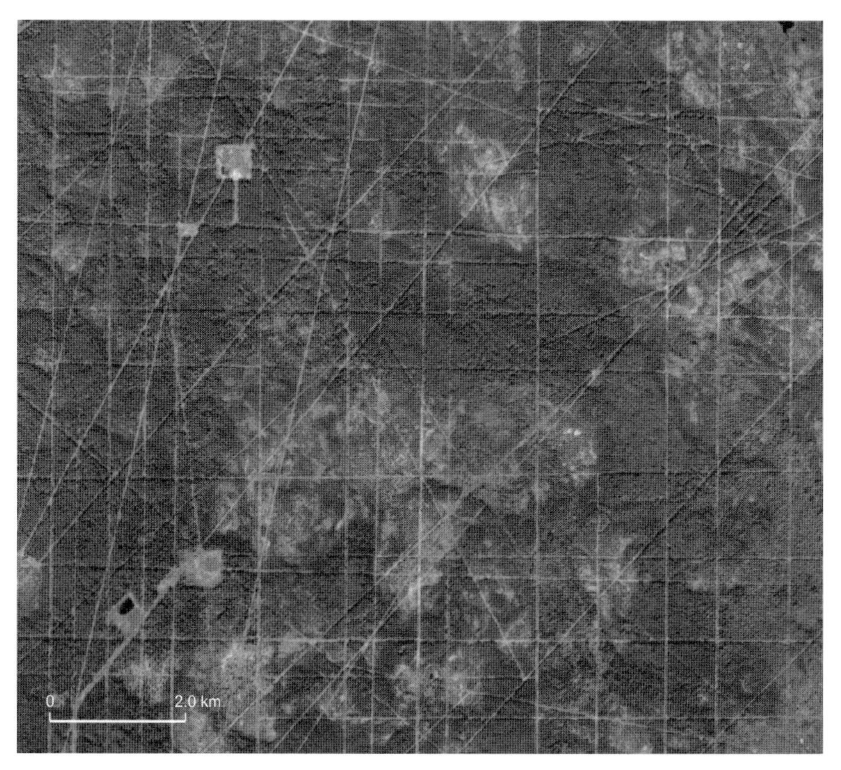

Figure 4. Air photo illustrating the high density of seismic lines (light-colored straight lines) in parts of the boreal forest in Alberta, Canada (source: AS4324, Photo 166, 1992). Reproduced with permission from Alberta Sustainable Resource Development, Air Photo Distribution. Image owned by the Government of Alberta and protected under the Copyright Act of Canada.

not the normal small-scale restoration practiced in recent decades for other human disturbances. Instead, a completely new landscape, not only of vegetation but also of hydrology, soil, and topography, has to be created. The present established area of mineable and deep tar sands in Alberta, Canada is 140,000 km^2. At present, no restoration has been undertaken, although research on how to do this is ongoing. There has also been serious criticism of the method of monitoring the direct and indirect environmental effects.[92] Downwind air pollution from the processing plants has serious impacts on lichens, the winter food of woodland caribou, and on the biodiversity of terrestrial and aquatic ecosystems on Precambrian bedrock. It is estimated that the restoration process to a "functioning ecosystem"[93] of tar sands mining will take hundreds of years. Nothing of this size and scope has taken place before in the boreal forest.

Shale gas development is in the exploration phase and is limited to sedimentary basin.[94] Shale gas requires deep wells and more wells than conventional gas, although directional drilling may reduce the above-ground footprint. The gas is produced by fracturing the sediment to provide passage for the natural gas. The fracturing is done hydraulically and with gas, chemicals, and sand to maintain the fractures. The process requires large amounts of fresh water. Additionally, the carbon footprint will be larger, as CO_2 is a natural impurity in some shale gas deposits. Some of this emission may be reduced by carbon capture techniques. At present, the most obvious impact will be the large use of surface and ground water. Water used will consist of stream and river water, shallow and deep groundwater, and recycled shale gas well water. Use of surface and ground water will certainly have effects on instream flow and the resulting aquatic ecosystems. The effects on hillslope hydrology are not yet clear. What is clear, however, is the conflict with already allocated water rights to other uses, both industrial and domestic.

Mining

Indigenous peoples everywhere in the boreal used rocks and minerals. Chert and other rocks were valued for making stone tools and were often being traded over long distances from specific sources. Native copper was mined 6,000 years ago in boreal Canada; however, mining impact by indigenous peoples would have been insignificant. In most parts of the boreal forest, hard rock mining started in the 1800s and was at first primarily for gold and silver and only later for nickel, asbestos, copper, uranium, iron, and diamonds. Most of this mining is on the Precambrian Shield with its deposits formed often billions of years ago; 80% of all mines in Canada are found in the boreal region.[95] Many mineral deposits in the southern boreal forest were discovered during development for agriculture, roads, and railroad construction. In recent decades, mineral deposits in the northern boreal have been found by more systematic surveys, exploration, and a better understanding of geology and use of geophysical techniques.

Hard rock mining is done by open pits and by underground excavation, producing large amounts of waste rock (tailings). For example, about 1 tonne of copper is produced for every 99 tonnes of waste. Waste rock is often remined later using more efficient and more toxic chemicals. The waste rock often contains sulfide, heavy metals, and other contaminants from the processing methods and deposits. In most mines, the waste is the major source of most heavy metal in streams and rivers. The large aboveground footprint of open pit mines is the most obvious environmental impact both locally and regionally, particularly on the above-ground and below-ground hydrology.

Igneous and metamorphic rocks containing sulfide minerals produce sulfuric acid when exposed to air and water with the help of *Thiobacillus ferroxidans* bacteria. The acid is then carried by rain and drainage into watersheds. This process can go on for long periods (e.g., thousands of years) until the sulfides are weathered out. The low pH and chemicals used in the mining process also release heavy metals from rocks.

Both open-pit and below-ground mining require large machinery and associated processing facilities. Isolated mines, as is the usual case in the boreal, require roads and rail access often over great distances. Additionally, considerable power is required to run the dragline, transport vehicles, and associated extraction facilities. Further, shipping heavy minerals such as copper and nickel require transport methods that can carry heavier loads. Mines also use large quantities of surface water in extraction and processing. The gross water use for the extraction stage of metal mining in Canada is estimated at 2,542 million m^3 per year.[95]

The environmental impact of hard rock mining, although limited to a smaller footprint, has a much longer term effect, even when the mines themselves have relatively short lifespans. Sudbury, Ontario, Canada has one of the longest running hard rock mining operations (starting in the 1880s) and also has research documentation on its impacts on vegetation, water, air, and soil. The mines around Sudbury cover an elliptical area with a width of 200 kilometers. The deposit was formed by a meteor impact 1.85 billion years ago. This deposit accounts for two thirds of the world's nickel-bearing sulfide ore. By the late 1960s, the impact of sulfur dioxide air pollution from the three Sudbury smelters was evidenced by an inner zone barren of vegetation with a steep gradient of resistant plants out from this zone.[96,97] Extensive research has been done on the soil chemistry, soil organisms, and vegetation effects across these sulfur dioxide-impacted areas; research has also been done on vegetation recovery and genetic selection for plants tolerant of the acidity.[98] The lesson learned from this example is that recovery is slow and it is better to provide strict regulation in order to minimize impact from the beginning.

Conclusions

The boreal region has probably always been a cultural landscape impacted by humans since the Pleistocene ice retreated. However, the impact has varied depending on the closeness to large population and commercial centers. What closeness means has depended on the cost and demand for the natural resources. The boreal cultural impact must be considered in the framework of its severe climate with associated short growing season, low soil nutrients, slow ecosystem recovery, and often permafrost.

In the first phase of cultural impact, only local indigenous populations were important, and these populations were small and dependent primarily on local subsistence hunting and gathering. There is possibly one exception to this statement in North

America, where the Pleistocene migration of humans from eastern Asia was into a landscape in which the resident fauna had not previously encountered humans. Although there is some debate on the exact impact of this human invasion, it is clear that many species of large mammals, both predators and herbivores, became extinct in the millennium after human arrival in North America.[99] The result of this extinction must have been a major change in the trophic cascade. Thus, in a very important way, the impact of the arrival of the first humans in boreal North America could have been as important as the impact that started after the arrival of Europeans.

The second phase of cultural impact is marked by the arrival of commercial enterprise and increasingly large population centers close enough to allow exploitation of natural resources. This phase is marked by the use of the boreal regions across the globe as hinterlands or sources of natural resources. This hinterlands–heartland continuum first occurred in Europe and eastern Russia and then in North America and western Russia. The boreal regions have never become heartlands because the climate is not conducive to agriculture except on its southern margin and not conducive to settlement because of the infrastructure required to deal with the cold climate. Further, the boreal region has generally been far from commercial markets, although this has diminished as demand for limiting resources has increased and offset the transportation cost. As a consequence, human populations in the boreal have consisted of two separate groups: those who tried to maintain their indigenous life ways and those that are there as government employees from the heartland or for resource exploitation.[100] Up to now, this second group has been the source of most of the ecological and environmental impact in the last three centuries.

The boreal has always had a boom or bust economy during this second phase, starting with the fur trade and continuing today with timber, minerals, oil, and gas. This has been primarily due to the fluctuation in price of these natural resources in distant markets and changing demand (e.g., beaver hats and in recent years the collapse of the fur coat market). Additionally, hard rock mines have a fairly short lifespan of decades, as do some hydroelectric reservoirs. All of these economic effects marginalize indigenous populations and lead to an exploitation strategy that has not encouraged sustainability and care for ecosystem services. Further, the populations of boreal regions are small, so their political influence as commercial and population centers is not great.

The effect of these forces has resulted in an increasing rate of environmental impact on boreal regions. Unfortunately, we can probably not say that North America, being later in the development of its boreal region, will trace the same history as, for example, the Fennoscandian boreal, due to the increasing globalization of natural resource demand.

Of course, throughout this paper we have ignored the largest cultural impact on the boreal in the form of global warming by greenhouse gases.[101] Boreal regions of the world are, along with the Arctic, experiencing the largest change in climate from greenhouse gas warming. This and other atmospheric changes, such as increased acid precipitation, will create a new ecosystem that is not like any of the past in the boreal; in fact, the ecosystem will likely no longer be boreal.

Acknowledgments

We gratefully acknowledge Marie Puddister, Department of Geography, University of Guelph for preparation of the figures and D.R. Charlton and G.I. Fryer for supplying information and ideas on oil, gas, and mining. The research was supported by an NSERC discovery grant to EAJ.

Conflicts of interest

The authors declare no conflicts of interest.

References

1. Sauer, C.O. 1925. The morphology of landscape. *U. Cal. Publ. Geog.* **2:** 19–54.
2. Yaroshenko, A.Y., P.V. Potapov & S.A. Turubanova. 2001. *The Last Intact Forest Landscapes of Northern European Russia.* Greenpeace Russia and Global Forest Watch.
3. Hornberg, G., M. Ohlson & O. Zackrisson. 1995. Stand dynamics, regeneration patterns and long-term continuity in boreal old-growth Picea-abies swamp-forests. *J. Veg. Sci.* **6:** 291–298.
4. Segerström, U., G. Hornberg & R. Bradshaw. 1996. The 9000-year history of vegetation development and disturbance patterns of a swamp-forest in Dalarna, northern Sweden. *Holocene* **6:** 37–48.
5. Raup, H.M. 1941. Botanical problems in boreal America II. *Bot. Rev.* **7:** 209–248.
6. Johnson, E.A. & J.S. Rowe. 1976. *Fire and Vegetation Change in the Western Subarctic* Arctic Land Use Research Information. Canada.

7. Andersen, B.G. 1979. The deglaciation of Norway 15,000–10,000 BP. *Boreas* **8:** 59–87.
8. Berglund, B.E. 1979. The deglaciation of southern Sweden 13,500–10,000 BP. *Boreas* **8:** 89–117.
9. Broadbent, N.D. 1978. Prehistoric settlement in northern Sweden: a brief survey and a case study. In *The Early Post-glacial Settlement of Northern Europe*. P. Mellars, Ed.: 177–204. Duckworth. London, UK.
10. Robertson, A.-M. & U. Miller. 1984. Garaselet—Biostratigraphical studies of human impact during different periods of settlement from the mesolithic to medieval times. *Iskos* **5:** 127–140.
11. Nygaard, S.E. 1989. The Stone-Age of northern Scandinavia, a review. *J. World Prehist.* **3:** 71–116.
12. McMillan, A.D. 1995. *Native Peoples and Cultures of Canada: An Anthropological Overview*. Douglas & McIntyre. Vancouver, BC.
13. Pitul'ko, V. 2001. Terminal Pleistocene—early Holocene occupation in northeast Asia and the Zhokhov assemblage. *Quaternary Sci. Rev.* **20:** 267–275.
14. Baudou, E. 1990. The archaeological background. In *The Post-Glacial History of Vegetation and Agriculture in the Luleälv River Valley*. U. Segerström, Ed.: 9–13. University of Umeå. Umeå, Sweden.
15. Forsberg, L. 1985. Site variability and settlement patterns. An analysis of the hunter-gatherer settlement system in the Lule River Valley, 1500 B.C.-B.C./A.D. *Archaeol. Environ.* **5.**
16. Rogers, E.S. 1964. The fur trade, the government and the central Canadian Indian. *Arctic Anthropol.* **2:** 37–40.
17. Burney, D.A. & T.F. Flannery. 2005. Fifty millenia of catastrophic extinctions after human contact. *Trends Ecol. Evol.* **20:** 395–401.
18. Buck, C.E. & E. Bard. 2007. A calendar chronology for Pleistocene mammoth and horse extinction in North America based on radiocarbon calibration. *Quaternary Sci. Rev.* **26:** 17–18.
19. Yurtsev, B.A. 2001. The Pleistocene "Tundra-Steppe" and the productivity paradox: the landscape approach. *Quaternary Sci. Rev.* **20:** 165–174.
20. Lewis, H.T. 1982. *A Time for Burning*. Boreal Institute for Northern Studies, University of Alberta. Edmonton, AB.
21. Lewis, H.T. & T.A. Ferguson. 1988. Yards, corridors, and mosaics—how to burn a boreal forest. *Hum. Ecol.* **16:** 57–77.
22. Davidson-Hunt, I.J. 2003. Indigenous lands management, cultural landscapes and Anishinaabe people of Shoal Lake, northwestern Ontario, Canada. *Environments* **31:** 21–41.
23. Miller, A.M. & I. Davidson-Hunt. 2010. Fire, agency and scale in the creation of aboriginal cultural landscapes. *Hum. Ecol.* **38:** 401–414.
24. Carcaillet, C., I. Bergman, S. Delorme, *et al.* 2007. Long-term fire frequency not linked to prehistoric occupations in northern Swedish boreal forest. *Ecology* **88:** 465–477.
25. Rich, E.E. 1955. Russia and the colonial fur trade. *Econ. Hist. Rev.* **7:** 307–328.
26. Innis, H.A. 1956. *The Fur Trade in Canada: An Introduction to Canadian Economic History*. University of Toronto Press. Toronto, ON.
27. Bishop, C.A. 1974. *The Northern Ojibwa and the Fur Trade: A Historical and Ecological Study*. Reihnhardt and Winston of Canada. Toronto, ON, Canada.
28. Piper, L. & J. Sandlos. 2007. A broken frontier: ecological imperialism in the Canadian north. *Environ. Hist.* **12:** 759–795.
29. Ray, A. 1974. *Indians in the Fur Trade*. University of Toronto Press. Toronto, ON.
30. Ray, A. 1978. History and archaeology of the northern fur trade. *Am. Antiquity* **43:** 26–34.
31. Winterhalder, B.P. 1980. Canadian fur bearer cycles and Cree-Ojibwa hunting and trapping practices. *Am. Nat.* **115:** 870–879.
32. Boyce, M.S. 1981. Beaver life-history responses to exploitation. *J. Appl. Ecol.* **18:** 749–753.
33. Carlos, A.M. & F.D. Lewis. 1993. Indians, the beaver, and the Bay—the economics of depletion in the lands of the Hudsons-Bay Company, 1700–1763. *J. Econ. Hist.* **53:** 465–494.
34. Carlos, A.M. & F.D. Lewis. 1999. Property rights, competition, and depletion in the eighteenth-century Canadian fur trade: the role of the European market. *Can. J. Econ.* **32:** 705–728.
35. Carlos, A.M. & F.D. Lewis. 2001. Trade, consumption, and the native economy: Lessons from York Factory, Hudson Bay. *J. Econ. Hist.* **61:** 1037–1064.
36. Beschta, R.L. & W.J. Ripple. 2009. Large predators and trophic cascades in terrestrial ecosystems of the western United States. *Biol. Conserv.* **142:** 2401–2414.
37. Johnston, C.A. & Naiman, R.J. 1990. Aquatic patch creation in relation to beaver population trends. *Ecology* **71:** 1617–1621.
38. Gurnell, A.M. 1998. The hydrogeomorphological effects of beaver dam-building activity. *Prog. Phys. Geog.* **22:** 167–189.
39. Rosell, F., O. Bozsér, P. Collen & H. Parker. 2005. Ecological impact of beavers *Castor fiber* and *Castor canadensis* and their ability to modify ecosystems. *Mammal Rev.* **35:** 248–276.
40. Natcher, D.C., M. Calef, O. Huntington, *et al.* 2007. Factors contributing to the cultural and spatial variability of landscape burning by native peoples of interior Alaska. *Ecol. Soc.* **12:** Article 7. Available at: http://www.ecologyandsociety.org/vol12/iss1/art7/ Accessed 22 November 2011.
41. Granström, A. & M. Niklasson. 2008. Potentials and limitations for human control over historic fire regimes in the boreal forest. *Philos. T. Roy. Soc. B* **363:** 2353–2358.
42. Aronsson, K-A. 1991. Forest reindeer herding A.D. 1–1800. An archaeological and palaeoecological study in northern Sweden. *Archaeol. Environ.* **10.**
43. den Herder, M., M.M. Kytoviita & P. Niemela. 2003. Growth of reindeer lichens and effects of reindeer grazing on ground cover vegetation in a Scots pine forest and a subarctic heathland in Finnish Lapland. *Ecography* **26:** 3–12.
44. Nieminen, M. & U. Heiskari. 1989. Diets of freely grazing and captive reindeer during summer and winter. *Rangifer* **9:** 17–34.

45. Baskin, L.M. & K. Danell. 2003. *Ecology of Ungulates: A Handbook of Species in Eastern Europe and Northern and Central Asia*. Springer. Berlin, Germany.
46. Mörner, M. 1982. The colonization of Norrland by settlers during the 19th century in a broader perspective. *Scand. J. Hist.* **7:** 315–337.
47. Bowman, I. 1931. *The Pioneer Fringe*. American Geographical Society of New York. New York, NY.
48. Niklasson, M. & A. Granström. 2000. Numbers and sizes of fires: long-term spatially explicit fire history in a Swedish boreal landscape. *Ecology* **81:** 1484–1499.
49. Tchir, T., E.A. Johnson & K. Miyanishi. 2004. A model of fragmentation in the Canadian boreal forest. *Can. J. Forest Res.* **34:** 2248–2262.
50. Weir, J.M.H. & E.A. Johnson. 1998. Effects of escaped settlement fires and logging on forest composition in the mixedwood boreal forest. *Can. J. Forest Res.* **28:** 459–467.
51. Osborne, B.S. 1989. The Russian frontier on the west--16th century Belorussia. *Sov. Geogr.* **30:** 197–206.
52. Wyckoff, W. & G. Hausladen. 1989. Special issue on settling the Russian frontier—with comparisons to North America. *Sov. Geogr.* **30:** 179–188.
53. Gachet, S., A. Leduc, Y. Bergeron, *et al.* 2007. Understory vegetation of boreal tree plantations: differences in relation to previous land use and natural forests. *Forest Ecol. Manag.* **242:** 49–57.
54. Kryuchkov, V.V. 1993. Extreme anthropogenic loads and the northern ecosystem condition. *Ecol. Appl.* **3:** 622–630.
55. Turetsky, M., K. Wieder, L. Halsey & D. Vitt. 2002. Current disturbance and the diminishing peatland carbon sink. *Geophys. Res. Lett.* **29**. doi: 10.1029/2001GLO14000, 12 June 2002.
56. Wilson, D., J. Alm, J. Laine, *et al.* 2008. Rewetting of cutaway peatlands: Are we re-creating hot spots of methane emissions? *Restor. Ecol.* **17:** 796–806.
57. Karlsson, H., G. Hörnberg, G. Hannon & E-M. Nordström. 2007. Long-term vegetation changes in the northern Scandinavian forest limit: a human impact-climate synergy? *Holocene* **17:** 37–49.
58. Williams, M. 1989. *Americans and Their Forests: A Historical Geography*. Cambridge University Press. Cambridge, UK.
59. Boucher, Y., D. Arseneault & L. Sirois. 2009. Logging history (1820–2000) of a heavily exploited southern boreal forest landscape: Insights from sunken logs and forestry maps. *Forest Ecol. Manag.* **258:** 1359–1368.
60. Moen, J. & E.C.H. Keskitalo. 2010. Interlocking panarchies in multi-use boreal forests in Sweden. *Ecol. Soc.* **15:** Article 17. Available at: http://www.ecologyandsociety.org/vol15/iss3/art17/ Accessed 22 November 2011.
61. Hellberg, E., M. Niklasson & A. Granström. 2004. Influence of landscape structure on patterns of forest fires in boreal forest landscapes in Sweden. *Can. J. Forest Res.* **34:** 332–338.
62. Josefsson, T., B. Gunnarson, L. Liedgren, *et al.* 2010. Historical human influence on forest composition and structure in boreal Fennoscandia. *Can. J. Forest Res.* **40:** 872–884.
63. Bergen, K.M., *et al.* 2008. Changing regimes: forested land cover dynamics in central Siberia 1974 to 2001. *Photogramm. Eng. Rem. S.* **74:** 787–798.
64. Johnson, E.A., K. Miyanishi & J. Choczynska. 2003. Effects of fire, logging and settlement on the boreal forest landscape in Ontario. Project Report 2003/2004. Sustainable Forest Management Network. Edmonton, AB.
65. Saint-Germain, M. & D.F. Greene. 2009. Salvage logging in the boreal and cordilleran forests of Canada: Integrating industrial and ecological concerns in management plans. *Forest. Chron.* **85:** 120–134.
66. Lejon, A.G.C., B. Malm Renöfält & C. Nilsson. 2009. Conflicts associated with dam removal in Sweden. *Ecol. Soc.* **14:** Article 4. Available at http://www.ecologyandsociety.org/vol14/iss2/art4/ Accessed 22 November 2011.
67. Graf, W.L. 2003. *Dam Removal Research: Status and Prospects*. H. John Heinz III Center for Science, Economics and the Environment. Washington, D.C., USA.
68. Gregory, S., H. Li & J. Li. 2002. The conceptual basis for ecological responses to dam removal. *BioScience* **52:** 713–723.
69. Stanley, E. H. & M. W. Doyle. 2003. Trading off: the ecological effects of dam removal. *Front. Ecol. Environ.* **1:** 15–22.
70. McClelland, J.W., R.M. Holmes, B.J. Peterson & M. Stieglitz. 2004. Increasing river discharge in the Eurasian Arctic: consideration of dams, permafrost thaw, and fires as potential agents of change. *J. Geophys. Res.* **109:** D18102, doi:10.1029/2004JD004583.
71. Rawlins, M.A., H. Ye, D. Yang, *et al.* 2009. Divergence in seasonal hydrology across northern Eurasia: emerging trends and water cycle linkages. *J. Geophys. Res.* **114:** D18119, doi:10.1029/2009JD011747.
72. Assani, A.A., T. Biffin-Bélanger & A.G. Roy. 2002. Impacts of a dam on the hydrologic regime of the Matawin River (Québec, Canada). *Rev. Sci. de l'eau* **15:** 557–574.
73. Woo, M-K. & R. Thorne. 2009. Effects of reservoirs on streamflow in the boreal region. In *Threats to Global Water Security*. A.A. Jones, T.G. Vardanian & C. Hakopian, Eds.: 339–348. Springer. Dordrecht, The Netherlands.
74. Woo, M-K., R. Thorne, K. Szeto & D. Yang. 2008. Streamflow hydrology in the boreal region under the influences of climate and human interference. *Phil. T. Roy. Soc. B* **363:** 2249–2258.
75. Peters, D.L. & T.D. Prowse. 2001. Regulation effects on the lower Peace river, Canada. *Hydrol. Process* **15:** 3181–3194.
76. Jansson, R., C. Nilsson, M. Dynesius & E. Andersson. 2000. Effects of river regulation on river-margin vegetation: a comparison of eight boreal rivers. *Ecol. Appl.* **10:** 203–224.
77. Nilsson, C. & M. Svedmark. 2002. Basic principles and ecological consequences of changing water regimes: riparian plant communities. *Environ. Manag.* **30:** 468–480.
78. Johnson, P.T.J., J.D. Olden & M.J. Vander Zanden. 2008. Dam invaders: impoundments facilitate biological invasions into freshwaters. *Front. Ecol. Environ.* **6:** 357–363.
79. Renöfält, B.M., R. Jansson & C. Nilsson. 2010. Effects of hydropower generation and opportunities for environmental flow management in Swedish riverine ecosystems. *Freshwater Biol.* **55:** 49–67.
80. Bodaly, R.A., V.L.St. Louis, M.J. Paterson, *et al.* 1997. Bioaccumulation of mercury in the aquatic food chain in newly flooded areas. In *Metal Ions in Biological Systems, Mercury*

and its Effects on Environment and Biology. Vol. 34. A. Sigel & H. Sigel, Eds.: 259–287. Marcel Dekker. New York, NY.
81. Einsele, G. & M. Hinderer. 1997. Terrestrial sediment yield and the lifetimes of reservoirs, lakes, and larger basins. *Geol. Rundsch.* **86:** 288–310.
82. Friedl, G. & A. Wüest. 2002. Disrupting biogeochemical cycles—consequences of damming. *Aquat. Sci.* **64:** 55–65.
83. Petts, G.E. & A.M. Grunell. 2005. Dams and geomorphology: research progress and future directions. *Geomorphology* **71:** 27–47.
84. Church, M. 1995. Geomorphic response to river flow regulation: case studies and time-scales. *Regul. Rivers* **11:** 3–22.
85. Walter, R.C. & D.J. Merritts. 2008. Natural streams and the legacy of water-powered mills. *Science* **319:** 299–304.
86. Gomi, T., R.C. Sidle & J.S. Richardson. 2002. Understanding processes and downstream linkages of headwater systems. *BioScience* **52:** 905–916.
87. Wieder, R., D. Vitt & B. Benscoter. 2006. Peatlands and the boreal forest. Boreal Peatland ecosystems. *Ecol. Stud.* **188:** 1–8.
88. Varty, T. 2001. Basic concerns from a forest industry perspective. In *Oil and Gas Planning on Forested Lands in Alberta: Overview of CIF-RMS Technical Session, March 23, 2001*. Canadian Institute of Forestry. Edmonton, AB.
89. McCarthy, J. 2001. Gap dynamics of forest trees: a review with particular attention to boreal forests. *Environ. Rev.* **9:** 1–59.
90. Schneider, R. R., J. B. Stelfox, S. Boutin & S. Wasel. 2003. Managing the cumulative impacts of land uses in the Western Canadian sedimentary basin: a modeling approach. *Conserv. Ecol.* **7:** 8. Available at: http://www.consecol.org/vol7/iss1/art8.
91. Johnson, E.A. & K. Miyanishi. 2008. Creating new landscapes and ecosystems: The Alberta Oil Sands. *Ann. NY Acad. Sci.* **1134:** 120–145.
92. Kelly, E.N., J.W. Short, D.W. Schindler, *et al.* 2009. Oil sands development contributes polycyclic aromatic compounds to the Athabasca River and its tributaries. *P. Natl. Acad. Sci. USA* **106:** 22346–22351.
93. Alberta Environment. 1999. *Regional Sustainable Development Strategy for the Athabasca Oil Sands Area*. Alberta Environment, Government of Alberta. Edmonton, AB.
94. National Energy Board of Canada. 2009. *A Primer for Understanding Canadian Shale Gas*. Publications Office, National Energy Board. Calgary, Alberta.
95. MiningWatch Canada. 2008. The boreal below: mining issues and activities in Canada's boreal forest. Ottawa, ON. Available at: http://www.miningwatch.ca/sites/www.miningwatch.ca/files/Boreal_Below_2008_ES_web_0.pdf. Accessed 22 November 2011.
96. Gorham, E. & A.G. Gordon. 1960. The influence of smelter fumes upon the chemical composition of lake waters near Sudbury, Ontario, and upon the surrounding vegetation. *Can. J. Bot.* **38:** 477–487.
97. Gorham, E. & A.G. Gordon. 1963. Some effects of smelter pollution upon aquatic vegetation near Sudbury, Ontario. *Can. J. Bot.* **41:** 371–378.
98. Winterhalder, K. 1996. Environmental degradation and rehabilitation of the landscape around Sudbury, a major mining and smelting area. *Environ. Rev.* **4:** 185–224.
99. Martin, P.S. & D.W. Steadmam. 1999. Prehistoric extinctions on Islands and Continents. In *Extinctions in Near Time: Causes, Context and Consequences*. R. MacPhee, Ed.: 17–55. Kluwer Academic, Plenum Publishers. New York.
100. Bone, R.M. 1992. *The Geography of the Canadian North: Issues and Challenges*. Oxford University Press. Toronto, Ontario.
101. Olsson, R. 2010. *Boreal Forest and Climate Change—Regional Perspectives*. Air Pollution & Climate Secretariat. Reinhold Pape. Göteborg, Sweden.
102. Hare, F.K. & J.C. Ritchie. 1972. The boreal bioclimates. *Geog. Rev.* **62:** 333–365.
103. Global Forest Watch. World Resources Institute. Washington, DC. Available at: http://www.globalforestwatch.org. Accessed 22 November 2011.
104. Lee, P., D. Aksenov, L. Laestadius, *et al.* 2003. *Canada's Large Intact Forest Landscapes*. Global Forest Watch Canada. Edmonton, Alberta.
105. Aksenov D., D. Dobrynin, M. Dubinin, *et al.* 2002. *Atlas of Russia's Intact Forest Landscapes*. Global Forest Watch. Moscow, Russia.

Ann. N.Y. Acad. Sci. ISSN 0077-8923

ANNALS OF THE NEW YORK ACADEMY OF SCIENCES
Issue: *The Year in Ecology and Conservation Biology*

Climate change and the ecology and evolution of Arctic vertebrates

Olivier Gilg,[1,2,3] Kit M. Kovacs,[4] Jon Aars,[4] Jérôme Fort,[5] Gilles Gauthier,[6] David Grémillet,[7] Rolf A. Ims,[8] Hans Meltofte,[5] Jérôme Moreau,[1] Eric Post,[9] Niels Martin Schmidt,[5] Glenn Yannic,[3,6] and Loïc Bollache[1]

[1]Université de Bourgogne, Laboratoire Biogéosciences, UMR CNRS 5561, Equipe Ecologie Evolutive, Dijon, France. [2]Division of Population Biology, Department of Biological and Environmental Sciences, University of Helsinki, Finland. [3]Groupe de Recherche en Ecologie Arctique (GREA), Francheville, France. [4]Norwegian Polar Institute, FRAM Centre, Tromsø, Norway. [5]Department of Bioscience, Aarhus University, Roskilde, Denmark. [6]Département de Biology and Centre d'Études Nordiques, Université Laval, Québec, Québec, G1V 0A6, Canada. [7]FRAM Centre d'Ecologie Fonctionnelle et Evolutive, UMR CNRS 5175, Montpellier, France. [8]Department of Arctic and Marine Biology, University of Tromsø, Tromsø, Norway. [9]Department of Biology, Penn State University, University Park, Pennsylvania

Address for correspondence: Olivier Gilg, Laboratoire Biogéosciences, UMR CNRS 5561, Equipe Ecologie Evolutive, 6 Boulevard Gabriel, 21000 Dijon, France. olivier.gilg@gmail.com

Climate change is taking place more rapidly and severely in the Arctic than anywhere on the globe, exposing Arctic vertebrates to a host of impacts. Changes in the cryosphere dominate the physical changes that already affect these animals, but increasing air temperatures, changes in precipitation, and ocean acidification will also affect Arctic ecosystems in the future. Adaptation via natural selection is problematic in such a rapidly changing environment. Adjustment via phenotypic plasticity is therefore likely to dominate Arctic vertebrate responses in the short term, and many such adjustments have already been documented. Changes in phenology and range will occur for most species but will only partly mitigate climate change impacts, which are particularly difficult to forecast due to the many interactions within and between trophic levels. Even though Arctic species richness is increasing via immigration from the South, many Arctic vertebrates are expected to become increasingly threatened during this century.

Keywords: impacts; phenological changes; plasticity; range shifts; adaptations; threat; trophic interactions; mismatches; sea ice; tundra; parasites; geese; shorebirds; rodents; lemmings; large herbivores; seabirds; marine mammals; polar bear

Introduction

Several studies have already reviewed the ongoing and forecasted impacts of climate change on the ecology and evolution of vertebrates on Earth.[1,2] Overall, changes in phenologies, in distributional range, in evolutionary adaptations—on the long term—and, in the structure and the dynamics of communities, are forecast to be the main biological consequences of climate change. However, due to regional differences in climate change and in biotic characteristics, climate-induced impacts on species will differ in their nature and magnitude between regions and ecosystems.[3] Examples include the following: in Africa, water stress is expected to become the main constraint in the future, with arid and semi-arid land projected to increase; in eastern South America, drought and increasing temperatures will induce a gradual replacement of tropical forests by savanna; in boreal regions, increasing rainfall and temperature will in many places allow the forest to expand in altitude and latitude, replacing the Arctic and Alpine tundras; and in many coastal lowlands, the elevation of sea level will replace terrestrial ecosystems by marine ones. In the Arctic, physical changes will be dominated by changes in the cryosphere (glaciers, permafrost, sea ice, and snow layers), primarily due to increasing temperatures.

doi: 10.1111/j.1749-6632.2011.06412.x

To understand how Arctic vertebrates will respond to these changes, we have to consider their particular history, the extreme physical environment they currently inhabit, and the relatively simple structure and composition of the communities to which they belong.

Of course, changes in climate are not entirely new, and Arctic vertebrates have already lived through several periods of "historical climate changes"—although the speed and the extent of the current changes are probably unprecedented. However, unless we enter the realm of paleoecology, we know little about ecological responses of vertebrates to past changes in climate. Vibe's seminal study[4] is one notable exception. He clearly showed, nearly 50 years ago, the importance of climate, and more precisely the availability of specific ice types, in determining the distribution of marine mammals on the west coast of Greenland. More recent analyses of decadal patterns of sea ice and their influences on marine mammals can also be used to predict that changes in recent years are likely to impact resident marine mammal populations at both regional and hemispheric scales.[5]

The Arctic biome is already changing, and this change is rapid and severe compared to other regions of the globe. It is expected that the changes that will take place in the Arctic in the coming decades will surpass what many Arctic vertebrates have experienced in the past, despite the extreme variability that is normal in the climate of the Arctic. Indeed, while the global mean surface temperature has warmed by 0.6–0.8° C during the 20th century, it has increased *ca.* twice as much in the Arctic during the same period.[3] The consequences of this warming include decreases in snow, ice, and frozen ground layers, while precipitation has increased overall in the Northern hemisphere.[3] The strong environmental constraints that prevail in the Arctic—including the harsh climate, strong seasonality, low productivity, and numerous ecological barriers—are likely to exacerbate the impacts that are expected due to direct influences of climate change.[6]

Documented declines in Arctic sea ice are symptomatic of these ongoing changes. Its overall extent, thickness, seasonal duration, and the proportion of multiyear ice have all declined. Multiyear sea ice has disappeared three times faster during the past decade than during the previous three decades, and seasonal ice is thus becoming the dominant sea ice type in the Arctic. If these trends continue as predicted by models, the Arctic Ocean should become ice free in summer well before the end of this century,[7] and this would be a first for Arctic marine systems during the last 5+ million years.[8]

Arctic vertebrates possess many adaptations that have enabled their persistence in Arctic environments. These traits include behavioral (e.g., day torpor, seasonal migration), physiological (e.g., seasonal metabolic adjustments, fasting ability), and morphological (e.g., white coats, short extremities, extensive blubber layers, thick fur, lack of dorsal fins among ice whales) mechanisms. But despite their extreme and varied adaptations to coping with the harsh and highly variable Arctic climate, these species are considered to be particularly vulnerable to ongoing climate changes. Their peculiar life-history strategies have served them well in a low-competition situation but leave them vulnerable to competitive stress from temperate species that are already invading their ranges. Arctic endemics have little potential for northward retraction to avoid such overlap. In addition, they may have limited capacity to deal with exposure to disease because of their limited exposure in the past. Furthermore, the heavy lipid-based dietary dependence of some Arctic species leaves them disproportionately vulnerable to the effects of lipophilic contaminants.

Arctic vertebrates can adjust to cope with climate-related changes (i.e., through their phenotypic plasticity) or adapt (via the differential selection of some genotypes) in many different ways. The aim of our paper is to review these different responses and, whenever possible, to present examples of the impacts of climate change at the community level, explicitly taking into account interspecific interactions. Given the wide range of possible interactions, predictions made at this level usually remain speculative. On the other hand, predictions that overlook interaction processes would lack realism. To avoid these pitfalls, the examples supporting our text are primarily chosen from the few biological models that have already been studied comprehensively in the Arctic in recent years—that is, rodents and their predators, geese and shorebirds, large herbivores, seabirds, and marine mammals and their prey.

Our paper is structured in nine sections and follows a process orientation, rather than the traditional species by species approach. In the first section, we present a general outline of how vertebrates will respond to changes in the cryosphere, because most species living in the Arctic biome are exceedingly dependent on snow and ice regimes for their reproduction and survival. Second, we give a brief overview of some important eco-physiological constraints that directly link climate and species and can hence result in very rapid responses. In the third and fourth sections, we provide an extensive review of the main responses of Arctic vertebrates to climate change: changes in phenology, that is, the timing of seasonal activities (e.g., migration, calving, egg-laying), and range shifts, that is, moving to track suitable habitat.[1,2,9,10] The fifth section aims to present the need for considering interactions, especially in relatively simple communities such as those found in the Arctic, in order to understand the complexity and diversity of the observed responses. A particular case study involving parasites is presented in section six as an example. The seventh section provides an evolutionary perspective for the phenotypic and genotypic responses that are likely to arise in Arctic vertebrates, highlighting how molecular ecology can help us to infer the fate of these species. In the eighth section, we list some principle knowledge gaps that currently prevent us from having a better understanding or being able to forecast climate change responses well. Finally, we summarize the general trends that emerged from our review in a concluding section.

Coping with a changing cryosphere

Ongoing changes in snow and sea ice conditions will have strong and wide impacts on Arctic vertebrates' habitats and food resources. These widespread and long-lasting layers of frozen water are far from homogenous. Dozens of different types of snow and ice can be distinguished according to their thickness, age, density, temperature, chemical composition, physical structure, location (e.g., latitude, which accounts for temporal differences in solar radiation), history, etc. Depending on the quality and quantity of snow or ice found in their habitats, Arctic vertebrates will react differently in order to optimize their fitness. Forecasted changes in the Arctic cryosphere will hence impact Arctic vertebrates in many different ways that will be presented in more details in the following sections, but let us just start here with a few examples chosen from among some of the most important vertebrate guilds found in the Arctic.

The winter is a key period in the annual cycle of Arctic small mammals as some—for example, the Arctic lemmings—live under the snow for up to nine months of the year at the most northerly sites; they even reproduce under the snow. The increase phase of lemming population cycles is almost always driven by reproduction under the snow,[11] and thus events occurring under the snow have a strong impact on their population dynamics. Lemmings clearly choose particular sites under the snow for their winter activities. The distribution of their winter nests show a strong association with deeper snow, and the greatest probability of occurrence is at snow depths from 60 to 120 cm.[12,13] A deep, dry snow cover reduces thermal stress on small mammals that overwinter in subnivean spaces, most notably by dampening the daily temperature fluctuations that they experience.[12,14] Many parts of the Arctic receive less than 40 cm of snowfall during a winter, but this is often redistributed heavily by wind, creating a mosaic of habitat patches differing substantially in snow depth, with snow being trapped by topography and vegetation. A recent manipulation experiment that altered snow cover showed that increasing the snow depth in marginal winter habitat increased habitat use by small mammals in these areas, as expected, but the impact on demography (reproductive rate or mortality due to predation) was less clear.[14] Nonetheless, it is likely that change in snow depth or snow quality (e.g., increased amount of wet snow, icing, or collapses in the subnivean spaces, which increase the energetic costs incurred in getting access to food plants or in severe instances totally obstruct access) caused by climate warming will have a strong impact on small mammal population dynamics. Such findings have been reported recently from some parts of Fennoscandia and Greenland (see section "Cascading changes").[15,16]

Among large mammals, variation and trends in snow cover, especially during late winter, will likely also affect foraging ecology and population dynamics. Evidence from long-term monitoring at Zackenberg in northeast Greenland indicates, for example, that biomass production and spatial synchrony in the growth of the main willow forage species for muskoxen *Ovibos moschatus* there, *Salix arctica*, are both negatively related to snow cover.[17] Increasing snowfall in the High Arctic with continued warming could therefore translate into increased spatial heterogeneity of willow growth, which might result in increases in local density-dependent competition for forage among muskoxen.[17] In the High Arctic Archipelago of Svalbard, population dynamics of Svalbard reindeer *Rangifer tarandus platyrhynchus* display a generally positive association with an index of winter ablation, or the energy available for snowmelt, suggesting that continued warming may promote an increase in that population through reduced winter starvation mortality.[18]

Most tundra breeding birds (shorebirds and passerines alike) are long-distance migrants, and many of them winter in rich intertidal flats in temperate and tropical regions. By spending only some weeks in the Arctic, they avoid the harsh Arctic winter but still take advantage of the summer boom of food on the tundra, particularly to raise their young.[19] Indeed, many tundra birds rely heavily on invertebrate food during the entire breeding season. For the adults, invertebrate food is particularly important for the development of eggs and for building up body stores to be used during the incubation period, while hatchlings either feed themselves (shorebirds) or are fed (passerines) with invertebrates during their growing period.[20,21] Both adult and juvenile shorebirds have to build up body stores rapidly during the postbreeding period, based on invertebrate food, for their long autumn migration, while passerines feed primarily on seeds that ripen at this time of the year. The first weeks following the shorebirds' arrival on the tundra (i.e., when eggs are formed) are particularly critical, because snow melt and weather conditions in general vary a lot from year to year in most parts of the Arctic.[21] Because global warming is expected to result in earlier snow melt and more benign summer weather, tundra birds may well benefit from more favorable breeding conditions in most years.[21] However, they might also suffer from more variable conditions, because the amount of snow falling in winter is expected to increase in most Arctic regions, a change that, in combination with cooler springs, could result in more chaotic—both in regularity and severity—breeding conditions (also see the following section).

When moving to ice loss–related threats to Arctic marine vertebrates, the polar bear *Ursus maritimus* is undoubtedly the first species people have in mind. During their short evolutionary history—they diverged from brown bears *Ursus arctos* less than 200,000 years ago[22]—polar bears have become highly adapted to life on the sea ice. Their dependency on sea ice is mainly via their extreme specialization in feeding on ice-associated seals, particularly ringed seals *Pusa hispida* and bearded seals *Erignathus barbatus*.[23] In addition, sea ice is important because it is a solid substrate on which bears can move and rest. Coastal sea ice habitats are particularly important to females with dependent young.[24] Some bears use sea ice in all seasons, while other bears stay on land for much of the year, but most feed at least seasonally on ice-associated seals. However, it is noteworthy that ice linkage varies geographically for bears among the different populations and also among individuals; different space-use strategies can be found even within polar bear subpopulations.[25] The observed loss of sea ice in recent decades has raised concerns about the future of polar bears in the Arctic.[26–28] In areas where sea ice is absent in summer, polar bears depend on fat reserves acquired in the peak feeding period to get them through long periods of fasting. Polar bears are unique among larger mammals in their ability to survive many months without feeding. Reproducing females may fast for up to eight months and still nurse cubs. There is, however, a balance between fat reserves and survival/breeding capabilities. In some areas where sea ice breaks up earlier in spring and access to ringed seal pups has been reduced, for example, in western Hudson Bay, survival of cubs, subadults, and even older bears is lower in years with early ice breakup, and recent longer ice-free summers are concomitant with a reduction in this subpopulation's size.[29] In the southern Beaufort Sea, survival of adult females was considerably higher in 2001–2003, when the mean ice-free period was 101 days, compared to 2004–2005, when it was 135 days. Breeding rates and cub survival also decreased during the latter period.[30] Another change in the Alaskan part of the southern Beaufort Sea is a profound decline in the proportion of females denning in the multiyear sea ice: 62% in 1985–1994 versus 37% in 1998–2004.[31] However, in most areas of the polar bear's range, denning occurs only on land. In Svalbard, many female bears denned on the isolated island Hopen decades ago. But in recent years, when sea ice has not extended south to Hopen in the late autumn, almost no denning has been documented on this island.[26] Reduction of sea ice has also been proposed as a causal factor behind increased intraspecific predation,[32] drowning of bears during storms,[33] and extraordinary long swims that may lead to loss of cubs.[34] In areas where polar bears have high levels of lipophilic pollutants, toxins and their metabolites are released into the blood stream under periods of nutritional stress. This is a concern given the longer fasting periods being experienced by populations with reduced ice conditions, although the exact effects this may have on survival and reproduction are not known.[35]

Other marine mammals will also face important environmental changes in the Arctic that will alter the communities in which they live,[36–38] particularly pagophilic ("ice-loving") species.[5,6,39–45] A summertime ice-free Arctic Ocean will have major implications for ocean circulation and our global climate system[3,46,47] and will also have impacts throughout marine food webs. Implications for the organisms that are residents of the unique Arctic sea ice habitat have been described as "transformative."[48] For its mammalian residents, Arctic sea ice has been a spatially extensive, virtually disease-free habitat that is to a large extent sheltered from open-water predators (i.e., killer whales *Orcinus orca*) and human impacts (e.g., oil development, shipping). It has been a low-competition environment that has provided a spatially predictable, seasonally rich food supply, particularly in the marginal ice-edge zone and at predictable polynyas.[49,50] It is furthermore sheltered from storm action for the mammals that have succeeded in dealing with the prevailing cold temperatures, risk of ice entrapment, dramatic seasonality, and other aspects of an ice-associated lifestyle.[41] Seven seal species and three whale species have evolved within the Arctic sea ice environment, or joined it, over the millions of years of its existence.[51] Loss of sea ice represents a reduction in available habitat for pagophilic mammals, and this is already affecting the distribution, body condition, survival rates, and reproductive output of some species.[41] In the longer term, it is expected that foraging success, fertility rates, mortality rates, and other population parameters will be impacted for additional populations and species. Northward range contractions are possible only within a limited scale before deep Arctic Ocean conditions replace the productive shelf habitats upon which most Arctic marine mammals depend. Generally speaking, specialist feeders are likely to be more heavily impacted by changes in Arctic food webs that will accompany sea ice losses compared to generalist feeders. In addition, widely distributed species will have greater chances of shifting to suitable regions than species with restricted ranges. Ice breeders that require long periods of stable ice late in the spring season are also likely to be impacted more rapidly than late winter ice breeders that require ice for shorter periods of time.[42,43,52] Other risks posed by climate change include the increased risk of disease in a warmer climate, the potential for increased effects of pollution, increased competition from temperate marine mammal species that are expanding northward, and stronger impacts of shipping and development in the north—in particular from the petroleum industry—in previously inaccessible areas.[41] In combination, these various changes are likely to result in substantial distributional shifts and abundance reductions for many endemic Arctic marine mammal species.

Although Arctic seabirds are highly adapted to extreme environmental conditions, their energy balance is nonetheless affected by harsh environments. Heat losses in cold air, water, or in stormy conditions can be extremely high despite the waterproof and well insulated plumage of adults.[53] Some species such as guillemots and little auks *Alle alle* can be found dead by the thousands in coastal areas following severe winter storms.[54] Because the intensity and frequency of winter storms are forecast to increase with climate warming,[6,55] important impacts are expected on seabird energetics and winter survival. The relative sensitivity of seabirds to climate changes is also obvious during the breeding season for their chicks, whose downy plumage is much less waterproof and a poorer insulator than that of the adults. Chicks are highly vulnerable to changes in wind speed and precipitation. Indeed, harsher wind conditions as well as more frequent heavy or freezing rain will most likely impact their energetics and result in higher mortality. Even for the extremely well-adapted High Arctic ivory gull *Pagophila eburnea*, a single day of rain accompanied by strong winds can destroy all clutches and broods in some colonies in some years (Gilg and Aebischer, personal communication 2011).

Ecophysiological constraints

The most direct link between climate and the ecology of most species is probably the physiological interplay between environmental and body temperatures. Each organism lives within a given range of temperature (i.e., "thermal window").[56] To minimize maintenance costs, thermal windows are thought to evolve to be as narrow as is possible around normal environmental temperature(s), which results in different temperature tolerance ranges between species, subspecies, or even populations of a given species living in different regions. When environmental temperatures shift closer to one edge of the individual's thermal window—for example, climate warming driving the temperature higher—the individual's physiological performance (e.g., growth, reproduction, foraging, immune competence, competitiveness) will be negatively impacted.[56] This is contrary to the layman's belief that an increase in environmental temperature will improve the "welfare" of Arctic vertebrates. Climate change is forecasted to have a stronger negative impact on species that have narrow thermal windows, long generation times, and limited genetic diversity.[56] Conversely and at a larger geographic scale, some vertebrate species currently living at the southern border of the Arctic region will benefit from climate warming and will be able to expand northward, because their thermal windows will soon match the new environmental temperatures found in the Arctic (see also section "Advanced phenologies"). Because aquatic habitats have more stable temperatures than terrestrial ones, most aquatic invertebrates and fish have narrower thermal windows than warm-blooded vertebrates and are probably more sensitive to changes in environmental temperatures (see examples for sockeye, Pacific salmon, Atlantic cod, and zooplankton).[57–59]

Because the energy budget of an individual integrates both biotic and abiotic constraints, a bioenergetic approach can be used to assess the impacts of climate change. For instance, seabirds wintering in the North Atlantic experience an energetic bottleneck in November and December due to low temperatures and strong winds.[53] A milder climate in this region would hence likely increase their winter survival and allow other less cold-adapted species to use these areas in winter. One of the main inferences from bioenergetic studies to date is that the impacts of climate change in the Arctic are likely to stem from the immigration of new colonizing species as much if not more than from the disappearance of current Arctic species. And some particular species, like hibernating mammals, should benefit—increase in abundance and distribution—more from climate change than others.[60]

Changes in hormonal responses are also expected in Arctic vertebrates as a result of climate change, with potentially very important outcomes on individual fitness. The stress axis in particular is believed to play a central role in the adaptation of birds and mammals to a changing environment. For many of them, breeding is only possible below a given stress response. An increasingly stressful environment, whether due to direct abiotic changes (e.g., temperature, snow, or ice cover) or to indirect biotic changes (e.g., new competitors, changing predation pressure), could negatively impact the fitness of Arctic vertebrates.[61] Expected range shifts will help some Arctic vertebrates maintain their stress levels within acceptable ranges, but changes in food resources and availability, which also impact the stress axis, also need to be taken into account to predict their overall responses. Finally, hormonal responses and immune status are also modulated to some extent by pollutants.[61,62] The future impact of contaminants will likely be correlated to the intensity of climate change;[63] pollutants make it even more difficult to untangle causes and consequences of climate-induced impacts on the ecophysiology of Arctic vertebrates.

Advanced phenologies and trophic match/mismatch

Because ongoing climate changes are relatively rapid while most Arctic vertebrates are long-lived animals with slow population growth rates and long generation times, selection for the most adapted genotypes to new environmental conditions will likely be too slow to guarantee the survival of these species in the short term (see section "Evolutionary responses"). Instead, phenotypic plasticity is likely to be the principle response mechanism that will permit Arctic vertebrates to respond to the rapid, ongoing changes.[64] If individuals can adjust rapidly to new environmental conditions by changing the timing of their migration or the initiation of their annual breeding events, then, in some cases at least, their populations could be little impacted by the climate-induced changes described earlier. Advanced phenologies are a widespread response of plants and animals to changes in climate. They have already been described in recent decades for organisms living in a wide variety of areas around the world, including the Arctic where they have been most pronounced (up to 20 day shifts during the past decade for some plants and invertebrates).[2,6,65,66] Unfortunately, phenologies are not plastic for all traits, nor among all species. For some vertebrates, the timing of migration or breeding is regulated by a rigid endogenous clock or by fixed environmental clues (e.g., photoperiod) and cannot therefore be adjusted to other environmental conditions that may be changing.[67] For such species, shifting their distribution range in order to track their optimal environmental conditions—abiotic but also biotic if interacting species also shift their range—might be a more relevant response. If they are not able to adjust the timing of their life cycle events (e.g., migration, breeding, molting, fattening) to temporal changes in the quality of their habitat and especially to their food resources, then they will face mismatches and will be left with no other choice than moving to different habitats, using different food resources (both responses allowing them to "rematch"), or their populations will decline.[68]

Mismatches have both trophic and dynamic implications. If temporal relationships are disrupted or altered, fitness will be affected,[69] leading to declines in population abundance.[70] Different trophic levels—plants, herbivores, predators—may respond differently to warming, which may lead to mismatches in the timing of events between trophic levels. For instance, gosling growth in herbivorous geese is dependent upon good synchrony between hatching and the seasonal change in plant nutritive quality, especially protein, an essential nutrient for growth.[71] In snow geese *Anser caerulescens* breeding in the High Arctic, there is evidence that in years with early and warm spring, gosling growth is reduced,[72] probably because they hatch too late to benefit from high-quality plants. This has a negative impact on the recruitment of young because their survival during the autumn migration is dependent on their mass at the end of the summer. Climate warming will increase this mismatch and may result in a reduction of recruitment in some populations.

Large terrestrial herbivores may also be susceptible to trophic mismatch derived from an advance in spring phenology associated with warming. In West Greenland, for instance, long time series of observational data indicate that April warming is associated with an advance in the emergence date and rate of emergence of species of plants that comprise important forage for caribou. In this case, despite an advance in the onset of spring plant growth, the timing of calving by caribou has not advanced to the same extent, with the result that calving is not well synchronized with the onset of the plant growing season in the warmest years, when calf production is lowest.[69] There is also a spatial component to trophic mismatch, in this case because during warm springs, the timing of plant growth is more highly synchronized across the landscape on the caribou calving grounds, reducing the advantage that spatial heterogeneity in plant phenology otherwise affords highly mobile herbivores such as caribou.[73] These observations suggest that warming may, in populations of caribou inhabiting regions with typically short plant growing seasons, adversely affect offspring production and survival if the timing of calving is inflexible and, thereby, unable to keep pace with advances in the timing of spring plant growth on calving grounds. Indeed, recent work indicates that caribou lack an internal circadian clock due to selection for a circ-annual clock and thus may display a physiologically "hard-wired" timing of parturition.[74]

In the Arctic, most mismatches are likely going to be related to the advancement of the spring.[66] Among Arctic shorebirds for example, initiation in egg laying is governed by a combination of spring cues, including snow cover and food availability during the egg-laying period, and it occurs in early June in most areas. If the snow melts early, invertebrate emergence could be affected. This is the strongest factor governing chick success, and shorebirds can adjust their laying date to these local conditions by up to four weeks.[75] In years with late snow melt, the birds have to wait for the snow to disappear, which might result in a mismatch between invertebrate availability and chick growth later in the season. However, in some regions, climate change could actually improve the match between the annual cycle of shorebirds and the phenology of their prey. There are indications that Arctic shorebirds may suffer from trophic mismatch on the High Arctic Siberian breeding grounds,[76] while they might benefit from a better temporal fit between their reproductive cycle and the availability of their food in other places—for example, in High Arctic Greenland.[21,77] The reason for this difference is that the duration of the peak in invertebrates on the High Arctic Siberian tundra is rather short, while invertebrates are plentiful for a more extended period in High Arctic Greenland. Hence, shorebird chicks in High Arctic Greenland may benefit from earlier appearance of sufficient amounts of invertebrates, while some of their later arriving Siberian counterparts—due to often later snow melt than in Greenland—may not be able to hatch in time for the peak.

Marine environments

Phenological adjustments also occur in marine ecosystems. The reduction of sea ice extent in the Arctic will allow earlier access to benthic feeding grounds and create new areas of open water earlier in the season for foraging, which could lead to changes in seabird breeding times. Falk and Møller[78] showed that the phenology of northern fulmars *Fulmarus glacialis* breeding in northeast Greenland was correlated to the sea ice dynamics of a nearby polynya. Similarly, Canadian northern fulmars as well as thick-billed murre *Uria lomvia* in West Greenland have been shown to respond to sea ice conditions by adjusting their timing of arrival at their colony, arriving later when the sea ice was more extensive.[79,80] A decrease of seasonal ice distribution should be beneficial to many Arctic seabird populations, as this may advance the start of their breeding season and therefore allow more time to raise their chicks, provided these species are not indirectly affected by other mismatches—for example, between plankton bloom or fish prey availability and the chick rearing period.[81] But sea ice declines will also be detrimental to some of the most typical High Arctic species, such as the ivory gull and several species of marine mammals (described later), that are sea ice specialists.[41,82] These contrasting effects are sometimes even found between populations of the same species. Thick-billed murres breeding in the Canadian Arctic experience a beneficial effect of a reduced sea ice season at the northern limit of the breeding range due a lengthening of the breeding season but a detrimental effect at its southern limit, because it leads to mismatches between the timing of fish prey availability around the colony and the chick rearing period.[83]

Along western Hudson Bay there has been a climate-driven increase in the overlap between nesting geese and polar bears.[84] Late ice formation in fall and early break-up has extended the terrestrial period for the bears. This means that the bears now forage on energy-rich goose eggs for a longer period, which can result in widespread failure of goose nests in the earliest ice breakup years. In this case, the increased match between bears and geese is beneficial to the predator but highly detrimental to the prey. In Svalbard, increased presence of polar bears at barnacle goose *Branta bernicla* colonies in recent years has contributed to a decline in goose numbers in some coastal areas.[85]

Beyond sea ice extent, the increase in sea surface temperatures also affects the breeding phenology of seabirds.[86] Despite the possibility for certain seabird species to adjust their breeding time in response to a change in their physical environment, most of them might face climate change–related modifications of the spatial and temporal distribution of their prey—that is, trophic mismatch between predators and resources.[68] For instance, an earlier melting of seasonal sea ice in the Arctic will advance the timing and modify the rate of nutrient supply in upper surface layers, therefore affecting the timing and species composition of the phytoplankton spring bloom. This change will in turn affect the zooplankton and fish communities[81] and modify their availability for predatory seabirds during their respective breeding seasons. To cope with these changes, seabirds have to breed earlier to follow the timing of prey availability. However, while this seems to be possible for some species—described earlier—it might not be the case for all of them. For example, seabird species nesting in crevices or burrows—little auks, auklets, puffins, etc.—can only access their nest when snow has melted and might therefore be limited in their response, as future climate changes are also expected to be associated with an increase in precipitation including snow fall.

Many ice-associated marine mammals use traditional breeding sites where these broadly dispersed populations congregate at very specific times—for example, harp seals *Phoca groenlandica.* These rendezvous match availability of their ice habitat, but also match food availability at the time of weaning of the young. As the sea ice season contracts, it is easy to envisage mismatches occurring between the timing of independence of the young and the lower trophic prey upon which they depend. Many temperate whale species migrate into the Arctic to feed in the marginal ice-edge zone, taking advantage of the spatially and temporally predictable late spring and early summer high productivity. In the future, it is likely that the Arctic marine ecosystem will be more productive in an overall sense, but it is also likely that prey will be more dispersed and hence more energetically costly to acquire.

As illustrated by the few examples described earlier, it will often be difficult or even sometimes totally impossible for Arctic vertebrates to adjust their phenologies in order to avoid mismatches. Responses to such challenges will be particularly difficult for species that use different habitats or food resources during their annual life cycles—for example, migratory species wintering outside the Arctic or species like salmonids using both oceanic and continental ecosystems; see also the section on "Cascading changes."[87] Indeed, the various mismatches these species will face in these different habitats throughout the year will be different in speed, intensity, and nature, and must therefore be solved by different—and sometimes antinomic—responses.

Poleward changes in habitat and species distributions: a large-scale paradox of enrichment

Due to various underlying processes already presented in other sections of this review (e.g., physiological constraints, specialization for some habitats and prey, limited phenological plasticity, interspecific interactions, biogeographical history, etc.), every species uses a given geographical area (delimited by its distribution range) characterized by a specific climate (sometime used to define species' "climatic envelopes").[88] As seen in the previous section, many species try to adjust to their changing environment (e.g., by changing the timing of their annual cycle) in order to stay in their current distribution range, but the phenotypic plasticity of individuals is finite and such adjustments are therefore limited.[89] Hence, shifting the latitude or altitude in which they live in order to track their optimal environment appears to be a better strategy in the long term, especially for Arctic vertebrates that are extremely well adapted to their harsh environment but that have limited evolutionary capabilities on short time scales (see section "Evolutionary responses").

Poleward shifts have been documented for many species in recent decades, even for animals that live only within the Arctic region.[90–93] As for phenological changes, given the intensity of climate change in the Arctic, range shifts are expected to be more pronounced here than in other biomes. In a strict sense, range shifts describe situations when both the southern (or lowest) and northern (or highest) borders of distribution ranges are constrained by climate and can hence move as a consequence of climate change. There are situations, however, when only one of these borders is constrained by climate—the other being constrained by geographical barriers, competitors, etc. In these cases, climate warming can either induce a decrease in distribution range—if the southern border is constrained by climate—or an increase. Several predictive models based on habitat utilization by feeding or nesting geese follow this rationale.[94,95] Jensen *et al.*[96] recently attempted to predict the future distribution of pink-footed geese *Anser brachyrhynchus* in Svalbard, considering that warm temperatures will create a longer summer season, thereby increasing the probability that geese will be able to complete their breeding cycle, and also experience increased food availability. According to their model, a 2°C increase in summer temperature should lead to an expansion in goose distribution and ultimately an enhanced population growth (but also see contrasting results based on seasonal changes in food quality in the previous section).

Another peculiarity of range shifts in the Arctic is related to the geophysics of the Earth. Any poleward shift in species distributions will mechanistically lead to a decline in the size of these ranges—a "polar squeeze"—the surface of latitudinal belts decreasing northward (e.g., a 1° Lat. by 1° Long. area covers *ca.* 6200 km^2 at 60° Lat. but only 4200 km^2 at 70° Lat. and 2100 km^2 at 80° Lat.). The most severe losses should therefore concern habitats that already reach the highest possible (90° Lat.) or potential (see later) latitudes, because surface lost in the South of their range cannot be fully compensated by the colonization of new areas in the North.

Sea ice is probably the best example of a habitat that can only decline under current climate change scenarios, because it already occupies the northernmost seas on Earth. Poleward shifts will be impossible for species already distributed in these extreme regions. In addition, retractions of sea ice that occur to the degree that the remaining ice is no longer over productive shelf seas, but instead over the deep, unproductive Arctic Ocean, are likely to have extreme impacts and may induce tipping points. The well-documented polar bear case has already been presented, but this will also be the case for less well-known species such as the ivory gull and the ice-dependent seals that require sea ice on a year-round basis.[41,82,97] In the long term, the disappearance of summer sea ice in most of the Polar Basin may restrict ice-dependent species to the north coasts of Greenland and Ellesmere Island (Canada), where multiyear ice is predicted to last longest or even drive some of these species to total extinction.

Some terrestrial habitats like the High Arctic polar deserts are similarly threatened because there is virtually no new area to colonize for the inhabitants of these areas, including the vertebrate species they host—for example, the Peary caribou *Rangifer tarandus Pearyi*. This "dead end" problem can be a concern even in low-Arctic regions. Here, narrow belts of open tundra are currently colonized at their southern borders by shrub communities, while the typical open tundra is prevented from extending northward due to bordering seas (e.g., in northern Alaska and on the continental part of northern Russia).

Species and subspecies that already have small distribution ranges, small population sizes, or that are to some extent dependant on typical Arctic habitats like sea ice (e.g., polar bear, walrus *Odobenus rosmarus*, narwhal *Monodon monoceros*, bowhead whale *Balaena mysticetus*, ivory gull, and a few others) or High Arctic tundra (e.g., several wildfowl and shorebird species, muskox, several reindeer subpopulations, etc.) will hence face a growing risk of extinction during the coming decades.[21,98,99]

Despite this dark picture for endemic Arctic species, many "meridional" species are colonizing the Arctic and their marginal Arctic populations are increasing. Due to their ability to cross-oceanic barriers, many of these species are fish, seabirds, seals, and cetaceans, with the highly visible colonial seabird being the best documented group. Climate change is expected to induce northward shifts in the distribution of most seabirds and some marine mammals in response to the appearance of new suitable habitats and to the modification of their prey distribution. In northeast Greenland, for example, the coastal waters between 72 and 80°Lat. North were until recently covered year round by multiyear drift ice transported from the Polar Basin by the East Greenland Current. Historically, this coast only hosted a few, small breeding seabird colonies.[100] Recent changes in sea ice regimes (both in extent and duration) are providing opportunities for this region to host larger seabird populations (e.g., kittiwakes *Rissa tridactyla* and common eiders *Somateria mollissima*) and for the establishment of new breeding species—like black-backed gulls *Larus marinus* and *Larus fuscus*.[101–104] The kittiwake nicely illustrates the ongoing changes in range shifts for widespread species: its populations are increasing and its distribution range is expanding in the Arctic, while they are simultaneously declining in Europe at the southern border of its range.[102,103,105] The first records of gray seals *Halichoerus grypus* in southern Greenland, harbor porpoise *Phocoena phocoena* in Svalbard and well-documented shifts in gray whale *Eschrichtius robustus* distribution suggest that range shifts are also occurring quite broadly among marine mammals that are resulting in traditionally more southerly species spending more time in the Arctic as well as regular Arctic migrants staying for longer periods.[41]

Another consequence of the reduction of sea ice is that many islands currently connected to the mainland by ice bridges for most of the year will become ice free for a longer period. This loss of connectivity will prevent access for predator species like the Arctic fox *Vulpes lagopus* to some bird colonies during the breeding period and will strongly benefit birds nesting on these islands, as already seen for common eiders in Canada.[106] On a larger spatial and temporal scale, it is also likely that the small Arctic fox populations historically found on Jan Mayen and Bjørnøya (Norway)—two oceanic islands holding important seabird colonies in summer but no alternative food resources to sustain fox populations during the winter period—declined and eventually became extinct in recent decades due to the loss of sea ice connectivity, making regular immigration from large nearby populations—that is, northeast Greenland and Spitzbergen—difficult.[107,108] Although extensive sea ice periods have promoted population exchanges among Arctic fox, reindeer, and other terrestrial mammal populations, they have actually limited population exchange within marine mammals and conversely, of course, periods with little sea ice have promoted population linkages and genetic exchange for marine mammals.[109] Less sea ice in the future is almost certain to promote more extensive and regular population exchange among various marine mammal species, particularly among those cetaceans that have extensive migrations as a normal part of their annual cycles.[110]

Simultaneously, the warming of the oceans has also led to important changes in zooplankton and fish communities, and several studies predict these modifications will continue.[111] The overall trend is a northward movement of species along with a replacement of Arctic species by more temperate ones with different energetic values—for example, the replacement of the cold adapted, lipid-rich copepods *Calanus hyperboreus* and *C. glacialis* by southern, low-lipid *C. finmarchicus* being a particularly well-studied example.[81] These sorts of changes will indirectly impact breeding seabirds and marine mammals through bottom–up control processes.[112] Because these Arctic marine top predators are often specialized, their fate will depend on their ability to modify their diet according to new prey conditions, a plasticity that has already been shown in several Arctic species but within given limits.[113,114]

For the long-distance migratory Arctic shorebirds, climate change effects on their staging and wintering areas may be just as—or even more—important than conditions in the Arctic.[21] This is because most of them stop over and winter along marine coasts, on a small number of exceedingly important intertidal areas, where sea level rise or other climate-related changes may modify their current feeding grounds (e.g., in the West European Wadden Sea, Banc de Arguin in West Africa, the Yellow Sea in East Asia, and Copper River Delta in southern Alaska).[115]

For all these reasons, overall distribution ranges of most Arctic vertebrates are predicted to shift northward, and population sizes of many of them are likely to decline as a consequence of the reduction of their habitats and immigration of new competitors. However, in the Arctic, gains and losses will not be similar to other biomes,[116] because only a few vertebrate taxa are likely to disappear while many "new" ones will colonize from the South. The paradox here is that species richness in the Arctic will undoubtedly increase in the future—more species colonizing than species disappearing—but Arctic biodiversity, that is, its contribution to the biological diversity on a worldwide scale as defined by its original sense,[117,118] can only decline. Indeed, the many taxa colonizing from the South will often just expand or shift their ranges northward (see the kittiwake example described earlier), while the few Arctic ones, only present in this biome and that have no alternate region to colonize to the North, will see their populations decline and in some cases even go extinct.

Cascading changes and feedbacks: from population dynamics to ecosystem functioning

Arctic ecosystems are hypothesized to have little functional redundancy. Changes in the phenology, distribution, or behavior of species—as already described in previous sections—may therefore have disproportional impacts on the abundance and dynamics of species, and in turn on the structure and functioning of their communities and ecosystems. Therefore, the impacts of climate change cannot be inferred or predicted using either single-species approaches or simple empirical relationships between processes and driving variables. To understand these changes and predict their consequences, one must instead explore them in a multitrophic and multispecies perspective, particularly in the Arctic.[119–121]

Climate change has already been shown to affect life-history traits like survival and reproductive success of many Arctic vertebrates.[122–124] Climate-driven impacts on dynamics are, however, often more difficult to understand because impacts also vary in direction and strength depending on how trophic interactions are disrupted.[125] Small rodents in general and lemmings in particular provide good examples of such interspecific interactions across several trophic levels. Lemmings are central to the multitrophic interactions in the Arctic tundra and play a key role in shaping the structure and dynamics of these ecosystems. They impact reproduction and abundance of many terrestrial Arctic predators,[126,127] which in turn creates feedbacks on lemming cyclic dynamics as well as on the dynamics of other vertebrate prey.[128,129] In Greenland and in Scandinavia, lemming cycles have recently been fading out.[15,16,130] This situation is not particularly surprising given the evidence from both paleo- and current records showing that rodent population dynamics are highly influenced by climate variables in the Arctic, in particular by snow characteristics.[11,12,15,16,129,131,132] Not only does snow provide insulation from the harsh Arctic weather, but a thick snow cover also prevents most predator access to the subnivean rodents.[133] For lemmings in particular, sufficient snow cover is important for winter breeding.[11,12] The ongoing and future changes in temperature and snow conditions are therefore expected to have important implications for the population dynamics of lemmings, in terms of both increased length of the lemming cycle and reduced peak densities.[16,129] Such changes in rodent dynamics will have immediate effects on the predator community, not only in terms of reduced abundance and productivity,[16] but also in terms of more indirect effects on the predation pressure on alternative prey species, such as ground-nesting birds.[134–138] Ultimately, this may even affect the geographical distribution of shorebirds and other species sensitive to predation.[139] The fading out of the small mammal population cycles will have far-reaching, cascading effects, ultimately impacting the structure and functioning of the entire terrestrial Arctic ecosystem.[130,140] As lemmings are key grazer and browsers of Arctic plants, loss of cyclic lemming outbreaks will also feedback on important structural components of the vegetation like tall shrubs.[141] Consequently, negative effects of climate warming on lemmings may create a positive feedback on the ongoing greening of the Arctic owing to shrubs encroaching on tundra plant communities.[142] Finally, shrub encroachment is predicted to provide a positive feedback on the climate system itself through more absorption of sunlight that is converted to heat.[143]

Changes in distribution ranges—see the previous section—can also impact the population dynamics of competing species and deeply modify the structure of communities. In Scandinavia, there is now extensive empirical evidence that the larger red fox *Vulpes vulpes* can expel its smaller congener, the arctic fox, from food resources in winter[144] and from breeding territories[145] and dens[146] in summer. During the last century, the red fox has extended its range northwards into the tundra zone at the expense of the Arctic fox, which has retracted its southern distribution limits.[147,148] Although it has been suggested that the range expansion of the red fox is related to climate warming and concomitant increased primary productivity of the tundra,[147] the link to increased secondary productivity and hence food for carnivores has yet to be demonstrated.[149] In this context, it is important to keep in mind that the Arctic is changing not only due to climate change, but also for other reasons owing to local anthropogenic activities—e.g., intensified land use, infrastructure, settlements—which may pave the way for highly adaptive generalist predators like the red fox and corvids.[144]

Bylot Island provides another interesting example of altered community function. Reproduction of snow geese is strongly affected by weather conditions prevailing in spring at this location. When the spring is early and warm, the probability of laying eggs increases, that is, the nest density is higher, laying is early, and individuals lay larger clutches, resulting in a higher reproductive effort at the population level.[72,150,151] But climate also drives primary production. In wetlands used by geese, production has increased by 85% during the period of 1990–2009, most likely as a direct consequence of the warming temperatures. The cumulative number of thawing degree-days during the summer is an important determinant of plant biomass at the end of the summer.[152] Although this suggests that snow geese should benefit from climate warming, other factors such as increasing mismatch between the timing of breeding and plant phenology (see section earlier) could mitigate these effects. Moreover, in recent decades, goose population increases have largely been driven by events occurring during winter—mostly food subsidy provided by feeding in agricultural lands—and this has had negative impacts with snow geese overgrazing their Arctic breeding habitat in several localities.[153,154] Thus, climatic effects cannot be fully understood without considering all factors affecting the dynamics of a population.

The well-known influences of large, mammalian herbivores on primary production and plant community composition may play a role in ecosystem responses to climate change in areas with increasing or stable populations of large herbivores. Experimental evidence indicates, for example, that grazing by caribou and muskoxen in West Greenland suppresses the positive response of woody, deciduous shrubs to warming, maintaining the plant community in a graminoid-dominated state.[155] Because woody shrubs have a much greater carbon uptake potential than do graminoids, this suggests mammalian herbivores may have the potential in some parts of the Arctic to constrain carbon uptake by ecosystems in response to warming. Indeed, experimental warming resulted in a threefold increase in ecosystem carbon uptake in West Greenland, but only where caribou and muskoxen had been excluded; in sites where these herbivores grazed warmed plots, there was no net increase in carbon uptake in response to warming.[156]

As seen earlier, climate change will impact the dynamics of Arctic multitrophic systems whether they are bottom-up or top-down controlled. Other important clues to understanding these impacts are space and time. Even a significant impact of climate on breeding success can take several years to show effects in many species. For instance, it takes the young several years to reach sexual maturity and return to the breeding sites in many seabird or marine mammal species.[157] In addition, one needs to account for regional climate change variation, particularly for animals that use areas experiencing different degrees of change during their annual cycles, in order to comprehensively assess changes in population dynamics. Arctic terns *Sterna paradisaea* winter in sub-Antarctic waters while little auks remain in Arctic waters, despite the fact that both breed in the High Arctic. These two species will of course be differently exposed and differently impacted by climate change.[112,158] Even the lemming predator community described in the previous example is indirectly linked to pelagic intertropical climates, via two skua species that are highly specialized on lemming prey spending the summer in the Arctic but the rest of the year in the south.[159,160] But despite the recent development of year-round tracking devices, our knowledge is still mostly restricted to the summer season for many Arctic migrants, while spatial distribution or foraging behavior are still poorly known during the nonbreeding season, which is of course an essential period in these animals' life cycles.[53]

The cryptic parasites

The overall health of an individual is the result of many interactions between its immune status, body condition, pathogens, exposure to pollutants, and other environmental conditions that influence these factors.[161] Because climate is one of the main factors driving the diversity, distribution, and abundance of pathogens, understanding how climate changes might affect host–pathogen interactions is a major, though often overlooked, challenge.[162] As the Arctic climate becomes warmer and wetter, profound changes are expected in host–parasite interactions; many pathogens are sensitive to temperature, rainfall, and humidity. The transmission rates of pathogens and their period of transmission will become longer in warmer conditions.[163–165] This change could have profound consequences on Arctic vertebrates, because parasites are known to influence individual fitness and in turn population dynamics.[166–170]

Several climate-driven changes in host–parasite interactions are already obvious in the Arctic, and several reviews have recently been devoted to the effects of climate change on parasite infections and the ecological and evolutionary consequences for Arctic vertebrates.[162,171–173] Overall, these studies suggest that climate warming can increase pathogen development and survival rates, disease transmission, and host susceptibility through at least three main mechanisms.

First, global warming could change the abundance as well as spatial and temporal distribution of existing parasites. Indeed, the life cycles of numerous parasites are more efficient in warmer conditions, and global warming is hence expected to increase the success of infective stages. One of the best examples of this phenomenon is the infection of muskoxen by the nematode *Umingmakstrongylus pallikuukensis*.[163] Empirical field and laboratory data have demonstrated that the life cycle of this parasite has become shorter due to temperature increases, resulting in an increased parasite burdens for their host.[163] Increases in pathogen infections linked to warmer temperature have also been documented for other parasites in muskox,[174] reindeer,[175] marine mammals,[176] and seabirds.[170] Second, global warming might also induce northward range shifts of parasites due to the relaxation of the environmental constraints that affect their survival and developmental rates. For example, in Canada, the northern limit of the moose winter tick *Dermacentor albipictus* has already significantly expanded.[173] Finally, temperature increase will facilitate the arrival and establishment of new pathogen species previously unknown in the North and for which Arctic hosts have no previous exposure and hence no immunity.[177] A number of studies have shown that new parasites are currently emerging in northern ungulates due to climate change, modified movements of animals, and population density.[178] Direct impacts of new parasites on naive host population could lead to epidemic disease outbreaks (e.g., *Elaphostrongylus rangiferi* within caribou in Newfoundland).[179]

There is no doubt that climate change is already altering the dynamics of host–parasite interactions in the Arctic. Such changes will have profound impacts on population dynamics of hosts and for the ecology and evolution of Arctic vertebrates, although the consequences of these changes have not yet been properly evaluated.

Evolutionary responses

Most studies to date that have attempted to forecast the impact of climate change on species and communities have used phenomenological models based on species niche, largely ignoring the species' adaptive potential and biotic interactions.[180] Many examples have already been given throughout this review to emphasize the importance of biotic interactions, but evolutionary responses are likely also important to consider because evolution is a permanent and sometimes rapid process that can mitigate the effects of climate change.[181,182]

Change in phenology: adjust where you live

Phenology is a key aspect of the adaptation of Arctic vertebrates. The short growing season in the Arctic represents a challenge for most animals, and life-history strategies of successful Arctic vertebrates have evolved to enable individuals to fulfill their life cycles under time constraints and high environmental unpredictability.[6,183] In such a highly seasonal environment, offspring production is timed to coincide with the annual peak of resource availability—see section "Advanced phenology." The mechanisms that allow organisms to cope with climate-induced phenological changes are likely to be of two basic kinds.

Phenotypic plasticity—changes within individuals—allows organisms to rapidly cope with short-term changes in the environment without a change in genotypes. Some individuals in a population can accommodate large amounts of environmental variation while other individuals can tolerate only a narrow range of environmental variation.[184] Across the circumpolar Arctic, species are regionally exposed to varying sets of environmental conditions in different parts of their range and thus may demonstrate considerable plasticity.[66] But plasticity will also depend on species–specific sensitivity to climate change that may vary according to population size, geographic range, habitat specificity, diet diversity, migration, site fidelity, sensitivity to changes in sea ice, sensitivity to changes in the trophic web, and maximum population growth potential. This series of factors allowed Laidre *et al.*[43] to quantify the sensitivity of seven Arctic and four subarctic marine mammal species. According to their sensitivity index, the hooded seal *Cystophora cristata*, the polar bear, and the narwhal appear to be the three most sensitive Arctic marine mammal species, primarily due to reliance on sea ice and specialized feeding. The least sensitive species were the ringed seal and bearded seal, primarily due to large circumpolar distributions, currently large population sizes, and flexible habitat requirements. Although, the very specific breeding requirements of ringed seals (for snow availability on sea ice that must last for a period of months) probably warrants their inclusion in the most sensitive category.[41]

There are limits to plastic responses, and they are unlikely to provide long-term solutions for the challenges currently facing populations of Arctic vertebrates, especially for specialized species that are experiencing continued directional environmental change.[185,186] Microevolutionary changes—that is, changes in gene frequencies between generations—allow populations to cope with longer-term environmental changes through permanent modifications of phenotypes and will reflect selection on phenotypic traits.[187] As climate warming continues, shifting optima for phenological traits—for example, the timing of development, reproduction, migration, or fall dormancy—will at some point exceed the limits of individual plasticity and selection for genetic change in populations will occur. Indeed, when faced with long-term directional changes in environmental conditions, evolutionary adaptation becomes essential for the long-term persistence of populations.

Evolutionary response to selection will depend on the presence of heritable variation—that is, standing genetic variation within populations and additional variation generated by mutation and immigration.[188–190] However, there is growing evidence that populations of Arctic species have relatively low genetic diversity, a pattern observed for such diverse species as polar bears,[22] muskoxen,[191] collared lemmings *Dicrostonyx groenlandicus*,[132] and the wolf *Canis lupus*.[192] Most studies suggest that the Last Glacial Maximum (25.0–18.0 cal. kyr BP), which was strongly associated with a cold and dry climate, had strong effects on the genetic variation of Arctic vertebrates. High Arctic shorebird species have probably been through similar genetic bottlenecks since the last interglacial, in contrast with low-Arctic and sub-Arctic species that show extensive polymorphism.[21] This likely indicates that High Arctic shorebirds have been repeatedly exposed to "Arctic squeeze"—that is, their southern limit of distribution being pushed northwards, while they could not move farther North due to the Arctic Ocean barrier (see also section "Poleward shifts").

Although one may therefore expect a constant decrease of genetic diversity for current Arctic populations and low evolutionary potential, it should be mentioned that most studies done to date have used neutral markers (e.g., mitochondrial DNA) to evaluate the level of genetic variation within and between populations. Hence, they may have little bearing on real measures of genetic variation in morphology and life-history traits, which are directly exposed to selection. Few Arctic vertebrates have been studied using a quantitative genetic approach that addresses the potential for rapid adaptation to climatic change. Using such a quantitative genetic approach, Réale *et al.*[193] showed that northern boreal red squirrels *Tamiasciurus hudsonicus* were able to respond to increased temperatures both plastically and genetically within a decade. Pedigree analysis showed that about 13% of the 18-day advance in mean parturition date was genetic and that 62% of the change in breeding dates was a result of phenotypic plasticity. This study emphasizes the need for extensive, long-term ecological and quantitative genetic data for estimating the relative roles of evolution and plasticity on response to climate change. Given that species with short generation times tend to have higher rates of molecular evolution,[194] we can expect that long-generation-time species such as bowheads or the monodont whales should have less evolutionary potential to respond to new selective pressures than shorter generation–time species such as shorebirds or rodents.[183]

Range shifts: tracking suitable habitats

Global warming over the past 40 years has shifted the ecological niche of many species spatially—that is, towards the poles or higher altitudes (see section "Poleward shifts"). An ability to track shifting habitat is critically influenced by individual dispersal, which is key to the dynamics of genetic diversity in time and space and has profound effects on the ecological structure and dynamics of populations.[180,195] In the meantime, however, global warming might cause an overall contraction and fragmentation of the distribution range of Arctic vertebrates, eventually below the critical level of metapopulation persistence.[196,197] Future decreases of sea ice extent will lead to a loss of connectivity between Arctic regions, preventing or impeding dispersal of pagophilic species in summer. Even for some terrestrial mammals like the Arctic fox, sea ice is necessary for long-distance dispersal.[198,199] Genetic structure of fox populations is therefore likely to change dramatically in the future, both at the circumpolar scale and within populations, as a consequence of the ongoing sea ice declines (see also section "Poleward shifts").[108,200,201]

To improve our knowledge of the evolutionary consequences of climate change, we can also learn from the past. The Quaternary period (the past 2.6 Myr) was marked by glacial–interglacial phases and can be used as a model system to infer how Arctic species have responded to past climatic variation. In order to track their preferred habitats, species should increase their distribution ranges during periods of suitable climate and conversely contract them when climate conditions become unsuitable. The lesson we have learned from the past is that for temperate species, warming periods are often indicative of range expansions, while for Arctic species (e.g., muskox), populations are seen to increase during global climatic cooling and decline during the warmer and climatically unstable interglacial period.[191]

Finally, a different approach to study past demographic changes is to use ancient DNA, which allows the study of dynamic genetic variation over time. Thanks to well-preserved samples buried in permafrost, this approach has been used in several Arctic species—e.g., muskoxen,[191] collared lemming,[132] and Arctic fox.[202] The results show that for the Arctic fox, populations in middle Europe became extinct at the end of the Pleistocene—*ca.* 18 Kyr ago—and did not track its habitat when it shifted to the north. These results suggest that some populations, despite high specific dispersal capabilities,[198,199] might be unable to track their habitats.[202] For collared lemming, results also showed that previous climate events have strongly influenced genetic diversity and population size.[132] Due to its already reduced genetic diversity, a further decrease would strongly impede the evolutionary potential of this species. Local extinctions of collared lemmings would in turn have severe effects on the entire terrestrial ecosystem, and perhaps even beyond it into the marine environment (see section "Cascading changes").[16,126,152]

Missing pieces

Due to low species richness, relatively simple functioning, and high exposure to climate changes, Arctic ecosystems are already being strongly impacted by climate-driven changes and are among the best candidates to study ecological impacts of this phenomenon.[6,123,140] Arctic ecologists often advertise their studies for their inferential potential—that is, for the overall knowledge that can be gained from their study to understand and possibly mitigate the impact of climate change in other ecosystems that are not yet, or not so strongly, impacted. Even if we agree with these statements in general, we also call for great caution when discussing the results of such studies on larger taxonomic or geographical scales. For the time being, most studies claiming to show impacts of climate change—in the Arctic and elsewhere—are species oriented and only present correlative evidence. They are therefore speculative because they only consider a small fraction of the interacting species and have a limited predictive value since they often do not explicitly study the factors driving these population dynamics. Meta-analyses, experiments, and modeling studies can partly address these problems, but they also have weaknesses—for example, low predictive power at the species level,[90] narrow scope for the second because only a few parameters can be tested simultaneously, and too many nontestable assumptions. Below, we address a few of these missing pieces.

Cryptic indirect interactions

As seen in the previous sections, interspecific interactions strongly influence how climate change affects organisms—for example, competition, parasitism, and predation can impact behavior, individual fitness, geographic range, and ultimately the structure and dynamics of the community—but are too often overlooked.[120] Predators, for example, are both impacting and in turn impacted by their prey—see section "Cascading changes." But interspecific interactions also often produce indirect and unpredictable impacts.[203,204] In Scandinavia, for instance, the impact of increasing red fox populations on the regionally endangered Arctic fox is thought to be mediated by their shared microtine prey—that is, "exploitative competition."[205] In the High Arctic, the breeding success of some waders and wildfowl is sometimes assumed to mirror changes in lemming densities mediated by shared predators—that is, "apparent competition."[136,137,139,206]

Precipitation can also lead to unforeseen changes in interspecific interactions. Lecomte *et al.*[207] recently showed that water availability—and rainfall—affects the interaction between snow geese and the Arctic fox. Egg predation is reduced in years of high rainfall because incubating females walk shorter distances from their nest to drink and feed and therefore have a better chance to defend their nests from predators. Because climate change should affect precipitation regimes in the Arctic,[3,6] this may impact nesting success of geese by changing water availability for incubating females. However, the direction of the effect is difficult to predict because although total precipitation should increase, it may be concentrated in fewer, more intense rainfall events.

Even though they are strong enough to shape entire vertebrate communities, such indirect interactions are often difficult to discern without long-term, large-scale, and multispecies time series. Such data sets are rare for Arctic systems. An additional challenge is to untangle climate-driven changes from other anthropogenic impacts. In Canada, for example, the predation of large gulls *Larus argentatus* and *Larus marinus* on kittiwake is sometimes linked to the availability of capelin *Mallotus villosus*, and when capelin are scarce, kittiwakes suffer higher predation from large gulls. But capelin availability is driven both by climate and fishing pressure.[208]

The winter gap

Until recently, and with the exception of a few species, our ecological knowledge of most Arctic vertebrates has been extremely limited for the winter period. Yet this time, season is very important to the understanding of the dynamics of species. Exchanges of resources across ecosystem boundaries are known to impact food webs, especially in low-productive Arctic systems.[152] Because most Arctic vertebrates are to some extent migratory and make use of different ecosystems during their annual cycles, they contribute to such exchanges. In recent years, satellite tracking has allowed major progress in our knowledge of nonbreeding distribution and migration flyways of many large Arctic vertebrates, and geolocators are currently filling the gaps for small, site fidel birds species. Thanks to these advances, we recently learned that several of the terrestrial predators that reproduce in the Arctic tundra in summer make extensive use of the sea ice ecosystem during the winter period.[199,209,210] This unexpected behavior—at least for the snowy owl and the gyrfalcon—undoubtedly impacts the fitness of these individuals and their subsequent impact on terrestrial prey in the tundra ecosystem during the breeding season. Consequently, the current changes in sea ice regimes may also increase the vulnerability of these predators to climate warming.[152] Further technological innovations will hopefully allow us to gain more knowledge regarding the winter ecology of many others species in the near future.

"Subsidies"

Anthropogenic food resources and animal carcasses have often been an overlooked resource in the past—except for typical scavenging species—when assessing energy flows and population dynamics of Arctic vertebrate communities. Yet these alternate food resources are currently increasing in the Arctic, both as a consequence of climate change—for example, herbivores suffer higher mortality due to icing events[140]—and increasing human activities—for example, reindeer management[149] and fishing waste.[211] Increasing numbers of ungulate carcasses, for example, are assumed to benefit red foxes in Scandinavia, which can in turn negatively impact the regionally endangered Arctic fox.[149,212] Similarly, it is assumed that spatio-temporal differences in muskox mortality impacts the population dynamics of Arctic foxes and stoats *Mustela erminea* in northeast Greenland, which could in turn explain regional differences in the amplitude and cycle length of lemming fluctuations.[16] Because the availability of such alternate food is likely to increase in the future or at least change in abundance and distribution, it is important to consider them in future studies aiming to assess population dynamics of Arctic vertebrates.

Tipping points

Ecosystems do not always respond in smooth, linear, and reversible ways to pressures. As seen in previous sections, many Arctic species will first be able to buffer the effects of climate change to some extent—for example, thanks to their behavioral plasticity. But such abilities are limited, and "points of no return" might be reached when conditions change sufficiently. Complex systems in particular are often characterized by nonlinear responses to changes and can have thresholds that once passed lead to abrupt and irreversible changes.[213] Such thresholds, or "tipping points," are highly unpredictable in biological systems, because they can be the result of a number of changes in the population dynamics of several interacting species. Any taxon that disappears from an assemblage or even just declines due to climate change will alter ecosystem functioning. This can lead to sudden, steep, and irreversible changes in Arctic ecosystems where vertebrates have low functional redundancy (see also section "Cascading changes").

When tipping points result from a single dominant driving force, their consequences can be easier to foresee. For example, thermal windows have given limits (see section "Ecophysiological constraints"). The dynamics of a species can remain relatively stable as long as the climate fluctuates within these limits, but might suddenly and rapidly decline once these limits are crossed. In addition, species can only adjust their phenologies to a changing environment within given limits—for example, long-distance migrants need a minimum period of time to build up fat reserves on their wintering and staging grounds and then to travel all the way to their breeding grounds. Hence, their arrival in the Arctic cannot be brought forward indefinitely. Similarly, plasticity of diets has limits, which once crossed—that is, when the more profitable food items become too "expensive" to get—will rapidly impact fitness. Finally, some biotic tipping points can just mechanistically mirror abiotic tipping point. The reduction of sea ice, for instance, is believed to have geophysical tipping points according to some climate models,[214] which would induce correlated tipping points for pagophilic species.

Where are the Arctic vertebrates heading?

Due to the large variance in climate models, inherent complexity of natural systems, and the difficulties of implementing large-scale, long-term, and multi-trophic studies in the Arctic, predictions of climate-induced ecological impacts are difficult and remain highly speculative.[119,215] Even though predictions might be easier in the Arctic than elsewhere, the Arctic is nonetheless a complex system where climate conditions, physical environments, or food web structure can be markedly different. Different species or even individuals within a species can react differently to similar environmental modifications according to their location and their current life-history strategy. Predictions made for a spatially restricted area might not be applicable to other regions and could lead to wrong interpretations, predictions, and management plans.[216]

The best example of an Arctic vertebrate for which we have enough knowledge to reach this aim with relatively high confidence is the iconic polar bear. Several studies have tried to predict what may happen to polar bears and their habitat in the coming decades. Because the biology of polar bears in different subpopulations around the Arctic varies considerably[23] and vital demographic parameters vary in unpredictable ways with environmental change, the predicted outcomes must be considered cautiously. But given the future predictions regarding sea ice in most climate models, it is clear that the optimal habitat of polar bears will be severely reduced in the coming decades. This habitat where polar bears hunt and spend most of their time is mainly "not too old sea ice over not too deep waters, not too far from land."[216] In addition, multiyear sea ice is an important denning habitat north of Alaska, while in other areas sufficient snow fall in autumn is necessary to form snow drifts where maternity dens will not thaw or collapse. Sea ice around denning areas must be in place in time for the autumn to permit access and also remain long enough to provide hunting possibilities for ringed seals in the spring. A model based on the expected changes of polar bear habitat, in combination with demographic and environmental data, has predicted a strong decline in the number of polar bears, with total extirpation in areas currently holding two thirds of the world population by 2050.[217] However, Amstrup *et al.*[218] point out that slowing greenhouse gas emissions could reduce loss of polar bear habitat and stop their decline in large parts of the Arctic. The intimate connection between polar bear distribution and sea ice allows us to predict where polar bears will likely survive in the future. The decline in survival, reproduction, and population size of the western Hudson Bay population has given us some information on the maximum ice-free periods that polar bears can tolerate, though the precise length of this period will depend on the productivity in each area and access to alternative prey. High Arctic Canada and North Greenland are most likely the only regions where polar bear populations are likely to survive in the future if summer sea ice continues to decline as predicted by most sea ice models.

Although further work is needed before we can draw similarly precise predictive scenario for most other Arctic vertebrates, the current knowledge summarized in this review nonetheless indicates that climate change driven impacts on Arctic vertebrates follow some general trends.

- Because the annual cycle of most Arctic vertebrates is tightly linked with the cryosphere (i.e., snow cover for terrestrial species and sea ice for marine ones), they will have to adjust their phenologies to these new conditions on the short term, especially in order to avoid trophic mismatches.
- As vegetation belts and associated species move northward, Arctic vertebrates will eventually have to follow these range shifts in order to track their optimal living conditions. With a few exceptions, this will mechanistically lead to the reduction of their distribution ranges, particularly among High Arctic species.
- For the most pronounced High Arctic species, range shift may not be possible and selection for adaptive genotypes may be too slow, if possible at all. These species will face the highest risk of extinction during the coming decades.
- The speed of these range shifts will for the most part be species specific. Due to these specific differences, the structure of the vertebrate communities will necessarily change in the future. It is expected that climate warming will have stronger impacts on specialist feeders, which are less prone to respond to a rapid change in prey availability, distribution, and quality.
- Because many Arctic vertebrates have long generation times and sometimes highly variable dynamics, a climate-induced reduction in individual fitness may take significant periods of time to become perceptible in the population dynamics of Arctic vertebrates.
- New interspecific interactions will appear in Arctic communities—for example, between current competitors, predator–prey, or parasite–host arrangements of species—

with highly unpredictable cascading changes, including feedbacks, in the dynamics of populations and ultimately in the functioning of these modified ecosystems. Due to these complex relations, even small changes in climate can potentially lead to dramatic ecological changes—for example, top-down regulated populations could well become bottom-up regulated in some cases or vice versa.

- Many of these ecological changes that are likely to occur in the coming decades are nonlinear, both in slope and intensity, and some can reach tipping points. Small differences in environmental conditions can lead to very different impacts between otherwise similar populations or regions.
- Our current knowledge of the impacts of climate change on Arctic vertebrates is hampered by major knowledge gaps in the ecology of the species—for example, the winter period or on the nonbreeding grounds for migratory species—and does not generally consider other anthropogenic impacts—for example, see section "Subsidies."

How climate change will impact peoples through changes in the occurrence of Arctic vertebrates is a question that we did not implicitly address in this review. However, from most of our examples, one can ascertain that Arctic peoples depending on traditional lifestyles will also be strongly impacted. Impacts on reindeer distribution, especially in North America, will deeply impact the economy and the culture of some Arctic and sub-Arctic peoples. Changes in fish, marine mammal, and seabird availability, central species in the economy and culture of many Arctic peoples, will also impact their societies and will force them to adjust their relationship with their environment and perhaps adopt new environmental and management policies.[219,220] Most of the examples discussed in our review actually have cascading impacts on peoples.

Despite the uncertainties associated to the many empirical and theoretical works available in the literature forecasting the fate of Arctic vertebrates, their future is at risk, especially the Arctic endemic species, which are highly adapted to survive in harsh Arctic environments that have been restrictive to other species. What can be taken for granted at least is that they will undergo dramatic changes in the coming years and decades. But any precise, specific prediction of their future state, even at the circumpolar scale, remains highly speculative given our currently limited knowledge and the complexity of the processes involved. There is even some doubt about the widespread claim that Arctic species are more sensitive to climate change impacts than other species; some recent studies have suggested that they might actually be more resilient to climate change, based on the fact that they have already undergone more dramatic changes in the recent past than species from other biomes.[221] Given all the doubts and major knowledge gaps, a conservative approach to mitigation policies is warranted at this time.[218]

Acknowledgments

We are grateful to the Conseil Régional de Bourgogne and the French Polar Institute (IPEV) for its support to the project "1036-Interactions" (to OG, JM and LB) and to the latter for its support to the project "388-Adaclim" (to JF and DG). Climate change research in Norway has been funded largely by the Norwegian Research Council's IPY and NORKLIMA research programs. Some of the researches reviewed herein have also been supported by the Norwegian Polar Institute.

Conflicts of interest

The authors declare no conflicts of interest.

References

1. Walther, G.-R. *et al.* 2002. Ecological responses to recent climate change. *Nature*, **416:** 389–395.
2. Parmesan, C. 2006. Ecological and evolutionary responses to recent climate change. *Ann. Rev. Ecol. Evol. Syst.* **37:** 637–669.
3. IPCC. 2007. *Climate Change 2007. Synthesis report*. IPCC, Valencia.
4. Vibe, C. 1967. Arctic animals in relation to climatic fluctuations. *Meddelelser om Grønland* **170:** 1–227.
5. Barber, D.G. & J. Iacozza. 2004. Historical analysis of sea ice conditions in M'Clintock Channel and the Gulf of Boothia, Nunavut: implications for ringed seal and polar bear habitat. *Arctic* **57:** 1–14.
6. ACIA. 2005. *Arctic Climate Impact Assessment*. Cambridge University Press, Cambridge.
7. Overland, J.E. & M.Y. Wang. 2010. Large-scale atmospheric circulation changes are associated with the recent loss of Arctic sea ice. *Tellus Series a-Dyn. Meteorol. Oceanogr.* **62:** 1–9.

8. Polyak, L. *et al.* 2010. History of sea ice in the Arctic. *Quat. Sci. Rev.* **29:** 1757–1778.
9. Parmesan, C. & G. Yohe. 2003. A globally coherent fingerprint of climate change impacts across natural systems. *Nature* **421:** 37–42.
10. Root, T.L. *et al.* 2003. Fingerprints of global warming on wild animals and plants. *Nature* **421:** 57–60.
11. Ims, R.A., N.G. Yoccoz & S.T. Killengreen. 2011. Determinants of lemming outbreaks. *Proc. Nat. Acad. Sci. USA* **108:** 1970–1974.
12. Duchesne, D., G. Gauthier & D. Berteaux, 2011. Habitat selection, reproduction and predation of wintering lemmings in the Arctic. *Oecologia* **167:** 967–980.
13. Sittler, B. 1995. Response of stoat (*Mustela erminea*) to a fluctuating lemming (*Dicrostonyx groenlandicus*) population in North East Greenland: preliminary results from a long term study. *Ann. Zool. Fennici* **32:** 79–92.
14. Reid, D., F. Bilodeau, *et al.* Lemming winter habitat choice: a snow-fencing experiment. *Oecologia*, in press. doi:10.1007/s00442-011-2167-x
15. Kausrud, K.L. *et al.* 2008. Linking climate change to lemming cycles. *Nature* **456:** 93–97.
16. Gilg, O., B. Sittler & I. Hanski. 2009. Climate change and cyclic predator-prey population dynamics in the high-Arctic. *Global Change Biol.* **15:** 2634–2652.
17. Forchhammer, M. *et al.* 2005. Local-scale and short-term herbivore-plant spatial dynamics reflect influences of large-scale climate. *Ecology* **86:** 2644–2651.
18. Tyler, N.J.C., M.C. Forchhammer & N.A. Oritsland. 2008. Nonlinear effects of climate and density in the dynamics of a fluctuating population of reindeer. *Ecology* **89:** 1675–1686.
19. Meltofte, H. 1996. Are African wintering waders really forced south by competition from northerly wintering conspecifics? Benefits and constraints of northern versus southern wintering and breeding in waders. *Ardea* **84:** 31–44.
20. Lyon, B. & R. Montgomerie. 1995. Snow Bunting and McKay's Bunting (*Plectrophenax nivalis*, and *Plectrophenax hyperboreus*). In *Birds of North America* No. 198–199. A. Poole & F. Gill, eds. The Academy of Natural Sciences, Philadelphia, and the American Ornithologists' Union, Washington, D.C.
21. Meltofte, H. *et al.* 2007. Effects of climate variation on the breeding ecology of Arctic shorebirds. *Meddelelser om Grønland-Bioscience* **59:** 1–48.
22. Lindqvist, C. *et al.* 2010. Complete mitochondrial genome of a Pleistocene jawbone unveils the origin of polar bear. *Proc. Nat. Acad. Sci. USA* **107:** 5053–5057.
23. Amstrup, S.C. 2003. Polar Bear (*Ursus maritimus*). In *Wild Mammals of North America: Biology, Management and Conservation.* G.A. Feldhamer, B.C. Thopmson & J.A. Chapman, Eds.: 587–610. Johns Hopkins University Press, Baltimore, MD.
24. Freitas, C. *et al.* 2012. Importance of fast ice and glacier fronts for female polar bears and their cubs during spring in Svalbard, Norway. *Mar. Ecol. Prog. Ser.*, in press. doi:10.3354/meps09516
25. Mauritzen, M., A.E. Derocher & O. Wiig. 2001. Space-use strategies of female polar bears in a dynamic sea ice habitat. *Can. J. Zool.-Revue Canadienne De Zoologie* **79:** 1704–1713.
26. Wiig, O., J. Aars & E.W. Born. 2008. Effects of climate change on polar bears. *Sci. Prog.* **91:** 151–173.
27. Stirling, I. & A.E. Derocher. 1993. Possible impacts of climatic warming on polar bears. *Arctic* **46:** 240–245.
28. Derocher, A.E., N.J. Lunn & I. Stirling. 2004. Polar bears in a warming climate. *Integr. Comp. Biol.* **44:** 163–176.
29. Regehr, E.V. *et al.* 2007. Effects of earlier sea ice breakup on survival and population size of polar bears in western Hudson bay. *J. Wildlife Manag.* **71:** 2673–2683.
30. Regehr, E.V. *et al.* 2010. Survival and breeding of polar bears in the southern Beaufort Sea in relation to sea ice. *J. Animal Ecol.* **79:** 117–127.
31. Fischbach, A., S. Amstrup & D. Douglas. 2007. Landward and eastward shift of Alaskan polar bear denning associated with recent sea ice changes. *Polar Biol.* **30:** 1395–1405.
32. Amstrup, S.C. *et al.* 2006. Recent observations of intraspecific predation and cannibalism among polar bears in the southern Beaufort Sea. *Polar Biol.* **V29:** 997–1002.
33. Monnett, C. & J. Gleason. 2006. Observations of mortality associated with extended open-water swimming by polar bears in the Alaskan Beaufort Sea. *Polar Biol.* **29:** 681–687.
34. Durner, G.M. *et al.* 2011. Consequences of long-distance swimming and travel over deep-water pack ice for a female polar bear during a year of extreme sea ice retreat. *Polar Biol.* **34:** 975–984.
35. Jenssen, B.M. 2006. Endocrine-disrupting chemicals and climate change: a worst-case combination for arctic marine mammals and seabirds? *Environ. Health Perspect.* **114:** 76–80.
36. MacGarvin, M. & M.P. Simmonds. 1996. Whales and climate change. In *The Conservation of Whales and Dolphins, Science Practice.* M.P. Simmonds & J.D. Hutchinson, Eds.: 321–332. John Wiley and Sons. Chichester, UK.
37. Tynan, C.T. & D.P. DeMaster. 1997. Observations and predictions of Arctic climatic change: potential effects on marine mammals. *Arctic* **50:** 308–322.
38. Kelly, B.P. 2001. Climate change and ice breeding pinnipeds. In *"Fingerprints" of Climate Change*, G.-R. Walther, C.A. Burga & P.J. Edwards, Eds.: 43–55. Kluwer Academic/Plenum Publications, New York.
39. Johnston, D.W. *et al.* 2005. Variation in sea ice cover on the east coast of Canada from 1969 to 2002: climate variability and implications for harp and hooded seals. *Clim. Res.* **29:** 209–222.
40. Simmonds, M.P. & S.J. Isaac. 2007. The impacts of climate change on marine mammals: early signs of significant problems. *Oryx* **41:** 19–26.
41. Kovacs, K. *et al.* 2011. Impacts of changing sea-ice conditions on Arctic marine mammals. *Mar. Biodivers.* **41:** 181–194.
42. Kovacs, K.M. & C. Lydersen. 2008. Climate change impacts on seals and whales in the North Atlantic Arctic and adjacent shelf seas. *Sci. Prog.* **91:** 117–150.
43. Laidre, K. *et al.* 2008. Quantifying the sensitivity of Arctic marine mammals to climate-induced habitat change. *Ecol. Appl.* **18**(Suppl.): S97–S125.

44. Moore, S.E. & H.P. Huntington. 2008. Arctic marine mammals and climate change: impacts and resilience. *Ecol. Appl.* **18:** S157–S165.
45. Ragen, T.J., H.P. Huntington & G.K. Hovelsrud. 2008. Conservation of Arctic marine mammals faced with climate change. *Ecol. Appl.* **18:** S166–S174.
46. Kern, S., L. Kaleschke & G. Spreen. 2010. Climatology of the Nordic (Irminger, Greenland, Barents, Kara and White/Pechora) Seas ice cover based on 85 GHz satellite microwave radiometry: 1992–2008. *Tellus Series a-Dyn. Meteorol. Oceanogr.* **62:** 411–434.
47. White, J.W.C. *et al.* 2010. Past rates of climate change in the Arctic. *Quat. Sci. Rev.* **29:** 1716–1727.
48. Johannessen, O.M. & M.W. Miles. 2010. Critical vulnerabilities of marine and sea ice-based ecosystems in the high Arctic. *Region. Environ. Change* **11:** S239–S248.
49. Heide-Jorgensen, M.P. & K. Laidre. 2004. Declining extent of open-water refugia for top predators in Baffin Bay and adjacent waters. *Ambio* **33:** 487–494.
50. Stirling, I. 1997. The importance of polynyas, ice edges, and leads to marine mammals and birds. *J. Mar. Syst.* **10:** 9–21.
51. Harington, C.R. 2008. The evolution of Arctic marine mammals. *Ecol. Appl.* **18:** S23–S40.
52. Williams, T.M.W.T.M., S.R. Noren & M. Glenn. 2011. Extreme physiological adaptations as predictors of climate-change sensitivity in the narwhal, Monodon monoceros. *Mar. Mammal Sci.* **27:** 334–349.
53. Fort, J., W.P. Porter & D. Gremillet. 2009. Thermodynamic modelling predicts energetic bottleneck for seabirds wintering in the northwest Atlantic. *J. Exp. Biol.* **212:** 2483–2490.
54. Gaston, A.J. & I.L. Jones, Eds. 1998. *Bird Families of the World: the Auks.* Oxford University Press, New York.
55. Hoegh-Guldberg, O. & J.F. Bruno. 2010. The impact of climate change on the World's marine ecosystems. *Science* **328:** 1523–1528.
56. Pörtner, H.O. & A.P. Farrell. 2008. Physiology and climate change. *Science* **322:** 690–692.
57. Farrell, A.P. *et al.* 2008. Pacific Salmon in hot water: applying aerobic scope models and biotelemetry to predict the success of spawning migrations. *Physiol. Biochem. Zool.* **81:** 697–708.
58. Perry, A.L. *et al.* 2005. Climate change and distribution shifts in marine fishes. *Science* **308:** 1912–1915.
59. Helaouet, P. & G. Beaugrand. 2007. Macroecology of Calanus finmarchicus and C-helgolandicus in the North Atlantic Ocean and adjacent seas. *Mar. Ecol.-Prog. Ser.* **345:** 147–165.
60. Humphries, M.M., J. Umbanhowar & K.S. McCann. 2004. Bioenergetic prediction of climate change impacts on northern mammals. *Integr. Comp. Biol.* **44:** 152–162.
61. Boonstra, R. 2004. Coping with changing northern environments: the role of the stress axis in birds and mammals. *Integr. Comp. Biol.* **44:** 95–108.
62. Bustnes, J.O., K.O. Kristiansen & M. Helberg. 2007. Immune status, carotenoid coloration, and wing feather growth in relation to organochlorine pollutants in great black-backed gulls. *Arch. Environ. Contam. Toxicol.* **53:** 96–102.
63. Bustnes, J.O., G.W. Gabrielsen & J. Verreault. 2010. Climate variability and temporal trends of persistent organic pollutants in the Arctic: a study of Glaucous Gulls. *Environ. Sci. Technol.* **44:** 3155–3161.
64. Grémillet, D. & A. Charmantier. 2010. Shifts in phenotypic plasticity constrain the value of seabirds as ecological indicators of marine ecosystems. *Ecol. Appl.* **20:** 1498–1503.
65. Parmesan, C. 2007. Influences of species, latitudes and methodologies on estimates of phenological response to global warming. *Global Change Biol.* **13:** 1860–1872.
66. Høye, T.T. *et al.* 2007. Rapid advancement of spring in the High Arctic. *Curr. Biol.* **17:** R449–R451.
67. Carey, C. 2009. The impacts of climate change on the annual cycles of birds. *Philos. Trans. R. Soc. B-Biol. Sci.* **364:** 3321–3330.
68. Durant, J.M. *et al.* 2007. Climate and the match or mismatch between predator requirements and resource availability. *Clim. Res.* **33:** 271–283.
69. Post, E. & M.C. Forchhammer. 2008. Climate change reduces reproductive success of an Arctic herbivore through trophic mismatch. *Philos. Trans. R. Soc. B-Biol. Sci.* **363:** 2369–2375.
70. Miller-Rushing, A.J. *et al.* 2010. The effects of phenological mismatches on demography. *Philos. Trans. R. Soc. B-Biol. Sci.* **365:** 3177–3186.
71. Lepage, D., G. Gauthier & A. Reed. 1998. Seasonal variation in growth of greater snow goose goslings: the role of food supply. *Oecologia* **114:** 226–235.
72. Dickey, M.-E., G. Gauthier & M.-C. Cadieux. 2008. Climatic effects on the breeding phenology and reproductive success of an arctic-nesting goose species. *Global Change Biol.* **14:** 1973–1985.
73. Post, E. *et al.* 2008. Warming, plant phenology and the spatial dimension of trophic mismatch for large herbivores. *Proc. R. Soc. B-Biol. Sci.* **275:** 2005–2013.
74. Lu, W.Q. *et al.* 2010. A Circadian Clock is not required in an Arctic Mammal. *Curr. Biol.* **20:** 533–537.
75. Meltofte, H. *et al.* 2007. Differences in food abundance cause inter-annual variation in the breeding phenology of High Arctic waders. *Polar Biol.* **V30:** 601–606.
76. Tulp, I. & H. Schekkerman. 2008. Has prey availability for arctic birds advanced with climate change? Hindcasting the abundance of tundra arthropods using weather and seasonal variation. *Arctic* **61:** 48–60.
77. Vatka, E., M. Orell & S. Rytkönen. 2011. Warming climate advances breeding and improves synchrony of food demand and food availability in a boreal passerine. *Global Change Biol.* **17:** 3002–3009.
78. Falk, K. & S. Møller. 1997. Breeding ecology of the fulmar *Fulmarus glacialis* and the Kittiwake *Rissa tridactyla* in high-arctic northeastern Greenland, 1993. *Ibis.* **139:** 270–281.
79. Mallory, M.L. & M.R. Forbes. 2007. Does sea ice constrain the breeding schedules of high Arctic Northern Fulmars? *Condor* **109**: 894–906.
80. Laidre, K.L. *et al.* 2008. Latitudinal gradients in sea ice and primary production determine Arctic seabird colony size in Greenland. *Proc. R. Soc. B: Biol. Sci.* **275:** 2695–2702.
81. Reygondeau, G. & G. Beaugrand. 2011. Future climate-driven shifts in distribution of Calanus finmarchicus. *Global Change Biol.* **17:** 756–766.

82. Gilg, O. *et al.* 2010. Post-breeding movements of the northeast Atlantic ivory gull *Pagophila eburnea* populations. *J. Avian Biol.* **41:** 532–542.
83. Gaston, A.J., H.G. Gilchrist & J.M. Hipfner. 2005. Climate change, ice conditions and reproduction in an Arctic nesting marine bird: Brunnich's guillemot (Uria lomvia L.). *J. Anim. Ecol.* **74:** 832–841.
84. Rockwell, R.F., L.J. Gormezano & D.N. Koons. 2011. Trophic matches and mismatches: can polar bears reduce the abundance of nesting snow geese in western Hudson Bay? *Oikos* **120:** 696–709.
85. Drent, R.H. & J. Prop. 2008. Barnacle goose *Branta leucopsis* survey on Nordenskiöldkysten, west Spitsbergen 1975–2007: breeding in relation to carrying capacity and predator impact. *Circumpolar Stud.* **4:** 59–83.
86. Moe, B. *et al.* 2009. Climate change and phenological responses of two seabird species breeding in the high-Arctic. *Mar. Ecol.-Prog. Ser.* **393:** 235–246.
87. Taylor, S.G. 2008. Climate warming causes phenological shift in Pink Salmon, Oncorhynchus gorbuscha, behavior at Auke Creek, Alaska. *Global Change Biol.* **14:** 229–235.
88. Malcolm, J.R. *et al.* 2006. Global warming and extinctions of endemic species from biodiversity hotspots. *Conserv. Biol.* **20:** 538–548.
89. Dingemanse, N.J. *et al.* 2010. Behavioural reaction norms: animal personality meets individual plasticity. *Trends Ecol. Evol.* **25:** 81–89.
90. Chen, I.-C. *et al.* 2011. Rapid range shifts of species associated with high levels of climate warming. *Science* **333:** 1024–1026.
91. Huntley, B. *et al.* 2006. Potential impacts of climatic change upon geographical distributions of birds. *Ibis* **148:** 8–28.
92. Beaumont, L.J. *et al.* 2007. Where will species go? Incorporating new advances in climate modelling into projections of species distributions. *Global Change Biol.* **13:** 1368–1385.
93. Levinsky, I. *et al.* 2007. Potential impacts of climate change on the distributions and diversity patterns of European mammals. *Biodivers. Conserv.* **16:** 3803–3816.
94. Wisz, M.S. *et al.* 2008. Where might the western Svalbard tundra be vulnerable to pink-footed goose (*Anser brachyrhynchus*) population expansion? Clues from species distribution models. *Diversity Distrib.* **14:** 26–37.
95. Speed, J.D.M. *et al.* 2009. Predicting habitat utilization and extent of ecosystem disturbance by an increasing Herbivore population. *Ecosystems* **12:** 349–359.
96. Jensen, R.A. *et al.* 2008. Prediction of the distribution of Arctic-nesting pink-footed geese under a warmer climate scenario. *Global Change Biol.* **14:** 1–10.
97. Gilg, O. *et al.* 2009. Status of the endangered Ivory Gull, *Pagophila eburnea*, in Greenland. *Polar Biol.* **32:** 1275–1286.
98. Schwartz, M.W. *et al.* 2006. Predicting extinctions as a result of climate change. *Ecology* **87:** 1611–1615.
99. Thomas, C.D. *et al.* 2004. Extinction risk from climate change. *Nature* **427:** 145–148.
100. Boertmann, D. 1994. A annotated checklist to the birds of Greenland. *Meddelelser om Grønland-Bioscience* **38:** 1–63.
101. Boertmann, D. & R.D. Nielsen. 2010. Geese, seabirds and mammals in North and Northeast Greenland. Aerial surveys in summer 2009. *NERI. Technical Report no. 773.* NERI. Copenhagen.
102. Labansen, A.L. *et al.* 2010. Status of the black-legged kittiwake (Rissa tridactyla) breeding population in Greenland, 2008. *Polar Res.* **29:** 391–403.
103. Gilg, O. *et al.* 2005. *Ecopolaris – Tara 5 expedition to NE Greenland 2004.* GREA. Francheville.
104. Boertmann, D. 2008. The lesser black-backed gull, *Larus fuscus*, in Greenland. *Arctic* **61:** 129–133.
105. Sandvik, H., T.I.M. Coulson & B.-E. Saether. 2008. A latitudinal gradient in climate effects on seabird demography: results from interspecific analyses. *Global Change Biol.* **14:** 1–11.
106. Chaulk, K.G., G.J. Robertson & W.A. Montevecchi. 2007. Landscape features and sea ice influence nesting common eider abundance and dispersion. *Can. J. Zool.* **85:** 301–309.
107. Rouvinen-Watt, K. 2004. Biological premises for management of an Arctic fox (Alopex lagopus) population – The demise and future of the hare-footed fox on Jan Mayen. In *Jan Mayen Island in Scientific Focus*, S. Skreslet, Ed.: 207–218. Springer. Dordrecht.
108. Geffen, E.L.I. *et al.* 2007. Sea ice occurrence predicts genetic isolation in the Arctic fox. *Mol. Ecol.* **16:** 4241–4255.
109. Corry-Crowe, G. 2008. Climate change and the molecular ecology of Arctic marine mammals. *Ecol. Appl.* **18:** S56–S76.
110. Heide-Jørgensen, M.P. *et al.* 2011. The Northwest Passage opens for bowhead whales. *Biol. Lett.* in press. doi:10.1098/rsbl.2011.0731
111. Beaugrand, G., M. Edwards & L. Legendre. 2010. Marine biodiversity, ecosystem functioning, and carbon cycles. *Proc. Nat. Acad. Sci. USA* **107:** 10120–10124.
112. Karnovsky, N. *et al.* 2010. Foraging distributions of little auks *Alle alle* across the Greenland Sea: implications of present and future Arctic climate change. *Mar. Ecol.-Prog. Ser.* **415:** 283–293.
113. Karnovsky, N.J. *et al.* 2003. Foraging behavior of little auks in a heterogeneous environment. *Mar. Ecol.-Prog. Ser.* **253:** 289–303.
114. Barrett, R.T. & Y.V. Krasnov. 1996. Recent responses to changes in stocks of prey species by seabirds breeding in the southern Barents Sea. *ICES J. Mar. Sci.* **53:** 713–722.
115. van de Kam, J. *et al.*, eds. 2004. *Shorebirds. An Illustrated Behavioural Ecology*. KNNV Publishers. Utrecht.
116. Wardle, D.A. *et al.* 2011. Terrestrial ecosystem responses to species gains and losses. *Science*, **332:** 1273–1277.
117. Kaschner, K. *et al.* 2011. Current and future patterns of global marine mammal biodiversity. *PLoS One* **6:** E19653. doi:10.1126/science.1183010
118. UNEP, ed. 1995. *Global Biodiversity Assessment*. Cambridge University Press. Cambridge.
119. Wookey, P.A. *et al.* 2009. Ecosystem feedbacks and cascade processes: understanding their role in the responses of Arctic and alpine ecosystems to environmental change. *Global Change Biol.* **15:** 1153–1172.
120. Gilman, S.E. *et al.* 2010. A framework for community interactions under climate change. *Trends Ecol. Evol.* **25:** 325–331.
121. Van der Putten, W.H. *et al.* 2004. Trophic interactions in a changing world. *Basic Appl. Ecol.* **5:** 487–494.

122. Sandvik, H. *et al.* 2005. The effect of climate on adult survival in five species of North Atlantic seabirds. *J. Anim. Ecol.* **74:** 817–831.
123. Post, E. *et al.* 2009. Ecological dynamics across the Arctic associated with recent climate change. *Science*, **325:** 1355–1358.
124. Callaghan, T.V. *et al.* 2005. Arctic tundra and polar desert ecosystems. In *Arctic Climate Impact Assessment, ACIA*: 243–352. Cambridge University Press. Cambridge.
125. Van der Putten, W.H., M. Macel & M.E. Visser. 2010. Predicting species distribution and abundance responses to climate change: why it is essential to include biotic interactions across trophic levels. *Philos. Trans. R. Soc. B-Biol. Sci.* **365:** 2025–2034.
126. Gilg, O. *et al.* 2006. Functional and numerical responses of four lemming predators in high arctic Greenland. *Oikos* **113:** 196–213.
127. Gauthier, G. *et al.* 2004. Trophic interactions in a High Arctic snow goose colony. *Integr. Comp. Biol.* **44:** 119–129.
128. Gilg, O., I. Hanski & B. Sittler. 2003. Cyclic dynamics in a simple vertebrate predator-prey community. *Science* **302:** 866–868.
129. Schmidt, N.M. *et al.* 2008. Vertebrate predator-prey interactions in a seasonal environment. *Adv. Ecol. Res.* **40:** 345–370.
130. Ims, R.A., J.-A. Henden & S.T. Killengreen. 2008. Collapsing population cycles. *Trends Ecol. Evol.* **23:** 79–86.
131. Berg, T.B. *et al.* 2008. High-arctic plant-herbivore interactions under climate influence. *Adv. Ecol. Res.* **40:** 275–298.
132. Prost, S. *et al.* 2010. Influence of climate warming on Arctic mammals? New insights from ancient DNA studies of the collared Lemming *Dicrostonyx torquatus*. *PLoS One* **5:** e10447.
133. Hansson, L. & H. Henttonen. 1985. Gradients in density vriations of small rodents: the importance of latitude and snow cover. *Oecologia* **67:** 394–402.
134. McKinnon, L. *et al.* 2010. Lower predation risk for migratory birds at high latitudes. *Science* **327:** 326–327. doi:10.1126/science.1183010
135. Summers, R.W., L.G. Underhill & E.E. Syroechkovski. 1998. The breeding productivity of dark-bellied brent geese and curlew sandpipers in relation to changes in the numbers of arctic foxes and lemmings on the Taimyr Peninsula, Siberia. *Ecography* **21:** 573–580.
136. Bêty, J. *et al.* 2001. Are goose nesting success and lemming cycles linked? Interplay between nest density and predators. *Oikos* **93:** 388–400.
137. Bêty, J. *et al.* 2002. Shared predation and indirect trophic interactions: lemming cycles and arctic-nesting geese. *J. Anim. Ecol.* **71:** 88–98.
138. Blomqvist, S. *et al.* 2002. Indirect effects of lemming cycles on sandpiper dynamics: 50 years of counts from southern Sweden. *Oecologia* **133:** 146–158.
139. Gilg, O. & N.G. Yoccoz. 2010. Explaining bird migration. *Science* **327:** 276–277.
140. Ims, R.A. & E. Fuglei. 2005. Trophic interaction cycles in Tundra ecosystems and the impact of climate change. *BioScience* **55:** 311–322.
141. Ravolainen, V.T. *et al.* 2011. Rapid, landscape scale responses in riparian tundra vegetation to exclusion of small and large mammalian herbivores. *Basic Appl. Ecol.* **12:** 643–653.
142. Sturm, M., C. Racine & K. Tape. 2001. Climate change – Increasing shrub abundance in the Arctic. *Nature* **411:** 546–547.
143. Chapin, F.S. *et al.* 2005. Role of land-surface changes in Arctic summer warming. *Science* **310:** 657–660.
144. Killengreen, S.T. *et al.* How ecological neighbourhoods influence the structure of the scavenger guild in low arctic tundra. *Divers. Distrib.* in press. doi:10.1111/j.1472-4642.2011.00861.x
145. Tannerfeldt, M., B. Elmhagen & A. Angerbjörn. 2002. Exclusion by interference competition? The relationship between red and arctic foxes. *Oecologia* **132:** 213–220.
146. Rodnikova, A. *et al.* 2011. Red fox takeover of arctic fox breeding den: an observation from Yamal Peninsula, Russia. *Polar Biology* **34:** 1609–1614.
147. Hersteinsson, P. & D.W. Macdonald. 1992. Interspecific competition and the geographical distribution of red and arctic foxes Vulpes vulpes and Alopex lagopus. *Oikos* **64:** 505–515.
148. Killengreen, S.T. *et al.* 2007. Structural characteristics of a low Arctic tundra ecosystem and the retreat of the Arctic fox. *Biol. Conserv.* **135:** 459–472.
149. Killengreen, S.T. *et al.* 2011. The importance of marine vs. human-induced subsidies in the maintenance of an expanding mesocarnivore in the arctic tundra. *J. Animal Ecol.* **80:** 1049–1060.
150. Morrissette, M. *et al.* 2010. Climate, trophic interactions, density dependence and carry-over effects on the population productivity of a migratory Arctic herbivorous bird. *Oikos* **119:** 1181–1191.
151. Reed, E.T., G. Gauthier & J.-F. Giroux. 2004. Effects of spring conditions on breeding propensity of greater snow goose females. *Animal Biodivers. Conserv.* **27:** 35–46.
152. Gauthier, G. *et al.* 2011. The tundra food web of Bylot Island in a changing climate and the role of exchanges between ecosystems. *Ecoscience* **18:** 223–235.
153. Jefferies, R.L., R.F. Rockwell & K.F. Abraham. 2004. Agricultural food subsidies, migratory connectivity and large-scale disturbance in arctic coastal systems: a case study. *Integr. Comp. Biol.* **44:** 130–139.
154. Gauthier, G. *et al.* 2005. Interactions between land use, habitat use and population increase in greater snow geese: what are the consequences for natural wetlands? *Global Change Biol.* **11:** 856–868.
155. Post, E. & C. Pedersen. 2008. Opposing plant community responses to warming with and without herbivores. *Proc. Nat. Acad. Sci. USA* **105:** 12353–12358.
156. Cahoon, S.M.P. *et al.* 2012. Large herbivores limit CO_2 uptake and suppress carbon cycle responses to warming in West Greenland. *Global Change Biol.* **18:** 469–479. doi:10.1111/j.1365-2486.2011.02528.x
157. Thompson, P.M. & J.C. Ollason. 2001. Lagged effects of ocean climate change on fulmar population dynamics. *Nature* **413:** 417–420.
158. Egevang, C. *et al.* 2010. Tracking of Arctic terns *Sterna paradisaea* reveals longest animal migration. *Proc. Nat. Acad. Sci. USA* **107:** 2078–2081.

159. Furness, R.W. 1987. *The Skuas*. Calton. Poyser.
160. Sittler, B., A. Aebischer & O. Gilg. 2011. Post-breeding migration of four Long-tailed Skuas (*Stercorarius longicaudus*) from North and East Greenland to West Africa. *J. Ornithol.* **152:** 375–381.
161. Burek, K.A., F.M.D. Gulland & T.M. Hara. 2008. Effects of climate change on Arctic marine mammal health. *Ecol. Appl.* **18:** S126–S134.
162. Kutz, S.J., A.P. Dobson & E.P. Hoberg. 2009. Where are the parasites? *Science* **326:** 1187–1188.
163. Kutz, S.J. *et al.* 2005. Global warming is changing the dynamics of Arctic host-parasite systems. *Proc. R. Soc. B-Biol. Sci.* **272:** 2571–2576.
164. Harvell, C.D. *et al.* 2002. Climate warming and disease risks for terrestrial and marine biota. *Science* **296:** 2158–2162.
165. Polley, L. & R.C.A. Thompson. 2009. Parasite zoonoses and climate change: molecular tools for tracking shifting boundaries. *Trends Parasitol.* **25:** 285–291.
166. Hudson, P.J., A.P. Dobson & D. Newborn. 1998. Prevention of population cycles by parasite removal. *Science* **282:** 2256–2258.
167. Hudson, P.J., A.P. Dobson & D. Newborn. 1992. Do parasites make prey vulnerable to predation – Red grouse and parasites. *J. Animal Ecol.* **61:** 681–692.
168. Albon, S.D. *et al.* 2002. The role of parasites in the dynamics of a reindeer population. *Proc. R. Soc. Lond. Ser. B-Biol. Sci.* **269:** 1625–1632.
169. Jenkins, E.J., E.P. Hoberg & L. Polley. 2005. Development and pathogenesis of Parelaphostrongylus odocoilei (Nematoda : Protostrongylidae) in experimentally infected thinhorn sheep (Ovis dalli). *J. Wildlife Dis.* **41:** 669–682.
170. Gaston, A.J., J.M. Hipfner & D. Campbell. 2002. Heat and mosquitoes cause breeding failures and adult mortality in an Arctic-nesting seabird. *Ibis* **144:** 185–191.
171. Davidson, R. *et al.* 2011. Arctic parasitology: why should we care? *Trends Parasitol.* **27:** 238–244.
172. Hueffer, K., T.M. O'Hara & E.H. Follmann. 2011. Adaptation of mammalian host-pathogen interactions in a changing arctic environment. *Acta Veterinaria Scandinavica* **53:** 17. doi:10.1186/1751-0147-53-17
173. Kutz, S.J. *et al.* 2009. The Arctic as a model for anticipating, preventing, and mitigating climate change impacts on host-parasite interactions. *Vet. Parasitol.* **163:** 217–228.
174. Ytrehus, B. *et al.* 2008. Fatal pneumonia epizootic in musk ox (*Ovibos moschatus*) in a period of extraordinary weather conditions. *Ecohealth* **5:** 213–223.
175. Laaksonen, S. *et al.* 2010. Climate change promotes the emergence of serious disease outbreaks of Filarioid Nematodes. *Ecohealth* **7:** 7–13.
176. Jensen, S.K. *et al.* 2010. The prevalence of *Toxoplasma gondii* in polar bears and their marine mammal prey: evidence for a marine transmission pathway? *Polar Biol.* **33:** 599–606.
177. Hoberg, E.P. *et al.* 2008. Integrated approaches and empirical models for investigation of parasitic diseases in northern wildlife. *Emerg. Infect. Dis.* **14:** 10–17.
178. Kutz, S.J. *et al.* 2004. "Emerging" parasitic infections in arctic ungulates. *Integr. Comp. Biol.* **44:** 109–118.
179. Ball, M.C., M.W. Lankester & S.P. Mahoney. 2001. Factors affecting the distribution and transmission of Elaphostrongylus rangiferi (Protostrongylidae) in caribou (Rangifer tarandus caribou) of Newfoundland, Canada. *Can. J. Zool.* **79:** 1265–1277.
180. Lavergne, S. *et al.* 2010. Biodiversity and climate change: integrating evolutionary and ecological responses of species and communities. *Annual Review of Ecology, Evolution, and Systematics* **41:** 321–350. doi:10.1146/annurev-ecolsys-102209-144628
181. Skelly, D.K. *et al.* 2007. Evolutionary responses to climate change. *Conserv. Biol.* **21:** 1353–1355.
182. Hoffmann, A.A. & C.M. Sgro. 2011. Climate change and evolutionary adaptation. *Nature* **470:** 479–485.
183. Berteaux, D. *et al.* 2004. Keeping pace with fast climate change: can Arctic life count on evolution? *Integr. Comp. Biol.* **44:** 140–151.
184. Bradshaw, W.E. & C.M. Holzapfel. 2008. Genetic response to rapid climate change: it's seasonal timing that matters. *Mol. Ecol.* **17:** 157–166.
185. Pigliucci, M. 2001. *Phenotypic Plasticity; Beyond Nature and Nurture. Syntheses in Ecology and Evolution.* John Hopkins University Press, Baltimore, MD.
186. Gienapp, P. *et al.* 2008. Climate change and evolution: disentangling environmental and genetic responses. *Mol. Ecol.* **17:** 167–178.
187. Bradshaw, W.E. & C.M. Holzapfel. 2006. Climate change – Evolutionary response to rapid climate change. *Science* **312:** 1477–1478.
188. Barton, N. & L. Partridge. 2000. Limits to natural selection. *BioEssays* **22:** 1075–1084.
189. Bürger, R. & M. Lynch. 1995. Evolution and extinction in a changing environment: a quantitative-genetic analysis. *Evolution* **49:** 151–163.
190. Falconer, D.S. & T.F.C. Mackay. 1996. *Introduction to Quantitative Genetics.* 4th ed. Pearson Education. Essex, UK.
191. Campos, P.F. *et al.* 2010. Ancient DNA analyses exclude humans as the driving force behind late Pleistocene musk ox (Ovibos moschatus) population dynamics. *Proc. Nat. Acad. Sci. USA* **107:** 5675–5680.
192. Leonard, J.A. *et al.* 2007. Megafaunal extinctions and the disappearance of a specialized wolf ecomorph. *Curr. Biol.* **17:** 1146–1150.
193. Reale, D. *et al.* 2003. Genetic and plastic responses of a northern mammal to climate change. *Proc. R. Soc. Lond. Ser. B-Biol. Sci.* **270:** 591–596.
194. Bromham, L., A. Rambaut & P. Harvey. 1996. Determinants of rate variation in mammalian DNA sequence evolution. *J. Mol. Evol.* **43:** 610–621.
195. Hanski, I. 1999. *Metapopulation Ecology*. Oxford University Press. Oxford.
196. Opdam, P. & D. Wascher. 2004. Climate change meets habitat fragmentation: linking landscape and biogeographical scale levels in research and conservation. *Biol. Conserv.* **117:** 285–297.
197. Hanski, I. & O. Ovaskainen. 2000. The metapopulation capacity of a fragmented landscape. *Nature* **404:** 755–758.
198. Noren, K. *et al.* 2011. Pulses of movement across the sea ice: population connectivity and temporal genetic structure in the arctic fox. *Oecologia* **166:** 973–984.

199. Tarroux, A., D. Berteaux & J. Bêty. 2010. Northern nomads: ability for extensive movements in adult arctic foxes. *Polar Biol.* **33:** 1021–1026.
200. Noren, K. *et al.* 2009. Farmed arctic foxes on the Fennoscandian mountain tundra: implications for conservation. *Animal Conserv.* **12:** 434–444.
201. Noren, K., A. Angerbjorn & P. Hersteinsson. 2009. Population structure in an isolated Arctic fox, Vulpes lagopus, population: the impact of geographical barriers. *Biol. J. Linnean Soc.* **97:** 18–26.
202. Dalen, L. *et al.* 2007. Ancient DNA reveals lack of postglacial habitat tracking in the arctic fox. *Proc. Nat. Acad. Sci. USA* **104:** 6726–6729.
203. Ruscoe, W.A. *et al.* 2011. Unexpected consequences of control: competitive vs. predator release in a four-species assemblage of invasive mammals. *Ecol. Lett.* **14:** 1035–1042.
204. Roemer, G.W., C.J. Donlan & F. Courchamp. 2002. Golden eagels, feral pigs, and insular carnivores: how exotic species turn native predators into prey. *Proc. Nat. Acad. Sci. USA* **99:** 791–796.
205. Henden, J.-A. *et al.* 2010. Strength of asymmetric competition between predators in food webs ruled by fluctuating prey: the case of foxes in tundra. *Oikos* **119:** 27–34.
206. Sittler, B., O. Gilg & T.B. Berg. 2000. Low abundance of King eider nests during low lemming years in Northeast Greenland. *Arctic* **53:** 53–60.
207. Lecomte, N., G. Gauthier & J.F. Giroux. 2009. A link between water availability and nesting success mediated by predator-prey interactions in the Arctic. *Ecology* **90:** 465–475.
208. Massaro, M. *et al.* 2000. Delayed capelin (*Mallotus villosus*) availability influences predatory behaviour of large gulls on black-legged kittiwakes (*Rissa tridactyla*), causing a reduction in kittiwak breeding success. *Can. J. Zool.* **78:** 1588–1596.
209. Burnham, K.K. & I.A.N. Newton. 2011. Seasonal movements of Gyrfalcons Falco rusticolus include extensive periods at sea. *Ibis* **153:** 468–484.
210. Therrien, J.-F., G. Gauthier & J. Bêty. 2011. An avian terrestrial predator of the Arctic relies on the marine ecosystem during winter. *J. Avian Biol.* **42:** 363–369.
211. Votier, S.C. *et al.* 2004. Changes in fisheries discard rates and seabird communities. *Nature* **427:** 727–730.
212. Selås, V. & J.O. Vik. 2006. Possible impact of snow depth and ungulate carcasses on red fox (Vulpes vulpes) populations in Norway, 1897–1976. *J. Zool.* **269:** 299–308.
213. Duarte, C. 2011. Not so smooth: thresholds and abrupt changes in ecosystems under pressure. In *The Arctic in the Earth System Perspective: The Role of Tipping Points. Book of Abstracts.* 13. University of Tromso. Tromso.
214. Winton, M. 2006. Does the sea ice have a tipping point? *Geophys. Res. Lett.* **33:** 1–5.
215. Mustin, K., W.J. Sutherland & J.A. Gill. 2007. The complexity of predicting climate-induced ecological impacts. *Clim. Res.* **35:** 165–175.
216. Durner, G.M. *et al.* 2009. Predicting 21st-century polar bear habitat distribution from global climate models. *Ecol. Monogr.* **79:** 25–58.
217. Amstrup, S.C., B.G. Marcot & D.C. Dougles. 2008. A Bayesian network modeling approach to forecasting the 21st century worldwide status of polar bears, in Arctic Sea Ice Decline: observations, projections, mechanisms, and implications. *Geophys. Monogr. Ser.* **180:** 213–268.
218. Amstrup, S.C. *et al.* 2010. Greenhouse gas mitigation can reduce sea-ice loss and increase polar bear persistence. *Nature* **468:** 955–958.
219. Huntington, H.P. & S.E. Moore. 2008. Assessing the impacts of climate change on Arctic marine mammals. *Ecol. Appl.* **18:** S1–S2.
220. Metcalf, V. & M. Robards. 2008. Sustaining a healthy human-walrus relationship in a dynamic environment: challenges for comanagement. *Ecol. Appl.* **18:** S148–S156.
221. Beaumont, L.J. *et al.* 2011. Impacts of climate change on the world's most exceptional ecoregions. *Proc. Nat. Acad. Sci. USA* **108:** 2306–2311.

Ann. N.Y. Acad. Sci. ISSN 0077-8923

ANNALS OF THE NEW YORK ACADEMY OF SCIENCES
Issue: *The Year in Ecology and Conservation Biology*

Effects of organic farming on biodiversity and ecosystem services: taking landscape complexity into account

Camilla Winqvist, Johan Ahnström, and Jan Bengtsson
Department of Ecology, Swedish University of Agricultural Sciences, Uppsala, Sweden

Address for correspondence: Camilla Winqvist, Department of Ecology, Swedish University of Agricultural Sciences, Box 7044 750 07 Uppsala, Sweden. Camilla.winqvist@slu.se

The recent intensification of the arable landscape by modern agriculture has had negative effects on biodiversity. Organic farming has been introduced to mitigate negative effects, but is organic farming beneficial to biodiversity? In this review, we summarize recent research on the effects of organic farming on arable biodiversity of plants, arthropods, soil biota, birds, and mammals. The ecosystem services of pollination, biological control, seed predation, and decomposition are also included in this review. So far, organic farming seems to enhance the species richness and abundance of many common taxa, but its effects are often species specific and trait or context dependant. The landscape surrounding the focal field or farm also seems to be important. Landscape either enhances or reduces the positive effects of organic farming or acts via interactions where the surrounding landscape affects biodiversity or ecosystem services differently on organic and conventional farms. Finally, we discuss some of the potential mechanisms behind these results and how organic farming may develop in the future to increase its potential for sustaining biodiversity and associated ecosystem services.

Keywords AES; biological control; pollination; review

Introduction

Agriculture has historically enriched and diversified the arable landscape in Europe.[1] Increased food and energy demands from a growing world population have resulted in agricultural intensification, which has had effects at both the local field or farm scale and the larger landscape scale. Negative effects of agricultural intensification have been demonstrated for many organism groups and associated ecosystem services.[2,3] Organic farming is gaining ground because it is assumed to counteract the negative effects of modern farming on, for example, soil organic matter, nutrient balances, and biodiversity. However, the impacts of organic farming on biodiversity and abundance have appeared to vary between taxa and seem to be affected by the surrounding landscape.[3–5] In this paper, we review some of the effects of organic farming on biodiversity and ecosystem services in the agricultural landscape and discuss if organic farming is important as an alternative farming practice now and in the future.

Traditional agricultural landscapes are often very species rich, especially in Europe.[1] They include a number of habitats and succession stages, and centuries of management have enabled a large number of species to adapt to management disturbances and inhabit the arable landscape. The human population and per capita food consumption is growing, and consequently biodiversity in arable landscapes has become negatively affected by agricultural intensification. On the local scale, agricultural intensification includes a number of interrelated processes such as higher inputs of energy, fertilizers and pesticides, a reduction of the area devoted to seminatural habitats such as hedgerows and field islands, and increased field sizes. On the larger landscape scale, natural habitats such as grasslands, wetlands, and forest are converted to arable land, thus decreasing the amount of natural and less intensively managed habitats in the landscape.

Organic farming and other agri-environment schemes (AES) have been suggested as a means to

doi: 10.1111/j.1749-6632.2011.06413.x

mitigate the negative effects of agricultural intensification in the European Union (Box 1). In organic farming, no inorganic pesticides or chemical fertilizers are used, and animal husbandry should be integrated in farm management, thus ideally producing a farmland that is of higher habitat quality than modern high-intensity farmlands. Twenty-five percent of the world's organically managed land is in Europe, and many studies of organic farming are European, but organic farming is a worldwide phenomenon. In 2009, certified organic farming in various forms existed in 160 countries. In total, 37.2 million hectares were organically farmed arable land (two thirds of which were grasslands or grazing areas), which represents 0.9% of the world's agricultural land.[6]

Box 1

"Organic production is an overall system of farm management and food production that combines best environmental practices, a high level of biodiversity, the preservation of natural resources, the application of high animal welfare standards, and a production method in line with the preferences of certain consumers for products produced using natural substances and processes." Council Regulation (EC) no. 824/2007 of 28 June 2007 on organic production and labeling of organic products and repealing Regulation (EEC) 2092/91.

Organic farming means different things to different people; some people are interested in soil quality and nutrient recycling, others in carbon emissions or the quality of nutrient levels of foods produced, and still others are interested in effects on biodiversity or ecosystem services. In this review, we summarize some recent findings regarding the effect of organic farming on agricultural biodiversity and associated ecosystem services in the light of landscape complexity. It is more or less established that organic farming results in higher species richness and greater abundance of many taxa.[3,4] Results, however, vary greatly between organisms, between studies with different designs, and according to the region or country in which the studies have been conducted. The effectiveness of organic farming to enhance biodiversity and ecosystem services may be affected by the composition of the surrounding landscape,[7–9] but such studies are still rare.

Biodiversity and ecosystem services

Many earlier studies of organic farming and other AES focus on a single organism group; plants tend to show the strongest positive responses to AES, followed by invertebrates, with birds and mammals showing the smallest responses.[10] Recently, multiple trophic levels or networks are increasingly being studied. Macfadyen *et al.* found that organic farms have higher species richness at three trophic levels, plant, herbivore, and parasitoid, and in a later study it was concluded that network modules on conventional farms had fewer links than on organic farms, which may reduce the stability of these networks.[11,12]

The surrounding landscape is now sometimes incorporated in studies (Table 1A and 1B). In a recent field study across five regions in Europe, the abundance and species richness of birds and plants were higher on organic fields, but ground beetles (Coleoptera: *Carabidae*) were no more abundant, when compared to conventional fields.[9] At the same time, birds and plants decreased both in species richness and in abundance with an increase in the percentage of arable land in the surrounding landscape, whereas the activity density of ground beetles increased.

Other information about the large-scale effects of organic farming on diversity can be gained by partitioning biodiversity into alpha-diversity (mean diversity of a site), between site beta-diversity, and between region beta-diversity.[13] Farm management acts on both the local and landscape scale, and different taxa respond at different, often multiple, scales.[14]

The effects of organic farming on biodiversity have been thoroughly studied and have been subject to several meta-analyses.[3,15] However, the effects of organic farming on ecosystem services and ecosystem functions have not received as much attention, especially not in a landscape perspective (Table 2). Ecosystem services are the benefits that humans derive from ecosystems.[16] Some of the services that are produced and consumed in the arable landscape include pollination, biological control, respiration, decomposition, and weed seed predation. Some earlier studies of pollinators were the first to discover interactions between farming practices and the surrounding landscape,[5,17] and this has now also been detected for ecosystem services such as pollination,[18] biological control,[9] and seed removal.[7]

Table 1. Results from recent studies on the effect of organic farming, landscape complexity, and their interaction on the species richness (A) and the abundance (B) of organisms in the arable landscape. For information regarding the definition of landscape in each study, see the main text.

A

Organism	Organic	Landscape	Interaction	References
Decomposers	ns	ns	ns	Diekötter *et al.*[35]
Ground beetles	ns	ns	*	Diekötter *et al.*[35]
Spiders	ns	ns	ns	Diekötter *et al.*[35]
Plants	***	*	ns	Winqvist *et al.*[9]
Ground beetles	ns	ns	ns	Winqvist *et al.*[9]
Breeding birds	*	*	ns	Winqvist *et al.*[9]
Pollinators	ns	ns	ns	Brittain *et al.*[18]

B

Organism	Organic	Landscape	Interaction	References
Decomposers	ns	ns	*	Diekötter *et al.*[35]
Springtails	ns	ns	ns	Diekötter *et al.*[35]
Ground beetles	ns	ns	ns	Diekötter *et al.*[35]
Spiders	ns	ns	ns	Diekötter *et al.*[35]
Plants (cover)	***	*	ns	Winqvist *et al.*[9]
Ground beetles	ns	**	ns	Winqvist *et al.*[9]
Breeding birds	*	**	ns	Winqvist *et al.*[9]
Small mammals	ns	ns	*	Fischer *et al.*[7]
Slugs	ns	ns	ns	Fischer *et al.*[7a]

$P < 0.1$ (ns), $P < 0.05$ (*), $P < 0.01$ (**), $P < 0.001$ (***).
[a]Additional effects were significant in the model: distance to field edge (**).

Plants

Plants are the most studied organisms in the agricultural landscape, since they are important for many other organisms and vital for some functions and services such as improving microclimate and, of course, to produce oxygen. Plants in arable fields are often considered to be weeds and are the main targets of many management practices. Therefore, plants often benefit from organic farming and other AES,[2,3,9] and they seem to respond quickly to conversion to organic farming in terms of species richness.[19] The most straightforward mechanism to explain these findings is that organic farmers do not use herbicides. In a pan-European study, the negative effects on plant species richness of increased frequency of insecticide and herbicide applications and the applied amount of active ingredients of fungicides was found.[2] Furthermore, weed species richness may be reduced by increased cereal cover, at least on organic farms.[20] Other potential explanations may have more to do with differences in the surrounding landscape between organic and conventional farms. The number of years with grass crops in the rotation may increase both weed abundance and species richness in the field as well as weed abundance in the seed bank, whereas the number of winter crops and the amount of inorganic nitrogen (N) and herbicide applied may have a negative effect on weed abundance and species richness in the field.[21]

The responses of plants to organic farming are often species specific. In a Mediterranean study, four of the 51 weed species found were more abundant in conventional fields, even though both a higher abundance, cover, diversity, and species richness of weeds was found on organic fields.[21]

Table 2. Results from recent studies on the effect of organic farming, landscape complexity, and their interaction on the ecosystem services in the arable landscape. For information regarding the definition of landscape in each study, see the main text

Ecosystem service	Organic	Landscape	Interaction	References
Biological control (b)	ns	*	0.057	Winqvist *et al.*[9]
Pollination (a)	ns	ns	0.054	Brittain *et al.*[18]
Seed predation	*	ns	ns	Diekötter *et al.*[35]
Seed removal	ns	ns	**	Fischer *et al.*[7a]
Litter decomposition	ns	ns	ns	Diekötter *et al.*[35]

$P < 0.1$ (ns), $P < 0.05$ (*), $P < 0.01$ (**), $P < 0.001$ (***).
[a]Additional effects were significant in the model: seed species (***) and predator exclusion (***), (a) = abundance of visits, (b) = removal of aphids from plastic labels placed in the field for 24 h.

Landscape structure such as the proportion of arable fields, grasslands or forests, and field boundary type may be more important for plants than organic farming.[22] The properties of the soil such as the percentage of N and carbon (C), pH, or clay content may be more important than farm management.[21] Sometimes, both farm management and the surrounding landscape are important for biodiversity. Winqvist *et al.* found positive effects on both plant species richness and plant cover of organic farming as well as a lower percentage of arable fields in the surrounding landscape, but there was no interaction between the two measures.[9] Finally, contrary to expectations, plants did not respond faster to organic farming in heterogeneous landscapes with a lower percentage of arable land than in homogeneous landscapes with more arable land.[19] Hence, the role of the surrounding landscape in plant responses to organic farming remains ambiguous, even though it seems well established that weeds are more species rich and abundant under organic management.

Arthropods

Arthropods are abundant in most habitats, including agricultural landscapes. Some are pests of crops; others are beneficial as pollinators or as naturally occurring predators of pests. Arthropods are directly affected by insecticide use and indirectly affected if food resources become less abundant as a result of pesticide use.[23] A conversion to organic farming may accordingly increase arthropod abundance, and this appears to occur gradually. Butterfly (Lepidoptera: *Rhopalocera* and *Zygaenidae*) abundance had increased 100% 25 years after the transition from conventional to organic agriculture.[19] Butterfly species richness, on the other hand, was not affected by time since the transition and was much higher on organic farms directly after transition. However, such studies are too rare to allow more general conclusions about the temporal responses of arthropods to changes in farm management.

A recent development is to analyze different taxa or functional groups in more detail, since it has become clear that responses are not general to all pollinators or predators. In a study from the Mediterranean area, organic fields had a overall higher richness than conventional fields, but for important groups such as spiders (Araneae), beetles (Coleoptera), and flies (Diptera), there were no differences, indicating that they are affected by different factors than those that vary between organic and conventional fields.[20] Using meta-analysis, Garratt *et al.* found that most natural enemy groups, as well as pests, increased in numbers, impact, or performance on organic farms, although this was not true for beetles. Even if the abundance and species richness of some natural enemy groups is increased in organic systems, individual species responses are often more specific and may be determined by one or a few specific elements of farming practices.[23] Most studies do not take factors such as soil structure or the surrounding landscape into account. Many arthropods utilize different habitats during a year—for instance, for overwintering or reproduction—making the surrounding landscape important. Bee species richness and the number of brood cells in trap nests have been shown to be enhanced by high proportions of noncrop habitats when both nesting sites and floral resources are present in the nearby landscape. In contrast, wasp species richness and brood cell numbers were enhanced by connecting corridors.[24] In a study designed to minimize differences in landscape complexity between organic and conventional farms, no effect of organic farming on ground beetles was detected, but these organisms were more abundant in homogenous landscapes.[9]

It is also becoming popular to study functional groups as defined by various traits, since the perception of crop–noncrop boundaries or the availability of their food resources differs between such groups.[24] The sensitivity to agricultural intensity may, for instance, differ between specialist and generalist parasitoids,[24] and the percentage of arable fields in a landscape may influence species with different dispersal abilities differently.[25]

Not much work has been done regarding differences in, for instance, body condition or fecundity between organic and conventional farming. Wolf spider females (Araneae: *Lycosidae*) have been found to be in better condition in landscapes with large fields of annual crops, but fecundity was unaffected by the proportion of arable land, number of crops grown, or field size; additionally, their body condition and fecundity were unaffected by farming practice.[26] Predatory ground beetles have been shown to be in better condition on organic farms than on conventional farms and in fields with a high perimeter-to-area ratio.[27]

To conclude, arthropod taxa have, until recently, been clumped together as "beetles" or "bees," but in order to examine the mechanisms behind responses to organic farming or landscape complexity, species-specific traits must be taken into account. This may be especially important when understanding effects on ecosystem services.

Pollination. Pollination is one of the most well-known and important ecosystem services in the arable landscape, and many crops rely on naturally occurring pollinators or honeybees. Many studies of "pollination" primarily focus on pollinators or insect–plant networks, and both pollinators and pollinator networks seem to be more diverse on organic farms.[28,29] In a German study, organic fields had a hundred times greater abundance of pollinators than conventional fields, but no attempt to study actual pollination was made.[28] In Ireland, insect–flower networks on organic farms have been found to be larger and more asymmetrically structured than networks on conventional farms, which may increase pollination.[29] Bees (Hymenoptera: *Apidae*) were found to be more abundant on organic fields partly because of higher floral abundances. Since organic farms produced more flowers that attracted more pollinators, pollination was improved.[29] Hoverfly (Diptera: *Syrphidae*) evenness was greater in organic farms, but this was not related to floral abundance, suggesting that organic farms provide additional and more diverse resources for this organism.[29]

Both organic farming and the surrounding landscape have been shown to affect pollinators, but to assess pollination, studies on pollination—not just pollinators—are needed. In a study of vine fields in Italy, flower-visiting insect abundance species richness did not differ between organic and conventional fields.[18] Instead, the abundance of visitors to potted plants was negatively affected by the proportion of uncultivated land in conventional fields only. The fruit set, weight of seeds per plant, and seed weight were negatively affected by the proportion of uncultivated land in the surrounding landscape on both field types.

More studies of pollination and the outcome on seed quality, etc. are badly needed, especially studies taking the surrounding landscape into account. To secure or restore pollination services, it seems important to reduce local management intensity and to restore or maintain natural or seminatural vegetation, providing nesting habitat and food resources.[30]

Biological control. Biological control of pests by naturally occurring (arthropod) predators is another well-known ecosystem service. As with pollination, very often only potential predators are studied or predation is studied in an inactive way, not revealing actual predators. Already 10 years ago, Östman *et al.* showed that organic farming as well as landscapes with abundant field margins and perennial crops contributed to biological control of cereal aphids during their establishment phase. During the aphid population growth phase, biological control was only greater in landscapes with more arable land.[27] Cereal aphids have often been shown to have lower abundance on organic fields in Germany; at the same time, predator abundances and predator–prey ratios were higher, but no actual study of predation was performed.[28] Pesticide-sprayed conventional fields had a higher abundances of aphids and lower abundances of predators late in the season, indicating that spraying only had a short-term effect on aphids but affected predators negatively over a longer period.[31]

Sometimes molecular methods are used to pinpoint predators or prey. Using stable isotope ratios, generalist predators have been found to consume higher proportions of herbivore prey in organic systems.[32]

In a recent European study using data from five regions, organic farms in complex landscapes with a low proportion of arable land had the highest biological control, but on conventional farms, the biological control potential was the same in all landscapes.[9] The mechanism behind this remains unclear, but Crowder *et al.* argue that organic agriculture promotes evenness among predators,[33] which in turn may be negatively affected by landscape simplification. Another possibility is that different species are the main predators in different landscapes and on farms with different management. Omnivorous ground beetles have been shown to be relatively more abundant in homogeneous landscapes with a high proportion of arable land, but they did not differ in abundance between organic and conventional farms.[25] Until we know which predators to promote under which circumstances, conserving a diverse natural enemy community of species with different traits may be a good strategy.[32]

Seed predation. The ecosystem service of weed seed predation can be carried out by, for instance, birds,[34] small mammals, or invertebrates.[7] Weed seed predation borders on the ecosystem disservice of weed seed dispersal (burial, movement), where weed seeds are not being predated but dispersed. Some seed predation studies do not include the surrounding landscape[34] or find that it is not important,[35] whereas other studies have found it very important.[7] In a German study, seed predation and removal increased in conventional fields, but decreased in organic fields as the proportion of arable fields increased. Of the studied potential seed predators, this pattern corresponded most closely to the activity density of slugs.[7] Slugs do not always predate on seeds, only disperse them, thereby performing an ecosystem disservice. Small omnivorous and herbivorous mammals showed the opposite response as slugs to seed removal; they decreased in conventional fields but increased in organic fields as landscape complexity decreased, corresponding to more arable land. In this study, ground beetles were more abundant on organic farms than on conventional ones, and they were unaffected by the surrounding landscape. Fischer *et al.*[7] concluded that "weed seed predation is provided by small mammals and invertebrates, whereas the disservice of seed dispersal and movement is provided by slugs."

It is too soon to draw any general conclusions regarding seed predation, since results differ between weed species, depending both on, for instance, seed size and its nutritional value and on which potential predators are being studied.

Soil biota

Soil quality and soil biodiversity have been widely studied since the soil is the base of farming and the ability to sustain high yields.[36] Soil organisms are important for the functioning of the soil and recycling of nutrients, and soil biota are believed to benefit from a reduction in the intensity of use of mechanical and manufactured inputs and by integrating biological inputs.[36] Negative direct and indirect effects of tillage have previously been shown on soil microfauna,[37] and epigeic springtails (Collembola) have been found to be vulnerable to certain fungicides since they reduce their fungal food supply.[38] In a recent review by Gomiero *et al.*, it was concluded that alternative (for instance, organic) farming practices in most cases enhance soil biodiversity regardless of climate and soil conditions.[39] Earthworms are usually more abundant in organic farming;[3] in the Netherlands, they were 2–4 times more abundant on organic farms.[40] Earthworms have large effects on both soil structure and the resources available to other soil organisms, and a higher abundance of earthworms may therefore result in decreases among other soil organism groups.[41]

Even though soil biota may be less mobile and utilize fewer habitats during their lifetime, the surrounding landscape has been shown to be important.[35] In a recent German study, Flohre *et al.* found an interaction between organic farming and landscape complexity for earthworm species richness.[8] The microbial biomass showed the same response but on a smaller landscape scale. Respiration (microbial activity) was higher on conventional farms and dependent on landscape complexity, measured as the proportion of arable fields. Species richness of earthworms and microbial biomass were low in organic farms in complex landscapes but increased with an increase in arable land, and the opposite was found in conventional farms, where higher species richness in complex landscapes decreased with increasing arable land. For earthworms, this is believed to be a result of greater predation in organic fields in complex landscapes. Springtails, on the other hand, were more abundant on conventional farms, regardless of the surrounding landscape.

Soil biotas have been thoroughly studied in organic farming, but the effects vary between studies. The effects of the different groups of soil biota, and the effects on surrounding landscape and ecosystem services and functions provided by soil biota, should be given more attention.

Decomposition and other soil ecosystem services. Sustaining the quality of the soil in arable ecosystems can be considered a supporting ecosystem services and reactive N and water quality can be considered regulating services,[16,42] but there are still few studies simultaneously examining soil ecosystem services, organic farming, and landscape complexity. Additionally, it is not possible to translate findings from organisms to services; even if decomposers were found to be affected by interactions between organic farming and landscape complexity, the associated ecosystem service of litter decomposition was not.[35] Millipedes (Diplopoda) and woodlice (Isopoda) were less abundant on conventional fields surrounded by other conventional fields than in either conventional or organic fields surrounded by organic fields. Litter decomposition, on the other hand, did not differ between farming types or landscapes.

Snapp *et al.* studied some of the potential mechanisms behind the effects of organic farming on soil ecosystem services and found that organic management, but not crop diversity, sustained soil fertility, augmented soil carbon, enhanced N retention, and improved N-use efficiency.[42] At the same time, grain quality, quantity, and temporal yield stability were lower under organic management than in conventional, integrated management. Therefore, improving soil quality through organic farming may come with the cost of reduced yields and lower quality crops that need to be considered in the light of higher energy, nutrient, and pesticide input on conventional farms. However, it is clear that a more complete and nuanced picture of soil ecosystem services in relation to organic farming requires more research.

Farmland birds

Birds often benefit from organic farming,[3,9] and one likely explanation of lower species richness and abundance of birds on conventional farms is that herbicides and insecticides reduce food resources. Geiger found a higher soil–seed density and weed biomass on organic farms, and seed density correlated both to bird abundance and species richness.[43] There are other differences between organic and conventional farms that may affect birds; the often increased rate of mechanical weed control on organic farms and mowing of green manure crops may be detrimental to many species.[17,43,44] Lapwings (*Vanellus vanellus*) have been shown to have a higher nest density on organic farms, but they may suffer from lower nesting success due to farming activities.[45,46] Skylarks (*Alauda arvensis*) on organic farms initiated nests throughout the season, but in conventional farms they did so only early and late in the season due to differences in the height of crops between farm types.[46] Later in the season, skylarks initiated nests in vegetable fields, available on both farm types, thereby reducing the difference between farm types; this indicates that differences between organic and conventional fields may be seasonal. Organic farms have been shown to enhance the species richness of birds during the breeding season in Germany, but not in the winter.[47] The opposite was found in the Netherlands by Geiger; species richness and abundance was positively influenced by organic farming in the winter, but not in the summer, but responses were species specific.[43]

Since birds are very mobile and often use resources not just from a single field, many studies are taking the surrounding landscape into account. Landscape complexity affected the diversity of birds both in the summer and in the winter in a German study, and responses differed between groups of birds. Forest and farmland bird species abundance, species richness, and diversity decreased as the percentage of arable land in the landscape increased, presumably owing to the reduced availability of nesting and sheltering places in noncrop habitats. The abundance of birds known to breed and feed in fields increased with decreasing landscape complexity, and they seem to require high proportions of arable land.[47]

The effect of the surrounding landscape also differed between species in a study from the Netherlands. Some bird species such as skylarks were more abundant in larger fields, whereas others such as yellowhammers (*Emberiza citronella*) preferred smaller fields. An increase in habitat diversity increased the abundance of some species, whereas an increase in the percentage of arable land in the landscape enhanced, for instance, skylarks.[43] In Sweden, organic farms had higher species richness in homogenous landscapes with a high proportion of arable fields and larger fields, especially for passerine invertebrate feeders, indicating that organic farming improved foraging conditions in these landscapes.[48]

Birds are also important for some ecosystem services, but this has not yet been widely studied in the light of organic farming and ecosystem services. Navntoft *et al.* filmed seed predation on organic and conventional farms but did not take landscape complexity into account. They found that birds were the most important seed predators, but mice and slugs were also observed predating seeds.[34]

Mammals

A number of mammals, such as mice, deer, hares, and foxes can be found on arable fields or in the arable landscape. Some species of bat also hunt in the arable landscape. Mammals are not very often well studied in the context of organic farming.

Wickramasinghe *et al.* found that bat activities and feeding activity were higher on organic farms than conventional farms. Taller hedgerows and better water quality may explain the difference, as might greater prey availability.[49,50] A more recent study of Pipistrelle bats, on the other hand, found lower bat activity and prey availability on farms participating in AES associated with hedgerows and water margins.[51] This finding may indicate that organic farming is more important than more specific AESs for some organisms.

In a British study, Bates and Harris did not find a difference in small mammal (seven species of mice, voles, and shrews) abundance and diversity between organic and conventional farms when hedgerow management and structure were studied. They argue that increasing the area of noncrop habitats may be more important.[52] Many mammals are mobile and utilize different habitats throughout their lives and may therefore be very dependant on the quality or structure of the landscape. Small omnivorous and herbivorous mammals have been shown to decrease in conventional fields but increase in organic fields in landscapes with a higher proportion of arable fields.[7] In this study, small mammals showed the opposite response to the ecosystem service of seed removal, which was highest on organic fields in complex landscapes and decreased as landscapes grew more homogeneous, whereas on conventional fields it increased with increasing homogeneity.[7] Mammals such as bats may also be important for pollination and pest control of, for instance, moths, but this is largely unstudied in an organic farming and landscape context.

Discussion

Sometimes the issue of organic versus conventional farming turns into a very polarized and unproductive debate. Instead of arguing which farming system is the worst or the best, we should try to see the strengths and weaknesses in these as well as other farming systems and try to improve them and make them more sustainable. Organic farming is not the only solution to the problem caused by modern agriculture, but where biodiversity is concerned, organic farming generally performs better than conventional farming. Both conventional and organic farms are very diverse in their management, resulting in a huge variation and overlap in farm management.[53] Organic farmers cannot use chemical fertilizers and inorganic pesticides, but conventional farmers can decide not to use or minimize these external inputs, thus reducing the differences between the systems. Some farmers have both organic and conventional fields, and many farmers adopt other AES than organic farming.

One step forward may be a stronger focus on integrated farming and management practices, so that farmers willing to adjust the number of pesticide applications to pest pressure and economical thresholds can be compensated. Compared to organic farming, integrated farms can have a higher species richness of, for instance, weeds on the regional or landscape scale as a result of these farms having a greater range of crop types and cropping practices between fields than both organic and conventional farms.[21] Integrated farm management is thus one way to increase the variation in the landscape, but the questions of how the complexity of the surrounding landscapes should be taken into account in policies such as organic farming is yet unresolved.

Is organic farming most beneficial in heterogeneous or homogeneous landscapes? Heterogeneous landscapes with plenty of natural, seminatural, and traditionally managed habitats may already have high biodiversity, and thus organic farming may not result in higher biodiversity in these landscapes, suggesting that organic farming has the greatest effect in homogeneous landscapes.[5,17] Perfecto and Vandermeer suggested that managing the matrix between natural habitats and seminatural grasslands was important for both preserving biodiversity and sustainable farming based on ecosystem services, just as is being done in organic farming.[54]

In regions where the presence of agriculture increases rather than decreases habitat heterogeneity, the positive effects of agriculture on biodiversity are more likely to occur.[55] In many countries and regions, there is intensification in arable landscapes but extensification in more forested areas as unproductive farms close down. In the arable landscape, organic practices lower farming intensity, and in forested areas, the subsidy for organic farming and the higher payment for organic products can save farms from closing.

In a meta-analysis of landscape-moderated biodiversity effects on agri-environmental management, Batáry *et al.* found that such schemes significantly increased species richness in simple, but not in complex, landscapes.[15] Still, the question of actual management of the surrounding landscape has not yet been properly addressed. Kleijn *et al.* studied whether AES in five European countries benefited diversity, and they concluded that common species may be enhanced with relatively simple modifications, whereas endangered species require more elaborate conservation methods.[56] The benefit of more specific AES is that they can be more precisely targeted at a specific species or service. In a recent European study of agricultural intensification, the percentage of land under AES in the landscape had a beneficial impact on biological control potential, whereas no significant positive effects of organic farming in itself were found.[2]

In recent years, the importance of studying not only biodiversity per se, but also associated ecosystem services has been emphasized.[57] More importantly, it is time to include ecosystem services in regulations and schemes in order to promote them in a sufficient way. Studies on ecosystem services so far show diverse results, but most agree that the landscape aspect is important for ecosystem services as well (Table 2). There is also a need for more studies on the actual outcome of, for instance, pollination—e.g., seed set, seed quality, and germination, or the increase in crop yield and quality resulting from the predation of pests. Since a high yield and quality of crops is essential to farming, a functioning pollination service and biological control are vital. Any measures to increase pollination and predation of pests are worthwhile, and in this respect, organic farming seems to be a good start. But since most pollinators and some predators only visit arable fields during a short period of their lifetime, the surrounding landscape with alternative natural and seminatural habitats is very important, and this is not included in the regulation of organic farming.

Rather than just discussing organic farming, we should be talking about (organic) farmers, the main actors in the arable landscape. Farmers decide farm management practices, including what crops to grow, whether to use pesticides or not, and whether to convert to organic farming or not. Conversion to organic farming often depends on the opportunity cost of farmers (or the farms); what will be the net economic gain when the frequently lower yield but higher payment is considered? These factors are landscape specific; the uptake of organic farming is higher in less-intensive agricultural landscapes or low-productive and high-diversity fields. This can be called the selection effect, where AES are preferentially located on fields with high biodiversity.[58] Alternatively, schemes could be implemented in low-diversity locations to enhance biodiversity in areas where it is needed, which may result in no increase in species richness since species are not present in the surrounding landscape. In addition, responses can be hard to measure, since time-lags and long response times are common in nature. The duration of many studies is just one season, and we therefore fail to incorporate "background variation" or demographic effects of organic farming. In addition, it seems like policies are expected to give swift responses to be proven useful. It has been argued that AES should result in quick responses on biodiversity and ecosystem services to make validations of costs and benefits easier.[19] This is quite a harsh requirement, since the effect of organic farming can take a long time to emerge. For example, these effects have been shown to be rapid for butterfly and plant species richness, whereas butterfly abundance increased gradually with time since transition over a 25-year period.[19] We can assume that the latter is likely to be the case for ecosystem services as well.

Another opportunity cost is that of the farmer's identity and social capital.[59] The conversion to organic farming is a process in which the farmer has to accept weedier fields and maybe endure negative comments from other farmers regarding organic farming. Weeds are not desired by organic farmers and other methods for reducing weeds are used, but these seem to be less effective than the use of herbicides. In organic farming, but also more and more in conventional farming, there is a discussion on the importance of management, instead of eradication, of weeds; the intent is to have appropriate, in an economic and ecological sense, populations of weeds.[60]

There is also an opportunity cost associated with the need to learn about techniques, to buy new machinery, and to join new environmental subsidies and certification bodies.

GM technology may help overcome some of the negative aspects of modern agriculture. In Europe, genetically modified organisms and products from them are incompatible with the concept of organic farming.[61] There may be both pros and cons with GM technology, but studies so far have not come to a general consensus, and studies on potential effects on ecosystem services are missing.[62]

Concluding remarks

So, where do we go from here? We need to raise the production of food, feed, fuel, fiber, and ecosystem services in a sustainable and resilient manner that is also economically and socially valid without exhausting resources, biodiversity, or ecosystems. It has been claimed that "existing farming systems and the knowledge system that supports them are no longer fit-for-purpose and that a new approach is called for."[63] These authors also argued that future agricultural production systems will need to produce food and other products using much scarcer resources than today, in terms of inputs of energy and nutrients such as phosphorus.[63] Can organic farming play a role in the developing these future systems? Different ways forward to resolve this conundrum have been proposed.

One of the most important issues is that enough funding should be allocated to AES such as organic farming, not only toward sufficient subsidies to farmers, but also toward adaptive management[10] and, importantly, toward monitoring results. Baseline data must be collected, and improvements or failures must be monitored, documented, and analyzed after implementation. We also need to calibrate our AES and farm management to be up to date with recent scientific findings. To make AES fulfill their goals, they need to have been developed together with stakeholders at all levels.[64] Previous uptake of AES has been associated with AES that demand the least changes from current farm management; that is, farmers keep doing what they have always done. It is important that the connection between management and positive effects on biodiversity, ecosystem services, or other aspects of organic farming is stated clearly and widely communicated. Farmers may refuse to join schemes because they do not understand why the management measures should be done, and they might also doubt whether the measures really have the effects that they are said to have.

Many agri-environment policies are too short term to be useful. Time lags in responses to AES may be common, and long-term effects would need to be included in management recommendations and policies.[19] AES are not currently designed specifically to deliver improved ecosystem services, but to include payment for ecosystem services from, for instance, organic farming could be a way to achieve a more long-term approach.[10]

Another maybe more radical idea is to make some measures compulsory—for instance, protective zones without pesticides or fertilizers lining water bodies, seminatural, and natural habitats. In many countries, some basic measures are mandatory—for instance, in Finland, some basic measures covered 93% of arable land in 2002.[64] Perfecto and Vandermeer suggested the matrix quality model to achieve small-scale sustainable agriculture.[54] According to this idea, a high-quality matrix within which fragments of high-diversity native vegetation can persist along with biodiversity-friendly agroecosystems to form an integrated landscape. If this is to work, much more focus on the landscape as a whole is needed, both by farmers and by policy makers, and policies concerning the landscape level need to be developed. To be able to answer questions like "how much organic farming in homogenous landscapes will lead to a 30% increase in biodiversity?" we need to study organic farming on a wider landscape scale and also take the amount of organic farming in the surrounding landscape into account in study design (see, for instance, Diekötter *et al.*[35]).

Active management of noncrop habitats is often not included in organic farming, but this may be included in other AES. For pollinators, a deliberate manipulation of nesting resources may be important, not just floral resource management. The provision of dead wood, bare soil, and areas of natural flooding may also be important for many organisms in the modern arable landscape. Studies of organic farming and other AES have demonstrated the importance of landscape effects on other than the target environment to record the total benefits of AES on biodiversity. The first step could be to offer financial support to farmers linked to spatial arrangements of organic fields.[65] For researchers to give the best recommendations, we need multiyear, multiscale, multibiodiversity, and multifunctional studies to guide us. We must make sure to not only study biodiversity per se, but also include the associated services!

Scope and method

This review is an attempt to bring together recent findings regarding effects of organic farming on biodiversity and ecosystem services and studies on how these effects may be affected by the surrounding landscape. Our aim is to give a broad overview on this matter regarding terrestrial biodiversity and associated ecosystem services in the arable landscape. We focused mainly on studies published from 2009 and onward, but for some important reviews or meta-analyses we refer to older studies. For some organism groups, we also included older studies simply because there are only few recent studies. We have used keywords such as "organic or biological farming," "biodiversity," "ecosystem services," "landscape complexity," "pollination," and "predation" when searching for studies. Most studies have been conducted on arable crops such as cereals, whereas some have been conducted in grasslands such as ley. Additionally, because of our background, most studies are European, with a few exceptions covering, for instance, the United States.

Acknowledgments

The authors would like to thank The Swedish Research Council Formas, The Swedish Research Council to the EuroDiversity Agripopes project, and finally the Ekhaga Foundation for funding. The authors would also like to thank Richard Hopkins for help with improving the manuscript.

Conflicts of interest

The authors declare no conflicts of interest.

References

1. van Elsen, T. 2000. Species diversity as a task for organic agriculture in Europe. *Agr. Ecosyst. Environ.* **77:** 101–109.
2. Geiger, F., J. Bengtsson, F. Berendse, *et al.* 2010. Persistent negative effects of pesticides on biodiversity and biological control potential on European farmland. *Basic Appl. Ecol.* **11:** 97–105.
3. Bengtsson, J., J. Ahnström & A-C. Weibull. 2005. The effects of organic agriculture on biodiversity and abundance: a meta-analysis. *J. Appl. Ecol.* **42:** 261–269.
4. Hole, D.G., A.J. Perkins, J.D. Wilson, *et al.* 2005. Does organic farming benefit biodiversity? *Biol. Conserv.* **122:** 113–130.
5. Rundlöf, M. & H. Smith. 2006. The effect of organic farming on butterfly diversity depends on landscape context. *J. Appl. Ecol.* **43:** 1121–1127.
6. Willer, H. & L. Kilcher. Eds. 2011. *The World of Organic Agriculture—Statistics and Emerging Trends 2011.* IFOAM, Bonn, Germany, and FiBL, Frick, Switzerland.
7. Fischer, C., C. Thies & T. Tscharntke. 2011. Mixed effects of landscape complexity and farming practice on weed seed removal. *Perspect. Plant Ecol.* **13:** 297–303.
8. Flohre, A., M. Rudnick, G. Traser, *et al.* 2011. Does soil biota benefit from organic farming in complex vs. simple landscapes? *Agr. Ecosyst. Environ.* **141:** 210–214.
9. Winqvist, C., J. Bengtsson, T. Aavik, *et al.* 2011. Mixed effects of organic farming and landscape complexity on farmland biodiversity and biological control potential across Europe. *J. Appl. Ecol.* **48:** 570–579.
10. Wittingham, M.J. 2011. The future of agri-environment schemes: biodiversity gains and ecosystem service delivery? *J. Appl. Ecol.* **48:** 509–513.
11. Macfadyen, S., R. Gibson, A. Polaszek, *et al.* 2009. Do differences in food web structure between organic and conventional farms affect the ecosystem service of pest control? *Ecol. Lett.* **12:** 229–238.
12. Macfadyen, S., R. Gibson, W.O.C. Symondson, *et al.* 2011. Landscape structure influences modularity patterns in farm food webs: consequences for pest control. *Ecol. Appl.* **2:** 516–524.
13. Gabriel, D. & T. Tscharntke. 2007. Insect pollinated plants benefit from organic farming. *Agr. Ecosyst. Environ.* **118:** 43–48.
14. Gabriel, D., S.M. Sait, J.A. Hodgson, *et al.* 2010. Scale matters: the impact of organic farming on biodiversity at different spatial scales. *Ecol. Lett.* **13:** 858–869.

15. Batáry, P., A. Báldi, D. Kleijn, *et al.* 2011. Landscape-moderated biodiversity effects of agri-environment management: a meta-analysis. *Proc. R. Soc. B.* **278:** 1894–1902.
16. Millennium Ecosystem Assessment. 2003. *Ecosystems and Human Well-Being: Current State and Trends Assessment*. Island Press. Washington, D.C.
17. Holzschuh, A., I. Steffan-Dewenter, D. Kleijn, *et al.* 2007. Diversity of flower-visiting bees in cereal fields: effects of farming system, landscape composition and regional context. *J. Appl. Ecol.* **44:** 41–49.
18. Brittain, C., R. Bommarco, M. Vighi, *et al.* 2010. Organic farming in isolated landscapes does not benefit flower-visiting insects and pollination. *Biol. Conserv.* **143:** 1860–1867.
19. Jonason, D., G.K.S. Andersson, E. Öckinger, *et al.* 2011. Assessing the effect of the time since transition to organic farming on plants and butterflies. *J. Appl. Ecol.* **48:** 543–550.
20. Ponce, C., C. Bravo, D. Garcia de Léon, *et al.* 2011. Effects of organic farming on plant and arthropod communities: a case study in Mediterranean dryland cereal. *Agr. Ecosyst. Environ.* **141:** 193–201.
21. Hawes, C., G.R. Squire, P.D. Hallett, *et al.* 2010. Arable plant communities as indicators of farming practice. *Agr. Ecosyst. Environ.* **138:** 17–26.
22. Aavik, T. & J. Liira. 2010. Quantifying the effect of organic farming, field boundary and landscape structure on the vegetation of field boundaries. *Agr. Ecosyst. Environ.* **135:** 178–186.
23. Garratt, M.P.D., D.J. Wright & S.R. Leather. 2011. The effects of farming system and fertilizers on pests and natural enemies: a synthesis of current research. *Agr. Ecosyst. Environ.* **141:** 261–270.
24. Holzschuh, A., I. Steffan-Dewenter & T. Tscharntke. 2010. How do landscape composition and configuration, organic farming and fallow strips affect the diversity of bees, wasps and their parasitoids? *J. Anim. Ecol.* **79:** 491–500.
25. Winqvist, C. 2011. *Biodiversity and Biological Control. Effects of Agricultural Intensity at the Farm and Landscape Scale*. Doctoral Thesis, Swedish University of Agricultural Sciences. Uppsala, Sweden.
26. Öberg, S. 2009. Influence of landscape structure and farming practice on body condition and fecundity of wolf spiders. *Basic Appl. Ecol.* **10:** 614–621.
27. Östman, Ö., B. Ekbom & J. Bengtsson. 2001. Landscape complexity and farming practice influence biological control. *Basic Appl. Ecol.* **2:** 365–371.
28. Krauss, J., I. Gallenberger & I. Steffan-Dewenter. 2011. Decreased functional diversity and biological pest control in conventional compared to organic crop fields. *PLoS One* **6:** 1–9.
29. Power, E.F. & J.C. Stout. 2011. Organic dairy farming: impacts on insect-flower interaction networks and pollination. *J. Appl. Ecol.* **48:** 561–569.
30. Kremen, C., N.M. Williams & R.W. Thorp. 2002. Crop pollination from native bees at risk from agricultural intensification. *Proc. Natl. Acad. Sci. USA* **99:** 16812–16916.
31. Krauss, J., I. Gallenberger & I. Steffan-Dewenter. 2011. Decreased functional diversity and biological pest control in conventional compared to organic crop fields. *PLoS One* **6:** 1–9.
32. Birkhofer, K., A. Fliessbach, D.H. Wise, *et al.* 2011. Arthropod food webs in organic and conventional wheat farming systems of an agricultural long-term experiment: a stable isotope approach. *Agric. For. Entomol.* **13:** 197–204.
33. Crowder, D.W., T.D. Northfield, M.R. Strand, *et al.* 2010. Organic agriculture promotes evenness and natural pest control. *Nature* **466:** 109–112.
34. Navntoft, S., S.D. Wratten, K. Kristensen, *et al.* 2009. Weed seed predation in organic and conventional fields. *Biol. Control.* **49:** 11–16.
35. Diekötter, T., S. Wamser, V. Wolters, *et al.* 2010. Landscape and management effects on structure and function of soil arthropod communities in winter wheat. *Agr. Ecosyst. Environ.* **137:** 108–112.
36. Stockdale, E.A., L. Philipps & C.A. Watson. 2006. Impacts of farming practice within organic farming systems on below-ground ecology and ecosystem function. In *Aspects of Applied Biology 79*, C. Atkinson, B. Ball & D. H. K. Davies, *et al.*, Eds.: 43–46. Association of Applied Biologists. Warwick, UK.
37. Wardle, D.A. 1995. Impacts of disturbance on detritus food webs in agro-ecosystems of contrasting tillage and weed management practices. *Adv. Ecol. Res.* **26:** 105–185.
38. Frampton, G.K. & S.D. Wratten. 2000. Effects of Benzimidazole and Triazole fungicide use on epigeic species of collembola in wheat. *Ecotox. Environ. Safe* **46:** 64–72.
39. Gomiero, T., D. Pimentel & M.G Paoletti. 2011. Environmental impact of different agricultural management practices: conventional vs. organic agriculture. *Crit. Rev. Plant Sci.* **30:** 95–124.
40. Kragten, S., W.L.M. Tamis, E. Gertenaar, *et al.* 2011. Abundance of invertebrate prey for birds on organic and conventional arable farms in the Netherlands. *Bird Conserv. Int.* **21:** 1–11.
41. Eisenhauer, N. 2010. The action of an animal ecosystem engineer: identification of the main mechanisms of earthworm impact on soil microarthropods. *Pedobiologia* **53:** 343–352.
42. Snapp, S.S., L.E. Gentry & R. Harwood. 2010. Management intensity—not biodiversity—the driver of ecosystem services in a long-term row crop experiment. *Agr. Ecosyst. Environ.* **138:** 242–248.
43. Geiger, F. 2011. *Agricultural Intensification and Farmland Birds*. Doctoral Thesis, Wageningen University. Wageningen, the Netherlands.
44. Swagemakers, P., H. Wiskerke & J.D. Van Der Ploeg. 2009. Linking birds, fields and farmers. *J. Environ. Manage.* **90**(Suppl 2): 185–192.
45. Kragten, S. & G.R. de Snoo. 2007. Nest success of lapwings (*Vanellus vanellus*) on organic and conventional arable farms in the Netherlands. *Ibis* **149:** 742–749.
46. Kragten, S. & G.R. de Snoo. 2008. Field-breeding birds on organic and conventional arable farms in the Netherlands. *Agr. Ecosyst. Environ.* **126:** 270–274.
47. Fischer, C., A. Flohre, L.W. Clement, *et al.* 2011. Mixed effects of landscape structure and farming practice on bird diversity. *Agr. Ecosyst. Environ.* **141:** 119–125.
48. Smith, H.G., J. Dänhardt, Å. Lindström, *et al.* 2010. Consequences of organic farming and landscape heterogeneity for

species richness and abundance of farmland birds. *Oecologia* **162:** 1071–1079.
49. Wickramasinghe, L.P., S. Harris, G. Jones, *et al.* 2003. Bat activity and species richness on organic and conventional farms: impact of agricultural intensification. *J. Appl. Ecol.* **40:** 984–993.
50. Wickramasinghe, L.P., S. Harris, G. Jones, *et al.* 2004. Abundance and species richness of nocturnal insects on organic and conventional farms: effects of cultural intensification on bat foraging. *Conserv. Biol.* **18:** 1283–1292.
51. Fuentes-Montemayor, E., D. Goulson & K.J. Park. 2011. Pipistrelle bats and their prey do not benefit from four widely applied agri-environment management prescriptions. *Biol. Conserv.* **114:** 2233–2246.
52. Bates, F.S. & S. Harris. 2009. Does hedgerow management on organic farms benefit small mammal populations? *Agr. Ecosyst. Environ.* **129:** 124–130.
53. Ahnström, J. 2009. *Farmland Biodiversity—in the Hands and Minds of Farmers.* Doctoral Thesis, Swedish University of Agricultural Sciences. Uppsala, Sweden.
54. Perfecto, I. & J. Vandermeer. 2010. The agroecological matrix as alternative to the landsparing/agriculture intensification model. *Proc. Natl. Acad. Sci. USA* **107:** 5786–5791.
55. Kremen, C., N.M. Williams, M.A. Aizen, *et al.* 2007. Pollination and other ecosystem services produced by mobile organisms: a conceptual framework for the effects of land-use change. *Ecol. Lett.* **10:** 299–314.
56. Kleijn, D., R.A. Baquero, Y. Clough, *et al.* 2006: Mixed biodiversity benefits of agri-environment scheme in five European countries. *Ecol. Lett.* **9:** 243–254.
57. Letourneau, D.K. & S.G. Bothwell. 2008. Comparison of organic and conventional farms: challenging ecologists to make biodiversity functional. *Front. Ecol. Environ.* **6:** 430–438.
58. Kleijn, D. & W.J. Sutherland. 2003. How effective are European agri-environment schemes in conserving and promoting biodiversity? *J. Appl. Ecol.* **40:** 947–969.
59. Burton, R.J.F. & U.H. Paragahawewa. 2011. Creating culturally sustainable agri-environmental schemes. *J. Rural Stud.* **27:** 95–104.
60. Håkansson, S. 2003. *Weeds and Weed Management on Arable Land: An Ecological Approach.* CABI Publishing. Wallingford, UK.
61. Council Regulation (EC) no 824/2007 of 28 June 2007 on organic production and labeling of organic products and repealing regulation (EEC) 2092/91.
62. Wolfenbarger, L.L. & P.R. Phifer. 2000. The ecological risks and benefits of genetically engineered plants. *Science* **290:** 2088–2093.
63. European Commission- Standing Committee on Agricultural Research (SCAR). 2011. *Sustainable Food Consumption and Production in a Resource-Constrained World.* Retrieved January 17, 2012 from http://ec.europa.eu/research/agriculture/conference/pdf/feg2-report-web-version.pdf
64. Bonnieux, F., P. Dupraz & K. Latouche. 2006. *Experience with Agri-Environmental Schemes in EU and Non-EU Members.* Notre Europe. Available at: http://www.notre-europe.eu/fileadmin/IMG/pdf/Bonnieux-EN.pdf
65. Holzschuh, A., I. Steffan-Dewenter & T. Tscharntke. 2008. Agricultural landscapes with organic crops support higher pollinator diversity. *Oikos* **117:** 354–361.

Ann. N.Y. Acad. Sci. ISSN 0077-8923

ANNALS OF THE NEW YORK ACADEMY OF SCIENCES
Issue: *The Year in Ecology and Conservation Biology*

The ecology of *Anopheles* mosquitoes under climate change: case studies from the effects of deforestation in East African highlands

Yaw A. Afrane,[1,2] Andrew K. Githeko,[1] and Guiyun Yan[3]

[1]Climate and Human Health Research Unit, Centre for Global Health Research, Kenya Medical Research Institute, Kisumu, Kenya. [2]School of Health Sciences, Bondo University, Bondo, Kenya. [3]Program in Public Health, College of Health Sciences, University of California Irvine, Irvine, California

Address for correspondence: Dr. Yaw A. Afrane, Climate and Human Health Research Unit, Centre for Global Health Research, Kenya Medical Research Institute, P.O. Box 1578, Kisumu 40100, Kenya. yaw_afrane@yahoo.com

Climate change is expected to lead to latitudinal and altitudinal temperature increases. High-elevation regions such as the highlands of Africa and those that have temperate climate are most likely to be affected. The highlands of Africa generally exhibit low ambient temperatures. This restricts the distribution of *Anopheles* mosquitoes, the vectors of malaria, filariasis, and O'nyong'nyong fever. The development and survival of larval and adult mosquitoes are temperature dependent, as are mosquito biting frequency and pathogen development rate. Given that various *Anopheles* species are adapted to different climatic conditions, changes in climate could lead to changes in species composition in an area that may change the dynamics of mosquito-borne disease transmission. It is important to consider the effect of climate change on rainfall, which is critical to the formation and persistence of mosquito breeding sites. In addition, environmental changes such as deforestation could increase local temperatures in the highlands; this could enhance the vectorial capacity of the *Anopheles*. These experimental data will be invaluable in facilitating the understanding of the impact of climate change on *Anopheles*.

Keywords: climate change; *Anopheles* mosquitoes; deforestation; malaria

Introduction

Anopheles mosquitoes are responsible for the transmission of a number of diseases in the world, including malaria, lymphatic filariasis (*Wuchereria bancrofti* and *Brugia malayi*), and viruses such as the one that causes O'nyong'nyong fever, among others. In Africa, *Anopheles gambiae* is one of the best-known vector species because of its prominent role in the transmission of the most dangerous malaria parasite species—*Plasmodium falciparum*. *A. gambiae sensu lato* is a complex of at least seven morphologically indistinguishable sibling species. There are approximately 460 recognized *Anopheles* mosquito species worldwide, and over 100 of them are capable of transmitting human or animal diseases.

Anopheles mosquitoes, being poikilotherms, have life-history characteristics that are dependent on ambient temperatures. These life-history characteristics include their biting rates, the duration of their gonotrophic cycles, their fecundity, and the survival and development of the immature mosquitoes and the adult. Thus, any factor that can alter any of these characteristics has the capacity to affect the disease transmission potential of the mosquitoes. Environmental changes, climate variability, and climate change are such factors that could affect the biology and ecology of *Anopheles* vectors and their disease transmission potential.

Climate change is expected to lead to latitudinal and altitudinal temperature increases. Future global warming projections indicate that the best estimate of surface air warming for a "high scenario" is 4 °C, with a likely range of 2.4–6.4 °C by 2100.[1] Such a temperature increase will alter the biology and ecology of many mosquito

doi: 10.1111/j.1749-6632.2011.06432.x

vectors and, subsequently, the dynamics of the diseases they transmit. Because arthropods critically depend on ambient temperature for survival and development,[2] their distribution range is often limited by temperature. For example, in the high-elevation areas or highlands of Africa, ambient temperature is low, restricting the development and survival of *Anopheles* mosquitoes. However, climate warming or any factor that alters the microclimatic conditions of *Anopheles* mosquitoes (e.g., deforestation) in the highlands may facilitate the persistence of the mosquito population there.[3] Furthermore, changes in precipitation and humidity are also expected to occur under climate change scenarios; the synergistic effects between temperature and precipitation are expected to have major effects on the ecology of *Anopheles* mosquitoes and mosquito-borne diseases.[4] Increased precipitation may affect larval habitat availability and stability as well as habitat productivity.[5] The association between precipitation, vector abundance, and malaria prevalence has been well supported.[6]

Mosquito populations in the highlands may be more sensitive to climate change than those in the lowland areas of Africa. Climate warming can mediate mosquito physiology and metabolic rate because metabolic rate increases exponentially rather than linearly with temperature in ectotherms.[7] Therefore, *Anopheles* mosquitoes in highland areas are expected to experience a larger shift in metabolic rate due to the effects of climate warming. Climatic conditions affect *Anopheles* mosquitoes in a number of ways. The development rate of immature mosquitoes is highly dependent on temperature. Low-temperature conditions result in severe delays in larval development and can also result in high mortality. Below 16 °C, larval development of *A. gambiae*, the main malaria vector in most parts of Africa, will stop; the larvae will die in water temperature below 14 °C.[8] When water temperature rises, the larvae take a shorter time to mature[9] and, consequently, there is a greater capacity to produce more offspring. In the adult stage, an increase in ambient temperature will accelerate the digestion of blood meals taken by mosquitoes, leading to increased human biting frequency and faster parasite sporogonic development,[10] translating to an increased disease transmission efficiency.[11] Increased biting frequency and faster blood meal digestion also mean increased fecundity and better reproductive fitness.[12] However, temperature above 34 °C generally has a negative impact on the survival of vectors and parasites.[13]

Anthropogenic environmental changes such as deforestation, urbanization, and agricultural practice may have significant effects on mosquito habitat availability and microclimatic conditions of the aquatic habitats and human residences where adult mosquitoes rest. This review will focus on the impact of anthropogenic environmental changes on the ecology of *Anopheles* vectors and malaria disease transmission in African highlands. It gives the perspective of how anthropogenic changes in the environment could affect the microclimate of an area. This could be likened to the effect of climate change and is expected to increase our understanding of how climate change could affect the ecology of *Anopheles* mosquitoes.

Impact of land use and land cover changes on *Anopheles* mosquito biology and ecology

In many parts of the world, especially in Africa and South America, anthropogenic environmental changes such as deforestation have been linked to altered malaria transmission dynamics. Deforestation for the purposes of logging and self-subsistence agriculture is a serious problem in the tropical regions of Africa. For example, Malava forest, a tropical rainforest in Kakamega district, Kenya, shrank from 150 km^2 in 1965 to 86 km^2 in 1997.[14] In the East African highlands, 2.9 million hectares of forest were cleared between 1981 and 1990, representing an 8% reduction in forest cover in one decade.[15] Land use and land cover changes may modify the temperature and relative humidity of malaria vector habitats in the highlands. In the southwestern highlands of Uganda, maximum and minimum temperatures were shown to be significantly higher in communities bordering cultivated swamps than in those near natural swamps.[15] These changes in regional climate and in the microclimatic conditions of mosquito habitats cause abundant changes in the existing mosquito species and may make some areas permissive to the proliferation of new species. Land use and land cover changes have been linked to changes in vector ecology and malaria transmission. For instance, deforestation in Cameroon caused the introduction of *A. gambiae* into a habitat that was previously dominated by *A. moucheti*.[16] In northern Brazil, *A. marajoara*, a species previously of minor importance, became the principal malaria vector following changes in land use.[17]

A series of studies were conducted in the western Kenyan highlands to determine the influence of land use and land cover changes on the microclimatic condition of human residences, and subsequently on the ecology of *A. gambiae* mosquitoes, the primary malaria vector in the region.[5] The examined parameters included the duration of gonotrophic cycle, biting frequency, fecundity and survivorship, and sporogonic development of *P. falciparum* malaria parasite. The study area was originally forested but has experienced severe deforestation for agricultural development in the past three decades. Deforestation increased the indoor mean temperature by 1.8 °C. Mean maximum and minimum temperatures were increased by 2.3 and 1.5 °C, respectively. Mean maximum outdoor temperature was significantly higher by 1.4 °C in the deforested site than in the forested site (31.3 vs. 29.9 °C). Mean outdoor temperatures were significantly higher by 1 °C in the deforested site than in the forested site (19.9 vs. 18.9 °C). The mean indoor relative humidity in the deforested area was about 22.6% lower (79.88% vs. 57.29%) than in the forested area during the dry seasons.[3,10–12]

The changes in the microclimatic conditions in the human residences induced by deforestation significantly shortened the duration of the mosquitoes' gonotrophic cycle by 1.7 days (4.6 vs. 2.9 days).[11] The duration of the gonotrophic cycle is the period between the taking of a blood meal by a mosquito, including the digestion of the blood meal, until oviposition or egg laying.[18] The decreased duration of the gonotrophic cycles implies an increase in human biting frequency from an average of once every five days to once every three days. An increase in the biting frequency means that *A. gambiae* will feed more frequently on humans and enhance malaria transmission potential exponentially. The microclimatic changes also shortened the sporogonic development time from an average of 14 days down to 12.6 days.[10] Both oocyst and sporozoite development times were reduced by 1 and 1.4 days, respectively. Reduced parasite development time in mosquitoes indicates that the parasite took a shorter time to become infectious and, therefore, is transmitted much more efficiently.

However, these changes in microclimate due to deforestation did not favor the survival of *A. gambiae* adults. The effect of deforestation decreased median survival of *A. gambiae* by five to seven days. The *A. gambiae* mosquito prefers areas with high humidity; since deforestation caused a decrease in indoor humidity, it decreased the median survival of *A. gambiae*. However, despite the decreased survivorship of the mosquitoes due to the effects of deforestation, mosquitoes exhibited an enhanced reproductive fitness by 40% over the course of mosquito life span,[12] partly due to faster blood-meal digestion and more frequent blood-feeding. The implication of these findings is that *A. gambiae* could increase its population within a short time when breeding sites are available. This could potentially lead to an increase in malaria transmission when infected humans are available.

These findings in the western Kenyan highlands are consistent with findings by other investigators in other regions of Africa. In the highlands of Uganda, Lindblade *et al.*[19] compared mosquito density, biting rates, sporozoite rates, and entomological inoculation rates between eight villages located along natural papyrus swamps and eight villages located along swamps that have been drained and cultivated. They found that on average all malaria indexes were higher near cultivated swamps. Maximum and minimum temperatures were significantly higher in communities bordering cultivated swamps. The average minimum temperature of a village was significantly associated with the number of *A. gambiae* per house. Thus, replacement of natural swamp vegetation with agricultural crops led to increased temperatures and elevated malaria transmission risk in cultivated areas.

Land use and land cover changes also affected the microclimatic condition of mosquito larval habitats. Munga *et al.*[9] compared microclimatic conditions and *A. gambiae* larval development and survivorship in seminatural larval habitats under three land cover types (farmland, forest, and natural swamp). They found significantly higher water temperatures in farmland habitats as compared to the other land cover types. The mosquito pupation rate was significantly greater in farmland habitats than in swamp and forest habitats, while larval-to-pupal development times were significantly shorter. Land cover type may affect larval survivorship and habitat productivity through its effects on water temperature and nutrients in the aquatic habitats.

It is important to note that the effects of land use and land cover on malaria vectors discussed above may be specific to African highlands where low ambient temperature is the major limiting factor for vector development and reproduction and sporogonic development of malaria parasites. Meta-analysis on the impact of environmental changes on the development and reproduction of malaria vectors that include large number of study sites and various anopheline species may reveal general principles on the effects of environmental changes on malaria vectors and the underlying biological mechanisms.[22]

Proliferation of mosquito species to new areas

Global climate warming may render suitable the high-altitude areas previously unsuitable for proliferation of the mosquito vector population. Each mosquito species has its own minimum niche requirement, and one important limiting factor for the spatial distribution range is climate. For instance, *A. arabiensis,* the sibling species to *A. gambiae*, is either absent or shows a very low abundance in high-altitude areas. Chen *et al.*[20] reported *A. arabiensis* mosquitoes breeding in the central Kenyan highlands of elevation of 1,720–1,921 m above sea level for the first time, suggesting that local climate or ecological conditions have become conducive to the proliferation of malaria vector species. The consequence of new vector species persistence on malaria transmission may be significant and warrants careful and long-term vector and malaria monitoring.

Land use and land cover may modify the microclimatic conditions of mosquito vectors, which may further facilitate the establishment and persistence of populations in previously unsuitable areas. For example, Manga *et al.*[16] observed that deforestation from airport construction in Cameroon caused the introduction of *A. gambiae* into a habitat that was previously dominated by *A. moucheti*. In northern Brazil, *A. darlingi* is generally the dominant vector. However, land use and land cover changes made *A. marajoara*, a species previously of minor importance, the principal malaria vector.[17] This vector species is highly susceptible to malaria infection and exhibits anthropophilic biting behavior. Changes in the vectorial system pose novel and special challenges to malaria control due to the presence of various species with different resting and feeding behaviors and various extents in susceptibility to insecticides.

Afrane *et al.*[3] used life-table analysis to investigate whether climate conditions in the western Kenyan highlands were permissive to the development and survival of *A. arabiensis* and whether deforestation promoted *A. arabiensis* larval and adult survivorship. They found that the mean water temperature of aquatic habitats in the deforested area was 4.8–6.1 °C higher than that in the forested area, and *A. arabiensis* larval-to-adult survivorship was increased by 65–82%. They also noted that the larval-to-adult development time was shortened by eight–nine days as a result of deforestation. Deforestation is not solely responsible for such effects on microclimatic conditions and mosquito larval survival due to reduced canopy coverage. For example, Munga *et al.*[9] found that larval development time was significantly shortened in breeding sites in farmlands compared to breeding habitats in natural swamps in the western Kenyan highlands.

Deforestation also enhanced the survivorship and reproductive fitness of adult *A. arabiensis* mosquitoes in the highlands.[3] The average indoor temperature in the houses in the deforested area was 1.8° C higher than in the forested area, and the relative humidity was 22–25% lower. The median survival time of adult mosquitoes in the deforested area was 49–55% higher than those in the forested area, and the net reproductive rate of female mosquitoes in the deforested area was 1.7- to 2.6-fold higher than that in the forested area. As a result, *A. arabiensis* placed in the deforested area had better survival and laid more eggs to produce more offspring. The implications of these findings are that if the current trends of deforestation continue in the highlands, *A. arabiensis* could invade, inhabit, and proliferate in the highlands. The establishment and persistence of *A. arabiensis* in the highlands could worsen the malaria situation because of the resilience demonstrated by this vector species to control measures such as insecticide-impregnated bed-nets or indoor residual spray methods.

Climate warming and malaria resurgence in the East African highlands

The conventional wisdom is that climate plays a large role in malaria, especially in transition regions such as highlands and desert fringes where temperature and rainfall limit the abundance of mosquitoes. However, there have been strong debates over the last decade over whether climate warming has occurred in the East African highlands and whether climate is a driving force for a series of malaria epidemics observed in the 1990s in this region.[21] Hay *et al.*[21] concluded that mean temperature and rainfall have not changed significantly in the past century at four locations in the East African highlands, where malaria incidence has been increasing. Pascual *et al.*[22] reanalyzed the temperature trend data in the same East African sites and found evidence for a significant warming trend at all sites—a rise of approximately 0.5 °C in the last half of the 20th century. Omumbo *et al.*[23] used over 30 years of quality-controlled daily observations of maximum, minimum, and mean temperature in the analysis of trends at Kericho meteorological station, situated in Kenya's western highlands. They found that an upward trend of approximately 0.2 °C per decade could be observed in all three temperature variables. Mean temperature variations in the Kericho area were associated with large-scale climate variations including tropical sea surface temperatures. Local rainfall was found to have inverse effects on minimum and maximum temperature. They also used three versions of a spatially interpolated temperature data set, which showed markedly different trends when compared with each other and with their data. This study demonstrates that the increases in temperatures observed in the East African region could be attributed to the effects of climate change and environmental changes.

The effect of climate change on the epidemiology of malaria is less conspicuous. Given that malaria transmission depends on vector abundance, mosquito biting rate, vector survivorship, and parasite sporogonic rate, changes to the ambient temperature may affect many or all of these factors. Loevinsohn[24] assessed the contribution of the climate to a malaria epidemic in the late 1980s in Rwanda. In late 1987, malaria incidence in the area increased by 337% over the three previous years. He found that the increase was greatest in groups with little acquired immunity—children under two years and people in high-altitude areas. An autoregressive correlation analysis found that temperature, especially mean minimum, best predicted incidence at higher altitudes where malaria had increased most. Alonso *et al.*[25] developed a dynamic model that incorporated the population dynamics of the mosquito vector with the temperature time series and found that a small increase in ambient temperature has a major but nonlinear effect on malaria transmission. In parallel, climate variability, defined as short-term fluctuations around the mean climate state, is associated with clinical malaria incidence in the East African highlands despite a high spatial variation in the sensitivity of malaria incidence to climate fluctuations in the highlands.[26] Interestingly, temperature and rainfall exhibited nonlinear and synergistic effects on malaria incidence. Wanjala *et al.*[27] indicated that topographic and drainage features could explain the different malaria transmission rates in different western Kenyan highland areas with similar climate and elevation.

An often used but unreliable indicator of malaria incidence is clinical malaria cases taken from hospital records. The unreliability of this indicator stems from variations in health seeking behavior, health policy, human immunity, drug resistance, and other related factors. In order to measure the impact of climate warming on malaria transmission potential, one can use vectorial capacity (C). Vectorial capacity is defined as the daily rate at which future inoculations arise from a currently infective case;[28] this can be seen as a true measure of malaria transmission potential. Vectorial capacity is expressed as

$$C = \frac{ma^2 pn}{-\mathrm{Log}(P)},$$

where m is the relative density of vectors in relation to humans, a is the average number of humans bitten by one mosquito in one day, p is the proportion of vectors surviving per day, and n is the duration of sporogony in days. Afrane *et al.*[10] found that deforestation nearly doubled the vectorial capacity in the western Kenyan highlands (Table 1).

As observed, deforestation greatly facilitates malaria transmission in the western Kenyan highlands; one would expect reforestation or other methods of increasing canopy coverage for aquatic habitats to reduce malaria transmission. Indeed, Wamae *et al.*[29] showed that in the western Kenyan highlands, shading habitats by growing Napier grass around mosquito breeding habitats reduced the temperature of aquatic habitats by as much as 5 °C, which significantly reduced habitat productivity by more than 85%. Forests have been shown

Table 1. Estimated vectorial capacity of *Anopheles gambiae* in forested and deforested areas in western Kenya highland; adapted from Afrane *et al.*[10]

Land use type	m	a	n	P	Vectorial capacity
Forested	3.05	0.198	13.9	0.927	0.54
Deforested	4.64	0.233	12.8	0.917	0.96

Note: m is the relative density of vectors in relation to humans; P is the proportion of vectors surviving per day; a is the average number of men bitten by one mosquito in one day; n is the duration of sporogony in days.

to stabilize local temperatures and reforestation can help in mitigating the effects of climate change on malaria.

Conclusions

Climate change has multifaceted effects on malaria transmission in the African highlands, where temperature and rainfall limit the abundance of mosquitoes. Climate warming may make the areas previously too cool for vector population establishment now suitable, causing an expansion of vector species to higher altitude areas. Warm climate may facilitate larval development, enhance vector survivorship and reproductive fitness, and increase the blood feeding frequency and parasite sporogonic development rate in the previously cooler highlands. These effects have been explained using scenarios from anthropogenic environmental changes in east Africa and other parts of the world. Environmental changes, either natural or anthropogenic, alter the ecological balance and context within which vectors and their parasites breed, develop, and transmit diseases[30] and may facilitate or reduce malaria transmission. Deforestation in the western Kenyan highlands exhibited strong positive effects on the development and survival of *A. gambiae* and *A. arabiensis* mosquito larvae, enhanced the survivorship of adult mosquitoes, increased human feeding frequency, and shortened the development time of the parasites through effects on the microclimatic condition of the mosquitoes. It is important to note that climate change is also involved with abnormal rainfall patterns. Synergistic effects between temperature and rainfall on vector ecology and malaria transmission may be produced. Climate warming is expected to have continental or regional effects on the ecology of *Anopheles* mosquitoes and their malaria transmission potential, while the influence of environmental changes on malaria transmission is local and may be site specific. As climate warming and environmental changes may have a long-term effect on malaria transmission, it is imperative to conduct systematic, long-term surveillance in various sentinel sites in Africa and other continents on climate, vector dynamics, and community structure, and malaria transmission dynamics. These experimental data will be invaluable in helping to understand the impact of climate change on malaria and to develop an effective approach in controlling malaria.

Acknowledgments

The authors were supported by grants from the National Institute of Health (R01 AI094580, D43 TW001505, and R01 AI050243).

Conflicts of interest

The authors declare no conflicts of interest.

References

1. IPCC. 2007. *Climate Change 2007: Impacts, Adaptation, and Vulnerability*. Cambridge University Press. Cambridge, UK.
2. Lindsay, S.W. & M.H. Birley. 1996. Climate change and malaria transmission. *Ann. Trop. Med. Parasitol.* **90:** 573–588.
3. Afrane, Y.A., G. Zhou, B.W. Lawson, *et al.* 2007. Life-table analysis of *Anopheles arabiensis* in western Kenya highlands: effects of land covers on larval and adult survivorship. *Am. J. Trop. Med. Hyg.* **7:** 660–666.
4. Githeko, A.K., S.W. Lindsay , U.E. Confalonieri & J.A. Patz. 2000. Climate change and vector-borne diseases: a regional analysis. *Bull. World Health Organ.* **78:** 1136–1147.
5. Githeko, A.K. & W. Ndegwa. 2001. Predicting malaria epidemics in the Kenya highlands using climate data: a tool for decision makers. *Global Change Human Health* **2:** 54–63.
6. Dery, D.B., C. Brown, K.P. Asante, *et al.* 2010. Patterns and seasonality of malaria transmission in the forest-savannah transitional zones of Ghana. *Malar. J.* **9:** 314.
7. Gillooly, J.F., J.H. Brown, G.B. West, *et al.* 2001. Effects of size and temperature on metabolic rate. *Science* **293:** 2248–2251.
8. Koenraadt, C.J., K.P. Paaijmans, P. Schneider, *et al.* 2006. Low larval vector survival explains unstable malaria in the western Kenya highlands. *Trop. Med. Int. Health* **11**(8): 1195–1205.
9. Munga, S., N. Minakawa, G. Zhou, *et al.* 2007. Survivorship of immature stages of *Anopheles gambiae* s.l. (Diptera: Culicidae) in natural habitats in western Kenya highlands. *J. Med. Entomol.* **44:** 758–764.
10. Afrane, Y.A., T. Little, B.W. Lawson, *et al.* 2008. Deforestation increases the vectorial capacity of *Anopheles gambiae* giles to

transmit malaria in the western Kenya highlands. *Emerg. Infect. Dis.* **14**(10): 1533–1538.
11. Afrane, Y.A., B.W. Lawson, A.K. Githeko, & G. Yan. 2005. Effects of microclimatic changes due to land use and land cover on the duration of gonotrophic cycles of *Anopheles gambiae* giles (diptera: culicidae) in western Kenya highlands. *J. Med. Entomol.* **42:** 974–980.
12. Afrane, Y.A., G. Zhou, B.W. Lawson, *et al.* 2006. Effects of microclimatic changes due to deforestation on the survivorship and reproductive fitness of *Anopheles gambiae* in western Kenya highlands. *Am. J. Trop. Med. Hyg.* **74:** 772–778.
13. Rueda, L.M., K.J. Patel, R.C. Axtell & R.E. Stinner. 1990. Temperature-dependent development and survival rates of Culex quinquefasciatus and Aedes aegypti (Diptera: Culicidae). *J. Med. Entomol.* **27**(5): 892–898.
14. FAO. 1993. Forest Resources Assessment, 1990: Tropical Countries. FAO forestry Paper No. 112, Rome, Italy.
15. Lindblade, K.A., E.D Walker, A.W. Onapa, *et al.* 2000. Land use change alters malaria transmission parameters by modifying temperature in a highland area of Uganda. *Trop. Med. Int. Health.* **5:** 263–274.
16. Manga, L., J.C. Toto, & P. Carnevale. 1995. Malaria vectors and transmission in an area deforested for a new international airport in southern Cameroon. *Societes Belges Medicine Tropicale* **75:** 43–49.
17. Conn, J.E., R.C. Wilkerson, M.N. Segura, *et al.* 2002. Emergence of a new neotropical malaria vector facilitated by human migration and changes in land use. *Am. J. Trop. Med. Hyg.* **66:** 18–22.
18. Santos, R.L., O.P. Forattini & M.N. Burattini. 2002. Laboratory and field observations on duration of gonotrophic cycle of Anopheles albitarsis s.l. (Diptera: Culicidae) in southeastern Brazil. *J. Med. Entomol.* **39**(6): 926–930.
19. Lindblade, K.A., D.B. O'Neill, D.P. Mathanga, *et al.* 2000. Treatment for clinical malaria is sought promptly during an epidemic in a highland region of Uganda. *Trop. Med. Int. Health* **5:** 865–875.
20. Chen, H., A.K. Githeko, G. Zhou, *et al.* 2006. New records of *Anopheles arabiensis* breeding on the Mount Kenya highlands indicate indigenous malaria transmission. *Malar. J.* **7:** 17.
21. Hay, S.I., J. Cox, D.J. Rogers, *et al.* 2002. Climate change and the resurgence of malaria in the East African highlands. *Nature* **415:** 905–909.
22. Pascual, M., J.A. Ahumada, L.F. Chaves, *et al.* 2006. Malaria resurgence in the East African highlands: temperature trends revisited. *Proc. Natl. Acad. Sci. USA* **11:** 5635–5636.
23. Omumbo, J.A., B. Lyon, S.M. Waweru, *et al.* 2011. Raised temperatures over the Kericho tea estates: revisiting the climate in the East African highlands malaria debate. *Malar. J.* **10:** 12.
24. Loevinsohn, M.E. 1994. Climatic warming and increased malaria incidence in Rwanda. *Lancet* **343:** 714–718.
25. Alonso, D., M.J. Bouma & M. Pascual. 2011. Epidemic malaria and warmer temperatures in recent decades in an East African highland. *Proc. Biol. Sci.* **278:** 1661–1669.
26. Zhou, G., N. Minakawa, A.K. Githeko & G. Yan. 2004. Association between climate variability and malaria epidemics in the East African highlands. *Proc. Natl. Acad. Sci. USA* **101:** 2375–2380.
27. Wanjala, C.L., J. Waitumbi, G. Zhou & A.K. Githeko. 2011. Identification of malaria transmission and epidemic hotspots in the western Kenya highlands: its application to malaria epidemic prediction. *Parasit. Vectors* **4:** 81.
28. MacDonald, G. 1957. *The Epidemiology and Control of Malaria.* Oxford University Press. Oxford, UK.
29. Wamae, P.M., A.K. Githeko, D.M. Menya & W. Takken. 2010. Shading by napier grass reduces malaria vector larvae in natural habitats in Western Kenya highlands. *Ecohealth* **7:** 485–497.
30. Patz, J.A., T.K. Graczyk, N. Geller & A.Y. Vittor. 2000. Effects of environmental change on emerging parasitic diseases. *Int. J. Parasitol.* **30:** 1395–1405.

Ann. N.Y. Acad. Sci. ISSN 0077-8923

ANNALS OF THE NEW YORK ACADEMY OF SCIENCES
Issue: *The Year in Ecology and Conservation Biology*

Ecology and conservation biology of avian malaria

Dennis A. LaPointe,[1] Carter T. Atkinson,[1] and Michael D. Samuel[2]

[1]U.S. Geological Survey, Pacific Island Ecosystems Research Center, Hawaii National Park, Hawaii. [2]U.S. Geological Survey, Wisconsin Cooperative Wildlife Research Unit, University of Wisconsin, Madison, Wisconsin

Address for correspondence: Dennis A. LaPointe, U.S. Geological Survey, Pacific Island Ecosystems Research Center, Hawaii National Park, HI 96718. dennis_lapointe@usgs.gov

Avian malaria is a worldwide mosquito-borne disease caused by *Plasmodium* parasites. These parasites occur in many avian species but primarily affect passerine birds that have not evolved with the parasite. Host pathogenicity, fitness, and population impacts are poorly understood. In contrast to continental species, introduced avian malaria poses a substantial threat to naive birds on Hawaii, the Galapagos, and other archipelagoes. In Hawaii, transmission is maintained by susceptible native birds, competence and abundance of mosquitoes, and a disease reservoir of chronically infected native birds. Although vector habitat and avian communities determine the geographic distribution of disease, climate drives transmission patterns ranging from continuous high infection in warm lowland forests, seasonal infection in midelevation forests, and disease-free refugia in cool high-elevation forests. Global warming is expected to increase the occurrence, distribution, and intensity of avian malaria across this elevational gradient and threaten high-elevation refugia, which is the key to survival of many susceptible Hawaiian birds. Increased temperatures may have already increased global avian malaria prevalence and contributed to an emergence of disease in New Zealand.

Keywords: avian malaria; Hawaiian forest birds; *Culex* mosquito; *Plasmodium*; climate change

Introduction

Avian malaria, a disease caused by protozoan parasites in the genus *Plasmodium*, has played a seminal role as a model for human malarial infection since these common intraerythrocytic parasites of wild birds were first recognized.[1–5] Today, avian malaria is more frequently used as a model system to investigate general host–parasite interactions, coevolutionary processes, and the role of parasites in host life-history evolution. The pathogenicity of most avian species of *Plasmodium* is poorly understood but ranges from sublethal effects on host fitness to population decline and extinction. One species, *Plasmodium relictum*, plays an important role as a limiting factor in the distribution and abundance of native Hawaiian forest birds.[6–11] In this paper, we provide a review of the ecology and epidemiology of avian malaria, its impact on avian communities, and how climate and landscape change may alter this disease system. We rely on examples from the Hawaiian Islands, but many interactions are universal and applicable to insular and continental avian malaria, human malaria, and other mosquito-borne diseases enzootic in passerines, such as West Nile virus.

Natural history

Avian malaria is a common mosquito-transmitted disease of wild birds that has a world-wide distribution. The disease is caused by intracellular protozoan parasites in the genus *Plasmodium*, which share morphological and developmental features with closely related haemosporidian parasites in the genera *Haemoproteus* and *Leucocytozoon*.[12,13] Avian malaria is a complex, spatially heterogeneous host–parasite disease caused by more than 40 parasite species that differ widely in host range, geographic distribution, vectors, and pathogenicity.[12] Although the taxonomy of *Plasmodium* is currently in flux and supporting field studies have been limited,[10] diversity of mitochondrial genes suggests that genetic

doi: 10.1111/j.1749-6632.2011.06431.x

structure may underlie differences in host susceptibility, vector competence, and parasite virulence.[14]

The parasites occur in many avian species and families but primarily affect passerine birds.[5] The mosquito vectors of avian malaria are generalist blood feeders that likely transfer parasites among multiple avian species.[15] Although there are numerous reports of individual birds with acute, pathogenic infections with *Plasmodium*, the impacts of these blood parasites on host fitness and host population dynamics in wild birds is poorly understood.[16] Reports of epizootics are rare and mostly associated with birds in zoological collections, introductions of parasites or mosquito vectors to remote islands, or domestic birds that are exposed to sylvatic cycles of malaria outside of their geographic origins.[17–19] At one extreme, avian malaria causes high mortality in naive Hawaiian birds and has been implicated in the extinction, population decline, and restricted distribution of multiple Hawaiian bird species.[6,7,10,17,20] At the other extreme, there is little evidence of overt mortality in wild bird populations that have a long evolutionary association with the parasites. There is increasing evidence, however, that malaria may have significant effects on host fitness, including mate selection, reproductive success, and immune response.[16,21]

The species of *Plasmodium* that infect birds have a cosmopolitan distribution and are found in all major zoogeographic regions of the world with the exception of Antarctica, where mosquito vectors responsible for their transmission do not occur. Reports of *Plasmodium* from the Australian region are notably limited, although it is not clear whether this reflects inadequate sampling or a distributional anomaly.[5,22] Seven species of *Plasmodium* have a cosmopolitan distribution. *Plasmodium relictum* and *P. circumflexum* have the broadest geographic distribution and are reported from the Nearctic, Palearctic, Oriental, Ethiopian, Neotropical, and Australian regions. *P. vaughani*, *P. cathemerium*, *P. nucleophilum*, *P. rouxi*, and *P. elongatum* have been reported from every region except the Australian region.[22]

Plasmodium infections have been reported from all avian orders with the exception of the Struthoniformes (ostriches), the Coliiformes (mousebirds), and the Trogoniformes (trogons and quetzals), but only about half of all avian species have been examined. The greatest diversity of *Plasmodium* species are found in Galliformes, Columbiformes, and Passeriformes.[5] *P. relictum* has one of the widest host ranges, occurring in more than 400 species in 70 different avian families.[5,12,22] The relatively broad host range of *Plasmodium* is characteristic of this genus, but exceptions are common.[5,12] For example, *P. elongatum* is known from as few as 67 avian species and *P. hermani* is recognized only from North American wild turkeys (*Meleagris gallopavo*). Recently, molecular methods revealed a greater complexity in the genetic lineages of *Plasmodium* (and closely related *Haemoproteus*) that is currently difficult to reconcile with traditional morphological species.[23] Multiple lineages can occur in the same host individual, and their occurrence in a wide range of avian orders, families, and species is much broader than previously recognized.[24–27]

Mosquitoes in the genus *Culex* are believed to be the most common vectors of avian *Plasmodium*; however, only *Culex quinquefasciatus*, *Cx. tarsalis*, and *Cx. stigmatasoma* have been identified as natural vectors of *P. relictum* in California and only *Cx. quinquefasciatus* in Hawaii.[28,29] By contrast, more than 60 different species of culicine and anopheline mosquitoes can support experimental development of a variety of *Plasmodium* from avian hosts.[30] *P. relictum*, one of the best studied species of avian malaria, can complete development in at least 26 mosquito species from four different genera (*Culex*, *Aedes*, *Culiseta*, and *Anopheles*) in the laboratory. More recently, molecular assays for avian malaria have identified a number of potential vector species.[31–36] Although few of these studies have confirmed the presence of salivary gland sporozoites or demonstrated transmission, they support the notion that relatively little host specificity among most parasite–vector associations may facilitate switching of vertebrate hosts.[37]

Epidemiology

Although mosquito transmission of *Plasmodium* was discovered by Ross,[38] much of what we know about the life cycle of avian malaria was described 30–40 years later in a series of detailed studies of the blood and tissue stages of *P. elongatum*[39,40] and *P. gallinaceum*.[41,42] Other than scattered reports and descriptions of tissue or exoerythrocytic stages of infection, only a handful of additional species have been studied in any detail.[5] All have an initial cycle of asexual, preerythrocytic reproduction (preerythrocytic merogony) within host tissues immediately after sporozoites are inoculated by mosquitoes. This is followed by multiple cycles of asexual reproduction within circulating erythrocytes (erythrocytic merogony) and host tissues (exoerythrocytic merogony) that eventually leads to production of gametocytes within circulating blood cells (gametogony). Gametocytes of all species of avian *Plasmodium* remain in circulation and do not continue development until ingested by a vector. Once in the midgut of a mosquito vector, gametocytes undergo gametogenesis to form true gametes. Male gametocytes produce up to eight, flagellated microgametes. One microgamete will fertilize a macrogamete, and within 24 hours a motile zygote develops that is capable of penetrating the midgut wall to begin development as an oocyst. These initial stages of gametogenesis and fertilization exhibit little or no specificity for mosquito vectors and can be completed *in vitro*. It is only during invasion of the membrane surrounding the blood meal and subsequent penetration of the midgut epithelium that development can be blocked.[43]

Oocysts undergo a type of asexual reproduction called sporogony and eventually produce thousands of sporozoites. The duration of sporogony is dependent on temperature, and at optimal temperatures oocysts can mature within seven days. When oocysts reach a diameter of approximately 40 μm, they rupture, releasing sporozoites that move to the salivary glands, penetrate the glandular cells, and eventually gain access to the salivary ducts. At this point, the mosquito is considered infective, and upon subsequent blood feeding, sporozoites pass with the saliva into a new avian host to initiate infection.

Birds typically undergo an acute phase of infection where parasitemia increases steadily to a peak called the crisis, approximately 6–12 days after parasites first appear in the blood. After this acute phase, intensity appears to be influenced by the complex interplay of host immunity, seasonal photoperiod, and hormones associated with reproduction. Infected birds are typically anemic, lethargic, anorexic, and have ruffled feathers. Hematocrits may fall by more than 50%. Disease progression and clinical signs closely parallel the number of parasites in the peripheral blood circulation and the reticuloendothelial system.[44] The hallmark signs of acute infections with *Plasmodium* include thin, watery blood with enlargement and discoloration of

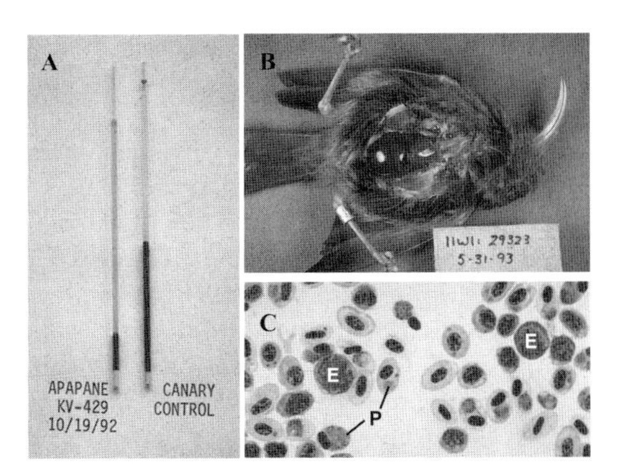

Figure 1. Hallmark signs of acute *Plasmodium* infections. (A) Severe anemia and low packed cell volume for an apapane infected with *P. relictum*. (B) Gross lesions (enlarged, darken liver) associated with a fatal malarial infection in an iiwi. (C) Blood smear of iiwi with an acute infection of *P. relictum*. Note parasitized erythrocytes (P) and erythroblasts (E).

the liver and spleen by deposition of malarial pigment in tissue macrophages (Fig. 1). Enlargement of these organs is due to hypercellularity and increased phagocytic activity of macrophages rather than edema.[45] Fatal pathologies may arise from anemia as infected erythrocytes rupture during asexual parasite reproduction and as both infected and uninfected erythrocytes are removed by the spleen. During the acute phase, indirect mortality may occur when birds weakened by anemia succumb to environmental stressors such as predation, starvation, or inclement weather.[46]

In birds that survive infection, the acute phase is followed by a rapid decline in intensity of infection to chronic levels as strong antibody and cell-mediated responses develop to the parasites.[44] Chronic infections likely persist for the lifetime of infected birds at extremely low intensities, and both circulating parasites and persistent exoerythrocytic meronts can serve as a source for recrudescing infections.[3,47,48] When chronically infected birds are rechallenged with homologous strains of the parasite, they may have only brief, low-intensity increases in peripheral parasitemia.[2,49,50] This concept of prolonged chronic infection that stimulates immunity to reinfection is fundamental to the epidemiology of avian malaria, yet supporting experimental evidence is limited to only a handful of studies in a few avian species.[48,51–53]

In temperate climates, a recrudescence, or spring relapse, occurs during the breeding season when increased corticosterone levels suppress the host's immune system.[54] Recrudescence of chronic infections is believed to facilitate seasonal transmission in temperate climates where vector populations may have corresponding seasonal population increases, but this may not apply to all species of avian *Plasmodium*. In Europe, absence of some lineages of *P. relictum* in hatch-year birds indicates that transmission occurs on the wintering rather than the breeding grounds.[55] Seasonal recrudescence has not been reported in Hawaii, but concomitant infections with immunosuppressing pox virus may trigger malaria recrudescence in the Hawaiian system.[7,12]

Population impacts

There is relatively little evidence that avian *Plasmodium* causes major epizootic die-offs in most natural bird hosts. The most significant reports of pathogenicity from avian *Plasmodium* occur in birds with acute infections, typically captive birds in zoological collections and the avifauna of isolated islands where new host–parasite associations become established. In a thorough review of over 5,000 papers on avian blood parasites, only about 4% reported mortality or pathogenicity, mostly in domestic birds or zoological collections.[22] Avian malaria is particularly pathogenic in captive penguins exposed to mosquito vectors from outside of their natural range.[56,57] At the Rotterdam Zoo, mortality from avian malaria occurred annually among Atlantic puffins (*Fratercula arctica*), common guillemots (*Uria aalge*), and black-legged kittewakes (*Rissa tridactyla*), as well as black-footed penguins (*Spheniscus demersus*).[58] Although mortality from *Plasmodium* has not been reported in wild penguins,[59,60] future spread of new mosquito vectors and potential effects of global climate change may place wild colonies at risk.[61,62]

There are only a handful of documented *Plasmodium* deaths of wild birds in the United States and other parts of the world, and virtually all are associated with acute infections.[5] For example, intense transmission of *Plasmodium* and other haemosporidians in the rookeries of wading birds (ciconiiformes) in Venezuela is suspected to be a cause of high levels of nestling mortality, but it is unclear if other factors are also involved.[5,63] Similarly, acute infections may lead to increased predation on infected hosts.[46,64]

The more significant impacts of these parasites may be subclinical and indirect, with long-term effects on the lifetime reproductive success of their avian hosts. For example, male white-crowned sparrows (*Zonotrichia leucophrys oriantha*) infected with *Plasmodium* sing fewer songs in response to experimental playbacks. Parasite-induced changes in song frequency or quality may affect female mate selection.[65] More direct effects of *Plasmodium* infection on reproductive success have been demonstrated in wild populations of blue tits (*Cyanistes caeruleus*) and great reed warblers (*Acrocephalus arundinaceus*).[66–70]

In contrast to malaria in continental species, introduced avian malaria poses a substantial threat to naive endemic birds on isolated islands. The accidental introduction of *P. relictum* and the southern house mosquito (*Culex quinquefasciatus*) to the Hawaiian Islands has devastated native Hawaiian forest birds[6,7] and continues to play a significant role in limiting the geographic and altitudinal distribution of remaining species.[71,72] Although numerous limiting factors have contributed to extinctions of Hawaiian avifauna, avian malaria is believed to be one of the key components driving population collapse in otherwise suitable, lowland (<900 m above sea level, asl) habitat. Mortality due to acute infection appears to be the major contributor to population effects.[6,7,20,46,71,73] Chronic infections may affect survival in adult Hawaii amakihi (*Hemignathus virens*) but do not affect reproductive success or prevent populations from growing.[74] Although many native species continue to decline in forests where avian malaria is prevalent, the Hawaii amakihi appears to be evolving tolerance to infection, and lowland populations have rebounded dramatically in recent years.[7,8,75]

Much less is known about the impact of avian malaria on the avifauna of New Zealand and the Galapagos Archipelago.[60,62,76,77] Although *Cx. quinquefasciatus* arrived in the North Island of New Zealand (1830s) about the same time as in the Hawaiian Islands (*circa* 1826) and malarial infections were reported in introduced birds as early as 1920,[78] significant deaths from avian malaria among captive native birds have only been reported in the last 15 years (New Zealand dotterel [*Charadrius obscures*], mohua [*Mohoua ochrocephala*], South Island saddlebacks [*Philesturnus carunculatus carunculatus*]).[79–82] Prevalence of *Plasmodium* by polymerase chain reaction (PCR) in wild nonnative birds ranges from 9% to 26%, suggesting the disease is more prevalent than previously thought.[60,62] Recent surveys of endemic birds have detected chronic malarial infections, as well as cases of acute mortality, in wild, endemic, and indigenous birds, and there is increasing concern that transmission of avian malaria may be limiting success of some reintroduction programs.[82] In the Galapagos, avian malaria has recently been reported in the endemic Galapagos penguins and, while there is no evidence of clinical illness or mortality, there is also concern that spread into the endemic avifauna of this archipelago might parallel what happened to native Hawaiian forest birds after introduction of *Plasmodium*.[60,76,77]

Ecological drivers of transmission

Several common drivers of transmission have been identified in continental studies of avian malaria, including host density, proximity to water, temperature, and host immunocompetence.[83–86] Comprehensive studies of host and vector dynamics and the environmental factors that influence transmission rates have been done in the Hawaiian Islands.[19] Although it has been argued that the Hawaiian system may not be applicable to continental areas because both the vector and parasites have been introduced,[69] its simplicity makes it highly relevant for understanding the complexities of mainland systems where multiple vectors and parasite species, as well as human influences, make it difficult to measure ecological effects on host fitness.[86,87]

The avian malaria system in Hawaii includes many factors that facilitate transmission of both endemic and epizootic disease across a geographically broad landscape, including a novel pathogen in highly susceptible naive hosts,[71,72] broad host range, a widely dispersed and highly competent vector,[7,29,95] favorable climate for vector and parasite,[11,88] and an efficient reservoir of chronically infected native birds.[20,73]

Habitat and human impacts on vector abundance

The geologically young, volcanic landscape of the Hawaiian Islands provides limited natural freshwater habitat for the mosquito vector of avian malaria. On the older islands (Kauai, Oahu, Maui, and Molokai) and volcanoes like Mauna Kea, erosive forces have exposed less permeable substrates, creating perched wetlands or rock pools along streambeds. As a result, Kauai's topography is characterized by numerous perennial streams and vast open bogs, while on the relatively young volcano of Mauna Loa, surface hydrology is limited to a few small bogs, ground pools, and intermittent streams. Depressions in exposed lava substrates along intermittent stream beds and elsewhere impound rainwater and surface drainage to create suitable larval mosquito habitat,[89,90] but there is little evidence of available natural tree holes or other phytotelmata in Hawaiian forests.[91,92] In Hawaii, forest fragmentation by natural and human activities increases the likelihood of vector and native bird interactions. On the slopes of the active Mauna Loa and Kilauea volcanoes, forests are naturally fragmented by lava flows into islands or *kipuka* where native birds and vectors may be concentrated in crucibles of malarial transmission.[93,94] In general, conservation areas in Hawaii exist within a matrix of residential and agricultural areas where roadways and open landscapes may enhance mosquito abundance and dispersal into native forest bird habitat.[95]

Other human activities and alterations of the environment can significantly enhance vector abundance, richness, and disease transmission.[96] Road construction, forestry, agriculture, ranching, and residential development have all contributed to vector abundance. *Cx. quinquefasciatus* is, primarily, a peridomestic mosquito and is found in great abundance in suburban and agricultural areas where artificial containers and impoundments are common larval habitats.[92,97–99] However, this mosquito also occurs in Hawaiian forests where feral pigs are managed as a game species. Feral pigs increase the abundance of mosquitoes by foraging on native tree ferns.[90,100] In forests with minimal surface water, pigs create cavities in tree fern trunks that fill with leaf litter and rainwater to form larval mosquito habitat (Fig. 2). More larval habitat leads to increased vector abundance, an important factor in determining transmission and prevalence rates of avian malaria throughout low- and midelevation (900–1500 m asl) forests.[20]

In New Zealand, there is a positive association between the distribution of *P. relictum* and introduced *Cx. quinquefasciatus*.[62] Anthropogenic landscape change may be facilitating the geographic spread of *Cx. quinquefasciatus*, as deforestation and

Figure 2. Tree fern (*Cibotium glacum*) cavities are created by the feeding behavior of feral pigs in Hawaiian wet forests. In forests where surface hydrology is limited, tree fern cavities are the primary larval habitat for the mosquito vector of avian malaria, *Culex quinquefasciatus.*

agriculture have created nutrient-rich larval habitat. The native New Zealand mosquito, *Cx. pervigilanus*, prefers aquatic habitat with medium detrital loads, leaving habitat with more detritus available for introduced species like *Cx. quinquefasciatus.*[101] In the Galapagos, populations of *Cx. quinquefasciatus* are closely associated with human residence and agriculture.[102] As human populations grow and move among the five inhabited islands, *Cx. quinquefasciatus* populations will likely increase in distribution, abundance, and seasonal occurrence unless peridomestic sources of water are managed for mosquito control.[102,103]

This positive relationship between forest disturbance, vector abundance, and avian malaria prevalence defines the emergent nature of *P. relictum* where *Cx. quinquefasciatus* has been introduced to parasite-depauperate island ecosystems. In the continental forests of Cameroon, West Africa, however, avian malaria prevalence in native rain forest birds may increase or decrease in response to forest disturbance.[104,105] There are many possible nonmutually exclusive explanations for these findings. For example, deforestation and encroaching agriculture may destroy or create larval habitat for specific vectors, affect the general health of hosts, or alter key vector–host interactions through native biodiversity loss or the introduction of livestock.[104]

Reservoirs of infection and the dilution effect

Human activities have greatly altered avian communities in Hawaii through the introduction of at least 17 avian species that have become established in forest habitats. The most abundant and widely dispersed species in Hawaiian forests is the Japanese white-eye (*Zosterops japonicas*), a species first released in 1929.[106] Our understanding of the role these introduced species play in the avian malaria disease system has changed over time. Warner[6] found malarial infections in two nonnative species, the house finch (*Carpodacus mexicanus*) and the Japanese white-eye, but not in native apapane or amakihi, suggesting that nonnative species were the reservoirs of malaria. Conversely, van Riper *et al.*[7] detected parasitemias of up to 30% in apapane and less than 3% in nonnative species, including the northern cardinal (*Cardinalis cardinalis*), Japanese white-eye, and red-billed leiothrix (*Leiothrix lutea*).[73] Recent archipelago-wide surveys using PCR and serological diagnostics[8,107–109] agree with the findings of van Riper *et al.*[7] with the possible exception that prevalence (based on PCR diagnostics) in Oahu amakihi (*Hemignathus chloris*) is lower than nonnative species.[110] Laboratory challenge experiments have shown that native birds surviving infection acquire immunity to reinfection but retain a chronic, low parasitemia, making these individuals efficient, lifetime reservoirs of disease.[7,72] Unfortunately, the malaria reservoir in native Hawaiian birds may well be the Achilles heel for these bird populations.[73] Efforts to control malaria transmission by removal of reservoir hosts would be counterproductive and may hinder natural selection for disease resistance in native species.[8,9]

In laboratory challenge experiments, Japanese white-eye and red-billed leiothrix develop transient malarial infections but are incapable of infecting mosquitoes beyond a brief period following the acute phase, thereby, making them poor reservoirs of disease. In contrast, the nonnative house sparrow (*Passer domesticus*) develops acute parasitemias that are intermediate in intensity and remain infectious to mosquitoes up to one year. House sparrows may serve as disease reservoirs for native birds inhabiting or foraging in the anthropogenic ecotones favored by this species. Isolated house sparrow populations may even allow for the evolution of more virulent strains of *P. relictum* that could ultimately spill over into the forest bird community.[97]

Because of their relative resistance to infection and short-lived parasitemias, Japanese white-eye and red-billed leiothrix are incompetent reservoirs that may serve to buffer or dilute disease transmission from infective mosquitoes.[99,111] Because Japanese white-eyes and other nonnatives make up a large proportion of some forest bird communities, one might expect a significant dilution effect. However, nonnative species, especially Japanese white-eyes and red-billed leiothrix, are more defensive to mosquito blood feeding and may thereby reduce the potential dilution effect. The potential dilution effect may be more significant when agriculture encroaches on forests and people and livestock become alternate blood hosts for generalist vectors.[104]

Climate

Although habitat and avian communities largely determine the geographic distribution of avian malaria, climatic factors are likely the strongest drivers of transmission.[20,87] Temperature and rainfall in Hawaii are highly variable across different temporal and spatial scales with significant variation across small geographical distances depending on elevation, aspect, and slope.[112] Rainfall occurs throughout the year with the greatest accumulation between October and March, but annual and seasonal rainfall can vary substantially between wet and dry years.[113] Rainfall can have direct and indirect effects on malarial transmission by influencing larval habitat availability and larval and adult survivorship. In Hawaiian rainforests, mean annual precipitation (2000–6000 mm) is sufficient to maintain larval mosquito habitats and keep relative humidity high for adult survival. However, the extended droughts associated with El Niño Southern Oscillation events will reduce available larval habitat and adult survivorship.[88,97,114] Extreme rainfall events (>200 mm/day) can also have a negative effect on larval and adult survivorship by flooding larval habitat and causing direct mortality to adults.[88,97,115]

Temperature effects on the disease system are not limited to extreme conditions. Seasonal temperatures have a direct effect on larval mosquito development that influences adult abundance. The link between human malaria transmission and climate has been recognized for a long time,[116] but the significance of temperature to the distribution of avian malaria is just being explored.[11,87] In a spatial analysis of avian malaria prevalence in the olive sunbird (*Cyanomitra olivacea*) across 28 sites in Central and West Africa, Sehgal *et al.*[87] found the maximum temperature of the warmest month to be the most important environmental factor influencing local prevalence. In Hawaii, where annual changes in mean temperature are minimal, elevation accounts for much of the variation in ambient temperatures. Temperature decreases with elevation, and lower temperature slows the development of mosquito larvae.[117] Lowland *Cx. quinquefasciatus* populations consist of year-round, overlapping generations that increase during the most favorable parts of the year. At higher elevations (≥900 m asl), populations consist of fewer cohorts which are seasonal and less abundant.[88,97] Temperature also influences survival of adults and duration of the gonotrophic cycle (time between blood meal and oviposition), factors that contribute to disease transmission.[20,97] Perhaps the most important temperature effect on avian malaria transmission is through control of the parasite extrinsic incubation period, the interval of parasite development in the vector from blood meal acquisition to salivary gland infection. This incubation period lasts 6 to 28 days, is inversely dependent on temperature, and ultimately restricts the altitudinal distribution of infectious mosquitoes and disease transmission.[11] *P. relictum* requires a minimum of 13° C for development; therefore, development is limited during cool seasons or at high elevations. Development is rapid above 28° C, slows considerably below 21° C, and is extended beyond 30 days at temperatures less than 17° C.[11]

Altitudinal, seasonal, and annual patterns of transmission

The intensity and seasonal patterns in avian malaria transmission vary dramatically among low, mid, and high elevations in Hawaiian rainforests.[20,73,97] Key patterns in the Hawaiian malaria–forest bird system are high malaria transmission in low-elevation forests with minor seasonal or annual variation in infection; episodic transmission in mid-elevation forests with site-to-site, seasonal, and annual variation; and disease refugia in high-elevation forests with slight risk of infection only during summer. These transmission patterns are driven primarily by climatic effects on the parasite extrinsic incubation period, mosquito population dynamics across an altitudinal gradient, and landscape variation in the type and abundance of larval habitat. A key factor in determining malaria transmission is the size of the vector population, which is largely dependent on temperature, rainfall, and availability of larval habitat. In low-elevation forests, climate is consistently favorable for mosquito and parasite, and larval habitat is sufficient to produce a high abundance of infectious mosquitoes, resulting in an absence of susceptible native species and a high prevalence of chronic infections in the native birds that remain. Under these conditions, susceptible juvenile birds are exposed to infection soon after hatch and adult birds represent the previous years' cohort of malaria survivors.

In midelevation forests, abundance of infectious mosquitoes is lower due to a decrease of larval habitat and cooler temperatures, and here seasonal patterns in malaria transmission are especially evident.[20,73,97] At midelevation, the number of susceptible juvenile birds build up until malaria transmission begins during the late summer and early fall, which coincides with peak mosquito abundance. Transmission continues through the fall and early winter, when temperature and rainfall remain favorable to mosquito survival and parasite development. In temperate regions, seasonal transmission starts in late spring and coincides with emergent populations of mosquitoes and recrudescing infections of immune-suppressed breeding birds.[54] This spring relapse in erythrocytic parasites bridges the gap in seasonal transmission. Transmission in Hawaii, however, occurs well after the breeding season because chronically infected native birds serve as a year-round reservoir of disease. In years or areas with lower transmission, some susceptible juveniles escape exposure and mature to become susceptible adults, allowing for a build-up of susceptibles across years.

At high elevation (> 1,700 m asl), larval mosquito habitat is typically scarce, and cooler temperatures restrict seasonal abundance of mosquitoes and slow parasite development. These elevations provide disease-free refugia, with only a brief risk of infection during summer. Factors driving year-to-year and site-to-site variation in intensity are less evident.

Annual variation in malaria infection in birds occurs primarily in midelevation forests, likely caused by the abundance of susceptible birds and annual shifts in weather patterns driving vector abundance. These annual disease patterns can vary from full-blown epizootics that infect nearly all of the susceptible population to lower intensity events that infect approximately 50% of the susceptible population.[10,73] This suggests that the severity of epizootics may be related to the build-up of susceptible adult birds, which are dramatically affected when annual rainfall and associated mosquito populations return to normal levels that cause high transmission. Susceptible birds from transmission-free elevations may also descend to a transmission zone while tracking the nectar of ohia lehua (*Metrosideros polymorpha*), the dominate flowering tree on the landscape. Hart *et al.*,[118] however, found little evidence to support this ecological trap hypothesis. Other factors that could produce annual variability in the severity of epizootics include the cycling of malarial variants with different pathogenicity[97] and concomitant avian pox infection.[73,119]

Management and future threats

Control of vectors

Because the parasite depends on vectors for development and transmission, mosquitoes provide a crucial link between infected and susceptible birds. However, many characteristics of the malaria system in Hawaii make vector control challenging. A highly competent vector, favorable climate, lifelong reservoir hosts, and a steady source of susceptible birds all make transmission of avian malaria so efficient that relatively few vectors are required to drive transmission. Therefore, to significantly reduce transmission, control of the vector has to be near complete and applied at large spatial scales across a landscape of natural, agricultural, and residential areas to target relatively small, unevenly dispersed vector populations.[90] General strategies to control mosquito-borne disease have focused on reducing the longevity or abundance of vectors.[90] One of the most effective ways to reduce mosquito abundance is removal of larval habitat. In some forests, significant vector reduction may require a combination of strategies including pig removal, destruction of larval mosquito habitat, and biological control. Other methods, including chemical control of larvae and adults, biological control using predators or microbial pathogens, genetic modification strategies, and sterile males, have also been successfully used to control some vector-borne diseases.[90] Many of these approaches, however, would be unacceptable in Hawaiian forests because of non-target effects to endemic invertebrates or impractical because of the extreme heterogeneity of island topography.

An important potential opportunity for controlling larval habitat lies in the residential and agricultural communities that encroach on natural areas. Eliminating or controlling artificial larval mosquito habitat in these areas requires considerable cooperation and vigilance from the public, which may require incentives and community involvement.[90] Reduction of tree fern cavities by elimination of feral pigs is also feasible but is expensive and controversial because feral pigs are considered a valuable game species.

Computer modeling of the Hawaiian malaria system indicates that larval habitat reduction in forested areas could provide effective mosquito control.[20,99] In midelevation forests with abundant larval habitat, substantial and sustained reductions of mosquitoes could reduce disease transmission and ultimately increase native bird populations. In midelevation forests with a low abundance of larval habitat and high-elevation forests, mosquitoes and disease transmission are already limited. Thus, further reduction in larval habitat may have minimal impact on native bird populations. In contrast, low-elevation forests typically have ample larval habitat and temperatures favorable for rapid development of larvae, producing a high and continuous abundance of adult mosquitoes. In these low-elevation forests, nearly complete removal of larval habitat may be required to substantially reduce mosquito abundance and subsequently increase native bird populations. Overall, these modeling results suggest that control actions that reduce mosquito larval habitat will be most beneficial to bird populations in midelevation forests where larval habitat is abundant. Unfortunately, these sites may require a permanent reduction of nearly all (>80%) of the larval habitat to produce a significant improvement in native Hawaiian bird abundance. Model simulations indicate that control efforts that substantially reduce larval habitat without reaching critical thresholds may not substantially reduce disease transmission or increase bird populations.

Climate change

Climate change, and global warming in particular, is expected to increase the occurrence, distribution, and intensity of vector-borne disease throughout the world.[116,120] Projected climate warming of 2–3 °C in Hawaii by 2100 will certainly increase the need for conservation measures to reduce malaria transmission but will also increase the challenge in developing successful strategies. Because malaria infection patterns in Hawaii are largely driven by the effects of temperature and rainfall on mosquito dynamics and parasite incubation rates, future climate change will have profound impacts on transmission patterns and undermine current conservation strategies to mitigate malaria transmission and mortality in Hawaiian forest birds.[10,11,20,121] Climate warming will mostly affect mid- to high-elevation bird communities where seasonal disease transmission will increase in both duration and intensity. Cooler, high-elevation forests, which are currently disease-free refugia, are likely to see seasonal transmission. In the long term, climate change poses a significant threat to the viability of Hawaiian forest bird populations because warming temperatures will facilitate upslope spread of mosquitoes and malaria.[121,122]

Recent analyses of climatic data from Hawaii indicate that mean temperatures have shown a significant upward trend over the past 80 years,[123] and recent studies suggest that climate change may already be increasing the risk of malaria infection in some upper elevation native bird communities.[109] Given current land use and a projected 2 °C temperature rise, disease-free, high-elevation forest habitat is predicted to decrease by 60–96% in some parts of Hawaii (Fig. 3).[121,122] Management of current conservation lands in disease-free, high-elevation habitat may not be sufficient for the long-term protection of many Hawaiian forest birds. Securing deforested lands adjacent and above the current refugia and restoring forest cover would provide additional benefit to birds by increasing carrying capacity.[90] On a small scale, disease-free islands of habitat might also be created in *kipuka* if mosquitoes can be eliminated or substantially reduced in these isolated areas.[90]

These imperiled high-elevation refugia exist in a narrow band of native forest that extends to tree line. If climatic changes occur rapidly, the upslope extension of suitable forest habitat could not occur before susceptible bird species are exposed to malaria.[121] These changes would effectively eliminate high-elevation refugia in Hawaii, likely driving remaining populations of threatened and endangered honeycreepers to extinction with severe declines in nonendangered species that exhibit high susceptibility to avian malaria.

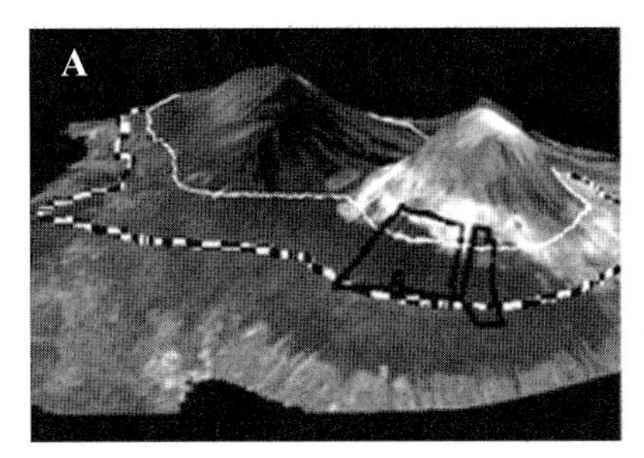

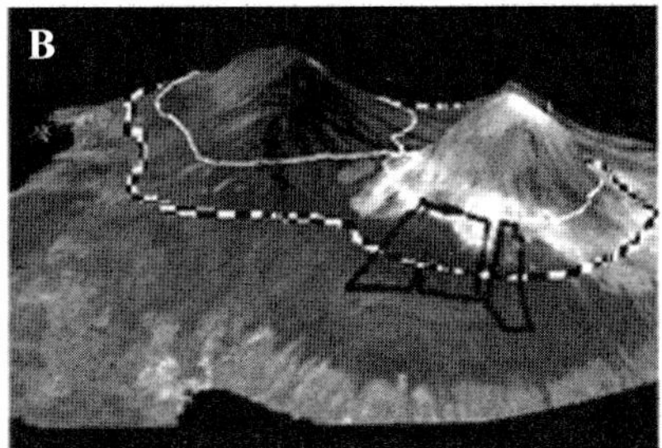

Figure 3. The current (A) and +2 °C projected (B) isotherms for the developmental threshold temperature for *P. relictum* (13 °C, solid while line) and the lower thermal boundary for seasonal transmission (17 °C, hatched line) at the Hakalau Forest National Wildlife Refuge island (black boundary) of Hawaii. The result of this warming scenario is the loss of 96% of disease-free refugia. Modified from Benning *et al.* (2002) with permission from the National Academy of Sciences.

Impacts from warming temperature, however, are only a part of the potential impact of climate change. Rainfall is also an important driver of mosquito population dynamics, and changes in seasonal or annual precipitation could alter transmission patterns across the landscape. The modest, predicted increase in precipitation during the dry season (summer) and decrease in the wet season (winter)[124] could lengthen seasonal availability of larval mosquito habitat and result in longer seasonal transmission.

Will climate change affect avian malaria dynamics and impact birds in continental disease systems? A recent analysis of global avian malaria prevalence and temperature anomalies over the last several decades suggests an increase in avian *Plasmodium* prevalence in Europe and Africa where significant increases in mean temperature have occurred.[125] Though exact mechanisms were not identified, geographical and seasonal expansion in vector populations and changes in parasite extrinsic incubation rate may increase prevalence. In temperate regions, where new infections depend on a spring relapse, higher temperatures may increase transmission by changing the phenology of vectors.[126,127] Climate change is also suspected as a driving factor in the southward expansion of *Cx. quinquefasciatus* in New Zealand.[62]

Protection/treatment of birds

Vaccines and antimicrobial agents have been successfully used to protect humans and domestic animals from infectious disease. Most wildlife vaccines have targeted mammalian reservoirs of zoonotic disease, but vaccines against protozoan pathogens have been difficult to develop. Although birds were some of the first experimental models for development of vaccines against *Plasmodium*, immunizing wild bird populations presents a significant challenge. A variety of experimental vaccines such as ultraviolet light-inactivated; formalin inactive and irradiated sporozoites, merozoites, and gametes; and synthetic vaccines based on parasite surface molecules have been used.[44] DNA vaccines based on the circumsporozoite protein of *P. gallinaceum* and *P. relictum* have been demonstrated to provide protection against *P. relictum* in penguins[128] and canaries,[129] but immunity was short lived and birds were susceptible to infection one year later.[12] Similarly, disease treatment by chemotherapy is typically limited to unique settings such as captive or closely managed flocks. Common antimalarial agents include chloroquine phosphate, primaquine phosphate, pyramethamine–sulfadoxine combinations, mefloquine, and atovaquone/proguanil combinations (Malarone™, GlaxoSmithKline, Research Triangle Park, NC), which have been used in treating canaries, other passerines, penguins, and raptors with avian malaria.[130,131] In general, the effectiveness of antimalarial agents in the treatment of wild birds is limited by the difficulties of delivery and potential development of drug-resistant parasites.[90]

Genetic evolution of resistance/tolerance

On the island of Hawaii, emergent populations of Hawaii amakihi have been found in low-elevation forests.[7,8] In spite of malaria prevalence ranging from 55% to 83%, these populations are expanding in range and densities that exceed those of Hawaii amakihi at elevations above 1500 m asl. Unlike the apparently *Plasmodium*-refractory Oahu amakihi,[110] the lowland Hawaii amakihi appear more tolerant to the pathologic effects of malaria. These birds have unique nuclear and mitochondrial haplotypes not found in amakihi from high elevation, suggesting their recent resurgence originated from pockets of surviving individuals with some natural disease tolerance, rather than recolonization of the lowlands by high-elevation birds.[9] Low-elevation Hawaii amakihi have also been found on Maui and Molokai where malaria prevalence exceeds 75%, suggesting that selection for disease resistance or tolerance has also occurred on other islands.[89]

Interaction with avian pox virus

The date when avian pox virus was introduced to the Hawaiian Islands remains unknown, but the disease was readily apparent by the late 19th century.[10,132] At that time, it was assumed that domestic poultry were the most likely route for virus introduction to Hawaii. Molecular analysis of isolates circulating in native forest birds, however, has determined that Hawaiian forest bird pox is distinct from fowl pox, has a higher genetic diversity than expected from a single introduction,[133] and is most similar to canarypox.[119] Although pox may be primarily transmitted by *Culex* mosquitoes, we know relatively little about its epizootiology and pathogenesis in Hawaiian forest birds.[134] Other than limited experimental studies,[6,119] information about the potential impacts of pox virus on Hawaiian forest bird populations is based on observations of pox-like lesions on captured wild birds.[10] VanderWerf[135] reported declines in some breeding cohorts of Hawaii elepaio (*Chasiempis sandwichensis*) that were correlated with the occurrence of pox epizootics. In addition, preliminary observations of Oahu elepaio (*C. ibidis*) suggest up to 40% annual mortality of birds with active lesions, while birds with mild infections involving only one or more toes frequently recover.[136] Overall, the prevalence of pox-like lesions in Hawaiian forest birds is substantially less than chronic malaria infections,[73,108,134] and birds with pox-like lesions had concurrent malarial infections more frequently than expected by chance.[7,108] However, we know little about whether this apparent interaction is caused by simultaneous transmission of the pathogens; differential mortality among pox, malaria, or pox-malaria infected birds; or results because older birds are more likely to have both chronic malaria and pox infections. Alternatively, concurrent infections may represent relapses of chronic malaria brought on by immune suppression from pox virus.[119] Whatever the etiology, the high frequency of concurrent pox and malaria infections makes it difficult to separate the demographic impacts of either agent alone. To further confound the issue, at least two variants of avian pox virus, with marked differences in virulence, circulate among passerines in the Hawaiian Islands.[119] How these variants interact with malaria may provide additional insight into the nature and drivers of malarial epizootics in Hawaiian birds. Avian pox virus has been present on the Galapagos Archipelago for over a century, and notable outbreaks and mortality have occurred among Galapagos mockingbirds (*Nesomimus parvulus parvulus*).[137–139] With the recent detection of *Plasmodium* in Galapagos penguins, similar interactions between pathogens may greatly increase the impact of avian disease in the Galapagos Archipelago.[77]

Conclusions, future concerns, and conservation recommendations

With the advent of PCR diagnostics, research in avian malaria has grown exponentially in the last 15 years, but it is unclear if these studies have revealed the dynamics of a rapidly emergent disease on the global scale or merely a leap in our understanding of the ecology of this disease. Certainly the evolving tolerance of once susceptible populations of Hawaii amakihi is a clear indication of the potential dynamics of bird–*Plasmodium* relationships. At the same time, some long-held preconceptions regarding the benign nature of evolutionarily old, host–parasite relationships are slowing giving way as larger population and modeling studies closely examine the survival and reproductive fitness in infected individuals. The growing impact of anthropogenic environmental change has also been documented at both the local and global scales. The effects of climate change, especially increased temperature, may be foretold by the prevalence of avian malaria across altitudinal gradients in the Hawaiian Islands and across the latitudinal gradients of continents. Perhaps the most alarming trend is the establishment and apparent spread of avian malaria in New Zealand and the Galapagos Archipelago, where globalization, deforestation, agriculture, and human encroachment have fostered an emergence of avian disease. Along with that of the Hawaiian Islands, the avifauna of these islands represents some of the most unique adaptive radiations of species in the world.

Whether the Hawaiian avifauna continue to adapt or succumb to avian malaria, globalization and regulatory limitations likely ensure future pathogen (i.e., West Nile virus)[140,141] and vector introductions. International transportation and commerce regulations that do not include improved disinsection and quarantine measures may lead to the further spread of vectors and avian pathogens. Considering the impacts of climate change, planning mitigation now for the loss of disease-free habitat on oceanic islands and fragmented continental landscapes may be necessary to save vulnerable species. Effective control of *Cx. quinquefasciatus*, the cosmopolitan mosquito vector of *P. relictum*, and identification and control of unknown vectors may be the key to disease management. Where traditional approaches to vector control would be difficult to apply across vast natural areas, innovative techniques will be needed to control avian malaria transmission if we want to successfully preserve many of the world's unique island avifaunas for future generations.

Acknowledgments

This research was funded through the U.S. Geological Survey's Wildlife, Invasive Species, and Natural Resource Protection Programs and a Biocomplexity grant from the National Science Foundation (DEB0083944). We also wish to thank our technical staff, postdoctoral researchers, and numerous research interns whose hard work and dedication made this research possible. The use of trade names or products does not constitute endorsement by the U.S. government.

Conflicts of interest

The authors declare no conflicts of interest.

References

1. Danilewsky, B. 1889. La parasitologie compareé du sang. 1. *Nouvelles recherches sur les parasites du sang des oiseaux.* Kharkov:Darre.
2. Hewitt, R.J. 1940. *Bird Malaria. Amer. J. Hyg. Monog.* ***Ser., No (15).*** The Johns Hopkins Press. Baltimore.
3. Garnham, P.C.C. 1966. *Malaria Parasites and Other Haemosporidia.* Blackwell Scientific. New York.
4. Harrison, G.A. 1978. *Mosquitoes, Malaria and Man: A History of the Hostilities since 1880.* E. P. Dutton. New York.
5. Valkiūnas, G. 2005. *Avian Malaria Parasites and Other Haemosporidia.* CRC Press. New York.
6. Warner, R.E. 1968. The role of introduced diseases in the extinction of the endemic Hawaiian avifauna. *Condor* **70:** 101–120.
7. van Riper, C., III, S.G. van Riper, M.L. Goff & M. Laird. 1986. The epizootiology and ecological significance of malaria in Hawaiian land birds. *Ecol. Monog.* **56:** 327–344.
8. Woodworth, B.L., C.T. Atkinson, D.A. LaPointe, *et al.* 2005. Host population persistence in the face of introduced vector-borne diseases: Hawaii amakihi and avian malaria. *Proc. Natl. Acad. Sci. USA* **102:** 1531–1536.
9. Foster, J.T., B.L. Woodworth, L.E. Eggert, *et al.* 2007. Genetic structure and evolved resistance in Hawaiian honeycreepers. *Mol. Ecol.* **22:** 4738–4746.
10. Atkinson, C.T. & D.A. LaPointe. 2009. Ecology and pathogenicity of avian malaria and pox. In *Conservation Biology of Hawaiian Forest Birds.* T. K. Pratt, C. T. Atkinson, P. C. Banko, J. Jacobi, & B. L. Woodworth, Eds.: 234–252. Yale University Press. New Haven.
11. LaPointe, D.A., M.L. Goff & C.T. Atkinson. 2010. Thermal constraints to the sporogonic development and altitudinal distribution of avian malaria *Plasmodium relictum* in Hawai'i. *J. Parasitol.* **96:** 318–324.
12. Atkinson, C.T. 2008. Avian malaria. In *Parasitic Diseases of Wild Birds.* C.T. Atkinson, N.J. Thomas & D.B. Hunter, Eds.: 35–53. Wiley-Blackwell. Ames, IA.
13. Forrester, D.J. & E.C. Greiner. 2008. Leucocytozoonosis. In *Parasitic Diseases of Wild Birds.* C.T. Atkinson, N.J. Thomas & D.B. Hunter, Eds.: 54–107. Wiley-Blackwell. Ames, IA.
14. Beadell, J.S., F. Ishtiaq, R. Covas, *et al.* 2006. Global phylogeographic limits of Hawai'i's Avian Malaria. *Proc. R. Soc., Ser. B: Biol. Sci.* **273:** 2935–2944.
15. Hellgren, O., J. Peréz-Tris & S. Bensch. 2009. A jack-of-all trades and still a master of some: prevalence and host range in avian malaria and related blood parasites. *Ecol. Soc. Am.* **90:** 2840–2849.
16. Lachish, S., S.C. Knowles, R. Alves, *et al.* 2011. Fitness effects of endemic malaria infections in a wild bird population: the importance of ecological structure. *J. Anim. Ecol.* **80:** 1196–1206.
17. Atkinson, C.T. & C. van Riper III. 1991. Pathogenicity and epizootiology of avian haemoatozoa: plasmodium, leucocytozoon, and haemoproteus. In *Bird-Parasite Interactions, Ecology, Evolution and Behavior*. J. E. Loye & M. Zuk Eds.: 19–48. Oxford University Press, New York.
18. Williams, R.B. 2005. Avian malaria: clinical and chemical pathology of *Plasmodium gallinaceum* in the domesticated fowl *Gallus gallus. Avian Path.* **34:** 29–47.
19. Palinauskas,V., G. Valkiūnas, V. Bolshakov & S. Bensch. 2008. *Plasmodium relictum* (lineage P-SGS1): effects on experimentally infected passerine birds. *Exp. Parasitol.* **120:** 372–380.
20. Samuel, M.D., P.H.F. Hobbelen, F. DeCastro, *et al.* 2011. The dynamics, transmission, and population impacts of avian malaria in native Hawaiian birds – a modeling approach. *Ecol. Appl.* **21:** 2960–2973.
21. Fallon, S.M., E. Bermingham & R.E. Ricklefs. 2003. Island and taxon effects in parasitism revisited: avian malaria in the Lesser Antilles. *Evolution* **57:** 606–615.
22. Bennett, G.F., M.A. Bishop & M.A. Peirce. 1993. Checklist of the avian species of *Plasmodium* Marchiafava & Celli, 1885 (Apicomplexa) and their distribution by avian family and Wallacean life zones. *Syst. Parasitol.* **26:** 171–179.

23. Bensch, S., J. Pérez-Tris, J. Waldenstrøm & O. Hellgren. 2004. Linkage between nuclear and mitochondrial DNA sequences in avian malaria parasites: multiple cases of cryptic speciation? *Evolution* **58:** 1617–1621.
24. Fallon, S.M., E. Bermingham & R.E. Ricklefs. 2005. Host specialization and geographic localization of avian malaria parasites: a regional analysis in the Lesser Antilles. *Am. Nat.* **165:** 466–480.
25. Ricklefs, R.E., B.L. Swanson, S.M. Fallon, *et al.* 2005. Community relationships of avian malaria parasites in Southern Missouri. *Ecol. Monog*.**75:** 543–559.
26. Szymanski, M.M. & I.J. Lovette. 2005. High lineage diversity and host sharing of malarial parasites in a local avian assemblage. *J. Parasitol.* **91:** 768–774.
27. Iezhova, T.A., M. Dodge, R.N.M. Sehgal, *et al.* 2011. New avian *Haemoproteus* species (Haemosporida: Haemoproteidae) from African birds, with a critique of the use of host taxonomic information in hemoproteid classification. *J. Parasitol.* **97:** 682–694.
28. Reeves, W.C., R.C. Herold, L. Rosen, *et al.* 1954. Studies on avian malaria in vectors and hosts of encephalitis in Kern County, California. II. Infections in mosquito vectors. *Am. J. Trop. Med. Hyg*.**3:** 696–703.
29. LaPointe, D.A., M.L. Goff, & C.T. Atkinson. 2005. Comparative susceptibility of introduced forest-dwelling mosquitoes in Hawai'i to avian malaria, *Plasmodium relictum*. *J. Parasitol.* **91:** 843–849.
30. Huff, C.G. 1965. The susceptibility of mosquitoes to avian malaria. *Exp.Parasitol.* **16:** 107–132.
31. Ejiri H., Y. Sato, E. Sasaki, *et al.* 2008. Detection of avian Plasmodium spp. DNA sequences from mosquitoes captured in Minami Daito Island of Japan. *J. Vet. Med. Sci.* **70:** 1205–1210.
32. Ishtiaq, F., L. Guillaumot, S.M. Clegg, *et al.* 2008. Avian haematozoan parasites and their associations with mosquitoes across Southwest Pacific Islands. *Mol. Ecol.* **17:** 4545–4555.
33. Ejiri, H., Y. Sato, R. Sawai, *et al.* 2009. Prevalence of avian malaria parasite in mosquitoes collected at a zoological garden in Japan. *Parasitol. Res.* **105:** 629–633.
34. Njabo, K.Y., A.J. Cornel, R.N.M. Sehgal, *et al.* 2009. Coquillettidia (Culicidae, Diptera) mosquitoes are natural vectors of avian malaria in Africa. *Malaria J.* **8:** 193. doi:10.1186/1475-2875-8-193.
35. Kimura M., J.M. Darbro & L.C. Harrington. 2010. Avian malaria parasites share congeneric mosquito vectors. *J. Parasitol.* **96:** 144–151.
36. Njabo, K.Y., A. J. Cornel, C. Bonneaud, *et al.* 2011 Nonspecific patterns of vector, host and avian malaria parasite associations in a central African rainforest. *Mol. Ecol.* **20:** 1049–1061.
37. Gager, A.B., J. Del Rosario Loaiza, D.C. Dearborn & E. Bermingham. 2008. Do mosquitoes filter the access of *Plasmodium* cytochrome *b* lineages to an avian host? *Mol. Ecol.* **17:** 2552–2561.
38. Ross, R. 1898. Report on the cultivation of *Proteosoma*, Labbé, in grey mosquitoes. *Ind. Med. Gaz.* **33:** 401–448.
39. Raffaele, G. 1934. Un ceppo italiano di *Plasmodium elongatum*. *Rivista di Malariogia* **13:** 3–8.
40. Huff, C. G. & W. Bloom. 1935. A malarial parasite infecting all blood and blood forming cells of birds. *J. Infect. Dis.* **57:** 315–336.
41. Huff, C.G. & F. Coulston. 1944. The development of *Plasmodium gallinaceum* from sporozoite to erythrocytic trophozoite. *J. Infect.Dis.***75:** 231–249.
42. Huff, C.G. 1951. Observations on the pre-erythrocytic stages of *P. relictum*, *P. cathemerium* and *P. gallinaceum* in various birds. *J. Infect.Dis.* **88:** 17–26.
43. Michel, K. & F.C. Kafatos. 2005. Mosquito immunity against *Plasmodium*. *Insect Biochem. Molec. Biology* **35:** 677–689.
44. van Riper C., III, C.T. Atkinson & T.M. Seed. 1994. Plasmodia of birds. In *Parasitic Protozoa*. J. P. Kreier Ed.: 73–140, Vol. 7. Academic Press. New York.
45. Al-Dabagh, M.A. 1966. *Mechanisms of Death and Tissue Injury in Malaria*. Shafik Press. Baghdad.
46. Yorinks, N. & C.T. Atkinson. 2000. Effects of malaria (*Plasmodium relictum*) on activity budgets of experimentally-infected juvenile Apapane (*Himatione sanguinea*). *Auk* **117:** 731–738.
47. Manwell, R.D. 1934. The duration of malarial infection in birds. *Am. J. Hyg*.**19:** 532–538.
48. Bishop, A., P. Tate & M.V. Thorpe. 1938. The duration of *Plasmodium relictum* in canaries. *Parasitology* **38:** 388–391.
49. Atkinson, C.T., R.J. Dusek & J.K. Lease. 2001. Serological responses and immunity to superinfection with avian malaria in experimentally-infected Hawaii Amakihi. *J. Wildlife Dis.* **37:** 20–27.
50. Paulman, A. & M.M. Mcallister. 2005. *Plasmodium gallinaceum*: clinical progression, recovery, and resistance to disease in chickens infected via mosquito bite. *Am. J. Trop. Med. Hyg.* **73:** 1104–1107.
51. Atkinson, C.T., J.K. Lease, B.M. Drake & N.P. Shema. 2001. Pathogenicity, serological responses, and diagnosis of experimental and natural malarial infections in native Hawaiian thrushes. *Condor* **103:** 209–218.
52. Jarvi, S.I., J.J. Schultz & C.T. Atkinson. 2002. PCR diagnostics underestimate the prevalence of avian malaria (*Plasmodium relictum*) in experimentally-infected passerines. *J. Parasitol.* **88:** 153–158.
53. Young, M.D., J.K. Nayar & D.J. Forrester. 2004. Epizootiology of *Plasmodium hermani* in Florida: chronicity of experimental infections in domestic turkeys and Northern Bobwhites. *J. Parasitol.* **90:** 433–434.
54. Applegate, J.E. & R.L. Beaudion. 1970. Mechanism of spring relapse in avian malaria: Effect of gonadotropin and corticosterone treatment. *J. Wildlife Dis.* **6:** 443–447.
55. Waldenström, J., S. Bensch, S. Kiboi, *et al.* 2002. Cross-species infection of blood parasites between resident and migratory songbirds in Africa. *Mol. Ecol.* **11:** 1545–1554.
56. Stoskopf, M.K. & J. Beier. 1979. Avian malaria in African Black-Footed Penguins. *J. Am. Assoc. Vet. Med.* **175:** 994–997.
57. Fix, A.S., C. Waterhouse, E.C. Greiner & M.K. Stoskopf. 1988. *Plasmodium relictum* as a cause of avian malaria in wild-caught Magellanic penguins (*Spheniscus magellanicus*). *J. Wildlife Dis.* **24:** 610–619.

58. Huijben, S., W. Schaftenaar, A. Wijsman, *et al.* 2007 Chapter 4. Avian malaria in Europe: an emerging infectious disease? In *Emerging Pests of Vector-Borne Diseases in Europe.* W. Takken & B. Knols, Eds.: 59–74. Wageningen Academic Publishers. Wageningen, Netherlands.
59. Jones, H.I. & G.R. Shellam. 1999. Blood parasites in penguins, and their potential impact on conservation. *Marine Ornithol.* **27:** 181–184.
60. Sturrock, H.J.W. & D.M. Tompkins. 2007. Avian malaria (*Plasmodium* spp.) in Yellow-eyed Penguins: investigating the cause of high seroprevalence but low observed infection. *New Zeal. Vet. J.* **55:** 158–160.
61. Miller, G.D., B.V. Hofkin, H. Snell, *et al.* 2001. Avian malaria and Marek's disease: potential threats to Galapagos penguins *Spheniscus mendiculus. Mar. Ornithol.* **29:** 43–46.
62. Tompkins, D.M. & D.M. Gleeson. 2006. Relationship between avian malaria distribution and an exotic invasive mosquito in New Zealand. *J. R. Soc. New Zealand* **36:** 51–62.
63. Gabaldon, A. & G. Ulloa. 1980. Holoendemicity of malaria: an avian model. *Trans. R. Soc. Trop. Med. Hyg.* **74:** 501–507.
64. Møller, A.P. & J.T. Nielsen. 2007. Malaria and risk of predation: a comparative study of birds. *Ecology* **88:** 871–881.
65. Gilman, S., D.T. Blumstein & J. Foufopoulos. 2007. The effect of hemosporidian infections on white-crowned sparrow singing behavior. *Ethology* **113:** 437–445.
66. Richner, H., P. Christe & A. Oppliger. 1995. Paternal investment affects prevalence of malaria. *Proc. Natl. Acad. Sci. USA* **92:**1192–1194.
67. Oppliger, A., P. Christe & H. Richner. 1996. Clutch size and malaria resistance. *Nature* **391:** 565.
68. Knowles, S.C.L., S. Nakagawa & B.C. Sheldon. 2009. Elevated reproductive effort increases blood parasitaemia and decreases immune function in birds: a meta-regression approach. *Funct. Ecol.* **23:** 405–415.
69. Knowles, S.C.L., M.J. Wood & B.C. Sheldon. 2010. Contest-dependent effects of parental effort on malaria infection in a wild bird population, and their role in reproductive trade-offs. *Oecologia* **164:** 87–97.
70. Asghar, M., D. Hasselquist & S. Bensch. 2011. Are chronic avian haemosporidian infections costly in wild birds? *J. Avian Biol.* **42:** 530–537.
71. Atkinson, C.T., K.L. Woods, R.J. Dusek, *et al.* 1995. Wildlife disease and conservation in Hawaii: pathogenicity of avian malaria (*Plasmodium relictum*) in experimentally infected Iiwi (*Vestiaria coccinea*). *Parasitology* **111:** S59–S69.
72. Atkinson, C.T., R.J. Dusek, K.L. Woods & W.M. Iko. 2000. Pathogenicity of avian malaria in experimentally infected Hawaii Amakihi. *J. Wildlife Dis.* **36:** 197–204.
73. Atkinson, C.T. & M.D. Samuel. 2010. Avian malaria (*Plasmodium relictum*) in native Hawaiian forest birds: epizootiology and demographic impacts on 'apapane (*Himatione sanguinea*). *J. Avian Biol.* **41:** 357–366.
74. Kilpatrick, A.M., D.A. LaPointe, C.T. Atkinson, *et al.* 2006. Effects of chronic avian malaria (*Plasmodium relictum*) infection of reproductive success of Hawaii Amakihi (*Hemignathus virens*). *Auk* **123:** 764–774.
75. Spiegel, C.S., P.J. Hart, B.L. Woodworth, *et al.* 2006. Distribution and abundance of native forest birds in low-elevation areas in Hawaii Island: evidence of range expansion. *Bird Conserv. Int.* **16:** 175–185.
76. Travis, E.K., F.H. Vargas, J. Merkel, *et al.* 2006. Hematology, serum chemistry, and serology of Galapagos penguins (*Spheniscus mendiculus*) in the Galapagos Islands, Ecuador. *J. Wildlife Dis.* **42:** 625–632.
77. Levin, I.I., D.C. Outlaw, F.H. Vargas & P.G. Parker. 2009. *Plasmodium* blood parasite found in endangered Galapagos penguins (*Spheniscus mendiculus*). *Biol. Conserv.* **142:** 3191–3195.
78. Dore, A. 1920. The occurrence of malaria in the native ground lark. *J. Sci. Technol.* **3:** 118–119.
79. Reed, C. 1997. Avian malaria in New Zealand dotterel. *Kokako* **4:** 3.
80. Alley, M.R., R.A. Fairley, D.G. Martin, *et al.* 2008. An outbreak of avian malaria in captive yellowheads/mohua (*Mohoua ochrocephala*). *New Zeal. Vet. J.* **56:** 345.
81. Alley, M.R., K.A. Hale, W. Case, *et al.* 2010. Concurrent avian malaria and avipox virus infection in translocated South island saddlebacks (*Philesturnus carunculatus carunculatus*). *New Zeal. Vet. J.* **58:** 218–223.
82. Howe, L.I., C. Castro, E.R. Schoener, *et al.* 2011. Malaria parasites (*Plasmodium* spp.) infecting introduced, native and endemic New Zealand birds. *Parasitol. Res.* **110:** 913–923. doi:10.1007/s00436–011-2577-z.
83. Wood, M.J., C.L. Cosgrove, T.A. Wilkin, *et al.* 2007. Within-population variation in prevalence and lineage distribution of avian malaria in blue tits, *Cyanistes caeruleus. Mol. Ecol.* **16:** 3263–3273.
84. Loiseau, C., T. Iezhova, G. Valkiūnas, *et al.* 2010. Spatial variation of haemosporidian parasite infection in African rainforest bird species. *J. Parasitol.* **96:** 21–29.
85. Lachish, S., S.C.L. Knowles, R. Alves, *et al.* 2011. Infection dynamics of endemic malaria in a wild bird population: parasite species-dependent drivers of spatial and temporal variation in transmission rates. *J. Anim. Ecol.* **80:** 1207–1216.
86. Knowles, S.C.L, M.J. Wood, R. Alves, *et al.* 2011. Molecular epidemiology of malaria prevalence and parasitemia in a wild bird population. *Mol. Ecol.* **20:** 1062–1076.
87. Sehgal, R.N.M., W. Buermann, R.J. Harrigan, *et al.* 2011. Spatially explicit predictions of blood parasites in a widely distributed African rainforest bird. *Proc. R. Soc. B.* **278:** 1025–1033.
88. Ahumada, J.A., D.A. LaPointe & M.D. Samuel. 2004. Modeling the population dynamics of *Culex quinquefasciatus* (Dipteria: Culicidae) along an altitudinal gradient in Hawaii. *J. Med. Entomol.* **41:** 1157–1170.
89. Aruch, S., C.T. Atkinson, A.F. Savage & D.A. LaPointe. 2007. Prevalence and distribution of pox-like lesions, avian malaria and mosquito vectors in Kipuhulu Valley, Haleakala National Park, Hawaii, USA. *J. Wildlife Dis.* **43:** 567–575.
90. LaPointe, D.A., C.T. Atkinson & S.I. Jarvi. 2009. Managing disease. In *Conservation Biology of Hawaiian Forest Birds.* T.K. Pratt, C.T. Atkinson, P.C. Banko, J. Jacobi, & B.L. Woodworth, Eds.: 405–424. Yale University Press. New Haven.
91. Laird, M. 1989. *The Natural History of Larval Mosquito Habitats.* Academic Press, Harcourt Brace Jovanovich. London.
92. Reiter, M.E. & D.A. LaPointe. 2009. Larval habitat for the avian malaria vector, *Culex quinquefaciatus* (Diptera: Culicidae), in altered mid-elevation mesic-dry forests in Hawai'i. *J. Vect. Ecol.* **34:** 208–216.

93. Goff, M.L. & C. van Riper III. 1980. Distribution of mosquitoes (Diptera: Culicidae) on the east flank of Mauna Loa Volcano, Hawaii. *Pac. Insects* **22:** 178–188.
94. van Riper, C., III, S.G. van Riper, M.L. Goff & M. Laird. 1982. The impact of malaria on birds in Hawaii Volcanoes National Park. Technical Report 47. Cooperative National Park Resources Studies Unit, Department of Botany, University of Hawaii, Honolulu.
95. LaPointe, D.A. 2008. Dispersal of *Culex quinquefasciatus* (Diptera: Culicidae) in a Hawaiian rain forest. *J. Med. Entomol.* **45:** 600–609.
96. Yanoviak, S.P., J.E. Ramírez Paredes, L.P. Lounibos & S.C. Weaver. 2006. Deforestation alters phytotelm habitat availability and mosquito production in the Peruvian Amazon. *Ecol. Appl.* **16:** 1854–1864.
97. LaPointe, D.A. 2000. Avian malaria in Hawai'i: the distribution, ecology and vector potential of forest-dwelling mosquitoes. Ph.D. dissertation, University of Hawai'i, Manoa. Honolulu, HI.
98. Reiter, M.E & D.A. LaPointe. 2007. Landscape factors influencing the spatial distribution and abundance of mosquito vector *Culex quinquefasciatus* (Diptera: Culicidae) in a mixed residential–agricultural community in Hawai'i. *J. Med. Entomol.* **44:** 861–868.
99. Ahumada, J.A., M.D. Samuel, D.C. Duffy, *et al.* 2009. Modeling the epidemiology of avian malaria and pox in Hawaii. In *Conservation Biology of Hawaiian Forest Birds.* T. K. Pratt, C. T. Atkinson, P. C. Banko, J. Jacobi, & B. L. Woodworth, Eds.: 331–335. Yale University Press. New Haven.
100. Baker, J.K. 1975. The feral pig in Hawaii Volcanoes National Park. *Cal.-Nevada Wildlife Soc.-Trans.* **11:** 74–80.
101. Leisnham, P.T., P.J. Lester, D.P. Slaney & P. Weinstein. 2004. Anthropogenic landscape change and vectors in New Zealand: effects of shade and nutrient levels on mosquito productivity. *EcoHealth* **1:** 306–316.
102. Whitehead, N.K., S.J. Goodman, B.J. Sinclair, *et al.* 2005. Establishment of the avian disease vector *Culex quinquefasciatus* Say, 1823 (Diptera: Culicidae) on the Galapagos Islands, Ecuador. *Ibis* **147:** 843–847.
103. Bataille, A., A.A. Cunningham, V. Cedeño, *et al.* 2009. Evidence for regular ongoing introductions of mosquito disease vectors into the Galapagos Islands. *Proc. R. Soc. B.* **276:** 3769–3775.
104. Bonneaud, C., I. Sepil, B. Mila, *et al.* 2009. The prevalence of avian *Plasmodium* is higher in undisturbed tropical forests of Cameroon. *J. Trop. Ecol.* **25:** 439–447.
105. Chasar, A., C. Loiseau, G. Valkiūnas, *et al.* 2009. Prevalence and diversity patterns of avian blood parasites in degraded African rainforest habitats. *Mol. Ecol.* **18:** 4121–4133.
106. Foster, J.T. 2009. The history and impact of introduced birds. In *Conservation Biology of Hawaiian Forest Birds.* T. K. Pratt, C. T. Atkinson, P. C. Banko, J. Jacobi, & B. L. Woodworth, Eds.: 312–330. Yale University Press. New Haven, CT.
107. Feldman, R.A., L.A. Freed & R.L. Cann. 1995. A PCR test for avian malaria in Hawaiian birds. *Mol. Ecol.* **4:** 663–673.
108. Atkinson, C.T., J.K. Lease, R.J. Dusek & M.D. Samuel. 2005. Prevalence of pox-like lesions and malaria in forest bird communities on leeward Mauna Loa Volcano, Hawaii. *Condor* **107:** 537–546.
109. Freed, L.A., R.L. Cann, M.L. Goff, *et al.* 2005. Increase in avian malaria at upper elevation in Hawaii. *Condor* **107:** 753–764.
110. Shehata, C., L. Freed & R.L. Cann. 2001. Changes in native and introduced bird populations on O'ahu: infectious diseases and species replacement. In *Evolution, Ecology, Conservation and Management of Hawaiian Birds: A Vanishing Avifauna. Stud. Avian Biol.* – ***Ser. 22.*** J. M. Scott, S. Conant, & C. van Riper III, Eds.: 264–273. Cooper Ornithological Society. Camarillo, CA.
111. Schmidt, K.A. & R.S. Ostfeld. 2001. Biodiversity and the dilution effect in disease ecology. *Ecology* **82:** 609–619.
112. Pratt, L.W. & J.D. Jacobi. 2009. Loss, degradation and persistence of habitats. In *Conservation Biology of Hawaiian Forest Birds.* T.K. Pratt, C.T. Atkinson, P.C. Banko, J.D. Jacobi & B.L. Woodworth, Eds.: 137–158. University of Hawaii Press. Honolulu, HI.
113. Giambelluca, T.W. & T.A.Schroeder. 1998. Climate. In Atlas of Hawaii. S.P. Juvik & J.O. Juvik, Eds.: 49–59. University of Hawaii Press. Honolulu, HI.
114. Alto, B.W. & S.A. Juliano. 2001. Precipitation and temperature effects on populations of *Aedes albopictus* (Diptera: Culicidae). *J. Med. Entomol.* **38:** 646–656.
115. Hayes, J. & T.D. Downs. 1980. Seasonal-changes in an isolated population of *Culex pipiens quinquefasciatus* (Diptera: Culicidae) – time-series analysis. *J. Med. Entomol.* **17:** 63–69.
116. Paaijmans, K.P., A.F. Reed & M.B. Thomas. 2009. Understanding the link between malaria risk and climate. *Proc. Natl. Acad. Sci. USA* **106:** 13844–13849.
117. Rueda L.M., K.J. Patel, R.C. Axtell & R.E. Stinner. 1990. Temperature-dependent development and survival rates of *Culex quinquefasciatus* and *Aedes aegypti* (Diptera: Culicidae). *J. Med. Entomol.* **27:** 829–898.
118. Hart, P.J., B.L. Woodworth, R.J. Camp, *et al.* 2011. Temporal variation in bird and resource abundance across an elevational gradient in Hawaii. *Auk* **128:** 113–126.
119. Jarvi, S.I., D. Triglia, A. Giannoulis, *et al.* 2008. Diversity, origins and virulence of *Avipoxviruses* in Hawaiian forest birds. *Conserv. Genetics* **9:** 339–348.
120. Sutherst, R.W. 2004. Global change and human vulnerability to vector borne diseases. *Clin. Microbiol. Rev.* **17:** 136–173.
121. Benning, T.L., D. LaPointe, C.T. Atkinson & P.M. Vitousek. 2002. Interactions of climate change with biological invasions and land use in the Hawaiian Islands: modeling the fate of endemic birds using a geographic information system. *Proc. Natl. Acad. Sci. USA* **99:** 14246–14249.
122. Atkinson, C.T. & D.A. LaPointe. 2009. Introduced avian disease, climate change, and the future of Hawaiian honeycreepers. *J. Avian Med. Surg.* **23:** 53–63.
123. Giambelluca, T.W., H.F. Diaz & M.S.A. Luke. 2008. Secular temperature in Hawaii. *Geophys.Res. Lett.* **35:** L12702, doi:10.1029/2008GL034377
124. Timm, O. & H.F. Diaz. 2009. Synoptic-statistical approach to regional downscaling of IPCC twenty-first-century climate projections: Seasonal rainfall over the Hawaiian Islands. *J. Clim.* **22:** 4261–4280.
125. Garamszegi, L.Z. 2011. Climate change increases the risk of malaria in birds. *Global Change Biol.* **17:** 1751–1759.

126. Møller, A.P. 2010. Host-parasite interactions and vector in the barn swallow in relation to climate change. *Global Change Biol.* **16:** 1158–1170.
127. Murdock, C.C. 2009. Studies on the ecology of avian malaria in an alpine ecosystem. PhD dissertation. University of Michigan. Ann Arbor, MI.
128. Grim, K.C., T. McCutchan, J. Li, *et al.* 2004. Preliminary results of an anticircumsporozoite DNA vaccine trial for protection against avian malaria in captive Africam black-footed penguins (*Spheniscus demersus*). *J. Zoo Wildlife Med.* **35:** 154–161.
129. McCutchan, T.F., K.C. Grim, J. Li, *et al.* 2004. Measuring the effects of an ever-changing environment on malaria control. *Infect. Immun.* **72:** 2248–2253.
130. Remple, J.D. 2004. Intracelluar hematozoa of raptors: a review and update. *J. Avian Med. Surg.* **18:** 75–88.
131. Palinauskas, V., G. Valkiūnas, A. Križanauskienė, *et al.* 2009. *Plasmodium relictum* (lineage P-SGS1): further observation of effects on experimentally infected passeriform birds, with remarks on treatment with Malarone™. *Exp. Parasitol.* **123:** 134–139.
132. Henshaw, H.W. 1902. *Birds of the Hawaiian Islands; Being a Complete List of the Birds of the Hawaiian Possessions, with Notes on their Habits.* Thos. G. Thrum. Honolulu, HI.
133. Tripathy, D.N., W.M. Schnitzlein, P.J. Morris, *et al.* 2000. Characterization of poxviruses from Hawaiian forest birds. *J. Wildlife Dis.* **36:** 225–230.
134. van Riper, C., III, S.G. van Riper & W.R. Hansen. 2002. Epizootiology and effect of avian pox on Hawaiian forest birds. *Auk* **119:** 929–942.
135. VanderWerf, E.A. 2001. Distribution and potential impacts of avian poxlike lesions in 'Elepaio at Hakalau Forest National Wildlife Refuge. In *Evolution, Ecology, Conservation and Management of Hawaiian Birds: A Vanishing Avifauna*. J. M. Scott, S. Conant, & C. van Riper III, Eds.: 247–253. Cooper Ornithological Society. Camarillo, CA.
136. VanderWerf, E.A., A. Cowell & J.L. Rohrer. 1997. Distribution, abundance, and conservation of O'ahu 'Elepaio in the southern leeward Ko'olau range. *Elepaio* **57:** 99–105.
137. Parker, P.G., E.L. Buckles, H. Farrington, *et al.* 2011. 110 Years of *Avipoxvirus* in the Galapagos Islands. *PLoS One* **6:** e15989. doi:10.1371/journal.pone.0015989
138. Vargas, H. 1987. Frequency and effect of pox-like lesions in Galapagos mockingbirds. *J. Field Ornith.* **58:** 101–264.
139. Thiel, T., N.K. Whiteman, A. Tirapé, *et al.* 2005. Characterization of canarypox-like viruses infecting endemic birds in the Galapagos Islands. *J. Wildlife Dis.* **41:** 342–353.
140. Kilpatrick, A.M., Y. Gluzberg, J. Burgett & P. Daszak. 2004. Quantitative risk assessment of pathways by which West Nile virus could reach Hawaii. *EcoHealth* **1:** 205–209.
141. LaPointe, D.A., E.K. Hofmeister, C.T. Atkinson, *et al.* 2009b. Experimental infection of Hawaii amakihi (*Hemignathus virens*) with West Nile virus and competence of a co-occurring vector, *Culex quinquefasciatus*: potential impacts on endemic Hawaiian avifauna. *J. Wildlife Dis.* **45:** 257–271.

Ann. N.Y. Acad. Sci. ISSN 0077-8923

ANNALS OF THE NEW YORK ACADEMY OF SCIENCES
Issue: *The Year in Ecology and Conservation Biology*

Dams in the Cadillac Desert: downstream effects in a geomorphic context

John L. Sabo,[1] Kevin Bestgen,[2] Will Graf,[3] Tushar Sinha,[4] and Ellen E. Wohl[5]

[1]School of Life Sciences, Arizona State University, Tempe, Arizona. [2]Larval Fish Laboratory, Department of Fish, Wildlife, and Conservation Biology, Colorado State University, Fort Collins, Colorado. [3]Department of Geography, University of South Carolina, Columbia, South Carolina. [4]Civil, Construction, and Environmental Engineering, North Carolina State University, Raleigh, North Carolina. [5]Department of Geosciences, Colorado State University, Fort Collins, Colorado

Address for correspondence: John L. Sabo, School of Life Sciences, Arizona State University, P. O. Box 874501, Tempe, AZ 85287. John.L.Sabo@asu.edu

This paper was motivated by the 25th anniversary of the publication of Marc Reisner's book, *Cadillac Desert: The American West and its Disappearing Water*. Dams are ubiquitous on rivers in the United States, and large dams and storage reservoirs are the hallmark of western U.S. riverscapes. The effects of dams on downstream river ecosystems have attracted much attention and are encapsulated in the serial discontinuity concept (SDC). In the SDC, dams create abrupt shifts in continua of downstream changes in physical and biotic properties. In this paper, we develop a framework for understanding how channel geometry and network structure influence how the physical components of habitat and the biota rebound from discontinuities set up by large dams. We apply this framework to data describing the flow regime, temperature, sediment flux, and fish community composition below Garrison Dam on the Missouri River, Glen Canyon Dam on the Colorado River, and Flaming Gorge Dam on the Green River. Sediment flux in dam tailwaters is under strong control by channel geometry. By contrast, dam-related changes in temperature and flow variation are not significantly modulated by channel geometry or tributary inputs if flow volumes are small (Missouri and Colorado River tributaries). Instead, small tributaries provide near-native conditions (flow and temperature variation) and, as such, provide key refuges for biota from novel habitats in mainstem rivers below large dams. Unregulated tributaries that are large relative to their respective mainstem (e.g., Yampa River) provide refuges as well as significant amelioration of flow and temperature effects from upstream dams. Finally, the proportion of native fish increases with distance from dam and exhibits sharp increases near tributary junctions. These results suggest that tributaries—even minor ones in terms of relative discharge—act as key refugia for native species in regulated river networks. Moreover, large, unregulated tributaries are key to restoring continuity in physical habitat and the biota in large regulated rivers.

Keywords: dam; river; valley form; network; tributary; sediment; temperature; flow anomaly; fish; nonnative species

Dams in the Cadillac Desert

This paper was motivated by the 25th anniversary of the publication of Marc Reisner's book, *Cadillac Desert: The American West and its Disappearing Water*.[1] This book is among the top 100 nonfiction titles published in the English language. One of the central theses of *Cadillac Desert* is that the vast quantity and storage capacity of dams and reservoirs in the West is simultaneously a testament to the success of human engineering of landscapes and a mirage that obscures the true costs of such projects. Water projects have greened large parts of western deserts and provide water for farms that generate high-quality produce, tap water to some of the fastest-growing U.S. cities, and are a critical cushion for this new society against regular droughts. The investment costs for these services were high and typically underwritten by taxpayers but were offset by the services they provide. However, the external costs of these projects were not fully considered and are still largely ignored. These externalities

doi: 10.1111/j.1749-6632.2011.06411.x

Table 1. Properties and data sources of three river case studies

River system	River	Gage location	Gage number	Location/ entry point relative) to dam (km)	Contributing area (km^2)	Mean annual discharge (m^3/s)	Years of flow (predam) (predam)	Years of flow record (postdam)	Years of temperature record	Years of sediment record	Years of fish record
Missouri	Mainstem	Garrison Dam	06338940	0	NA	NA	NA	NA	1971–1975	1954–1998[a]	1997–2009
	Knife River	Hazen, ND	06340500	24	5801.57	4.28	1938–1951	1954–1967	1974–1981	1954–1998[a]	NA
	Mainstem	Bismark, ND	06342500	122	482773.78	620.79	1929–1946	1954–1971	1971–1975	1954–1998[a]	NA
	Heart River	Mandan, ND	06349000	125	8572.86	7.05	NA	NA	NA	1954–1998[a]	NA
	Mainstem	Lake Oahe, ND	Extrapolated[b]	144	NA	NA	NA	NA	NA	1954–1998[a]	NA
Green	Mainstem	Green River, WY	09217000	Upstream	9740.00	45.11	NA	1966–1985	NA	1956–1959[d]	NA
	Mainstem	Greendale, UT	09234500	0	39082.92	55.84	NA	1966–1985	1991–2011	1956–1959[d]	2002–2009
	Yampa River	Maybell, CO[c]	09251000	105	8761.93	43.78	NA	1966–1985	1991–2011 (near Maybell)	1950–1958[d]	NA
	Mainstem	Jensen, UT	09261000	117	65785.70	118.02	NA	NA	NA	1947–1979[d]	NA
	White River	Mouth of Green	09306500	186	13260.74	17.90	NA	NA	NA	1974–1982[d]	NA
	Price River	Woodside, UT	09314500	270	3988.58	3.34	NA	NA	NA	1975–1982[d]	NA
	Mainstem	Green River, UT	09315000	290	116160.97	155.71	NA	NA	NA	1944–1982[d]	NA
Colorado	Mainstem	Lees Ferry, AZ	09380000	24	289560.67	387.03	1930–1960	1970–2000	1992–2010 (variable)	Pre: 1951–1955	2000–2009
	Paria River	Lees Ferry, AZ	09382000	24	3651.88	0.74	1930–1960	1970–2000	1997–2010	Post: 1964–1970	NA
	Little CO River	Cameron, AZ	09402000	123	69857.16[e]	11.98	NA	NA	1992–2010		NA
	Mainstem	Grand Canyon, AZ	09402500	165	366742.32	402.38	NA	NA			NA

[a]Estimated from sediment transport curves (Guy and Norman[9]) derived from more limited data from the same stations; see Macek-Rowland[10] for details.

[b]See Macek-Rowland[10] for details.

[c]Includes load from Little Snake River at Lily (1958–1964) NWIS: 09260000.

[d]From Andrews.[11]

[e]Measured at mouth at Desert View gauge.

include large-scale changes in the seasonality and variability of river discharge, in the distribution and fluxes of salt and sediment on the landscape, and in the distribution and viability of native riverine flora and fauna. Reduced seasonality and variability in discharge have facilitated invasion by nonnative species.[2] Salinity associated with irrigation and water transfers has already significantly reduced annual agricultural production and revenue.[2] Sediment loading to reservoirs constrains their lifespan, although for large reservoirs these life spans are still greater than several centuries.[2,3]

We review the downstream effects of dams on ecosystems with a focus on alterations to flow, sediment, temperature, and fish community composition in an attempt to synthesize data associated with the environmental externalities of dams. We develop a conceptual framework for how rivers respond to dams. This framework borrows from the river continuum and serial discontinuity concepts[4–6] but adds to it a geomorphic template that includes the effects of valley geometry[7] and river network structure.[8] We use three case studies in our analysis: the Garrison reach of the Missouri River (below Garrison Dam), the Grand Canyon reach of the Colorado River (below Glen Canyon Dam), and the Lodore Canyon reach of the Green River (below Flaming Gorge Dam). These rivers provide examples of systems with contrasting valley geometry and tributary density and volume (Table 1).

Geomorphic framework for downstream effects of dams on ecosystems

A strong tradition in stream ecology derives from the river continuum concept,[5] which holds that as the river channel widens (from headwaters to sea), energetic inputs change and the biota shift to utilize the energy sources available. This tradition has been enriched by consideration of discontinuities created by tributaries and dams. Dams create abrupt changes in physical parameters and biological properties that may perpetuate downstream, and these effects occur regularly, dam after dam, from headwaters to sea, creating serial discontinuity[4] in the river continuum. Case studies[12–17] and conceptual or synthesis papers[18–26] attest to the strong influence of dams on downstream ecosystems. We propose that the geomorphic setting, as reflected in valley geometry and river network structure, modulates the magnitude of downstream changes in water temperature, discharge variation, and sediment flux imposed by large dams. Because flow, sediment, and temperature are key components of the habitat template for fish, we further propose that the geomorphic setting also modulates the degree to which changes in master habitat variables (temperature, flow, sediment) alter fish community composition downstream of large dams. We define *community structure* as the proportion of nonnative species, measured either as relative number of nonnative to native species (richness) or individuals (abundance). We characterize valley geometry via the ratio of valley bottom width to channel width, and we characterize river network structure as the number and discharge (relative to the mainstem) of tributaries entering the river downstream from the dam of interest.

We ask four questions:

(1) How does valley form influence the spatial extent of dam-related discontinuities in flow variation, temperature, and sediment flux and community composition?
(2) How do tributary junctions influence the spatial extent of dam-related discontinuities in flow variation, temperature, and sediment flux and community composition?
(3) Does the relative importance of tributaries in restoring continuity in physical and biotic characteristics downstream of large dams vary depending on the valley form of the mainstem?
(4) Do tributaries provide physical conditions (flow variation, temperature) that deviate from the mainstem, thereby providing refugia from dam-related discontinuities in the mainstem?

A geomorphic framework for river ecosystem responses to dams

Physical processes act across varied spatial and temporal scales to influence the channel geometry and disturbance regime of riverine corridors. Channel geometry includes channel planform (straight, meandering, braided), width/depth ratio, dominant bedforms, bed and bank composition, and the abundance and diversity of riparian and aquatic habitat. Channel boundaries can be physically complex or relatively simple. Complex channel boundaries include variations in channel geometry among successive cross sections that create opportunities for flow separation and associated sediment and nutrient retention and habitat diversity. Complexity can derive from local features such as instream wood or from repetitive variations such as meander bends. Physical complexity is created and maintained by fluxes of water and sediment that vary in magnitude and duration.

Disturbance regime refers to the areal extent, magnitude, frequency, and predictability of disturbances, which in this context will be primarily changes in water and sediment discharge. Single disturbances caused by extreme events (e.g., exceptionally large floods) that drive reach-scale (channel lengths of 10^1–10^3 m) changes in sediment distribution and the establishment and mortality of plants. Such changes can persist for decades or be transient.[27] A disturbance regime and events can also be characterized by their predictability and timing, a characteristic of high or low flows that may have more significance for biota than sediment fluxes.[28,29] In this context, flow anomalies are unpredictable events that exceed the long-term trend in seasonal flow fluctuations by 1–2 standard deviations.[30]

The water and sediment moving down the river interact with the channel boundaries and the valley geometry to create channel geometry and disturbance regime. Characteristics of riparian and aquatic communities are likely governed by channel geometry and by the disturbance regime (measured as extreme events or anomalies), and dams alter both of these drivers. The magnitude of these alterations depends on a balance between the constraints valley geometry places on the ability of the floodplain to absorb fluxes or maintain pools of sediment or water and the magnitude of lateral inputs from tributaries. In particular, valley geometry directly influences channel geometry and response to disturbance by providing constraints on lateral and vertical channel adjustments (Fig. 1). Steep, narrow valleys, for example, limit channel sinuosity and development of broad, complex floodplains and also limit the physical complexity of the riverine environment. Changes in water and sediment discharge are likely to cause vertical channel adjustment in the form of incision or aggradation, and this can further reduce complexity. Sediment and water inputs from tributaries can exacerbate or mitigate mainstem response to changing water and sediment yield. Where declining water inputs limit mainstem sediment transport and tributaries continue to introduce sediment, the mainstem can aggrade more rapidly,[31,32] whereas tributaries that contribute primarily water inputs can limit

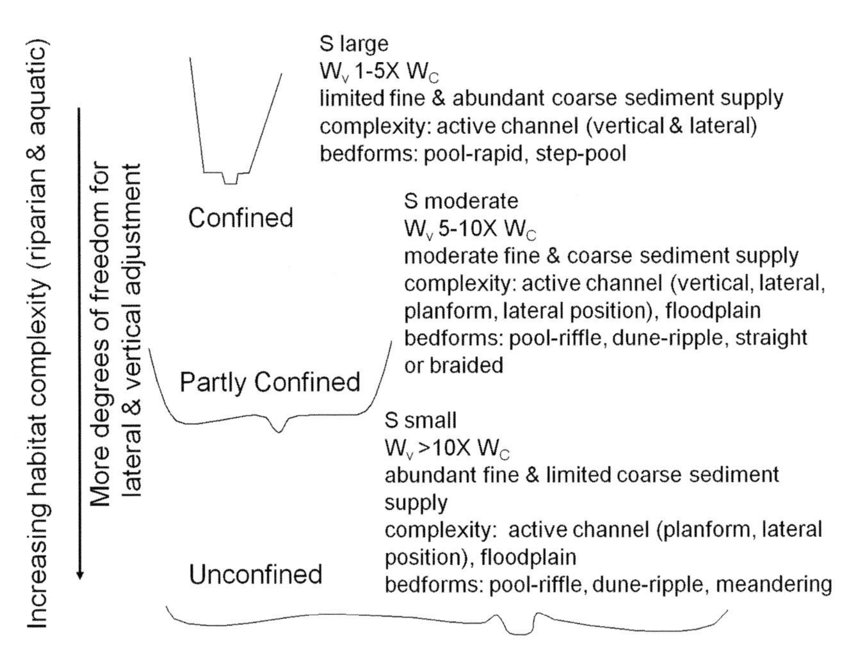

Figure 1. Schematic view of different types of valley geometry (view is upstream or downstream in each case). For each type of valley geometry, descriptions include characteristic values of ratio of valley-bottom width to channel width; relative supply of fine (pebble and smaller) sediment derived from upstream sources, lateral channel migration, or erosion of the floodplain; relative supply of coarse (gravel and larger) sediment from mass movements on the adjacent valley walls and floods on steep tributary channels; relative erosional resistance of channel boundaries and potential for channel change in response to changes in water and sediment discharge; sources of complexity; and characteristic bedforms.

mainstem aggradation. In contrast, the extensive floodplains more commonly present along gentle, wide valleys buffer the effects on the mainstem of changes in water and sediment discharge, as well as buffering the influence of tributary inputs.

Considerations such as these led us to develop a simple, two-way classification of the downstream effects of dams based on valley geometry and tributary junctures. We propose that certain valley geometries and tributaries offset discontinuities in physical conditions and the composition of the biota created by dams. The first axis of this classification scheme is valley geometry. Confined valleys are relatively deep and narrow, with steep valley walls and large downstream gradients. Channels in confined valleys have limited complexity associated with a single channel and narrow valley bottom. Adjustment to changing water and sediment supply occurs mainly through variation in width and depth of the main channel,[32–34] although limited adjustment in channel planform and floodplains may be possible.

In partly confined valleys, the river is less likely to be supply limited for fine sediment, which facilitates sediment storage in floodplains and complex patterns of sediment exchange between the active channel and floodplain. The thickness of alluvium below the contemporary channel is greater, creating more hyporheic exchange along the channel. Channels in partly confined valleys have greater complexity associated with the main channel and may have secondary channels, as well as complexity on the floodplain. Adjustments to changing water and sediment supply can occur solely via the width and depth of the main channel but are more likely to include changes in planform, lateral shifts of the main channel, and changes in secondary channels and floodplains.[17,35]

Unconfined valleys contain channels of lower gradient and greater sinuosity. As with partly confined valleys, complex exchanges of sediment, nutrients, and water among the floodplain, floodplain wetlands, secondary channels, and the main channel occur with changes in discharge. Channels in unconfined valleys typically have the greatest complexity because of downstream variations in the geometry of the main channel, as well as the presence of secondary channels and broad floodplains. Extreme high flows that create lateral channel movement and overbank flooding in unconfined valleys stimulate ecological processes and maintain diversity. Channel adjustment to altered water and sediment supply occurs mainly through lateral adjustment such as changes in sinuosity, width or location of the main channel, the configuration of secondary channels, or extent of overbank flooding.[36]

The second axis of our classification scheme is river network structure, specifically the number and spatial arrangement of entry points of tributaries downstream of the dam. Tributaries may create discontinuities in the physical and biotic properties of the mainstem rivers they enter that are distinct from upstream discontinuities set up by dams.[8] For example, large tributaries may contribute warmer water to the mainstem or deliver high flows during periods when the mainstem is regulated for low flow. Tributaries may also contribute sediment to sediment-starved mainstem channels downstream of dams. These effects depend on channel geometry and the relative size of tributaries. Small tributaries feeding canyon-bound mainstems may have little influence on physical habitat in the mainstem and merely provide small refuge catchments for the biota. We hypothesize that the influence of tributaries depends on the valley geometry of the mainstem and on the characteristics of water and sediment introduced by the tributaries (Fig. 2). Tributaries that introduce sufficient sediment to locally exceed transport capacity of the mainstem are point sources of sediment that enhance physical complexity on the mainstem by creating a constriction. The constriction may take the form of a boulder fan such as Crystal Rapids in the confined valley of the Grand Canyon, which results from episodic debris flows and flash floods, or the constriction may be an extensive lower velocity zone such as Peoria Lake on the unconfined Mississippi River, which results from more continual fluxes of finer sediment. The greater the valley confinement, the greater the magnitude of tributary-induced changes is likely to be because water, sediment, wood, and nutrient inputs from the tributaries are not distributed across a broad valley bottom or spread among multiple channels. Tributaries that primarily increase water discharge on the mainstem without adding appreciable sediment are much less common in the Cadillac Desert region and in rivers generally, but the presence of the tributary mouth can still enhance physical complexity on the mainstem.

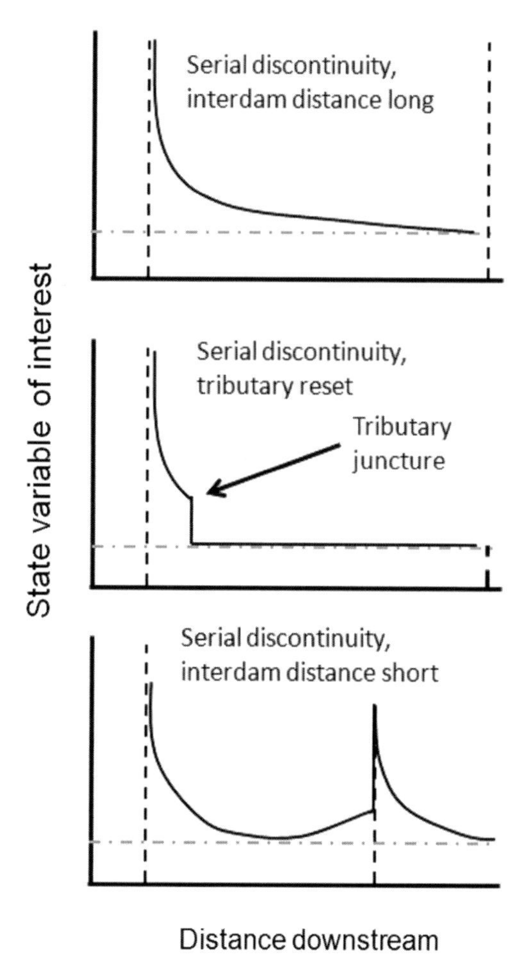

Figure 2. A conceptual model of the serial discontinuity concept (*sensu* Ref. 4; SDC) and variations of the SDC relevant to our case studies. Top panel: interdam segment of river with large interdam distance. Middle panel: interdam segment with large interdam distance, but entry of a significant tributary at midreach. Bottom panel: interdam segment with short interdam distance. Example of state variable used for this conceptual model was the proportion of nonnative species. Dashed line is location of dams. Dash-dotted line is reference condition (predam or upstream) for state variable.

A focus on valley geometry and tributary inputs facilitates understanding and prediction of the type and relative magnitude of channel response to altered water and sediment supply associated with flow regulation. If the primary effect of flow regulation is to reduce transport capacity below that required to mobilize sediment supplied from lateral sources downstream from the dam, the channel will aggrade or narrow, a common effect in confined valleys. If the primary effect is to trap sediment in the reservoir and reduce downstream sediment supply, the channel will widen, incise, or alter its planform to increase boundary erosion and sediment supply downstream from the dam. This can occur in partly confined or unconfined valleys; it can also be exacerbated by tributaries that substantially increase water discharge below the dam without contributing much sediment or can be offset by tributaries that introduce substantial volumes of sediment below the dam. A more typical scenario for the sediment-rich rivers of the Cadillac Desert is that reduced flow results in limited transport capacity even in partly confined and unconfined rivers, resulting in channel narrowing, aggradation, or change in planform (braided to meandering, or meandering to straight) geometry. Tributary sediment inputs only enhance accumulation of sediment downstream from the dam.

For the three case studies examined in this paper, the Colorado River below Glen Canyon Dam exemplifies a confined valley. The Green River below Flaming Gorge Dam is a partly confined valley, and the Missouri River below Garrison Dam is an unconfined river.

Rivers of the Cadillac Desert

As defined by Reisner,[1] the Cadillac Desert is approximately the area of the coterminous United States west of the 100th meridian, but not including the humid portions of the West Coast. The major drainages of this region are the Missouri, the Colorado, the interior Snake-Columbia, the Rio Grande, the Sacramento-San Joaquin, and the central valley rivers of California. The largest rivers head in the Rocky Mountains, where snowmelt sustains large spring–early summer peak flows. Rivers supplied primarily by snowmelt discharge more than 70% of their annual water budget during two to three months of snowmelt and have instantaneous discharges 10–100 times the mean low flow.[37] The sediment load of these rivers increases substantially as they leave the crystalline rocks of the mountains and cross more erodible substrates on the Great Plains, the interior desert basins, or the plateaus of the Southwest. Rain-fed ephemeral tributaries join the main channel along the middle and lower zones of these drainage basins, introducing large volumes of sediment. Although snowmelt can produce interannual variability in peak flow,[38] rainfall floods in the smaller subbasins have produced some of the largest peak unit discharges (flow volume per drainage area) recorded in the world.[39]

Lateral confinement and gradient of rivers in the Cadillac Desert also reflect physiographic provinces. Headwater rivers in the Rocky Mountains are likely to be confined (Fig. 1), as are the canyon rivers of the Colorado Plateau in the Southwest and the basalt plateaus of the interior Northwest.[40] Rivers of the western Great Plains and interior deserts are likely to be unconfined, and partly confined rivers occur throughout the Cadillac Desert.

Aquatic and riparian organisms of mainstem and smaller tributary rivers of the Cadillac Desert have been strongly affected by introduced species, changes in water and sediment yield to rivers in response to land use, and flow regulation. Of the 166 species of fishes in the Rio Grande basin, for example, at least 42 are vulnerable, threatened, endangered, or extinct.[41,42] Thirty-four unique species or subspecies of fish on the Missouri River are vulnerable, threatened, endangered, or extinct, as are at least 14 on the Colorado River. Many of these species are minnows and suckers (Family Cyprinidae and Catostomidae) adapted to a seasonal snowmelt peak or, in some cases, to access to extensive floodplains present discontinuously along the rivers. Elimination of the flood pulse and associated overbank flooding and nursery habitat limit natural recruitment of these fish,[43] as does sedimentation of spawning gravels because of reduced peak discharges.[44,45] Elimination of the flood pulse along the Rio Grande and fragmentation and isolation of the channel from the floodplain through flow regulation and water withdrawals have reduced establishment of native cottonwoods (*Populus* spp.) and willows (*Salix* spp.) and favored invasions by introduced tamarisk (*Tamarix* spp.) and Russian olive (*Elaeagnathus angustifolia*).[41]

Dams and diversions have been built along the rivers of the Cadillac Desert since the mid 19th century to provide water storage and flood control and, subsequently, hydroelectric power generation. Rivers of the Cadillac Desert are among those with the highest ratio of reservoir storage to mean annual runoff in the United States.[2,46]

Methods

Hydrology

Dams are well known to alter downstream hydrology of rivers by dampening seasonal and interannual discharge variation.[15,47] Specifically, peak discharge is muted and base flows are artificially elevated by extended release of captured peak flows. Dams also have altered flow variability by changing the frequency, magnitude, duration, and timing of extreme events.[47] Spectral methods provide a tool for quantifying the timing, magnitude, and frequency of low- and high-flow anomalies.[28–30] In this section, we use a spectral approach to analyze changes in the seasonality of low- and high-flow anomalies in average daily discharge. We used average daily discharge data from the USGS National Water Information System (NWIS) database to characterize changes in the timing of high and low flows before and after dam installation (or in one case upstream and downstream of an installed dam) at our three case study sites: Missouri River below Garrison Dam (USGS # 06440000; pre: 1929–1946; post: 1954–1971), Colorado River below Glen Canyon Dam (USGS # 09380000; pre: 1930–1960; post: 1970–2000), and Green River (upstream: 09217000; downstream: 09234500; postdam time period: 1987–2006). For all three case studies, there were adequate observations before and after dam construction/closure to compare flows during 15- to 30-year time windows bracketing the dam construction–filling interval. This was not the case for the Green River, where we used an upstream–downstream comparison in a space-for-time substitution. Such a substitution was justified as both mean and standard deviation of monthly spatial average precipitation (including rain and snow) and air temperature (one of the key drivers of snowmelt) of the upstream (09217000) and downstream (09234500) subbasins of the Green River were similar during 1987–2003 (mean = 25 mm and 24 mm, respectively; standard deviation at both subbasins = 11.9 mm). Similarly, mean and standard deviations of spatially averaged precipitation and air temperature were constant during pre- and postdam periods for the Colorado River below Glen Canyon Dam.

We also analyzed time series (during similar time periods) for one unregulated tributary to the reach downstream of the dam of each of our case studies. These tributaries include the Knife River (USGS # 06340500, entering the Missouri River 24 km below Garrison Dam), Paria River (USGS # 09382000, entering the Colorado River 25 km downstream of Glen Canyon Dam), and Yampa River (USGS # 09260050, entering the Green River 105 km downstream of Flaming Gorge Dam).

Our method involves three steps. First, we use a continuous 15- to 30-year predam (or upstream) record of daily average flows to quantify the long-term seasonal average discharge at each ordinal day of the year (following Ref. 29). Using this predam (or upstream) trend, we then estimate the standard deviation of high- and low-flow discharge anomalies (following Ref. 29). Second, we superimpose this long-term trend on a time series of daily discharge events of similar length but occurring 5–25 years after the closure (or downstream) of the dam. Third, we define catastrophic events in the postdam (or downstream) time series as anomalies with residual magnitude greater than or equal to 2σ of the long-term seasonal average. We index postdam (or downstream) events and their residual magnitude to the predam (or upstream) seasonal trend and σ of discharge anomalies. By referencing postdam (downstream) to predam (upstream) first and second moments, we amply quantify dam-related changes in the seasonality, magnitude, and frequency of anomalous flows (high and low).

In addition to this spectral analysis, we quantify changes in extreme events (i.e., 10-year high and low flows) using a similar approach. Specifically, we estimate the 10-year low- and high-flow magnitude using Log Pearson III methods[29,48] for the same predam (or upstream) record used in spectral analysis. We then search for 10-year high and low extreme events in the postdam (or downstream) record using the predam high- and low-flow magnitudes as thresholds for floods and droughts, respectively. In all of our discharge analyses, we use and present normalized discharge, actual average daily discharge divided by the long-term mean of average daily discharge and expressed on a log base 10 scale.

Temperature

We compiled water temperature data from our own primary records (Green River, U.S. Fish and Wildlife Service, Lakewood, Colorado; K. Bestgen, unpublished data) or from published sources. Temperature data for the Colorado River in the Grand Canyon and five tributaries of this reach were obtained from the Grand Canyon Monitoring and Research Center (http://www.gcmrc.gov/dasa/tabdata/). Temperature data for the Missouri River and one tributary (the Knife River) were obtained from the USGS NWIS (USGS # 06340500, 06338490, 06342500). Using this record, we estimate mean, standard deviation, minimum and maximum temperatures, and compare the average and the range of temperature conditions in the mainstem and nearby tributaries.

Sediment

We compiled time series of annual sediment loads for our three case studies. We attempted to identify time series before and after the upstream dam closure for locations directly below the dam and downstream of major tributary junctions on the mainstem. We also tried to identify time series before and after mainstem dam closure for major tributaries (Table 1). For the Missouri River, we identified data sets before the closure of the Garrison Dam at the USGS gaging station at Bismark, ND (downstream of the Heart River confluence, but upstream of the Knife River confluence). Data were more abundant after the closure of Garrison Dam such that we found adequate records for the mainstem below Garrison Dam, at Bismark and at the tail pool of Lake Oahe, as well as for both the Knife River (at Haven, ND) and the Heart River (at Mandan, ND). For the Colorado River in the Grand Canyon, we found adequate time series for sediment loads before and after the closure of Glen Canyon Dam at Lees Ferry, Paria River (at Lees Ferry); the Lower Colorado River (at Cameron); and at the gaging station at Grand Canyon, Arizona (downstream of both tributaries). For the Green River, data were less consistently available such that post-Flaming Gorge data for tributaries had to be used to estimate predam conditions for sediment fluxes from these tributaries to the mainstem.[11] Nevertheless, there were adequate data before and after the closure of Flaming Gorge Dam at Green River, Utah (downstream of several major tributary junctions). We also found records for the Yampa River (at Maybell, CO), the Little Snake River (which feeds the Yampa below Maybell), the White River (at the mouth), and the Price River (at Woodside, UT).

For all stations, we tallied annual sediment loads (in metric tonnes) from reported daily loads and then computed time averages (and standard errors) for all available years (pre- and postdam). We then constructed a full sediment budget for the reach between the dam and downstream-most gauge. We report the results from this sediment budget in terms of the percentage contribution of the mainstem at the upstream gauge (where available: Garrison for the Missouri River, and Lees Ferry for the Colorado River) and any downstream tributaries to the observed load in the mainstem at the downstream gauge. For the Missouri River, the Bismark gauge (intermediate to the Garrison and Oahe gauges) permitted us to assess the relative role of channel adjustments on the mainstem and tributaries (Heart River) to mainstem sediment supplies after the closure of Garrison Dam. We differentiate these two sources of sediment as "bank" (channel adjustment above Bismark) and "tributary" inputs. For all case studies, we were able to estimate the percentage contribution of all tributaries to the mainstem sediment load downstream of a large dam.

Fish

We compiled time series of fish community composition for multiple sites below dams at our case study sites: Garrison reach of the Missouri River (North Dakota Department of Fish and Game), Lodore-Yampa reach of the Green River (K. Bestgen, unpublished data), and Grand Canyon reach of the Colorado River (Arizona Game and Fish Department, USGS Grand Canyon Monitoring and Research Center).

All datasets were based on standardized electrofishing surveys that span multiple sites from below a dam to typically the tail pool of the next reservoir, ranging from 50–180 km longitudinally. All data sets are also temporally extensive, spanning 6–20 years. We present these data in two ways. First, we estimate the average proportion of nonnative fishes (hybrid species pooled with nonnatives) for each spatial location at each study site. The proportion of nonnative fishes provides an index of the longitudinal change in relative dominance of biodiversity by introduced fishes from the dam to the tail pool of the next reservoir (or significantly downstream from the dam). Second, we estimate the proportion of total fish abundance made up by nonnative species (expressed as abundance per unit sampling effort, or catch per unit effort, CPUE).

Results

Hydrology

Dams mute seasonal fluctuations in discharge in large rivers, dampening peak flows following snow melt and artificially maintaining much higher base flows during low-flow periods (Fig. 3A,C, and E). In all three river systems, the historical (or upstream) sinusoidal seasonal trend is replaced by a solid band of regulated flow observations. Observed flows after dam construction (or below dams) are significantly higher during historical base flow periods and can be (but are not always) lower than average peak flow during peak-flow season. Dams also change the pattern of stochastic variation in large rivers. Interannual variation is constrained into a narrow band of controlled flow levels.

Unregulated tributaries provide unique hydrologic conditions compared to the regulated large river systems they feed (Fig. 3). In particular, tributaries have much higher interannual variation (Fig. 3 B, D, and F) and show considerable seasonality (Fig. 3 B, D, and F). The Knife and Paria Rivers are flashy smaller river systems that show less (Knife) or none (Paria) of the snow melt peak exhibited by the mainstems they feed that were historically dominated by a snowmelt signal.

Dams eliminate extreme high flows and alter the timing of the season for extreme low flows in large regulated rivers (Fig. 4). Specifically, dams diminish the frequency of (Green River) or eliminate (Colorado and Missouri rivers) 2- and 10-year floods and enhance the frequency of 2- and 10-year low flows in the early spring when reservoirs are filling with snow melt (Flaming Gorge and Powell). In the Missouri River, Garrison Dam eliminates both extremes, not surprisingly, as flood control, recreation, and navigation are more central to the mission of the Garrison Dam than to the storage reservoirs on the Colorado River System. In contrast to Garrison Dam, Glen Canyon and Flaming Gorge dams exacerbate low-flow events. Specifically, the timing of 2- and 10-year low flow extremes is later, occurring during the period of historical peak flows from snow melt. Aseasonal low flows occur in dam tailwaters because of water storage. Reservoir operations recharge pools with snowmelt during peak discharge, and appear to reduce downstream flows below historical lows.

Tributaries provide a very different set of extreme events than the large regulated rivers they feed. Two-year high flows (via log-Pearson III methods) are still present in all three tributaries, even when referenced

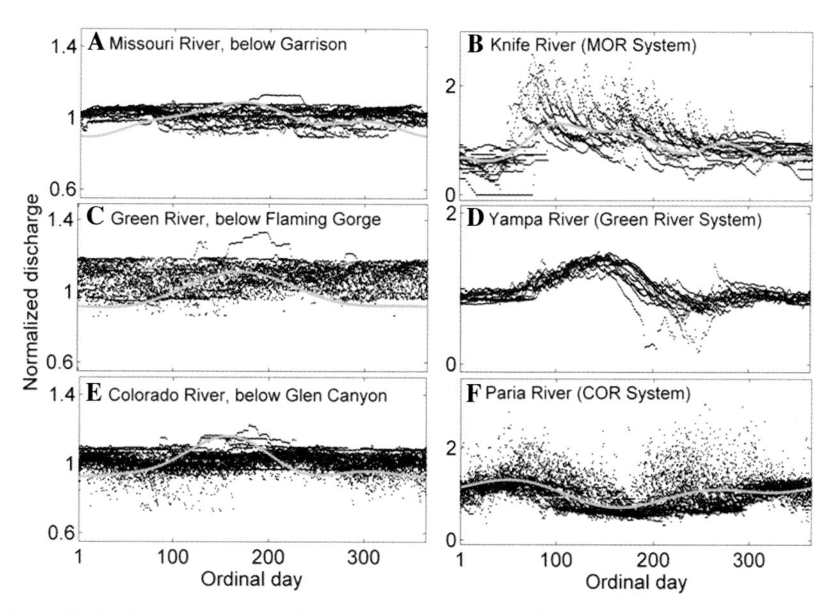

Figure 3. Hydrographs for three large rivers and one of their tributaries before and after dam construction on the mainstem. Panels show the predam or above-dam (Green River only) trend (gray line) estimated from 15–30 years of average daily discharge measurements using Fourier analysis compared with actual discharge (points) during a similar 15- to 30-year period postdam or below-dam (Green River only). Discharge is normalized to the average log-10 transformed value such that units reflect proportional change from average conditions. Tributaries (right-hand column) all enter river in below dam reach. Time period for trend and data for tributaries match that of mainstem except for Green river, where trend and data are both from the same period, following construction of Flaming Gorge Dam.

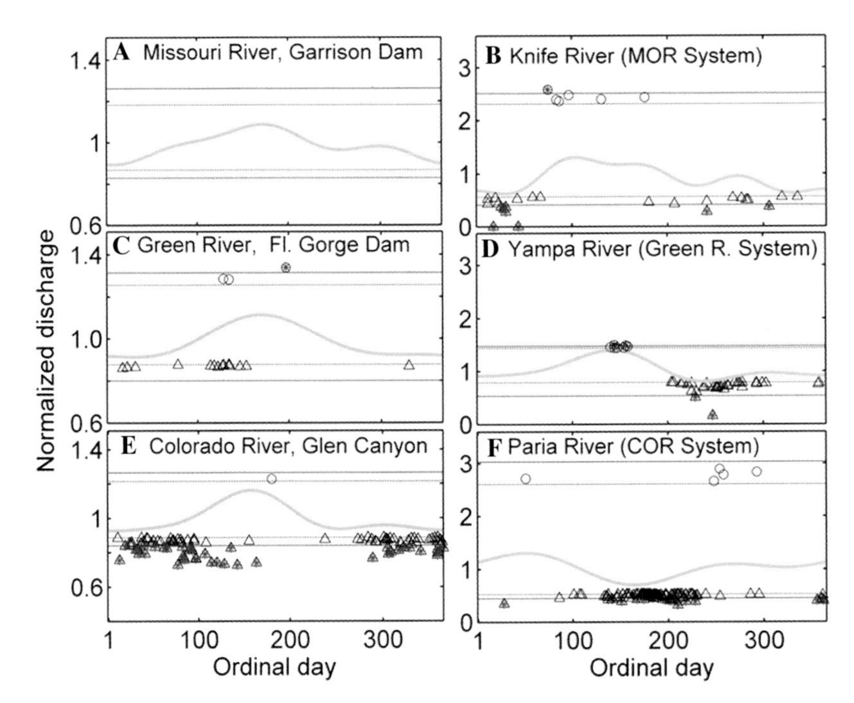

Figure 4. Extreme events for three large rivers and one of their tributaries before and after dam construction on the mainstem. Panels show predam extreme event magnitude thresholds (blue dashed and solid line: Q_h^2, Q_h^{10}; red dashed and solid line: Q_l^2, Q_l^{10}); circles and triangles show Q_h^2 and Q_l^2 of predam magnitude or higher, but occurring after dam construction; and circles filled with blue stars and triangles filled with red stars show Q_h^{10} and Q_l^{10} of predam, but occurring after dam construction. The predam signal from Fourier analysis is shown to illustrate the timing of extreme events (this figure) and catastrophic events (Fig. 5). Tributary extreme events (right-hand column) are analyzed in the same way as mainstem (e.g., postdam events referenced to extreme event threshold magnitudes from predam period), except for Yampa River (see methods for description).

to the 2-year flow magnitude from the matching discharge time series predating the dam on the corresponding mainstem (for Knife and Paria rivers). Ten-year high flows are likewise present in two of three tributary systems (except the Paria River). Ten-year high flows correspond better with seasonal peaks and 10-year low flows with base flow periods in tributary systems than in the mainstem. Thus, the timing of extreme events is altered in the regulated mainstem but not in tributaries of the river systems comprising our three case studies.

Two- and 10-year extreme events are estimated using a time series of maximum (or minimum) annual flows with no reference to the timing (ordinal day) of these events. By contrast, anomalous flows from spectral analysis provide an integrated measure of event timing, magnitude, and frequency.[29] Dams broaden or reverse the temporal pattern of high- and low-flow anomalies (events $>1\sigma$ in residual discharge magnitude) and increase the number of large magnitude low-flow anomalies or "catastrophic" flows (events $>2\sigma$ in residual discharge magnitude; Fig. 5). In all three large rivers, low-flow anomalies occur during historical peak-flow period, and high-flow anomalies occur during historical low-flow period. Thus, dams in these river systems smooth the hydrograph and reverse the historical pattern of high- and low-flow anomalies (hereafter, "anomaly reversal"). In addition, high- and low-flow anomalies are observed almost any time of the year in the Colorado River. Anomaly timing is spread out in the Green River as well, but to a lesser extent. Thus, water supply reservoirs in arid regions can broaden the temporal occurrence of positive and negative anomalies (hereafter, "anomaly broadening").

The temporal pattern of high- and low-flow anomalies in tributaries contrasts with that of the larger, regulated rivers they feed, but the way in which these patterns contrasts is idiosyncratic among case studies. In the Knife River, high- and low-flow anomalies occur across much of the

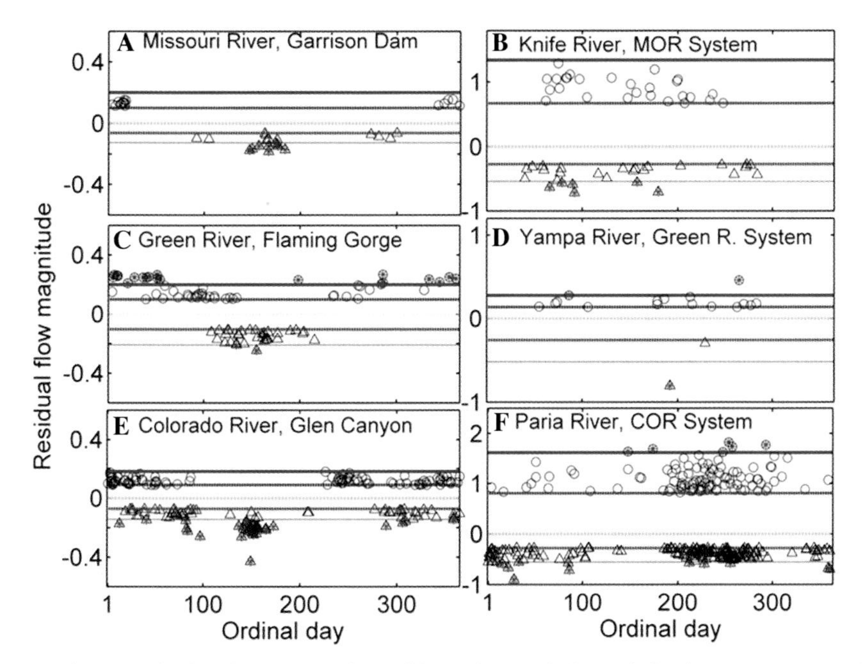

Figure 5. Catastrophic events for three large rivers and one of their tributaries before and after dam construction on the mainstem. Panels show predam catastrophic flow magnitude (blue dotted and solid lines: σ_{hf} and 2 × σ_{hf}; red dotted and solid lines: $\sigma_{lf}$and 2 × σ_{lf}), and high- and low-flow anomalies > σ (circles and triangles) or > 2σ (circles or triangles filled with blue or red stars, respectively) occurring after dam construction but referenced to long predam long-term signal from Fourier analysis (cyan line).

calendar year in contrast to the anomaly reversal observed in the Missouri River below Garrison Dam. In the Yampa River, high- and low-flow anomalies occur during periods that correspond to peak- and base-flow conditions in contrast to anomaly broadening and reversal observed in the Green River. Finally, the Paria has a much higher number of catastrophic high flows than the Colorado River as a result of a stronger response to monsoon rains (ordinal days 250–300). Moreover, extreme low flows and anomalies occur during the same timeframe in the Colorado River but occur in nearly nonoverlapping time periods in the Paria River.

Hydrologic changes can directly influence riverine environments by altering the depth and velocity of flow (hydraulics), the extent of wetted perimeter in the channel, and the duration of flow. Hydrologic changes can indirectly influence riverine environments by altering water temperature, water chemistry, and the transport of nutrients and sediment. In the context of valley geometry, hydraulic changes are likely to be greatest in confined valleys, whereas extent of the wetted perimeter may change the most in partly and unconfined valleys. Partly confined and unconfined valleys will also be most subject to changes in duration of flow in secondary channels and across floodplains. The indirect effects of hydrologic changes are likely to be lesser in partly confined and unconfined valleys, where hyporheic exchange, slower surface flow paths, biological processing of nutrients, and sediment storage in or removal from floodplains can each potentially buffer the effects caused by a dam. The greater the inputs from unregulated tributaries below a dam, the lesser the magnitude of direct and indirect hydrologic changes in the mainstem are likely to be.

Temperature

In many Western rivers, large dams create deep reservoir pools that exhibit stratified temperatures such that deep (hypo- or metalimnetic) release point of impounded water results in a significant change in water temperature below the dam. This discontinuity may be offset by downstream tributaries, depending on their relative discharge and temperature. Dam tailwater originates from the metalimnion in the Lake Sakajawea (above Garrison Dam) and in Lake Powell (above Glen Canyon dam), whereas a selective withdrawal mechanism allows for releases at variable depth from Flaming Gorge Reservoir. Average annual discharge contributions from downstream tributaries are ~3% for the Missouri River, 56% for the Green River (above Jensen), and 4.5% in the Colorado River (above the Grand Canyon gauge). Based on this information, we would expect the temperature discontinuity to be strongest in the Colorado River, but we would expect tributaries to have the strongest effects in restoring (warmer) temperatures in the Green River given the larger relative discharge of its tributary (the Yampa River).

Temperatures increase downstream of dams in all three of our large river case studies. Colder water prevails below dams that release colder water from depths below the warmer surface waters of the epilimnion, and more generally water temperatures increase downstream as snowmelt from headwaters enters warmer semi-arid or desert biomes at lower elevation. Temperature increased 0.005, 0.015, and 0.01° C/km with ranges of 0.6, 2.3, and 4.7°C in the Missouri, Green, and Colorado rivers, respectively (Fig. 6). In all three case studies, mean water temperatures in the mainstem appear to increase linearly; however, water temperature ranges (maximum–minimum) exhibit steeper increases at tributary mouths, especially in the Colorado River (Fig. 6).

Tributaries themselves, offer novel temperature ranges for biota. For the Green–Yampa Rivers, temperature ranges are similar in the tributary and mainstem near tributary junction. These similarities perhaps overshadow finer-scale seasonal differences in mean monthly temperature. The Yampa is warmer during the summer and colder during the winter on average than the Green River at the mouth of the Yampa (K. Bestgen, unpublished data). For the Colorado and Missouri rivers, however, mean annual water temperature and its range are both higher in tributaries than in the mainstem near tributary junctions. For the Colorado River, mean temperature of tributaries approaches the annual maximum temperature of the mainstem near tributary junctions.

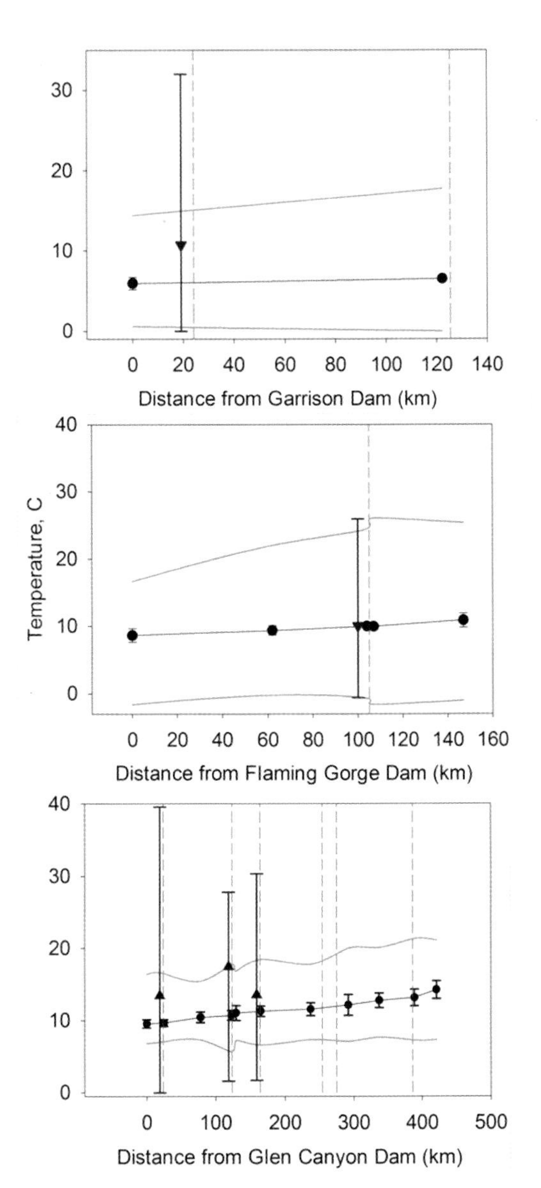

Figure 6. Mean ± 1 standard deviation annual water temperatures (circles ± error bar) for mainstem river locations. Gray lines indicate maximum (top panel) and minimum (bottom panel) annual water temperatures across entire period of record, smoothed by spline. Dotted vertical lines are entry points of major tributaries and mean and range (maximum and minimum) annual temperatures of tributaries are given for comparison (triangles and error bars).

In the context of valley geometry, confined valleys are likely to be the most sensitive to dam-induced changes in water temperature. Partly confined and unconfined valleys typically have greater surface–hyporheic exchange, as well as greater storage and downstream transmission times as a result of flow in secondary channels and across floodplains. The greater the unregulated tributary inputs below a dam, the lesser the influence of dam-induced temperature alterations.

A special case for large tributaries

Of our three case studies, the Green River has the largest tributary in terms of relative mean daily discharge, the Yampa River. Large unregulated tributaries may have disproportionate effects in terms of resetting discontinuities in discharge variation and temperature created by dams. Mean monthly temperatures fluctuate much more significantly below the mouths of the Yampa River (Green River) than upstream below Flaming Gorge Dam (Fig. 6). This is also true for the Colorado River below the confluence of the Little Colorado River compared to below Glen Canyon Dam. More importantly, summer temperatures are much lower directly below the dam in both rivers than below downstream tributaries. This shift in mean monthly temperatures eliminates most native fishes from reaches of river directly below dams with cold water release by creating thermal conditions conducive for nonnative salmonids. The Yampa River also restores the semblance of a "native" hydrograph—a snow melt peak and variation in daily discharge more similar to the Yampa River than the Green River below Flaming Gorge Dam (Fig. 3 and Fig. 7, bottom panels).

Sediment

Predam sediment conditions for both case studies for which we had predam data (Colorado and Green rivers) were dominated in the mainstem by upstream supply (Fig. 8). Tributaries contributed very little sediment compared to the watershed upstream of their junctions. This trend changed following dam closure, but to varying degrees depending on confinement. In the Missouri, the postdam load at Garrison declined significantly, but tributaries continued to contribute a small fraction of the relative load at Oahe. The relative contributions of the Knife and Heart rivers at Oahe ranged from 0% to 18% and 0% to 40%, respectively; on average, however, both tributaries contribute only 12%

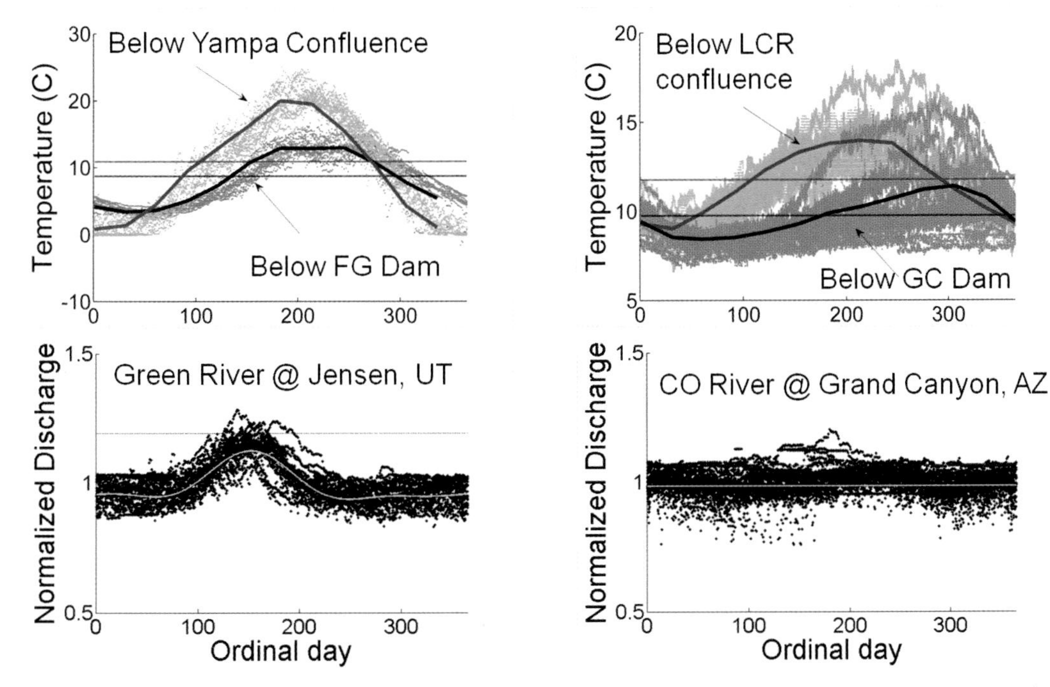

Figure 7. Effects of tributaries on temperature regimes (top row) and daily flow variation (bottom row) for the Green River below the Yampa River confluence (left-hand side) and the Colorado River below the confluence of the Little Colorado River (right-hand side). Top panels show water temperature (° C) for an ordinal day over the period of record including 1991–2010. In each panel, we plot temperature directly below a dam (gray dots) and downstream of a major tributary (rose dots). Dotted lines are mean temperatures across the period of record (black, downstream of dam; red downstream of tributary), and solid lines are the average monthly temperature (unsmoothed). Bottom panels show normalized daily discharge for an ordinal day over the same period of record as upstream (Green) or after dam (Colorado) records in Figures 6–8.

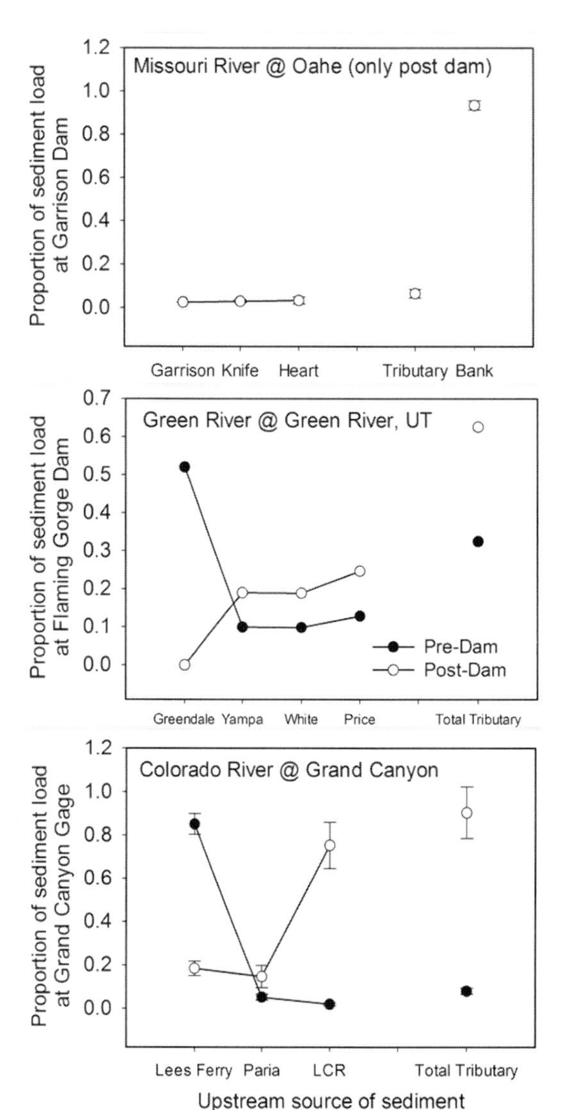

Figure 8. Sediment loads attributable to tributaries below three major western U.S. dams. Proportional annual sediment loads for tailwater (at dam) and several downstream tributaries. Proportional loads are measured relative to total annual load at a downstream site on the mainstem: Lake Oahe (for Missouri River), Greendale Utah (for Green River), and Grand Canyon (for Colorado River).

of the total sediment load at Oahe. Channel adjustment and land use downstream of Garrison (and above Bismark) produced significant sediment inputs that in most years dwarf tributary inputs.

Postdam tributary inputs increase systematically with confinement (Fig. 8). Total tributary inputs are higher in our partially confined case study (Green River) and dominate the mainstem sediment load in our confined case study (Grand Canyon). The total relative contributions of the Paria and Little Colorado rivers at Grand Canyon, AZ, ranged from 2% to 9% before the closure of the Glen Canyon Dam and 27% to 130% after the dam closure. The importance of tributaries as a source of mainstem sediment below large dams increases with confinement in the mainstem. Large rivers in steep canyons lack floodplain sources of sediment that dominate downstream sediment budgets in such systems below large dams.

Fish

Nonnative fishes thrive below dams and the relative dominance of nonnative fish fauna declines with distance from dam (Fig. 9). Nonnative fish make up a higher proportion of the fish fauna directly below dams in the Colorado and Green rivers than in the Missouri River due to higher regional native species richness in the Missouri River. The decline in dominance by nonnative species in all three river systems is more pronounced when measured in terms of abundance (CPUE) than richness alone. The decline in dominance by nonnative species is also more precipitous by tributary mouths. This pattern is most evident for the Missouri and Green rivers. This pattern is less obvious for Colorado River, but a stronger pattern is likely masked by binning data from surveys from variable locations into 10-km blocks. These 10-km blocks do not match up precisely with the entry points of tributaries in the Grand Canyon.

In the context of valley geometry, the greater the physical complexity of the river, the more likely fishes are to be able to find refugia from hydrologic variations and nonnative species. As noted in the context of changes in sediment supply, water temperature, and other parameters, partly confined and unconfined valleys can effectively create buffers against change by providing more diverse and extensive habitats for fish. Tributaries can also serve as critical refugia, as illustrated by the Little Colorado River and other tributaries to the Colorado River in Grand Canyon.[49]

Discussion

Large dams that create vast water storage reservoirs are the hallmark of Western U.S. riverscapes.[2] These reservoirs have had significant effects on downstream water temperature,[50] sediment flux,[11,15,51]

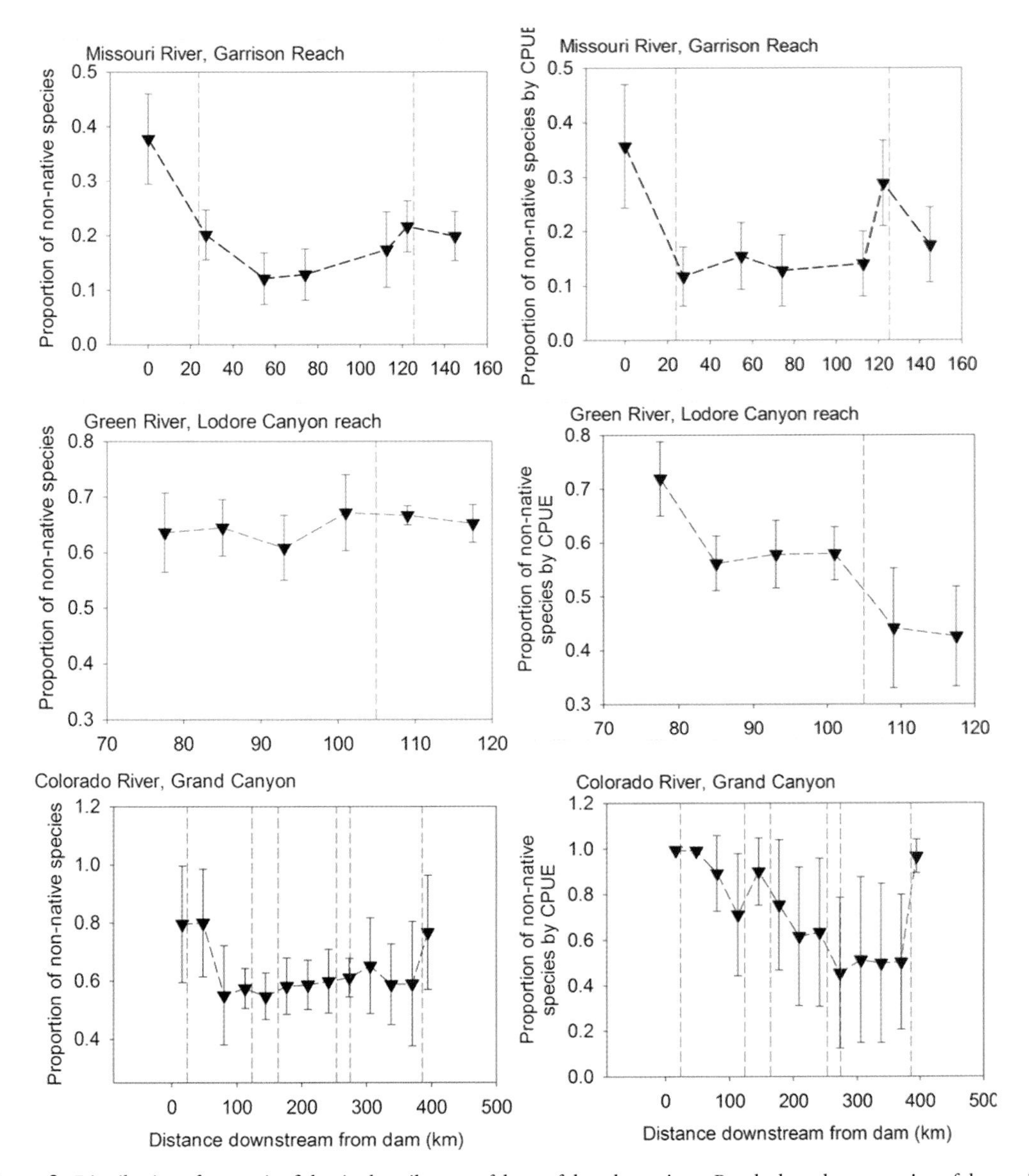

Figure 9. Distribution of nonnative fishes in the tailwaters of dams of three large rivers. Panels show the proportion of the species pool comprised by nonnative species (left-hand panels) and proportion of total community abundance of nonnative species as indexed by catch per unit effort (right-hand panels). Dotted lines show entry points of major tributaries.

and flow regime,[47] as well as on the composition of the fish community.[2,52,53] In this paper we ask how the geomorphic setting—valley form and river network structure—influences these patterns downstream of dams. We find that valley form has a strong effect on the way the downstream channel responds to the discontinuation of the upstream sediment supply below a dam. Sediment flux in confined channels like the Grand Canyon comprises primarily lateral inputs from tributaries, whereas, in unconfined channels, downstream sediment supply is dominated by channel adjustment and inputs from eroding banks of a sediment-hungry mainstem.

In contrast to this story about sediment, we find the relative influence of tributaries on the flow or temperature regime of the mainstem (across valley forms) is proportional to tributary size. All tributaries provide flow and thermal refugia for native species, but large ones (e.g., Yampa River) provide significant amelioration of effects of upstream flow and water temperature alteration,[54,55] especially if tributaries are far downstream of the dam. Perhaps as a result of the near-native conditions presented by tributaries (large and small), the composition of the fish community changes nonlinearly at tributary junctions in a way that enhances the relative richness and abundance of native species. For example, the Green River upstream of the Yampa River has a substantial cold water salmonid community, increasing in an upstream direction. In contrast, the Green River downstream of the Yampa supports few salmonids and nearly all warm water fish species because the upper ranges of water temperature excludes cold water fishes while enhancing warm water kinds. Similarly, the Little Colorado River supports one of the only populations of the endangered humpback chub; this tributary is the epicenter for the species in the mainstem Colorado River.[56] Thus, significant tributary effects on fish communities may arise from subtle variation in temperature or flow variation at seasonal (monthly) time scales that goes unmeasured when comparing annual mean values for these abiotic factors.

Tributaries are a significant factor in influencing physical habitat conditions downstream from dams, but broad generalizations on the subject are not well defined except for the role of sediment. Previous studies have shown that the amount of sediment and the size of sedimentary particles in the main channel are adjusted by contributions from tributaries. First, the amount of sediment contributed by each tributary partly replaces the sediment that has been subtracted by sedimentation in the upstream reservoir, and eventually most of the sediment lost to reservoir sedimentation is replaced in the downstream reaches in some cases (e.g., Ref. 57). Second, particle size of sediment is often changed by tributary contributions. In the case of canyon rivers, tributaries are capable of moving larger particles than those moved by the main channel because the tributaries have much steeper gradients (e.g., Ref. 57). The most striking examples are in canyons where alluvial fans of boulders are constructed by tributaries in the main channel, but even in unconstrained rivers the tributary contributions may be more coarse that sediments in the main channel. Our study extends our understanding of tributary influences beyond sediment to other important controls on key components of fish habitats and fish community composition. In this context, undammed tributaries that feed mainstem channels below large Western dams have significant conservation value for the long-term viability of native fauna in highly regulated river systems.

Dams and discontinuity in the physical habitat template

Dams that create storage reservoirs on many of the largest reservoirs in the western United States disrupt a more natural continuum of change in physical characteristics that comprise a habitat template for organisms in river ecosystems. We quantify this discontinuity for three such components of the habitat template for fishes (stream temperature, sediment transport, flow variability) and evaluate the degree to which geomorphology rectifies the discontinuity. Our results are rich but complex, highlighting the difficulties in replicating spatial patterns of habitat for rivers as large as the Missouri and Colorado. Nevertheless, three salient patterns emerge. First, dam operations produce engineered hydrographs downstream of dams that simultaneously reduce peak flows and augment base flows. Formerly sinusoidal seasonal variation in daily average discharge is supplanted by a "square wave pattern" in which dam operations shift abruptly between constant discharge that deviates from long-term seasonal average discharge. This new hydrograph presents uncharacteristic high flows during the historical low-flow season and mutes the snow melt signal (high spring flows with snow melt) in all three river systems. Dams also change the nature of daily variation in discharge. Extreme high flows, especially those with a 10-year recurrence interval, are rare in the postdam hydrograph, whereas extreme low flows are more common. Similarly, the temporal pattern of spectral low- and high-flow anomalies is much different in the postdam hydrograph. All three river systems exhibit reversal in the pattern of 2σ anomalies—high-flow anomalies occurring during the historic low-flow period and low-flow anomalies occurring during the historic snow melt peak. In the Green and Colorado rivers, the season for high- and low-flow anomalies is also broadened. Hence, the engineered hydrograph presents highs and lows that are unexpected relative to the historical patterns that shaped the evolution and life histories of organisms in these systems. It is important to note that tributaries to all three systems present hydrographs that are virtually unaltered compared to predam conditions. These more natural flow regimes likely provide refugia for native species that may have better adapted to local variation than some nonnative species that benefit from flow regulation in the mainstem.

Second, water temperature is reduced downstream of dams by periodic deep release from large reservoirs, but lowered temperatures increase steadily with distance from the dam. In the Grand Canyon, although mean annual stream temperatures increase more or less linearly, the range of stream temperatures (i.e., minimum and maximum) appears to be exaggerated around key tributary junctures, matching the wider temperature ranges of the stream water in tributaries themselves. Hence, although tributaries appear not to alter mainstem mean water temperatures, they do drive seasonal variation in the mainstem water temperature profile. Moreover, average temperatures are higher and their ranges are broader in tributaries, again suggesting that tributaries offer refuge from novel habitat created downstream of mainstem dams.

Third, dams disconnect downstream rivers from upstream sources of sediment that historically determined the presence, density, and life span of lateral habitats. These lateral habitats included braided channel systems in unconstrained valleys such as the Missouri River below the current location of Garrison Dam. Even in canyon-bound systems, lateral habitats such as backwaters may be important rearing habitats for juvenile native fishes[58] and are maintained by upstream sediment sources.[59] These lateral habitats are maintained by upstream channel adjustments and bank erosion from the mainstem in large unconfined river systems, whereas they are maintained by tributary inputs in confined systems. Moreover, these habitats are increasingly rare given reduced sediment input from upstream sources.

Dams and discontinuity in fish community structure

In all three rivers, the proportion of native species is very low near the dam and increases with distance downstream. This pattern is particularly pronounced for the relative abundance data (CPUE). In the Colorado River system, nonnative abundance and richness increase at the last sampling station. This increase is most likely due to proximity to a new nonnative species pool in nearby Lake Mead. Similarly, in the Missouri River, nonnative richness and abundance increase relative to natives in the second-to-last downstream station (Bismark) and decline again following the entrance of the Heart River just before the tail of Lake Oahe. The increase in abundance and richness of nonnatives at Bismark may be related to urban fisheries enhancement programs (i.e., stocking of bass), proximity to a new nonnative species pool in Lake Oahe, or both.

Patterns of relative abundance of native fishes also suggest that tributaries contribute to the downstream decline in nonnative relative abundance in a nonlinear fashion. Nonnatives are in lower relative abundance immediately downstream from tributary junctures. This effect is likely reinforced by the habitat template of tributaries providing a refuge from novel habitats in the mainstem.

Caveats

The data we present in this paper are comprehensive in spatial and temporal scope and cover a wide range of physical and biological detail, but the data have clear limitations. First, there are many holes in the dataset. This is particularly true for the Missouri River, where we have notably fewer temperature and sediment data. Here, the pattern we observe may not hold at finer resolution or broader extent. Second, replication of large basins like the rivers we analyze here is difficult and so many of our inferences about the effects of valley form and tributaries could be artifacts of locations reflecting random site variation. This data limitation underscores the importance of good monitoring of regulated rivers. Complete datasets that include observations of multiple components of physical processes and multiple groups of organisms (fishes, plants, macroinvertebrates) with adequate spatial and temporal resolution are nearly nonexistent except where mandated by law. More funding and effort should be allocated to compiling datasets like those that exist for the Grand Canyon.

Downstream effects of dams in the Cadillac Desert

In this paper, we propose that the geomorphic setting modulates the magnitude of downstream changes in master variables (water temperature, discharge variation, and sediment flux) imposed by large dams. Because flow, sediment, and temperature are key components of the habitat template for fish, we further propose that the geomorphic setting also modulates the degree to which changes in master habitat variables alter fish community composition downstream of large dams. Here, we define the geomorphic setting as valley form in the mainstem river and the number, size, and entry points of tributaries downstream of dams (i.e., network structure). We find that valley form has a significant effect on changes to sediment budgets downstream of dams but does not modulate changes in temperature or flow variation. In contrast, tributary junctions (network structure) can ameliorate downstream changes to temperature and flow regimes and sediment budgets, but only when tributaries are large relative to the mainstem. Finally, undammed tributaries, regardless of size and entry point, regularly present animals with predam or "native" conditions in terms of temperature, flow, and sediment. As a result, tributaries are refugia for native fishes in watersheds with heavily regulated mainstems, and large, undammed tributaries are exceptionally important in this regard because they can restore continuity in master variables of large rivers below dams.

The rivers and reservoirs we use as case studies in this paper are some of the largest and are certainly representative of regulated rivers in the Cadillac Desert of the western United States. Lake Powell and Lake Sakakawea are the second and third largest reservoirs in the United States, respectively, and Flaming Gorge Reservoir ranks 16th by storage volume. These three reservoirs store nearly 65 km^3 at full capacity.[60] Lake Powell and Flaming Gorge reservoir are principally water storage reservoirs, while Garrison Dam and many of the Missouri River dams serve as flood control impoundments (among other purposes). All three rivers were historically characterized by a spring snow melt peak originating from mountain headwaters and are fed by smaller, much flashier tributaries lower in the river network. Consequently, we think our story is broadly relevant to any western river in a semi-arid or arid catchment, where large dams that store one or more years of annual streamflow repeatedly fragment a river network. In this way, many of our observations are also relevant to reaches of the Snake, San Joaquin, Rio Grande, Owens, Arkansas, and South Platte rivers.

Acknowledgments

We thank D. Fryda of the North Dakota Game and Fish Department; R. H. Meade of the U.S. Geological Survey; and T. Kennedy and W. Persons of the Grand Canyon Monitoring and Research Center, USGS, and the field crews at Colorado State University for providing data and advice. D. Galat, S. Trimble, and R.H. Webb helped shape the ideas in the manuscript. This project was supported by National Science Foundation Grant EAR-0756817 (to J.L.S.) and ASU's Office of the Vice President for Research and Economic Affairs. This work was conducted as a part of the "Sustainability of Freshwater Resources in the United States" Working Group supported by the National Center for Ecological Analysis and Synthesis, a center funded by National Science Foundation Grant EF-0553768; the University of California, Santa Barbara; and the State of California.[10,9]

Conflicts of interest

The authors declare no conflicts of interest.

References

1. Reisner, M. 1986. *Cadillac Desert: The American West and its Disappearing Water*. Penguin Books. New York.
2. Sabo, J.L. *et al.* 2010. Reclaiming freshwater sustainability in the Cadillac Desert. *Proc. Natl. Acad. Sci. USA* **107:** 21263–21270.
3. Graf, W.L. *et al.* 2010. Sedimentation and sustainability of western American reservoirs. *Water Resources Res.* **46:** W12535. doi:1029/2009WR008856.
4. Ward, J.V. & J.A. Stanford. 1983. The serial discontinuity concept of lotic ecosystems. In *Dynamics of Lotic Ecosystems*. T. D. Fontaine & S. M. Bartell, Eds. Ann Arbor Science. Ann Arbor, MI.
5. Vannote, R.L. *et al.* 1980. River continuum concept. *Can. J. Fish. Aquat. Sci.* **37:** 130–137.
6. Stanford, J.A. & J.V. Ward. 2001. Revisiting the serial discontinuity concept. *Regul. Rivers: Res. Manage.* **17:** 303–310.
7. Ward, J.V. & J.A. Stanford. 1995. The serial discontinuity concept: extending the model to floodplain rivers. *Regul. Rivers: Res. Manage.* **10:** 159–168.
8. Rice, S.P., M.T. Greenwood & C.B. Joyce. 2001. Tributaries, sediment sources, and the longitudinal organisation of macroinvertebrate fauna along river systems. *Can. J. Fish. Aquat. Sci.* **58:** 824–840.
9. Guy, H.P. & V.W. Norman. 1970. Field methods for measurement of fluvial sediment. In *U.S. Geological Survey*

Techniques of Water-Resources Investigations. U.S. Geological Survey. Reston, VA.

10. Macek-Rowland, K.M. 2000. Suspended-sediment loads from major tributaries to the Missouri River between Garrison Dam and Lake Oahe, North Dakota, 1954–98. U.S. Geological Survey Water-Resources Investigations Report 00-407224.
12. Lagasse, P. 1981. Geomorphic response of the Rio Grande to dam construction. *New Mexico Geol. Soc. Spec. Publ.* **10:** 27–46.
13. Johnson, W. 1988. Dams and riparian forests: case study from the Upper Missouri River. In *Restoration, Creation, and Management of Wetland and Riparian Ecosystems in the American West*. K.M. Mutz, Ed. Society of Wetland Scientists. Denver, CO.
14. Stanford, J. & F. Hauer. 1992. Mitigating the impacts of stream and lake regulation in the Flathead River catchment, Montana, USA: an ecosystem perspective. *Aquat. Conserv.: Mar. Freshwat. Ecosyst.* **2:** 35–63.
15. Collier, M., R.H. Webb & J.C. Schmidt. 1996. Dams and rivers: primer on the downstream effects of dams US Geological Survey. *Circular* **1126:** 94.
16. Collier, M., R. Webb & E. Andrews. 1997. Experimental flooding in Grand Canyon. *Sci. Am.* **276:** 82–89.
17. Van Steeter, M. & J. Pitlick. 1998. Geomorphology and endangered fish habitats of the upper Colorado River. 1. Historic changes in streamflow, sediment load, and channel morphology. *Water Resources Res.* **34:** 287–302.
18. Williams, G. & M. Wolman. 1984. Downstream effects of dams on alluvial rivers. U.S. Geological Survey Professional Paper 128683.
19. Ligon, F., W. Dietrich & W. Trush. 1995. Downstream ecological effects of dams. *BioScience* **45:** 183–192.
20. Power, M., W. Dietrich & J. Finlay. 1995. Dams and downstream aquatic biodiversity: potential food web consequences of hydrologic and geomorphic change. *Environ. Manage.* **20:** 887–895.
21. Poff, N. *et al.* 1997. The natural flow regime: a paradigm for conservation and restoration of river ecosystems. *BioScience* **47:** 769–784.
22. Graf, W.L. 2006. Downstream hydrologic and geomorphic effects of large dams on American rivers. *Geomorphology* **79:** 336–360.
23. Nilsson, C. & K. Berggren. 2000. Alterations of riparian ecosystems caused by river regulation. *BioScience* **50:** 783–792.
24. Nilsson, C. *et al.* 2005. Fragmentation and flow regulation of the world's large river systems. *Science* **308:** 405–408.
25. Moyle, P. & J. Mount. 2007. Homogenous rivers, homogenous faunas. *Proc. Nat. Acad. Sci. USA* **104:** 5711–5712.
26. Braatne, J.H., R.S., L.A. Goater, C.L. Blair. 2008. Analyzing the impacts of dams on riparian ecosystems: a review of research strategies and their relevance to the Snake River through Hells Canyon. *Environ. Manage.* **41:** 267–281.
27. Brunsden, D. & J. Thornes. 1979. Landscape sensitivity and change. *Trans. Inst. Br. Geog* **4:** 463–484.
28. Sabo, J.L. *et al.* 2010. The role of discharge variation in scaling of drainage area and food chain length in rivers. *Science* **330:** 965–967.
29. Sabo, J.L. & D.M. Post. 2008. Quantifying periodic, stochastic, and catastrophic environmental variation. *Ecol. Monogr.* **78:** 19–40.
30. Grossman, G. & J. Sabo. 2010. Incorporating environmental variation into models of community stability: examples from stream fish. *Am. Fish. Soc. Symp.* **73:** 407–426.
31. Webb, R. *et al.* 2000. Sediment delivery by ungaged tributaries of the Colorado River in Grand Canyon, Arizona. USGS Water-Resources Investigations Report 00-405567.
32. Dubinski, I. & E. Wohl. 2007. Assessment of coarse sediment mobility in the Black Canyon of the Gunnison River, Colorado. *Environ. Manage.* **40:** 147–160.
33. Allen, P., R. Hobbs & N. Maier. 1989. Downstream impacts of a dam on a bedrock fluvial system, Brazos River, central Texas. *Bull. Assoc. Eng. Geol.* **26:** 165–189.
34. Larsen, I., J. Schmidt & J. Martin. 2004. Debris-fan reworking during low-magnitude floods in the Green River canyons of the eastern Uinta Mountains, Colorado and Utah. *Geology* **32:** 309–312.
35. Grams, P. & J. Schmidt. 2002. Streamflow regulation and multi-level flood plain formation: channel narrowing on the aggrading Green River in the eastern Uinta Mountains, Colorado and Utah. *Geomorphology* **44:** 337–360.
36. Phillips, J., M. Slattery & Z. Musselman. 2005. Channel adjustments of the lower Trinity River, Texas, downstream of Livingston Dam. *Earth Surf. Process. Landforms* **30:** 1419–1439.
37. Hauer, F. *et al.* 1997. Assessment of climate change and freshwater ecosystems of the Rocky Mountains, USA and Canada. *Hydrol. Process.* **11:** 903–924.
38. Pitlick, J. 1994. Relation between peak flows, precipitation, and physiography for five mountainous regions in the western USA. *J. Hydrol.* **158:** 219–240.
39. Costa, J.E. 1987. A comparison of the largest rainfall-runoff floods in the United States with those of the People's Republic of China and the world. *J. Hydrol.* **96:** 101–115.
40. Wohl, E.E. 2002. Rivers. In *The Physical Geography of North America*. E.E. Wohl & A.R. Orme, Eds.: 199–216. Oxford University Press. Oxford, UK.
41. Dahm, C., R. Edwards & F. Gelwick. 2005. Gulf Coast rivers of the southwestern United States. In *Rivers of North America*. A. Benke & C. Cushing, Eds.: 181–228. Elsevier Academic Press. Amsterdam.
42. Jelks, H. *et al.* 2008. Conservation status of imperiled North American freshwater and diadromous fishes. *Fisheries* **33:** 372–407.
43. Modde, T., K. Burnham & E. Wick. 1996. Population status of the razorback sucker in the middle Green River. *Conserv. Biol.* **10:** 110–119.
44. O'Brien, J.S. 1987. A case study of minimum streamflow for fishery habitat in the Yampa River. In *Sediment Transport in Gravel-Bed Rivers*. C. Thorne, J. Bathurst & R.D. Hey, Eds.: 921–946. John Wiley and Sons. Chichester, UK.
45. Wick, E.J. 1997. *Physical Processes And Habitat Critical to the Endangered Razorback Sucker on the Green River, Utah*. Doctoral dissertation. Colorado State University, Ft. Collins.

46. Graf, W.W. 1999. Dam nation: a geographic census of American dams and their large-scale hydrologic impacts. *Water Resources Res.* **35:** 1305–1311.
47. Poff, N.L. *et al.* 2007. Homogenization of regional river dynamics by dams and global biodiversity implications. *Proc. Nat. Acad. Sci. USA* **104:** 5732–5737.
48. Kottegoda, N. 1980. *Stochastic Water Resources Technology*. John Wiley. New York.
11. Andrews, E.D. 1986. Downstream effects of Flaming Gorge Reservoir on the Green River, Colorado and Utah. *Geol. Soc. Am. Bull* **97:** 1012–1023.
49. Gloss, S.P. & L.G. Coggins. 2005. Fishes of Grand Canyon. In *The State of the Colorado River Ecosypostem in Grand Canyon.* S.P. Gloss, J.E. Lovich & T.S. Mellis, Eds.: 33–56. U.S. Geological Survey Circular 1282.
50. Wright, S.A., C.R. Anderson & N. Voichick. 2009. A simplified water temperature model for the Colorado River below Glen Canyon Dam. *River Res. Appl.* **25:** 675–686.
51. Topping, D.J., D.M. Rubin & L.E. Vierra. 2000. Colorado River sediment transport: 1. Natural sediment supply limitation and the influence of Glen Canyon Dam. *Water Resources Res.* **36:** 515–542.
52. Johnson, P.T.J., J.D. Olden & M.J. vander Zanden. 2008. Dam invaders: impoundments facilitate biological invasions into freshwaters. *Front. Ecol. Environ.* **6:** 359–365.
53. Seegrist, D.W. & R. Gard. 1972. Effects of floods on trout in Sagehen Creek, California. *Trans. Am. Fisheries Soc.* **101:** 478.
54. Bestgen, K.R. *et al.* 2006. Response of the Green River fish community to changes in flow and temperature regimes from Flaming Gorge Dam since 1996 based on sampling conducted from 2002 to 2004. Final report submitted to the Biology Committee, Upper Colorado Endangered Fish Recovery Program. [Larval Fish Laboratory Contribution.]
55. Muth, R.T., L.W. Crist, K.E. LaGory, *et al.* 2000. Flow and temperature recommendations for endangered fishes in the Green River downstream of Flaming Gorge Dam. Final report to the Upper Colorado River Endangered Fish Recovery Program.
56. Coggins, L.G.J. *et al.* 2006. Abundance trends and status of the Little Colorado River population of humpback chub. *North Am. J. Fisheries Manage.* **26:** 233–245.
57. Graf, W.L. 1979. Rapids in canyon rivers. *J. Geol.* **87:** 533–551.
58. U.S. Department of Interior. 2008. Final biological opinion for the operation of Glen Canyon Dam: Phoenix, Arizona. AESO/SE 22410-1993-F-167R188.
59. Grams, P.E., J.C. Schmidt & M.E. Andersen. 2010. 2008 high-flow experiment at Glen Canyon Dam: morphologic response of eddy-deposited sandbars and associated aquatic backwater habitats along the Colorado River in Grand Canyon National Park U.S. Geological Survey Open-File Report 2010-103273.
60. U.S. Army Corps of Engineers. 1999. National Inventory of Dams. 2011.